# Mathematics

## ANALYSIS AND APPROACHES SL

FOR THE
**IB DIPLOMA**

IN COOPERATION WITH

# Mathematics
## ANALYSIS AND APPROACHES SL

Paul Fannon
Vesna Kadelburg
Ben Woolley
Stephen Ward

**HODDER**
EDUCATION
AN HACHETTE UK COMPANY

Although every effort has been made to ensure that website addresses are correct at time of going to press, Hodder Education cannot be held responsible for the content of any website mentioned in this book. It is sometimes possible to find a relocated web page by typing in the address of the home page for a website in the URL window of your browser.

Hachette UK's policy is to use papers that are natural, renewable and recyclable products and made from wood grown in well-managed forests and other controlled sources. The logging and manufacturing processes are expected to conform to the environmental regulations of the country of origin.

Orders: please contact Hachette UK Distribution, Hely Hutchinson Centre, Milton Road, Didcot, Oxfordshire, OX11 7HH. Telephone: +44 (0)1235 827827. Email education@hachette.co.uk Lines are open from 9 a.m. to 5 p.m., Monday to Friday. You can also order through our website: www.hoddereducation.com

ISBN: 9781510462359

© Paul Fannon, Vesna Kadelburg, Ben Woolley, Stephen Ward 2019

First published in 2019 by

Hodder Education,
An Hachette UK Company
Carmelite House
50 Victoria Embankment
London EC4Y 0DZ

www.hoddereducation.com

Impression number      10 9 8 7 6 5

Year        2023

Cover photo © glifeisgood - stock.adobe.com

Illustrations by Integra Software Services Pvt. Ltd., Pondicherry, India and also the authors

Typeset in Integra Software Services Pvt. Ltd., Pondicherry, India

Printed in India

A catalogue record for this title is available from the British Library.

# Contents

# Introduction

Welcome to your coursebook for Mathematics for the IB Diploma: analysis and approaches SL. The structure and content of this coursebook follow the structure and content of the 2019 IB Mathematics: analysis and approaches guide, with headings that correspond directly with the content areas listed in the guide.

This is also the first book required by students taking the higher level course. Students should be familiar with the content of this book before moving on to Mathematics for the IB Diploma: analysis and approaches HL.

## Using this book

The book begins with an introductory chapter on the 'toolkit', a set of mathematical thinking skills that will help you to apply the content in the rest of the book to any type of mathematical problem. This chapter also contains advice on how to complete your mathematical exploration.

The remainder of the book is divided into two sections. Chapters 1 to 10 cover the core content that is common to both Mathematics: analysis and approaches and Mathematics: applications and interpretation. Chapters 11 to 21 cover the remaining SL content required for Mathematics: analysis and approaches.

Special features of the chapters include:

## ESSENTIAL UNDERSTANDINGS

Each chapter begins with a summary of the key ideas to be explored and a list of the knowledge and skills you will learn. These are revisited in a checklist at the end of each chapter.

## CONCEPTS

The IB guide identifies 12 concepts central to the study of mathematics that will help you make connections between topics, as well as with the other subjects you are studying. These are highlighted and illustrated with examples at relevant points throughout the book.

## KEY POINTS

Important mathematical rules and formulae are presented as Key Points, making them easy to locate and refer back to when necessary.

## WORKED EXAMPLES

There are many Worked Examples in each chapter, demonstrating how the Key Points and mathematical content described can be put into practice. Each Worked Example comprises two columns:

On the left, how to **think** about the problem and what tools or methods will be needed at each step.

On the right, what to **write**, prompted by the left column, to produce a formal solution to the question.

## Exercises

Each section of each chapter concludes with a comprehensive exercise so that students can test their knowledge of the content described and practise the skills demonstrated in the Worked Examples. Each exercise contains the following types of questions:

- **Drill questions:** These are clearly linked to particular Worked Examples and gradually increase in difficulty. Each of them has two parts – **a** and **b** – desgined such that if students get **a** wrong, **b** is an opportunity to have another go at a very similar question. If students get **a** right, there is no need to do **b** as well.
- **Problem-solving questions:** These questions require students to apply the skills they have mastered in the drill questions to more complex, exam-style questions. They are colour-coded for difficulty.

**1** Green questions are closely related to standard techniques and require a small number of processes. They should be approachable for all candidates.

**2** Blue questions require students to make a small number of tactical decisions about how to apply the standard methods and they will often require multiple procedures. They should be achievable for SL students aiming for higher grades.

**3** Red questions often require a creative problem-solving approach and extended technical procedures. They will stretch even advanced SL students and be challenging for HL students aiming for the top grades.

**4** Black questions go beyond what is expected in IB examinations, but provide an enrichment opportunity for the most advanced students.

The questions in the Mixed Practice section at the end of each chapter are similarly colour-coded, and contain questions taken directly from past IB Diploma Mathematics exam papers. There is also a review exercise halfway through the book covering all of the core content, and two practice examination papers at the end of the book.

Answers to all exercises can be found at the back of the book.

 Throughout the first ten chapters it is assumed that you have access to a calculator for all questions. A calculator symbol is used where we want to remind you that there is a particularly important calculator trick required in the question.

 A non-calculator icon suggests a question is testing a particular skill for the non-calculator paper.

 The guide places great emphasis on the importance of technology in mathematics and expects you to have a high level of fluency with the use of your calculator and other relevant forms of hardware and software. Therefore, we have included plenty of screenshots and questions aimed at raising awareness and developing confidence in these skills, within the contexts in which they are likely to occur. This icon is used to indicate topics for which technology is particularly useful or necessary.

 **Making connections:** Mathematics is all about making links. You might be interested to see how something you have just learned will be used elsewhere in the course and in different topics, or you may need to go back and remind yourself of a previous topic.

## Be the Examiner

These are activities that present you with three different worked solutions to a particular question or problem. Your task is to determine which one is correct and to work out where the other two went wrong.

**LEARNER PROFILE**
Opportunities to think about how you are demonstrating the attributes of the IB Learner Profile are highlighted at the beginning of appropriate chapters.

### TOOLKIT

There are questions, investigations and activities interspersed throughout the chapters to help you develop mathematical thinking skills, building on the introductory toolkit chapter in relevant contexts. Although the ideas and skills presented will not be examined, these features are designed to give you a deeper insight into the topics that will be. Each toolkit box addresses one of the following three key topics: proof, modelling and problem solving.

### Proof

Proofs are set out in a similar way to Worked Examples, helping you to gain a deeper understanding of the mathematical rules and statements you will be using and to develop the thought processes required to write your own proofs.

### International mindedness

These boxes explore how the exchange of information and ideas across national boundaries has been essential to the progress of mathematics and to illustrate the international aspects of the subject.

### You are the Researcher

This feature prompts you to carry out further research into subjects related to the syllabus content. You might like to use some of these ideas as starting points for your mathematical exploration or even an extended essay.

### TOK Links

Links to the interdisciplinary Theory of Knowledge element of the IB Diploma programme are made throughout the book.

**Tip**

There are short hints and tips provided in the margins throughout the book.

### Links to: Other subjects

Links to other IB Diploma subjects are made at relevant points, highlighting some of the real-life applications of the mathematical skills you will learn.

Topics that have direct real-world applications are indicated by this icon.

There is a glossary at the back of the book. Glossary terms are **purple**.

## About the authors

The authors are all University of Cambridge graduates and have a wide range of expertise in pure mathematics and in applications of mathematics, including economics, epidemiology, linguistics, philosophy and natural sciences.

Between them they have considerable experience of teaching IB Diploma Mathematics at Standard and Higher Level, and two of them currently teach at the University of Cambridge.

The 'In cooperation with IB' logo signifies the content in this coursebook has been reviewed by the IB to ensure it fully aligns with current IB curriculum and offers high quality guidance and support for IB teaching and learning.

# The toolkit and the mathematical exploration

Mathematics is about more than just arithmetic, geometry, algebra and statistics. It is a set of skills that are widely transferable. All of the IB Diploma Programme Mathematics courses allocate time to the development of these skills, collectively known as the 'toolkit', which will help you formulate an approach to any mathematical problem and form a deeper understanding of the real-life applications of mathematics.

In this chapter, we will look at four of these skills:

- problem solving
- proof
- modelling
- technology.

For each, we have provided some background information and activities to help you develop these skills. There will be many more activities interspersed at appropriate places throughout the book. We will then also look at how these skills can be demonstrated in your exploration.

For some students, part of the additional 'toolkit' time might be usefully spent practising their basic algebra skills, so we have also provided an exercise to assist with this.

This chapter has been designed as a useful resource of information and advice that will give you a good grounding in the required skills at the start of the course, but that you can also refer back to at appropriate times throughout your studies to improve your skills.

## Problem Solving

Some people think that good mathematicians see the answers to problems straight away. This could not be further from the truth.

 **TOOLKIT: Problem Solving**

Answer the following problem as fast as you can using your mental maths skills.

A bottle and a cork cost $1.10. The bottle costs $1 more than the cork. How much is the cork?

The human brain is highly evolved to do things such as spot predators and food. There has not been a lot of evolutionary pressure to do mental maths, so most people get this type of problem wrong when they try to do it quickly (the most common answer is usually $0.10; the correct answer is actually $0.05). Good mathematicians do not just try to spot the correct answer, but work through a clear, analytical process to get to the right answer.

One famous mathematician, George Polya, tried to break down the process mathematicians go through when solving problems. He suggested four steps:

1 Understand the problem.
2 Make a plan.
3 Carry out the plan.
4 Look back on your work and reflect.

Most people put all their efforts into step 3, but often the others are vital in developing your problem-solving skills.

Polya also came up with a list of problem-solving strategies, called heuristics, which he used to guide him when he was stuck. For example, when dealing with a difficult problem; answer a simpler one.

---

 **TOOLKIT: Problem Solving**

A strange mathematical prison has 100 cells each containing 1 prisoner. Initially they are all locked.

- On day 1 the guard turns the key in every cell door, therefore all the doors are unlocked.
- On day 2 the guard turns the key on the door of every cell that is a multiple of 2. This locks all the even numbered cells.
- On day 3 the guard turns the key on the door of every cell that is a multiple of 3.
- On day 4 the guard turns the key on the door of every cell that is a multiple of 4.

This continues in this way until the 100th day, on which the guard turns the key on every cell that is a multiple of 100 (i.e. just the 100th cell.) All prisoners whose cell doors are then opened are released. Which prisoners get released?

---

You might find that the cycle of mathematical inquiry below is a useful guide to the process you have to go through when tackling the problem above, along with many of the other problems in this section.

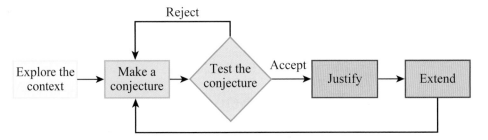

---

 **TOOLKIT: Problem Solving**

What is the formula for the sum of the angles inside an *n*-sided polygon?

**a** Explore the context:

What is a polygon? Is a circle a polygon? Does it have to be regular? Can it have obtuse angles? What about reflex angles? What is the smallest and largest possible value of *n*?

**b** Make a conjecture:

Using your prior learning, or just by drawing and measuring, fill in the following table:

| Shape | *n* | Sum of angles, *S* |
|---|---|---|
| Triangle | | |
| Quadrilateral | | |
| Pentagon | | |

Based on the data, suggest a rule connecting *n* and *S*.

**c** Test your conjecture:

Now, by drawing and measuring, or through research, see if your conjecture works for the next polygon. If your conjecture works, move on to **d**, otherwise, make a new conjecture and repeat **c**.

> **d** Justify your conjecture:
>
> If you connect one corner of your polygon to all the other corners, how many triangles does this form? Can you use this to justify your conjecture?
>
> **e** Extend your conjecture:
>
> Does the justification still work if the polygon has reflex angles? If the polygon is regular, what is each internal angle? What whole-number values can the internal angle of a regular polygon be? What is the external angle of a regular *n*-sided polygon? Which regular polygons can tessellate together?

One way of exploring the problem to help you form a conjecture is to just try putting some numbers in.

**TOOLKIT: Problem Solving**

Simplify $\cos^{-1}(\sin(x))$ for $0° < x < 90°$.

With a lot of problem solving, you will go through periods of not being sure whether you are on the right track and not knowing what to do next. Persistence itself can be a useful tool.

**TOOLKIT: Problem Solving**

Each term in the look-and-say sequence describes the digits of the previous term:

1, 11, 21, 1211, 111221, 312211, …

For example, the fourth term should be read as '1 two, 1 one', describing the digits in the third term. What is the 2000th term in this sequence? How would your answer change if the first element in the sequence was 2?

One of the key challenges when dealing with a difficult problem is knowing where to start. It is often useful to look for the most constrained part.

**TOOLKIT: Problem Solving**

The problem below is called a KenKen or Calcudoku. The numbers 1 to 5 are found exactly once in each row and column. The numbers in the cells connected by outlines can be used, along with the operation given, to form the stated result. For example, the box labelled 2÷ gives you the desired result and the operation used to get it. The two numbers must be able to be divided (in some order) to make 2, so it could be 1 and 2 or 2 and 4.

| 60× | 2÷ |  | 1− |  |
|---|---|---|---|---|
|  |  | 4+ |  | 2 |
| 4− | 12× |  | 2÷ | 4+ |
|  | 10+ |  |  |  |
| 2÷ |  |  | 4− |  |

Sometimes your immediate response to a hard problem is total panic, as you do not know where to begin. Always remember that the problems you will face will have a solution; it just might require some patience and careful thinking to get there.

---

 **TOOLKIT: Problem Solving**

In the multiple-choice quiz below, each question refers to the quiz as a whole:

1   How many answers are A?

    **A**   0

    **B**   1

    **C**   2

    **D**   3

    **E**   4

2   The first question whose answer is A is question

    **A**   1

    **B**   2

    **C**   3

    **D**   4

    **E**   There are no 'A's

3   The previous answer is

    **A**   C

    **B**   D

    **C**   E

    **D**   A

    **E**   B

4   The only repeated answer is

    **A**   C

    **B**   B

    **C**   A

    **D**   E

    **E**   D

---

# Proof

## ■ What makes a good proof?

In many ways, proof is the defining feature of mathematics. It is steps of logical reasoning beginning at a clear starting point (called an axiom). This requires precise, unambiguous communication so mathematical proofs shared between mathematicians often look scarily formal. For example, overleaf is a short section of Bertrand Russell's 1910 *Principia Mathematica* in which he used fundamental ideas about sets to show that 1 + 1 = 2.

**∗54·43.**   $\vdash :. \, \alpha, \beta \,\epsilon\, 1 \,.\, \supset \,:\, \alpha \cap \beta = \Lambda \,.\, \equiv \,.\, \alpha \cup \beta \,\epsilon\, 2$

*Dem.*

$\vdash . \ast 54{\cdot}26 . \supset \vdash :. \, \alpha = \iota{}^{\prime}x \,.\, \beta = \iota{}^{\prime}y \,.\, \supset \,:\, \alpha \cup \beta \,\epsilon\, 2 \,.\, \equiv \,.\, x \neq y \,.$

$[\ast 51{\cdot}231] \qquad\qquad\qquad\qquad\qquad\qquad \equiv \,.\, \iota{}^{\prime}x \cap \iota{}^{\prime}y = \Lambda \,.$

$[\ast 13{\cdot}12] \qquad\qquad\qquad\qquad\qquad\qquad \equiv \,.\, \alpha \cap \beta = \Lambda \qquad\qquad (1)$

$\vdash . (1) . \ast 11{\cdot}11{\cdot}35 . \supset$

$\vdash :. (\exists x, y) \,.\, \alpha = \iota{}^{\prime}x \,.\, \beta = \iota{}^{\prime}y \,.\, \supset \,:\, \alpha \cup \beta \,\epsilon\, 2 \,.\, \equiv \,.\, \alpha \cap \beta = \Lambda \qquad (2)$

$\vdash . (2) . \ast 11{\cdot}54 . \ast 52{\cdot}1 . \supset \vdash . \text{Prop}$

From this proposition it will follow, when arithmetical addition has been defined, that $1 + 1 = 2$.

However, you should not get too hung up on making proofs look too formal. A good proof is all about convincing a sceptical peer. A famous study by Celia Hoyles and Lulu Healy in 1999 showed that students often thought totally incorrect algebraic proofs were better than well-reasoned proofs that used words or diagrams. Good proof is about clearly communicating ideas, not masking them behind complicated notation.

**TOOLKIT: Proof**

Does the following diagram prove Pythagoras' Theorem?

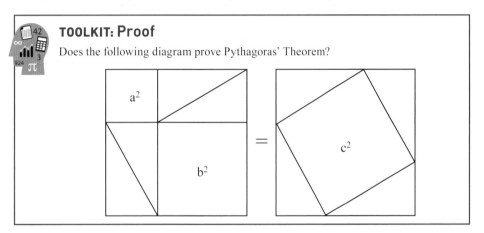

One of the problems you might encounter in developing proofs is not knowing what axioms you can start from. There is a certain amount of 'community knowledge' in knowing what acceptable starting points are. For example, if you were asked to prove that the area of a circle is $\pi r^2$, it would not be acceptable to say this is a well-known fact found in the formula book. However, if you were asked to prove that a circle is the shape with the largest area for a given perimeter then you could probably use $A = \pi r^2$ without proof.

The other issue is that too often in maths you are asked to prove obvious facts that you already know, for example proving that the sum of two odd numbers is even. However, the methods used to do that are helping with the development of precise reasoning which can be used to prove facts that are much less obvious.

**TOOLKIT: Proof**

Are there more positive even numbers or positive whole numbers?

Imagine creating a list of all the positive whole numbers alongside a list of all the positive even numbers:

$$1 \rightarrow 2$$

$$2 \rightarrow 4$$

$$3 \rightarrow 6$$

$$\vdots$$

$$n \rightarrow 2n$$

For every positive whole number, $n$, we can associate with it exactly one even number, $2n$. Therefore, there are the *same* number of whole numbers and even numbers.

Most people find this fact counter-intuitive. You might want to see if you can extend this method to ask:

a   Are there more positive whole numbers or prime numbers?

b   Are there more positive whole numbers or fractions between 0 and 1?

c   Are there more positive whole number or decimals between 0 and 1?

**You are the Researcher**

Reasoning with infinity is famously problematic. There is a famous proof that the sum of all positive whole numbers is $-\frac{1}{12}$. Although massively counter-intuitive and usually considered invalid, this result has found applications in quantum theory. You might want to research this result, along with other astounding results of the great Indian mathematician Ramanujan.

One way of developing your own skills in proof is to critically appraise other attempted proofs. Things can look very plausible at first but have subtle errors.

**TOOLKIT: Proof**

Find the flaw in the following proof that $7 = 3$.

Start from the true statement:

$$-21 = -21$$

Rewrite this as:

$$49 - 70 = 9 - 30$$

This is equivalent to:

$$7^2 - 10 \times 7 = 3^2 - 10 \times 3$$

Add on 25 to both sides:

$$7^2 - 10 \times 7 + 25 = 3^2 - 10 \times 3 + 25$$

This can be expressed as:

$$7^2 - 2 \times 5 \times 7 + 5^2 = 3^2 - 2 \times 5 \times 3 + 5^2$$

Both sides are perfect squares and can be factorized:

$$(7 - 5)^2 = (3 - 5)^2$$

Square rooting both sides:

$$7 - 5 = 3 - 5$$

Therefore $7 = 3$.

**You are the Researcher**

There is a famous proof that all triangles are equilateral. See if you can find this proof and determine the flaw in it.

**TOOLKIT: Proof**

Simplify each of the following expressions:

$\mathbb{N} \cup \mathbb{Z}^+$

$\mathbb{N} \cap \mathbb{Z}^+$

$\mathbb{Q} \cap \overline{\mathbb{Q}}$

$\mathbb{Q} \cup \overline{\mathbb{Q}}$

# ■ Sets, logic and the language of proof

Unfortunately, the precision of the notation used in mathematics can sometimes be intimidating. For example, the different notation for number sets:

$\mathbb{Z}$: Integers: $\{\ldots -1, 0, 1, 2, 3 \ldots\}$

$\mathbb{Z}^+$: Positive integers: $\{1, 2, 3 \ldots\}$

$\mathbb{N}$: Natural numbers: $\{0, 1, 2 \ldots\}$

$\mathbb{Q}$: Rationals: $\{-\frac{1}{3}, 0, 0.25, 4 \ldots\}$

$\overline{\mathbb{Q}}$: Irrationals : $\{-\pi, \sqrt{2}, 1+\sqrt{3} \ldots\}$

$\mathbb{R}$: Reals : $\{-3, 0, \frac{1}{5}, \sqrt{2}, 8 \ldots\}$

In common English usage, zero is *neither* positive nor negative. However, in some countries, for example, some parts of France, zero is considered *both* positive and negative. How much of maths varies between countries?

Thinking about sets (possibly visualizing them using Venn diagrams) is a very powerful way of describing many mathematical situations.

One of the main purposes for working on sets and proof is to develop logic.

**TOOLKIT: Proof**

A barber in a town shaves all those people who do not shave themselves, and only those people. Does the barber shave himself?

This is an example of a paradox often attributed to the mathematician and philosopher Bertrand Russell. Can you see why it causes a paradox?

**TOOLKIT: Proof**

A set of cards has numbers on one side and letters on the other. James claims that if there is a vowel on one side there must be an even number on the other side. Which of the following cards must be turned over to test James's claim?

| A | B | 1 | 2 |

This question, called the Watson selection test, formed part of a psychological study into deductive reasoning and very few people gave the correct answer. However, when they were asked the following question, which is logically identical, nearly everybody gave the correct answer.

Four people are drinking in a bar. If someone is drinking beer, they must be aged over 21. Which of the following people need to be investigated further to check that everybody is complying with this rule?

| Beer drinker | Coke drinker | 16-year-old person | 28-year-old person |

One of the conclusions from this study was that people are generally weak at abstract reasoning. The important message for you is, where possible, take abstract ideas and try to put them into a context.

(The answer in both cases is the yellow and pink cards.)

The previous example demonstrates that many people do not have an intuitive approach to formal logic. To help develop this, it is useful to have some terminology to work with.

A logical statement, $A$, is something which is either true or false, e.g. $A$ = 'The capital of Italy is Rome'.

The negation of a logical statement, $\overline{A}$, is something which is true when $A$ is false and vice versa, e.g. $\overline{A}$ = 'The capital of Italy is not Rome'. Sometimes you have to be careful here – if the statement is 'the cup is full', the negation is 'the cup is not full', rather than 'the cup is empty'.

An implication is the basic unit of logic. It is of the form 'if $A$ then $B$'. People are often not precise enough with this logic and confuse this implication with other related statements:

The converse: 'if $B$ then $A$'

The inverse: 'if $\overline{A}$ then $\overline{B}$'

The contrapositive: 'if $\overline{B}$ then $\overline{A}$'.

---

**TOOLKIT: Proof**

For each of the following statements, decide if the converse, inverse or contrapositive are always true:
- **If** a shape is a rectangle with equal sides, **then** it is a square.
- **If** $x$ is even, **then** $x^2$ is even.
- **If** two shapes are similar, **then** all corresponding angles are equal.
- **If** two numbers are both positive, **then** their product is positive.
- **If** a shape is a square, **then** it is a parallelogram.
- **If** a number is less than 5, **then** it is less than 10.

---

Logic gets even harder when it is taken out of abstract mathematical thought and put into real-life contexts. This is often implicitly presented in the form of a syllogism:

Premise 1: All IB students are clever.

Premise 2: You are an IB student.

Conclusion: Therefore, you are clever.

The above argument is logically consistent. This does not necessarily mean that the conclusion is true – you could dispute either of the premises, but if you agree with the premises then you will be forced to agree with the conclusion.

---

**TOOLKIT: Proof**

Decide if each of the following arguments is logically consistent.
- If Darren cheats he will get good grades. He has got good grades; therefore, he must have cheated.
- If you work hard you will get a good job. Therefore, if you do not work hard you will not get a good job.
- All HL Maths students are clever. No SL Maths students are HL Maths students; therefore, no SL maths students are clever.
- All foolish people bought bitcoins. Since Jamila bought bitcoins, she is foolish.
- All of Paul's jokes are either funny or clever but not both. All of Paul's maths jokes are clever, therefore none of them are funny.

---

**TOK Links**

Is it more important for an argument to be logically consistent or the conclusion to be true?

# Modelling

## ■ Creating models

Mathematics is an idealized system where we use strict rules to manipulate expressions and solve equations, yet it is also extremely good at describing real-world situations. The real world is very complicated, so often we need to ignore unnecessary details and just extract the key aspects of the situation we are interested in. This is called creating a model. The list of things we are assuming (or ignoring) is called the modelling assumptions. Common modelling assumptions include:

- Treating objects as just being a single point, called a particle. This is reasonable if they are covering a space much bigger than their size, for example, modelling a bird migrating.

- Assuming that there is no air resistance. This is reasonable if the object is moving relatively slowly, for example, a person walking.

- Treating individuals in a population as all being identical. This is reasonable if the population is sufficiently large that differences between individuals average out.

**TOOLKIT: Modelling**

None of the following statements are always true, but they are all sometimes useful. When do you think these would be good modelling assumptions? When would they be a bad idea to use?
- The Earth is a sphere.
- The Earth is flat.
- Parallel lines never meet.
- The sum of the angles in a triangle add up to 180 degrees.
- Each possible birthday of an individual is equally likely.
- The average height of people is 1.7 m.
- The Earth takes 365 days to orbit the Sun.
- The Earth's orbit is a circle.
- The more individuals there are in a population, the faster the population will grow.

One of the most difficult things to do when modelling real-world situations is deciding which variables you want to consider.

**TOOLKIT: Modelling**

What variables might you consider when modelling:
- the population of rabbits on an island
- the height reached by a basketball thrown towards the net
- the average income of a teacher in the UK
- the optimal strategy when playing poker
- the volume of water passing over the Niagara Falls each year
- the profit a company will make in 3 years' time
- the result of the next national election
- the thickness of cables required on a suspension bridge.

Can you put the variables in order from most important to least important?

Once you have decided which variables are important, the next step is to suggest how the variables are linked. This can be done in two different ways:

**TOK Links**

Is theory-led or data-led modelling more valid?

- **Theory-led:** This is where some already accepted theory – such as Newton's laws in physics, supply–demand theory in economics or population dynamics in geography – can be used to predict the form that the relationship should take.
- **Data-led:** If there is no relevant theory, then it might be better to look at some experimental data and use your knowledge of different functions to suggest an appropriate function to fit the data.

## ■ Validating models

Once you have a model you need to decide how useful it is. One way of doing this is to see how well it fits the data. Unfortunately, there is a balance to be struck between the complexity of the model and how well it fits – the more parameters there are in a model, the better the fit is likely to be. For example, a cubic model of the form $y = ax^3 + bx^2 + cx + d$ has four parameters ($a$, $b$, $c$ and $d$) and will fit any data set at least as well as a linear function $y = ax + b$ with two parameters ($a$ and $b$).

**TOK Links**

There is a very important idea in philosophy known as Occam's razor. This basically says that, all else being equal, the simpler explanation is more likely to be the correct one.

**TOOLKIT: Modelling**

Consider the following set of data, showing the speed of a ball (in metres per second) against the time since the ball was dropped.

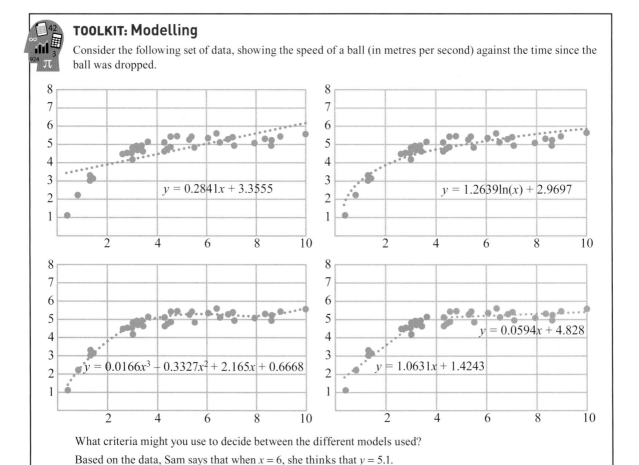

$y = 0.2841x + 3.3555$

$y = 1.2639\ln(x) + 2.9697$

$y = 0.0166x^3 - 0.3327x^2 + 2.165x + 0.6668$

$y = 1.0631x + 1.4243$

$y = 0.0594x + 4.828$

What criteria might you use to decide between the different models used?

Based on the data, Sam says that when $x = 6$, she thinks that $y = 5.1$.

Olesya says that when $x = 6$, she thinks that $y$ is between 4.5 and 5.5.

What are the benefits or drawbacks of each of their ways of reporting their findings?

## ■ The modelling cycle

In real-life applications you are unlikely to get a suitable model at the first attempt. Instead, there is a cycle of creating, testing and improving models.

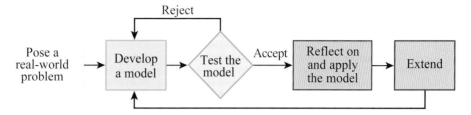

**TOOLKIT: Modelling**

Carlos is creating a business plan for his bicycle repair business. Based on a survey, he believes that he will have 10 customers each week if he charges $30 for a repair and 20 customers each week if he charges $20.

**a** Pose a real-world problem:

Can you predict how many customers Carlos will have each week ($N$) if he charges $d$ for a repair?

**b** Develop a model:

Use a formula of the form $N = ad + b$ to model the number of customers Carlos will have.

**c** Test the model:

In his first week, Carlos charges $25 and gets 14 customers. Evaluate your model. Should it be improved? If so suggest a better model.

**d** Apply the model:

Carlos believes he can cope with up to 30 customers each week. What is the least amount of money he should charge?

**e** Reflect on the model:

Does it matter that your original model predicts a negative number of customers if Carlos charges $100 for a repair? How accurate does the model have to be? Is it more important for a new business to make a lot of money or to have a lot of customers? Is a single price for any repair a sensible business proposition?

**f** Extend:

Use your model to predict the total income Carlos will get if he charges $d$. What price would you advise Carlos to charge?

# Using technology

The main technology that you need to master is your Graphical Display Calculator (GDC), which you should use as much as possible early in the course to get accustomed to all its features. However, outside of examinations, there are many more technological tools which you might find helpful at various points, especially when writing your exploration. The following questions provide starting points to develop your skills with different types of software.

## ■ Computer algebra systems (CAS)

Most professional mathematicians make extensive use of CAS to avoid the more tedious elements of algebraic simplification. Various websites, such as Wolfram Alpha, and some calculators will do a lot of manipulation for you. This reflects the fact that, in modern mathematics, the skill is not algebraic manipulation, but more deciding which equation needs to be solved in the first place.

a   Use CAS to simplify $\sin(\cos^{-1} x)$.

b   Simplify $(\sin x)^4 + (\cos x)^4 + \dfrac{1}{2}(\sin(2x))^2$.

## ■ Spreadsheets

a   Spreadsheets can be used to automate calculations.

A teacher uses the following boundaries to assign grades to internal tests.

| Percentage, $p$ | Grade |
|---|---|
| $p < 11$ | Fail |
| $11 \leqslant p < 22$ | 1 |
| $22 \leqslant p < 33$ | 2 |
| $33 \leqslant p < 44$ | 3 |
| $44 \leqslant p < 55$ | 4 |
| $55 \leqslant p < 66$ | 5 |
| $66 \leqslant p < 77$ | 6 |
| $p \geqslant 77$ | 7 |

Design a spreadsheet that will allow a teacher to input raw marks on Paper 1 (out of 40) and Paper 2 (out of 50) and automatically calculate an overall percentage (reported to the nearest whole number) and grade.

b   Spreadsheets are also useful for answering questions about modelling.

The change in speed of a ball dropped off a cliff with air resistance is modelled using the following rules.

The speed (in metres per second) at the end of each second is given by $0.5 \times$ previous speed $+ 10$. The distance travelled in each second is given by the average of the speed at the beginning of the second and the speed at the end. The initial speed is zero.

How far has the ball travelled after 20 seconds? What speed is achieved?

c   Spreadsheets are very useful for investigating sequences.

i   The first term of a sequence is 2. All subsequent terms are found by doing $1 - \dfrac{1}{\text{previous term}}$. Find the 100th term. What do you notice about this sequence?

ii   The first term of a sequence is $n$. The subsequent terms are found using the following rules:

■ If the previous term is odd, the next term is 1 plus 3 times the previous term.

■ If the previous term is even, the next term is half the previous term.

Investigate the long-term behaviour of this sequence for various different values of $n$.

### TOK Links

The sequence in **c ii** leads to a famous mathematical idea called the Collatz conjecture. It has been tested for all numbers up to about $87 \times 2^{60}$ and no counterexamples have been found, but this is still not considered adequate for a proof in mathematics. What are the differences between 'mathematical proof' and 'scientific proof'?

## ■ Dynamic geometry packages

Dynamic geometry packages are useful for exploring different geometric configurations by drawing a diagram using the given constraints, then seeing what stays the same while the unconstrained quantities are changed. They often allow you to make interesting observations that can then be proved; however, it is a skill in itself to turn a problem in words into a diagram in a dynamic geometry package.

**a** A triangle *ABC* has a right angle at *C*. The midpoint of *AB* is called *M*. How is the length of *MC* related to the length of *AB*? How could you prove your conjecture?

**b** Two touching circles share a common tangent which meets the first circle at *A*. The opposite end of the diameter of the first circle at *A* is called *B*. The tangent from *B* meets the other circle tangentially at *C*. Find the ratio *AB* : *BC*.

## ■ Programming

Programming is a great way to provide clear, formal instructions, that requires planning and accuracy. All of these are great mathematical skills. There are many appropriate languages to work in, such as Java, Python or Visual Basic. Alternatively, many spreadsheets have a programming capability.

**a** Write a program to find the first 1000 prime numbers.

**b** Write a program that will write any positive number as the sum of four square numbers (possibly including zero).

**c** Write a program that will decode the following message:

```
Jrypbzr gb Zngurzngvpf sbe gur Vagreangvbany
Onppnynherngr. Jr ubcr gung lbh rawbl guvf obbx
naq yrnea ybgf bs sha znguf!
```

**Tip**

You might want to do a frequency analysis on the message and compare it to the standard frequencies in the English language.

# The exploration

The purpose of this maths course is much more than just to prepare for an examination. We are hoping you will develop an appreciation for the beauty of mathematics as well as seeing a wide range of applications. 20% of the final mark is awarded for an exploration – a piece of independent writing, approximately 12 to 20 pages long, that explores a mathematical topic. There are many different types of mathematical exploration, but some of the more common types include:

■ creating a mathematical model for a situation

■ using statistics to answer a question

■ exploring the applications of a mathematical method

■ solving a pure mathematics puzzle

■ exploring the historical development of a mathematical idea.

Your exploration will be marked according to five criteria:

■ presentation

■ mathematical communication

■ personal engagement

■ reflection

■ use of mathematics.

**Tip**

There are many successful explorations in each of these categories, but you should be aware that the first three of these types listed, involving models, statistics and applications of a method, are often felt by students to be easier to get higher marks, particularly in relation to the personal engagement and reflection assessment criteria.

## ■ Presentation

This is about having a well-structured project which is easy to follow. If someone else reads your presentation and can summarize what you have written in a few sentences, it is probably well-structured.

It needs to be:

■ **Organized:** There should be an introduction, conclusion and other relevant sections which are clearly titled.

■ **Coherent:** It should be clear how each section relates to each other section, and why they are in the order you have chosen. The aim of the project should be made clear in the introduction, and referred to again in the conclusion.

■ **Concise:** Every graph, calculation and description needs to be there for a reason. You should not just be repeating the same method multiple times without a good reason.

## ■ Mathematical communication

As well as the general communication of your ideas, it is expected that by the end of this course you will be able to communicate in the technical language of mathematics.

Your mathematical communication needs to be:

■ **Relevant:** You should use a combination of formulae, diagrams, tables and graphs.

All key terms (which are not part of the IB curriculum) and variables should be defined.

■ **Appropriate:** The notation used needs to be at the standard expected of IB mathematicians. This does not mean that you need to use formal set theory and logical implications throughout, but it does means that all graphs need to labelled properly and computer notation cannot be used (unless it is part of a screenshot from a computer output).

Common errors which cost marks include:

☐ $2^\wedge x$ instead of $2^x$

☐ $x*y$ instead of $x \times y$

☐ 2E12 instead of $2 \times 10^{12}$

☐ writing $\frac{1}{3} = 0.33$ instead of $\frac{1}{3} = 0.33$ (2 d.p.) or $\frac{1}{3} \approx 0.33$.

■ **Consistent:** Your use of appropriate notation needs to be across all of the exploration, not just in the mathematical proof parts. You also need to make sure that your notation is the same across all of the exploration. If you define the volume of a cone to be $x$ in one part of the exploration it should not suddenly become $X$ or $V_c$ later (unless you have explained your good reason for doing so!).

One other part of consistency is to do with accuracy. If you have measured something to one significant figure as part of your exploration, it would not then be consistent to give your final answer to five significant figures.

## ■ Personal engagement

Your exploration does not need to be a totally new, ground-breaking piece of mathematics, but it does need to have a spark of originality in it to be likely to score well in personal engagement. This might be using novel examples to illustrate an idea, applying a tool to something from your own experience or collecting your own data to test a model.

**TOK Links**

Is mathematics a different language? What are the criteria you use to judge this?

**Tip**

One common error from people using calculus is to call every derivative $\frac{dy}{dx}$. If you are looking at the rate of change of $C$ over time, the appropriate notation is $\frac{dC}{dt}$.

**TOK Links**

Although personal engagement is difficult to describe, it is often easy to see. Can you think of other situations in which there is wide consensus about something despite no clear criteria being applied?

**Tip**

Beware overcomplicated sources, such as many internet discussion boards. Personal engagement marks are often lost by people who clearly have not understood their source material.

Also, try to avoid the common topics of the golden ratio and the prisoner's dilemma. It is quite tricky to show personal engagement with these, although it is not impossible to do so.

The key to writing an exploration with great personal engagement is to write about a topic which really interests you. We recommend that from the very first day of your course you keep a journal with mathematical ideas you have met that you have found interesting. Some of these will hopefully develop into an exploration. Importantly, do not just think about this during your maths lessons – we want you to see maths everywhere! If you are struggling for inspiration, the following sources have been fruitful for our students in the past:

- podcasts, such as More or Less: Behind the Stats
- YouTube channels, such as Numberphile
- websites, such as Underground Mathematics, NRICH or Khan Academy
- magazines, such as *Scientific American* or *New Scientist*.

One of our top tips is to find an overlap between mathematics and your future studies or career. As well as looking good on applications, it will give you an insight that few people have. Every future career can make use of mathematics, sometimes in surprising ways!

## Links to: Other subjects

Some topics we have seen that show great applications of mathematics relevant to other subjects include:
- **Psychology:** Do humans have an intuitive understanding of Bayesian probability?
- **Art:** Did Kandinsky use the golden ratio in his abstract art?
- **Modern languages:** Can you measure the distance between two languages? A phylogenetic analysis of European languages.
- **English literature:** Was Shakespeare sexist? A statistical analysis of the length of male and female character speeches.
- **Medicine:** How do doctors understand uncertainty in clinical tests?
- **Law:** What is the prosecutor's fallacy?
- **Economics:** Does the market model apply to sales in the school cafeteria?
- **Physics:** How accurate are the predictions made by Newton's laws for a paper aeroplane?
- **Chemistry:** Using logarithms to find orders of reactions.
- **Biology:** Did Mendel really cheat when he discovered genetics? A computer simulation.
- **History:** Was it worth it? A statistical analysis of whether greater mortality rates led to more land gained in WW1.
- **Politics:** How many people need to be included in a poll to predict the outcome of an election?
- **Geography:** Creating a model for the population of a city.

## ■ Reflection

Reflection is the point where you evaluate your results. It should be:
- **Meaningful:** For example, it should link to the aim of the exploration. It could also include commentary on what you have learned and acknowledge any limitations of your results.
- **Critical:** This might include looking at the implications of your results in other, related contexts; a consideration of how the assumptions made affect the ways in which the results can be interpreted; comparing the strengths and weaknesses of different mathematical methods (not all of which need to have been demonstrated in the exploration); and considering whether it is possible to interpret the results in any other way.
- **Substantial:** The reflection should be related to all parts of the exploration. The best way to do this is to include subsections on reflection in every appropriate section.

## ■ Use of mathematics

This is the area where your actual mathematical skills are assessed – things such as the technical accuracy of your algebra and the clarity of your reasoning. It is the only criterion that has a slight difference between the Standard Level and Higher Level.

Your work has to be:

- **Commensurate with the level of the course:** This means that it cannot just be on material from the prior knowledge if you want to score well.
- **Correct:** Although a few small slips which are not fundamental to the progress of the exploration are acceptable, you must make sure that you check your work thoroughly.
- **Thorough:** This means that all mathematical arguments are well understood and clearly communicated. Just obtaining the 'correct' answer is not sufficient to show understanding.

In addition, if you are studying Higher Level Mathematics then to get the top marks the work should be:

- **Sophisticated:** This means that the topic you choose must be sufficiently challenging that routes which would not be immediately obvious to most of your peers would be required.
- **Rigorous:** You should be able to explain why you are doing each process, and you should have researched the conditions under which this process holds and be able to demonstrate that they hold. Where feasible, you should prove any results that are central to your argument.

If you pick the right topic, explorations are a really enjoyable opportunity to get to know a mathematical topic in depth. Make sure that you stick to all the internal deadlines set by your teacher so that you give yourself enough time to complete all parts of the exploration and meet all of the assessment criteria.

# Algebra and confidence

Mathematics, more than many other subjects, splits people into those who love it and those who hate it. We believe that one reason why some people dislike mathematics is because it is a subject where you can often be told you are wrong. However, often the reason someone has got the wrong answer is not because they misunderstand the topic currently being taught, but because they have made a simple slip in algebra or arithmetic which undermines the rest of the work. Therefore, for some mathematicians, a good use of toolkit time might be to revise some of the prior learning so that they have the best possible foundation for the rest of the course. The exercise below provides some possible questions.

**Tip**

Although it might sound impressive to write an exploration on the Riemann Hypothesis or General Relativity, it is just as dangerous to write an exploration on too ambitious a topic as one which is too simple. You will only get credit for maths which you can actually use.

## TOK Links
What are the rules of algebra? Can you know how to do something without having the technical language to describe it?

## Algebra Practice

1 Simplify the following expressions.

 a $3x + 5y + 8x + 10y$
 b $3x^2 + 5xy + 7x^2 + 2xy$
 c $5 + 2(x - 1)$

 d $(2x + 1) - (x - 3)$
 e $(xy) \times (yz)$
 f $7xy \times 9x$

 g $(2x)^3$
 h $\dfrac{x}{2} - \dfrac{x}{3}$
 i $\dfrac{3 + 6x}{3y}$

 j $\dfrac{7xy}{21xz}$
 k $\dfrac{x^2 + 3x}{2x + 6}$
 l $\dfrac{x - y}{y - x}$

 m $\dfrac{x}{3} \times \dfrac{x}{5}$
 n $\dfrac{x + 1}{x + 2} \times \dfrac{2x + 4}{5}$
 o $\dfrac{x}{3} \div \dfrac{y}{6}$

2 Expand the following brackets.

 a $2x(x - 3)$
 b $(x + 3)^2$
 c $(x - 4)(x + 5)$

 d $x(x + 1)(x + 2)$
 e $(x + y + 1)(x + y - 1)$
 f $(x + 1)(x + 3)(x + 5)$

3 Factorize the following expressions.

 a $12 - 8y$
 b $3x - 6y$
 c $7x^2 - 14x$

 d $5yx^2 + 10xy^2$
 e $2z(x + y) + 5(x + y)$
 f $3x(x - 2) + 2 - x$

4 Solve the following linear equations and inequalities.

 a $5x + 9 = 2 - 3x$
 b $3(x + 2) = 18$
 c $4 - (3 - x) = 9 + 2x$

 d $\dfrac{x - 1}{2} + 3 = x$
 e $4 - 2x > 17$
 f $10 - 2x \leqslant 5 + x$

5 Solve the following simultaneous equations.

 a $x + y = 10$
 $x - y = 2$

 b $2x + y = 7$
 $x - y = 2$

 c $2x + 5y = 8$
 $3x + 2y = 1$

6 Evaluate the following expressions.

 a $3x + 4$ when $x = -2$
 b $(3 - x)(5 + 2x)$ when $x = 2$
 c $3 \times 2^x$ when $x = 2$

 d $x^2 - x$ when $x = -1$
 e $\dfrac{2}{x} + 3x$ when $x = \dfrac{1}{6}$
 f $\sqrt{x^2 + 9}$ when $x = 4$

7 Rearrange the following formulae to make $x$ the subject.

 a $y = 2x + 4$
 b $y(3 + x) = 2x$
 c $y = \dfrac{x + 1}{x + 2}$

 d $y = \dfrac{ax}{x - b}$
 e $a = \sqrt{x^2 - 4}$
 f $y = \dfrac{1}{x} - \dfrac{3}{2x}$

8 Simplify the following surd expressions.

 a $(1 + 2\sqrt{2}) - (-1 + 3\sqrt{2})$
 b $2\sqrt{3}\sqrt{5}$
 c $\sqrt{2} \times \sqrt{8}$

 d $\sqrt{3} + \sqrt{12}$
 e $(1 - \sqrt{2})^2$
 f $(2 + \sqrt{5})(2 - \sqrt{5})$

9 **HL only:** Rationalize the following denominators.

 a $\dfrac{2}{\sqrt{2}}$
 b $\dfrac{1}{\sqrt{2} - 1}$
 c $\dfrac{1 + \sqrt{3}}{3 + \sqrt{3}}$

10 **HL only:** Solve the following quadratic equations.

 a $x^2 + 5x + 6 = 0$
 b $4x^2 - 9 = 0$
 c $x^2 + 2x - 5 = 0$

11 **HL only:** Write the following as a single algebraic fraction.

 a $\dfrac{1}{a} - \dfrac{1}{b}$
 b $\dfrac{3}{x} + \dfrac{5}{x^2}$
 c $x + \dfrac{13}{x - 1}$

 d $\dfrac{1}{1 - x} + \dfrac{1}{1 + x}$
 e $\dfrac{5 + a}{a} - \dfrac{a}{5 - a}$
 f $\dfrac{x + 1}{x - 1} + \dfrac{x + 2}{x - 2}$

## ESSENTIAL UNDERSTANDINGS

■ Number and algebra allow us to represent patterns, show equivalences and make generalizations which enable us to model real-world situations.
■ Algebra is an abstraction of numerical concepts and employs variables to solve mathematical problems.

## In this chapter you will learn…

■ how to use the laws of exponents with integer exponents
■ how to perform operations with numbers in the form $a \times 10^k$, where $1 \leq a < 10$ and $k$ is an integer
■ about the number e
■ about logarithms
■ how to solve simple exponential equations.

## CONCEPTS

The following concepts will be addressed in this chapter:
■ Different **representations** of numbers enable quantities to be compared and used for computational purposes with ease and accuracy.
■ Numbers and formulae can appear in different, but equivalent forms, or **representations**, which can help us establish identities.

### Tip

The notation $k \in \mathbb{Z}$ means that $k$ is an integer.

## PRIOR KNOWLEDGE

Before starting this chapter, you should already be able to complete the following:

1   Evaluate the following:

   a   $3^4$                  b   $5 \times 2^3$

2   Write the following values in the form $a \times 10^k$ where $1 \leq a < 10$ and $k \in \mathbb{Z}$.

   a   342.71            b   0.00856

3   Express 64 in the form $2^k$, where $k \in \mathbb{Z}$.

■ **Figure 1.1** How do we compare very large numbers?

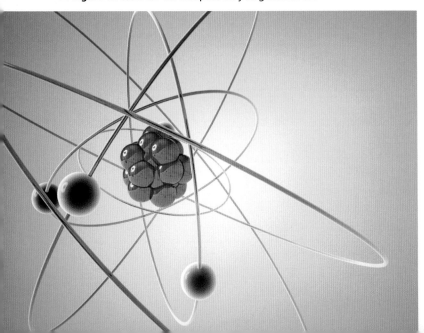

When you first study maths, it is easy to think it is a subject that is all about numbers. But actually, the study of numbers is just one area in which we apply the logic of maths. Once you have experience of numbers, you will be looking to make links: for example, $3^2 + 4^2 = 5^2$, but is this true in general for consecutive numbers? Of course the answer is 'no', but $3^2 \times 4^2 = 12^2$ gives you a rule that works when applied to multiplications.

In this chapter we will be generalizing some patterns you are probably already aware of. You will then be applying these rules to numbers represented in different ways. You might wonder why we need a different form for writing numbers, but try writing down the number of atoms in the sun. When numbers are very large or very small, it is useful to have a more convenient form to express them in.

Finally, you will see how to solve equations where the unknown is in the exponent, such as $10^x = 7$. In so doing you will meet a number that has wide-ranging applications in fields including science, economics and engineering.

## Starter Activity

Look at the pictures in Figure 1.1. In small groups discuss whether there are more atoms in a jug of water than there are jugs of water in the Atlantic Ocean.

**Now look at this problem:**

A model suggests that the level of carbon dioxide ($CO_2$) in the atmosphere in parts per million is given by $400 \times 1.05^n$, where $n$ is the number of years after 2018.

Use a spreadsheet to estimate when this model predicts the level of $CO_2$ in the atmosphere will reach 1000 parts per million.

**LEARNER PROFILE** – Inquirers
Is mathematics invented or discovered? Is mathematics designed to mirror reality?

# 1A Laws of exponents

### TOK Links

In expressions like $a^n$, $a$ is called the **base** and $n$ is called the **exponent**, although you may see it referred to as power, index or indice. Does having a label help you understand it? Would it be better if everybody used the same label?

In your previous work you have may have noticed that $2^3 \times 2^2 = (2 \times 2 \times 2) \times (2 \times 2) = 2^5$. Based on specific examples such as this you can generalize to a formula.

## KEY POINT 1.1

- $a^m \times a^n = a^{m+n}$
- $\dfrac{a^m}{a^n} = a^{m-n}$
- $(a^m)^n = a^{mn}$

 To formally prove these rules requires a method called mathematical induction, which you will encounter if you are studying the Mathematics: analysis and approaches HL course.

If we set $m$ and $n$ equal in the second law in Key Point 1.1 then it follows that $a^0 = 1$ for any (non-zero) value of $a$.

 **TOOLKIT:** Problem Solving

Anything raised to the power 0 is 1, but 0 to any power is 0. So, what is the value of $0^0$?

## WORKED EXAMPLE 1.1

Simplify $x^6 \times x^3$.

Use $a^m \times a^n = a^{m+n}$ $\qquad\qquad x^6 \times x^3 = x^{6+3}$

$= x^9$

## WORKED EXAMPLE 1.2

Simplify $\dfrac{y^8}{y^2}$.

Use $\dfrac{a^m}{a^n} = a^{m-n}$ $\qquad\qquad \dfrac{y^8}{y^2} = y^{8-2}$

$= y^6$

**WORKED EXAMPLE 1.3**

Simplify $\left(7^3\right)^5$. You do not need to evaluate your result.

Use $\left(a^m\right)^n = a^{mn}$ ............... $\left(7^3\right)^5 = 7^{3\times5}$

$= 7^{15}$

If the expression has numbers as well as algebraic values, just multiply or divide the numbers separately. This follows because we can do multiplication in any convenient order.

**WORKED EXAMPLE 1.4**

Simplify $2b^3 \times 7b^4$.

You can reorder the ............... $2b^3 \times 7b^4 = 14b^3b^4$
multiplication as $2 \times 7 \times b^3 \times b^4$

Then $b^3 \times b^4 = b^{3+4}$ ............... $= 14b^7$

**WORKED EXAMPLE 1.5**

Simplify $\dfrac{6c^9}{2c^4}$.

You can use your knowledge of fractions to ............... $\dfrac{6c^9}{2c^4} = \dfrac{6}{2} \times \dfrac{c^9}{c^4}$
write the expression as a numeric fraction
multiplied by an algebraic fraction

Use $\dfrac{c^9}{c^4} = c^{9-4}$ ............... $= 3c^5$

The laws in Key Point 1.1 work for negative as well as positive integers. But what is the meaning of a negative exponent? In the second law in Key Point 1.1 we can set $m = 0$ and use the fact that $a^0 = 1$ to deduce a very important rule.

**KEY POINT 1.2**

● $a^{-n} = \dfrac{1}{a^n}$

**WORKED EXAMPLE 1.6**

Without a calculator write $2^{-4}$ as a fraction in its simplest terms.

Use $a^{-n} = \dfrac{1}{a^n}$ ............... $2^{-4} = \dfrac{1}{2^4}$

$= \dfrac{1}{16}$

**WORKED EXAMPLE 1.7**

Write $4x^{-3}$ as a fraction.

Apply the power $-3$ to $x$ (but not to 4) .................................... $4x^{-3} = 4 \times \dfrac{1}{x^3}$

$$= \dfrac{4}{x^3}$$

You will often need to apply an exponent to a product or fraction. We can use the fact that multiplication can be reordered to help suggest a rule. For example,

$$(ab)^3 = (ab)(ab)(ab) = (a \times a \times a) \times (b \times b \times b) = a^3 b^3$$

This suggests the following generalization.

**KEY POINT 1.3**

- $(ab)^n = a^n \times b^n$
- $\left(\dfrac{a}{b}\right)^n = \dfrac{a^n}{b^n}$

**WORKED EXAMPLE 1.8**

Simplify $\left(2x^2 y^{-4}\right)^5$.

Apply the power 5 to each term in the product .................... $\left(2x^2 y^{-4}\right)^5 = 2^5 \left(x^2\right)^5 \left(y^{-4}\right)^5$

$$= 32x^{10} y^{-20}$$

**WORKED EXAMPLE 1.9**

Simplify $\left(\dfrac{2u}{3v^2}\right)^{-3}$.

Apply the power $-3$ to each part of the fraction ................ $\left(\dfrac{2u}{3v^2}\right)^{-3} = \dfrac{2^{-3} u^{-3}}{3^{-3} \left(v^2\right)^{-3}}$

$$= \dfrac{\dfrac{1}{8} \times \dfrac{1}{u^3}}{\dfrac{1}{27} \times \dfrac{1}{v^6}}$$

$$= \dfrac{\left(\dfrac{1}{8u^3}\right)}{\left(\dfrac{1}{27v^6}\right)}$$

Four level fractions are easiest dealt with by dividing two normal fractions. Flip the second fraction and multiply ............ $= \dfrac{1}{8u^3} \div \dfrac{1}{27v^6}$

$$= \dfrac{1}{8u^3} \times \dfrac{27v^6}{1}$$

$$= \dfrac{27v^6}{8u^3}$$

**Tip**

Perhaps an easier way to do Worked Example 1.9 is to use the last rule from Key Point 1.1 to write the expression as $\left(\left(\dfrac{2u}{3v^2}\right)^3\right)^{-1}$, but it is good to practise working with four level fractions too!

## Be the Examiner 1.1

Simplify $\left(2x^2 y^3\right)^4$.

Which is the correct solution? Identify the errors made in the incorrect solutions.

| Solution 1 | Solution 2 | Solution 3 |
|---|---|---|
| $\left(2x^2y^3\right)^4 = 2^4\left(x^2\right)^4\left(y^3\right)^4$ | $\left(2x^2y^3\right)^4 = 2^4\left(x^2\right)^4\left(y^3\right)^4$ | $\left(2x^2y^3\right)^4 = 2^4\left(x^2\right)^4\left(y^3\right)^4$ |
| $= 16\,x^6y^7$ | $= 8\,x^8y^{12}$ | $= 16\,x^8y^{12}$ |

More complicated expressions may have to be simplified before the laws of exponents are applied.

### WORKED EXAMPLE 1.10

Simplify $\dfrac{16a^5b^2 - 12ab^4}{4ab^2}$.

You can split a fraction up if the top is a sum or a difference ............ $\dfrac{16a^5b^2 - 12ab^4}{4ab^2} = \dfrac{16a^5b^2}{4ab^2} - \dfrac{12ab^4}{4ab^2}$

Simplify numbers, $a$s and $b$s separately ················· $= 4a^4 - 3b^2$

In Chapters 9 and 10 you will need to use the laws of exponents to simplify expressions before they can be differentiated or integrated.

You can only apply the laws of exponents when the bases are the same. Sometimes you can rewrite one of the bases to achieve this.

### WORKED EXAMPLE 1.11

Express $3^4 \times 9^5$ in the form $3^k$, for some integer $k$.

Write 9 as $3^2$ ······················ $3^4 \times 9^5 = 3^4 \times \left(3^2\right)^5$

$= 3^4 \times 3^{10}$

$= 3^{14}$

$\therefore k = 14$

This technique can be used to solve some equations.

An equation like this with the unknown ($x$) in the power is called an **exponential equation**. In Section 1C, you will see how to solve more complicated examples using logarithms.

### WORKED EXAMPLE 1.12

Solve $2^{x+6} = 8^x$.

Write 8 as $2^3$ ···················· $2^{x+6} = 8^x$

$2^{x+6} = \left(2^3\right)^x$

$2^{x+6} = 2^{3x}$

Since the bases are the same, the exponents must be equal ············· $\therefore x + 6 = 3x$

$2x = 6$

$x = 3$

The laws of exponents and your prior knowledge of algebra can be applied to modelling real-life situations.

---

**WORKED EXAMPLE 1.13**

The length of a baby fish is modelled by $L = 2t^2$ where $t$ is the age in days and $L$ is the length in cm. Its mass in grams is modelled by $M = 4L^3$.

a  Find and simplify an expression for $M$ in terms of $t$.
b  Find the age of the fish when the model predicts a mass of 1000 g.
c  Explain why the model is unlikely to still hold after 100 days.

Substitute $L = 2t^2$ into the expression for $M$ ⋯⋯⋯ **a** $M = 4L^3$

$$= 4(2t^2)^3$$

Apply the exponent 3 to 2 and to $t^2$ ⋯⋯⋯⋯ $= 4\left(2^3\left(t^2\right)^3\right)$

$$= 4\left(8t^6\right)$$

$$= 32t^6$$

You need to find the value of $t$ when $M = 1000$ ⋯⋯⋯ **b** When $M = 1000$,

$$1000 = 32t^6$$

$$t^6 = \frac{1000}{32}$$

Take the sixth root of both sides to find $t$ ⋯⋯⋯⋯⋯ $t = \sqrt[6]{\frac{1000}{32}} = 1.77$ days

**c** The model predicts that the fish will continue growing, whereas in reality it is likely that after 100 days the fish will be growing far more slowly, if at all.

---

## Exercise 1A

For questions 1 to 4, use the method demonstrated in Worked Example 1.1 to simplify the expressions. If numerical, you do not need to evaluate your result.

1  a  $x^2 \times x^4$
   b  $x^5 \times x^7$

2  a  $y^3 \times y^3$
   b  $z^5 \times z^5$

3  a  $a^6 \times a$
   b  $a^{10} \times a$

4  a  $5^7 \times 5^{10}$
   b  $2^{12} \times 2^{12}$

For questions 5 to 8, use the method demonstrated in Worked Example 1.2 to simplify the expressions. If numerical, you do not need to evaluate your result.

5  a  $\dfrac{x^4}{x^3}$
   b  $\dfrac{x^8}{x^5}$

6  a  $y^8 \div y^4$
   b  $z^9 \div z^3$

7  a  $\dfrac{b^7}{b}$
   b  $\dfrac{b^9}{b}$

8  a  $11^{12} \div 11^4$
   b  $7^{10} \div 7^5$

For questions 9 to 12, use the method demonstrated in Worked Example 1.3 to simplify the expressions. If numerical, you do not need to evaluate your result.

9  a  $\left(x^3\right)^5$
   b  $\left(x^4\right)^8$

10  a  $\left(y^4\right)^4$
    b  $\left(z^5\right)^5$

11  a  $\left(c^7\right)^2$
    b  $\left(c^2\right)^7$

12  a  $\left(3^5\right)^{10}$
    b  $\left(13^7\right)^4$

For questions 13 to 15, use the method demonstrated in Worked Example 1.4 to simplify the expressions.

13    a    $12x^2 \times 4x^5$       14    a    $a \times 3a^2$       15    a    $5x^2yz \times 4x^3y^2$

     b    $3x^4 \times 5x^3$            b    $b^2 \times 5b^2$               b    $6x^7yz^2 \times 2xz^3$

For questions 16 to 19, use the method demonstrated in Worked Example 1.5 to simplify the expressions.

16    a    $\dfrac{10x^{10}}{5x^5}$       17    a    $\dfrac{8x}{16x^4}$       18    a    $15x^2 \div 9x^4$       19    a    $14x^3y^5 \div 2xy^2$

     b    $\dfrac{9x^9}{3x^3}$             b    $\dfrac{5x^2}{20x^3}$           b    $21x^5 \div 28x^3$       b    $6x^5yz^2 \div 3x^2y^2z$

For questions 20 to 23, use the method demonstrated in Worked Example 1.6 to write the expression as a fraction in its simplest terms.

20    a    $10^{-1}$       21    a    $3^{-3}$       22    a    $\left(\dfrac{3}{4}\right)^{-1}$       23    a    $\left(\dfrac{2}{3}\right)^{-2}$

     b    $7^{-1}$             b    $5^{-2}$              b    $\left(\dfrac{5}{7}\right)^{-1}$         b    $\left(\dfrac{2}{5}\right)^{-3}$

For questions 24 to 26, use the method demonstrated in Worked Example 1.7 to write the expression as a fraction in its simplest terms.

24    a    $7 \times 3^{-2}$       25    a    $6x^{-1}$       26    a    $3^{-2} \div 2^{-3}$

     b    $5 \times 2^{-4}$           b    $10x^{-4}$          b    $4^{-3} \div 3^{-4}$

For questions 27 to 29, use the method demonstrated in Worked Example 1.8 to simplify each expression.

27    a    $\left(3u^{-2}\right)^3$       28    a    $\left(2a^{-5}\right)^{-2}$       29    a    $\left(5x^2y^3\right)^2$

     b    $\left(2v^{-3}\right)^5$        b    $\left(3b^{-7}\right)^{-3}$        b    $\left(3a^2b^{-2}\right)^4$

For questions 30 to 32, use the method demonstrated in Worked Example 1.9 to simplify each expression.

30    a    $\left(\dfrac{x}{3}\right)^3$       31    a    $\left(\dfrac{3x^2}{2y^3}\right)^3$       32    a    $\left(\dfrac{3u}{4v^2}\right)^{-2}$

     b    $\left(\dfrac{5}{x^2}\right)^2$         b    $\left(\dfrac{5uv^3}{7b^4}\right)^2$        b    $\left(\dfrac{2a^3}{3b^2}\right)^{-3}$

For questions 33 to 35, use the method demonstrated in Worked Example 1.10 to simplify each expression.

33    a    $\dfrac{6x^2 - 21x^7}{3x}$       34    a    $\dfrac{15u^3v + 18u^2v^3}{3uv}$       35    a    $\dfrac{10p^3q - 6pq^3}{2p^2q}$

     b    $\dfrac{15y^4 + 25y^6}{5y^3}$      b    $\dfrac{20a^5b^7c^2 - 16a^4b^2c^3}{4a^3b^2c}$      b    $\dfrac{14s^4t^3 + 21s^5t^7}{7s^2t^5}$

For questions 36 to 39, use the method demonstrated in Worked Example 1.11 to express each value in the form $a^b$ for the given prime number base $a$.

36    a    Express $9^4$ as a power of 3.                   b    Express $27^8$ as a power of 3.

37    a    Express $2^5 \times 4^2$ as a power of 2.          b    Express $5^4 \times 125^2$ as a power of 5.

38    a    Express $4^7 \div 8^3$ as a power of 2.            b    Express $27^5 \div 9^2$ as a power of 3.

39    a    Express $8^3 \times 2^7 + 4^8$ as a power of 2.      b    Express $16^2 - 4^8 \div 8^3$ as a power of 2.

For questions 40 to 43, use the method demonstrated in Worked Example 1.12 to solve each equation to find the unknown value.

40    a    Solve $3^x = 81$.       41    a    Solve $2^{x+4} = 8$.       42    a    Solve $7^{3x-5} = 49$.       43    a    Solve $3^{2x+5} = \dfrac{1}{27}$.

     b    Solve $5^x = 125$.          b    Solve $3^{x-3} = 27$.        b    Solve $4^{2x-7} = 16$.        b    Solve $2^{3x+5} = \dfrac{1}{16}$.

**44** Simplify $\dfrac{4x^2 + 8x^3}{2x}$.    **45** Simplify $\dfrac{\left(2x^2 y\right)^3}{8xy}$.    **46** Write $\left(2ab^{-2}\right)^{-3}$ without brackets or negative indices.

**47** The number of people suffering from a disease 'D' in a country is modelled by $D = 1\,000\,000n^{-2}$ where $n$ is the amount spent on prevention (in millions of dollars).

   a   Rearrange the equation to find $n$ in terms of $D$.

   b   According to the model, how many people will have the disease if $2 million dollars is spent on prevention?

   c   How much must be spent to reduce the number of people with the disease to $10\,000$?

**48** A computer scientist analyses two different methods for finding the prime factorization of a number. They both take a time $T$ microseconds that depends on the number of digits ($n$).

Method A: $T_A = k_A n^3$

Method B: $T_B = k_B n^2$

Both methods take 1000 microseconds to factorize a five digit number.

   a   Find the values of $k_A$ and $k_B$.

   b   Find and simplify an expression for $\dfrac{T_A}{T_B}$.

   c   Which method would be quicker at factorizing a 10 digit number? Justify your answer.

**49** Solve $10 + 2 \times 2^x = 18$.    **50** Solve $9^x = 3^{x+5}$.    **51** Solve $5^{x+1} = 25 \times 5^{2x}$.

**52** Solve $8^x = 2 \times 4^{2x}$.    **53** Solve $25^{2x+4} = 125 \times 5^{x-1}$.

**54** The pressure ($P$) in a gas is equal to $0.8T$ where $T$ is the temperature measured in kelvin. The air resistance, $R$, in newtons, of an aeroplane is modelled by $R = 5P^2$.

   a   Find an expression for $R$ in terms of $T$.

   b   If the air resistance has a magnitude of $200\,000$ newtons, find the temperature in kelvin.

**55** A boat travels 3 km at a speed of $v$ km per hour.

   a   Find an expression for the time taken.

The boat uses up $0.5v^2$ litres of petrol per hour.

   b   Find an expression for the amount of petrol used in the 3 km journey.

The boat has 60 litres of fuel.

   c   Find the maximum speed the boat can travel at if it is to complete the 3 km journey.

**56** Solve the simultaneous equations:

$$8^x 2^y = 1 \text{ and } \frac{4^x}{2^y} = 32$$

**57** Solve $6^x = 81 \times 2^x$.    **58** Solve $32 + 2^{x-1} = 2^x$.

**59** Find all solutions to $(x-2)^{x+5} = 1$.

**60** Determine, with justification, which is larger out of $2^{7000}$ and $5^{3000}$.

**61** What is the last digit of $316^{316} + 631^{631}$?

**Tip**

The notation $k \in \mathbb{Z}$ means that $k$ is an integer.

# 1B Operations with numbers in the form $a \times 10^k$, where $1 \leq a < 10$ and $k$ is an integer

You already know that it can be useful to write very large or very small numbers in the form $a \times 10^k$ where $1 \leq a < 10$ and $k \in \mathbb{Z}$. This is often referred to as **standard index form**, standard form or scientific notation.

**You are the Researcher**

Extremely large numbers require other ways of representing them. You might want to research tetration and the types of number – such as Graham's Number – which require this notation for them to be written down.

You now need to be able to add, subtract, multiply and divide numbers in this form.

**WORKED EXAMPLE 1.14**

Without a calculator, write $(3 \times 10^7) \times (4 \times 10^{-3})$ in the form $a \times 10^k$ where $1 \leq a < 10$ and $k \in \mathbb{Z}$.

Reorder so that the respective parts of each number are together. $\qquad (3 \times 10^7) \times (4 \times 10^{-3}) = (3 \times 4) \times (10^7 \times 10^{-3})$

$10^7 \times 10^{-3} = 10^{7+(-3)} \qquad\qquad = 12 \times 10^4$

$12 \times 10^4 = (1.2 \times 10) \times 10^4 \qquad\qquad = 1.2 \times 10^5$

**WORKED EXAMPLE 1.15**

Show that if $x = 3 \times 10^7$ and $y = 4 \times 10^{-2}$, then $\frac{x}{y} = 7.5 \times 10^8$.

Split off the powers of 10 $\qquad\qquad \dfrac{x}{y} = \dfrac{3 \times 10^7}{4 \times 10^{-2}}$

$$= \frac{3}{4} \times \frac{10^7}{10^{-2}}$$

$$= 0.75 \times 10^{7-(-2)}$$

$$= 0.75 \times 10^9$$

Change the number into standard form $\qquad\qquad = (7.5 \times 10^{-1}) \times 10^9$

$$= 7.5 \times 10^{-1+9}$$

$$= 7.5 \times 10^8$$

## TOK Links

In Worked Example 1.15, we made explicit the rules of indices used, but just assumed that $\frac{3}{4} = 0.75$ was obvious. When you were 11 years old you might have had to explain this bit too, but part of mathematics is knowing the mathematical level and culture of your audience. If you continue with mathematics, then in a few years' time you would not be expected to explain the rules of indices anymore as everybody reading your explanations will probably know them. Is this unique to mathematics or do explanations in every area of knowledge depend on the audience?

With a 'show that' question, like that in Worked Example 1.15, you need to be able to explain each step in the calculation; most of the time, however, you will be able to do this type of arithmetic on a calculator, as shown below.

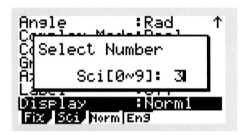

Note that in this mode, the output 1.59E+05 shown on the right means $1.59 \times 10^5$.

You can choose the number of significant figures to which the number is rounded (here 3 was chosen, as shown on the left).

---

**WORKED EXAMPLE 1.16**

Show that $3.2 \times 10^{19} + 4.5 \times 10^{20} = 4.82 \times 10^{20}$.

Write $3.2 \times 10^{19}$ as $0.32 \times 10^{20}$ so that there is clearly a factor of $10^{20}$ ·········· $3.2 \times 10^{19} + 4.5 \times 10^{20} = 0.32 \times 10^{20} + 4.5 \times 10^{20}$

$$= (0.32 + 4.5) \times 10^{20}$$

Change the number into standard form ·················· $= 4.82 \times 10^{20}$

---

In the 'show that' questions you need to be able to explain all the steps without referring to a calculator. Most of the time, however, you will be able to do this type of arithmetic on a calculator.

---

## CONCEPTS – REPRESENTATION

$5^{20}$ and $9.5367... \times 10^{13}$ are both **representations** of the same number. However, they have different uses. The second probably gives you a better sense of the scale of the number, but the former might be more useful in solving $5^{2x} = 5^{20}$ or comparing it to $3^{18}$. One major skill in mathematics is deciding which representation to use in which problem.

## Exercise 1B

Questions 1 to 4 are designed to remind you of your prior learning.

1 a   Write $3.2 \times 10^4$ as a normal number.

   b   Write $6.92 \times 10^6$ as a normal number.

2 a   Write $4.8 \times 10^{-2}$ as a decimal.

   b   Write $9.85 \times 10^{-4}$ as a decimal.

3 a   Write the value 612.07 in standard index form.

   b   Write the value 3076.91 in standard index form.

4 a   Write the value 0.003 061 7 in standard index form.

   b   Write the value 0.022 19 in standard index form.

For questions 5 to 7, use the method demonstrated in Worked Example 1.14 to write the given expressions in standard index form. Do not use a calculator.

5 a   $(2 \times 10^4) \times (3.4 \times 10^3)$     6 a   $(5 \times 10^4) \times (2 \times 10^{-5})$     7 a   $\left(5 \times 10^{10}\right)^2$

  b   $(3.2 \times 10^5) \times (3 \times 10^6)$       b   $(3 \times 10^{-2}) \times (4 \times 10^{-4})$       b   $\left(6 \times 10^6\right)^2$

For questions 8 to 10, use the method demonstrated in Worked Example 1.15 to write the given expressions in standard index form. Do not use a calculator.

8 a   $\dfrac{6 \times 10^2}{3 \times 10^4}$            9 a   $\dfrac{1 \times 10^5}{2 \times 10^4}$         10 a   $\left(8.4 \times 10^{14}\right) \div \left(4 \times 10^3\right)$

  b   $\dfrac{8 \times 10^6}{2 \times 10^{10}}$          b   $\dfrac{2 \times 10^5}{8 \times 10^2}$           b   $\left(9.3 \times 10^{15}\right) \div \left(3 \times 10^6\right)$

For questions 11 to 13, use the method demonstrated in Worked Example 1.16 to write the given expressions in standard index form. Do not use a calculator.

11 a   $1 \times 10^4 + 2 \times 10^5$      12 a   $8 \times 10^5 - 4 \times 10^4$      13 a   $(2.1 \times 10^3) + (3.8 \times 10^4)$

   b   $3 \times 10^8 + 2 \times 10^6$        b   $9 \times 10^{14} - 9 \times 10^{12}$       b   $(5.7 \times 10^{13}) + (4.3 \times 10^{12})$

14   Show that if $a = 4 \times 10^6$, $b = 5 \times 10^{-3}$, then $a \times b = 2 \times 10^4$.

15   Show that if $c = 1.4 \times 10^3$, $d = 5 \times 10^8$, then $c \times d = 7 \times 10^{11}$.

16   Show that if $a = 4 \times 10^6$, $b = 5 \times 10^{-3}$, then $\dfrac{a}{b} = 8 \times 10^8$.

17   Show that if $c = 1.4 \times 10^3$, $d = 2 \times 10^8$, then $\dfrac{c}{d} = 7 \times 10^{-6}$.

18   Show that if $a = 4.7 \times 10^6$, $b = 7.1 \times 10^5$, then $a - b = 3.99 \times 10^6$.

19   Show that if $c = 3.98 \times 10^{13}$, $d = 4.2 \times 10^{14}$, then $d - c = 3.802 \times 10^{14}$.

20   Let $p = 12.2 \times 10^7$ and $q = 3.05 \times 10^5$.

   a   Write $p$ in the form $a \times 10^k$ where $1 \leqslant a < 10$ and $k \in \mathbb{Z}$.

   b   Evaluate $\dfrac{p}{q}$.

   c   Write your answer to part b in the form $a \times 10^k$ where $1 \leqslant a < 10$ and $k \in \mathbb{Z}$.

21   The number of atoms in a balloon is approximately $6 \times 10^{23}$. Theoretical physics predicts that there are approximately $10^{80}$ atoms in the known universe. What proportion of atoms in the known universe are found in the balloon?

22   12 grams of carbon contains $6.02 \times 10^{23}$ atoms. What is the mass (in grams) of one atom of carbon? Give your answer in the form $a \times 10^k$ where $1 \leqslant a < 10$ and $k \in \mathbb{Z}$.

23   The diameter of a uranium nucleus is approximately 15 fm where 1 fm is $10^{-15}$ m.

   a   Write the diameter (in metres) in the form $a \times 10^k$ where $1 \leqslant a < 10$ and $k \in \mathbb{Z}$.

If the nucleus is modelled as a sphere, then the volume is given by $V = \dfrac{1}{6}\pi d^3$ where $d$ is the diameter.

   b   Estimate the volume of a uranium nucleus in metres cubed.

24  The area of Africa is approximately $3.04 \times 10^{13}\,\text{m}^2$. The area of Europe is approximately $1.02 \times 10^{13}\,\text{m}^2$. The population of Africa is approximately 1.2 billion and the population of Europe is 741 million.

    a  How many times bigger is Africa than Europe?

    b  Write the population of Europe in the form $a \times 10^k$ where $1 \leq a < 10$ and $k \in \mathbb{Z}$.

    c  Does Africa or Europe have more people per metre squared? Justify your answer.

25  You are given that
$$\left(3 \times 10^a\right) \times \left(5 \times 10^b\right) = c \times 10^d$$
where $1 \leq c < 10$ and $d \in \mathbb{Z}$.

    a  Find the value of $c$.

    b  Find an expression for $d$ in terms of $a$ and $b$.

26  You are given that
$$\left(2 \times 10^a\right) \div \left(5 \times 10^b\right) = c \times 10^d$$
where $1 \leq c < 10$ and $d \in \mathbb{Z}$.

    a  Find the value of $c$.

    b  Find an expression for $d$ in terms of $a$ and $b$.

27  $x = a \times 10^p$ and $y = b \times 10^q$ where $4 < a < b < 9$. When written in standard form, $xy = c \times 10^r$. Express $r$ in terms of $p$ and $q$.

# 1C Logarithms

## ■ Introduction to logarithms

If you want to find the positive value of $x$ for which $x^2 = 5$ you can use the square root function: $x = \sqrt{5} \approx 2.236$.

There is also a function that will let you find the value of $x$ such that, say, $10^x = 5$. That function is logarithm with base 10: $x = \log_{10} 5 \approx 0.699$.

Although base 10 logarithms are common, any positive base other than 1 can be used.

**Tip**

Usually $\log_{10} x$ will just be written as $\log x$.

---

**KEY POINT 1.4**

$a = b^x$ is equivalent to $\log_b a = x$.

---

### TOK Links

The study of logarithms is usually attributed to the Scottish mathematician John Napier. Do you think he discovered something which already existed or invented something new?

As $10^x$ is always positive, note that there is no answer to a question such as '10 raised to what exponent gives $-2$?' Therefore, you can only take logarithms of positive numbers.

---

**WORKED EXAMPLE 1.17**

Without a calculator, calculate the value of log 1000.

Let the value of log 1000 be $x$ ·········· $x = \log 1000$

Use $a = b^x$ is equivalent to $\log_b a = x$ ·········· $\therefore 10^x = 1000$

You can see (or experiment to find) that $x = 3$ ·········· $\log 1000 = 3$

---

**WORKED EXAMPLE 1.18**

Find the exact value of $y$ if $\log_{10} (y + 1) = 3$.

Use $a = b^x$ is equivalent to $\log_b a = x$ ·········· $\log_{10} (y + 1) = 3$

$$y + 1 = 10^3 = 1000$$

$$y = 999$$

The number e has many applications in real-world systems due to the fact that it has a very special rate of change. You will learn much more about it in Chapter 20 in relation to calculus.

Another very common base for logarithms is the number $e \approx 2.718\,28$, which is an irrational number a bit like $\pi$.

The logarithm with base e is called the natural logarithm and is written as $\ln x$.

---

**WORKED EXAMPLE 1.19**

Make $x$ the subject of $\ln(3x - 2) = y$.

Use $a = b^x$ is equivalent to $\log_b a = x$, with $b = e$ ·········· $\ln(3x - 2) = y$

$$3x - 2 = e^y$$

$$x = \frac{e^y + 2}{3}$$

---

Logarithms can be treated algebraically with all the usual rules applying.

---

**WORKED EXAMPLE 1.20**

Simplify $\dfrac{3\ln x - \ln x}{\ln x}$.

Since $3y - y = 2y$ we can simplify the numerator ·········· $\dfrac{3\ln x - \ln x}{\ln x} = \dfrac{2\ln x}{\ln x}$

Divide top and bottom by $\ln x$ ·········· $= 2$

---

Since the process of taking a logarithm with base $b$ reverses the process of raising $b$ to an exponent, you have the following important results.

> **KEY POINT 1.5**
>
> For base 10:
> - $\log 10^x = x$
> - $10^{\log x} = x$
>
> For base e:
> - $\ln e^x = x$
> - $e^{\ln x} = x$

---

**WORKED EXAMPLE 1.21**

Without using a calculator, find the exact value of $e^{3\ln 2}$.

| | |
|---|---|
| Use the law of exponents $a^{mn} = (a^m)^n$ | $e^{3\ln 2} = \left(e^{\ln 2}\right)^3$ |
| $e^{\ln 2} = 2$ | $= 2^3$ |
| | $= 8$ |

## ■ Numerical evaluation of logarithms using technology

You need to be able to use your graphical display calculator (GDC) to evaluate logarithms.

---

**WORKED EXAMPLE 1.22**

Using a calculator, find the value of $\ln(345678)$, giving your answer in standard index form to three significant figures.

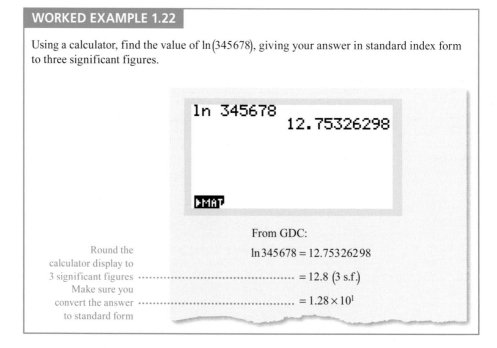

|  |  |
|---|---|
| | From GDC: |
| | $\ln 345678 = 12.75326298$ |
| Round the calculator display to 3 significant figures | $= 12.8 \ (3 \text{ s.f.})$ |
| Make sure you convert the answer to standard form | $= 1.28 \times 10^1$ |

## ■ Solving exponential equations

You can use technology to find approximate solutions to exponential equations.

> **WORKED EXAMPLE 1.23**
>
> Solve the equation $e^x = 4$, giving your answer to three significant figures.
>
> Use $a = b^x$ is equivalent to $\log_b a = x$, with $b = e$ ·············· $x = \ln 4$
>
> Evaluate using ············· $= 1.39 \ (3 \text{ s.f.})$
> your GDC

Sometimes you need to find exact solutions.

> **WORKED EXAMPLE 1.24**
>
> Solve $10^{x-2} = 21$, giving your answer in an exact form.
>
> Use $a = b^x$ is equivalent to $\log_b a = x$ ·············· $x - 2 = \log 21$
>
> Rearrange ·············· $x = 2 + \log 21$

You might have to rearrange an equation first.

> **WORKED EXAMPLE 1.25**
>
> Solve $e^{3x+2} = 4e^x$, giving your answer correct to three significant figures.
>
> This equation is not of the form $a = b^x$ so you ·············· $e^{3x+2} = 4e^x$
> cannot just introduce logarithms immediately.
> Instead, aim to express everything as a power of e $\qquad e^{3x+2} = e^{\ln 4}e^x$
>
> Use the $a^m a^n = a^{m+n}$ on the RHS ·············· $e^{3x+2} = e^{\ln 4 + x}$
>
> Since the bases are the same, the ·············· $3x + 2 = \ln 4 + x$
> exponents must be equal as well
>
> $2x = \ln 4 - 2$
>
> $x = \dfrac{\ln 4 - 2}{2}$
>
> $= -0.307 \ (3 \text{ s.f.})$

## Tip

There are many possible approaches to answering the question in Worked Example 1.25. For example, you could also have divided both sides by $e^x$ and then taken natural logs of both sides.

## Exercise 1C

For questions 1 to 4, without using a calculator, find the exact value of each of the given logarithms, using the method demonstrated in Worked Example 1.17.

1   a   $\log 10$
    b   $\log 100$

2   a   $\log 100\,000$
    b   $\log 1\,000\,000$

3   a   $\log 1$
    b   $\log 0.1$

4   a   $\log 0.01$
    b   $\log 0.0001$

For questions 5 to 8, without using a calculator, find the exact value of each of the given logarithms, using the method demonstrated in Worked Example 1.17. Note that these logs are to a base other than 10, but practising these will enhance your understanding.

5   a   $\log_2 2$
    b   $\log_2 4$

6   a   $\log_3 27$
    b   $\log_3 81$

7   a   $\log_5 1$
    b   $\log_5 0.2$

8   a   $\log_2 0.25$
    b   $\log_2 0.125$

For questions 9 to 11, without using a calculator, find the exact value of $x$ in each of the given logarithms, using the method demonstrated in Worked Example 1.18.

9   a   $\log x = 2$
    b   $\log x = 3$

10  a   $\log(2x - 4) = 5$
    b   $\log(3x + 4) = 3$

11  a   $\log(2x - 3) = -1$
    b   $\log(x + 2) = -2$

For questions 12 to 14, rearrange each equation to find an expression for $x$, using the method demonstrated in Worked Example 1.19.

12  a   $\ln x = 2$
    b   $\ln x = 5$

13  a   $\ln x = y + 1$
    b   $\ln x = y^2$

14  a   $\ln(2x + 4) = y - 3$
    b   $\ln\left(\frac{1}{2}x - 6\right) = 2y + 1$

For questions 15 to 17, simplify each expression using the method demonstrated in Worked Example 1.20.

15  a   $\log x + 4\log x$
    b   $10\log x - 5\log x$

16  a   $\dfrac{2(\log 3x)^2}{\log 3x}$
    b   $\dfrac{\log 2x}{(\log 2x)^2}$

17  a   $\dfrac{\ln x + (\ln x)^2}{\ln x} - \ln x$
    b   $\dfrac{\ln x - (\ln x)^3}{(\ln x)^2} + \ln x$

For questions 18 to 25, simplify each expression using the method demonstrated in Worked Example 1.21 and Key Point 1.5.

18  a   $\log 10^{-15}$
    b   $\log 10^{17}$

19  a   $\ln e^{4.5}$
    b   $\ln e^{-1.5}$

20  a   $10^{\log 13}$
    b   $10^{\log 7}$

21  a   $e^{\ln 3}$
    b   $e^{\ln 7}$

22  a   $10^{3\log 2}$
    b   $10^{2\log 3}$

23  a   $e^{2\ln 5}$
    b   $e^{4\ln 3}$

24  a   $10^{-\log 3}$
    b   $10^{-\log 6}$

25  a   $e^{-5\ln 2}$
    b   $e^{-3\ln 4}$

For questions 26 to 28, use the method demonstrated in Worked Example 1.22 (that is, use technology) to evaluate the following to three significant figures.

26  a   $\log 124.7$
    b   $\log 1399.8$

27  a   $\ln 245.3$
    b   $\ln 17.9$

28  a   $\log 0.5$
    b   $\log 0.04$

For questions 29 to 31, use the method demonstrated in Worked Example 1.23 to solve the given equations, giving an exact answer.

29  a   $10^x = 5$
    b   $10^x = 7$

30  a   $10^x = 0.2$
    b   $10^x = 0.06$

31  a   $e^x = 3$
    b   $e^x = 7$

For questions 32 to 34, use the method demonstrated in Worked Example 1.24 to solve the given equations to find $x$, giving your answer in an exact form.

32  a   $10^{x-2} = 7$
    b   $10^{x+4} = 13$

33  a   $10^{x-2} = 70$
    b   $10^{x+4} = 1300$

34  a   $e^{x+2} = k - 2$
    b   $e^{x+2} = 2k + 1$

For questions 35 and 36, use the method demonstrated in Worked Example 1.25 to solve the given equations, giving your answers correct to three significant figures.

35  a  $10^x = 5 \times 10^{-x}$

    b  $10^{2x+1} = 4 \times 10^x$

36  a  $e^{2x} = 6e^{2-x}$

    b  $e^{3x-1} = 4e^{1-x}$

37  Find the exact solution of $1 + 2\log x = 9$.

38  Find the exact solution of $\log(3x+4) = 3$.

39  Simplify $\ln(e^a e^b)$.

40  Use technology to solve $10^x = 5$.

41  Use technology to solve $3 \times 10^x = 20$.

42  Use technology to solve $2 \times 10^x + 6 = 20$.

43  Find the exact solution of $5 \times 20^x = 8 \times 2^x$.

44  One formula for the pH of a solution is given by pH $= -\log[H^+]$ where $[H^+]$ is the concentration of $H^+$ ions in moles per litre.

    a  A solution contains $2.5 \times 10^{-8}$ moles per litre of $H^+$ ions. Find the pH of this solution.

    b  A solution of hydrochloric acid has a pH of 1.9. Find the concentration of $H^+$ ions.

45  The radioactivity ($R$) of a substance after a time $t$ days is modelled by $R = 10 \times e^{-0.1t}$.

    a  Find the initial ($t = 0$) radioactivity.

    b  Find the time taken for the radioactivity to fall to half of its original value.

46  The population of bacteria ($B$) at time $t$ hours after being added to an agar dish is modelled by $B = 1000 \times e^{0.1t}$

    a  Find the number of bacteria

      i  initially

      ii  after 2 hours.

    b  Find an expression for the time it takes to reach 3000. Use technology to evaluate this expression.

47  The population of penguins ($P$) after $t$ years on two islands is modelled by:

First island: $P = 200 \times e^{0.1t}$

Second island: $P = 100 \times e^{0.25t}$.

How many years are required before the population of penguins on both islands is equal?

48  The decibel scale measures the loudness of sound. It has the formula $L = 10\log(10^{12}I)$ where $L$ is the noise level in decibels and $I$ is the sound intensity in watts per metre squared.

    a  The sound intensity inside a car is $5 \times 10^{-7}$ watts per metre squared. Find the noise level in the car.

    b  The sound intensity in a factory is $5 \times 10^{-6}$ watts per metre squared. Find the noise level in the factory.

    c  What is the effect on the noise level of multiplying the sound intensity by 10?

    d  Any noise level above 90 decibels is considered dangerous to a human ear. What sound intensity does this correspond to?

49  a  Write 20 in the form $e^k$.

    b  If $20^x = 7$ find an exact expression for $x$ in terms of natural logarithms.

50  Solve the simultaneous equations

$\log(xy) = 3$ and $\log\left(\dfrac{x}{y}\right) = 1$

51  Evaluate $\log 10x - \log x$ where $x$ is a positive number.

## Checklist

■ You should know how to use the laws of exponents with integer exponents:

□ $a^m \times a^n = a^{n+r}$          □ $(a^m)^n = a^{mn}$          □ $(ab)^n = a^n \times b^n$

□ $\dfrac{a^m}{a^n} = a^{m-n}$          □ $a^{-n} = \dfrac{1}{a^n}$          □ $\left(\dfrac{a}{b}\right)^n = \dfrac{a^n}{b^n}$

■ You should be able to apply these rules of exponents to numbers in standard form.

■ You should know the definition of a logarithm and be able to work with logarithms including those with base e:
□ $a = b^x$ is equivalent to $\log_b a = x$

■ For base 10:
□ $\log 10^x = x$                    □ $10^{\log x} = x$

■ For base e:
□ $\ln e^x = x$                      □ $e^{\ln x} = x$

■ You should be able to use logarithms to solve simple exponential equations.

---

## ■ Mixed Practice

**1** A rectangle is 2680 cm long and 1970 cm wide.
   **a** Find the perimeter of the rectangle, giving your answer in the form $a \times 10^k$, where $1 \leqslant a < 10$ and $k \in \mathbb{Z}$.
   **b** Find the area of the rectangle, giving your answer correct to the nearest thousand square centimetres.

Mathematical Studies SL May 2009 TZ2 Paper 1 Q1

**2** Simplify $\dfrac{(3xy^2)^2}{(xy)^3}$ .

**3** Write $(3x^2 y^{-3})^{-2}$ without brackets or negative indices.

**4** Zipf's law in geography is a model for how the population of a city $(P)$ in a country depends on the rank of that city $(R)$, that is, whether it is the largest $(R = 1)$, second largest $(R = 2)$, and so on. The suggested formula is $P = kR^{-1}$.

In a particular country the second largest city has a population of $2\,000\,000$.
   **a** Find the value of $k$.
   **b** What does the model predict to be the size of the fourth largest city?
   **c** What does the model predict is the rank of the city with population $250\,000$?

**5** Find the exact solution of $8^x = 2^{x+6}$.

**6** Show that if $a = 3 \times 10^8$ and $b = 4 \times 10^4$ then $ab = 1.2 \times 10^{13}$.

**7** Show that if $a = 1 \times 10^9$ and $b = 5 \times 10^{-4}$ then $\dfrac{a}{b} = 2 \times 10^{12}$.

**8** Show that if $a = 3 \times 10^4$ and $b = 5 \times 10^5$ then $b - a = 4.7 \times 10^5$.

**9** The speed of light is approximately $3 \times 10^8 \, \text{m s}^{-1}$. The distance from the Sun to the Earth is $1.5 \times 10^{11} \, \text{m}$. Find the time taken for light from the Sun to reach the Earth.

**10** Find the exact solution of $\log(x + 1) = 2$.

**11** Find the exact solution to $\ln(2x) = 3$.

**12** Use technology to solve $e^x = 2$.

**13** Use technology to solve $5 \times 10^x = 17$.

**14** Rearrange to make $x$ the subject of $5e^x - 1 = y$.

**15 a** Given that $2^m = 8$ and $2^n = 16$, write down the value of $m$ and of $n$.
   **b** Hence or otherwise solve $8^{2x+1} = 16^{2x-3}$.

Mathematics SL May 2015 TZ1 Paper 1 Q3

**16** You are given that $(7 \times 10^a) \times (4 \times 10^b) = c \times 10^d$
   where $1 \leqslant c < 10$ and $d \in \mathbb{Z}$.
   **a** Find the value of $c$.
   **b** Find an expression for $d$ in terms of $a$ and $b$.

**17** You are given that $(6 \times 10^a) \div (5 \times 10^b) = c \times 10^d$

   where $1 \leqslant c < 10$ and $d \in \mathbb{Z}$.
   **a** Find the value of $c$.
   **b** Find an expression for $d$ in terms of $a$ and $b$.

**18** The Henderson–Hasselbach equation predicts that the pH of blood is given by:

$$\text{pH} = 6.1 + \log\left(\frac{\left[\text{HCO}_3^-\right]}{\left[\text{H}_2\text{CO}_3\right]}\right)$$

where $\left[\text{HCO}_3^-\right]$ is the concentration of bicarbonate ions and $\left[\text{H}_2\text{CO}_3\right]$ is the concentration of carbonic acid (created by dissolved carbon dioxide). Given that the bicarbonate ion concentration is maintained at 0.579 moles per litre, find the range of concentrations of carbonic acid that will maintain blood pH at normal levels (which are between 7.35 and 7.45).

**19** In attempting to set a new record, skydiver Felix Baumgartner jumped from close to the edge of the Earth's atmosphere. His predicted speed ($v$) in metres per second at a time $t$ seconds after he jumped was modelled by: $v = 1350\left(1 - e^{-0.007t}\right)$.
   **a** Find the predicted speed after one second.
   **b** Baumgartner's aim was to break the speed of sound (300 ms$^{-1}$). Given that he was in free fall for 600 seconds, did he reach the speed of sound? Justify your answer.

**20** The Richter scale measures the strength of earthquakes. The strength ($S$) is given by $S = \log A$, where $A$ is the amplitude of the wave measured on a seismograph in micrometres.
   **a** If the amplitude of the wave is 1000 micrometres, find the strength of the earthquake.
   **b** If the amplitude on the seismograph multiplies by 10, what is the effect on the strength of the earthquake?
   **c** The 1960 earthquake in Chile had a magnitude of 9.5 on the Richter scale. Find the amplitude of the seismograph reading for this earthquake.

**21** Solve $3 \times 20^x = 2^{x+1}$.

**22** Solve the simultaneous equations:
$$9^x \times 3^y = 1$$
$$\frac{4^x}{2^y} = 16$$

**23** Solve the simultaneous equations:
$$\log(xy) = 0$$
$$\log\left(\frac{x^2}{y}\right) = 3$$

# 2 Core: Sequences

## ESSENTIAL UNDERSTANDINGS

- Number and algebra allow us to represent patterns, show equivalences and make generalizations which enable us to model real-world situations.
- Algebra is an abstraction of numerical concepts and employs variables to solve mathematical problems.

### In this chapter you will learn...

- how to use the formula for the $n$th term of an arithmetic sequence
- how to use the formula for the sum of the first $n$ terms of an arithmetic sequence
- how to use the formula for the $n$th term of a geometric sequence
- how to use the formula for the sum of the first $n$ terms of a geometric sequence
- how to work with sigma notation for sums of arithmetic and geometric sequences
- how to apply arithmetic and geometric sequences to real-world problems.

### CONCEPTS

The following concepts will be addressed in this chapter:
- **Modelling** real-life situations with the structure of arithmetic and geometric sequences and series allows for prediction, analysis and interpretation.
- Formulae are a **generalization** made on the basis of specific examples, which can then be extended to new examples.
- Mathematical financial models such as compounded growth allow computation, evaluation and interpretation of debt and investment both **approximately** and accurately.

■ Figure 2.1 The early development of a fertilized egg.

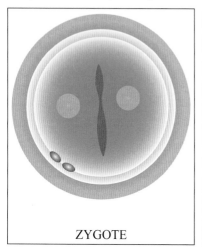
ZYGOTE

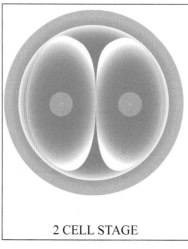
2 CELL STAGE

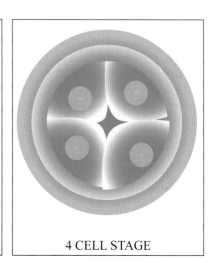
4 CELL STAGE

One area of overlap between mathematics, science and art is patterns. While these might seem purely aesthetic in art, some artists have made use of mathematics in their drawings and sculptures, from the ancient Greeks to Leonardo da Vinci, to in modern times the graphic artist Escher.

In science, patterns appearing in experimental data have led to advances in theoretical understanding.

In mathematics, an understanding of simple relationships can help in modelling more complex patterns in the future.

## Starter Activity

Look at Figure 2.1. In small groups discuss whether these images represent doubling or halving.

**Now look at this problem:**

In early 2018, a Canadian woman won the jackpot on the lottery after buying a scratch card for the first time on her 18th birthday. She was given the choice of taking a C$ 1 million lump sum or receiving C$ 50 000 a year for life – both options would be tax free.

Which option would you choose in this situation?

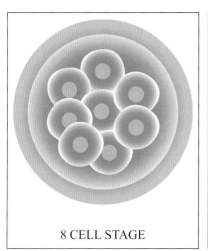

8 CELL STAGE

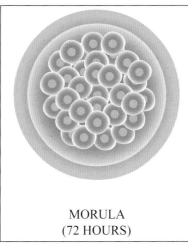

MORULA
(72 HOURS)

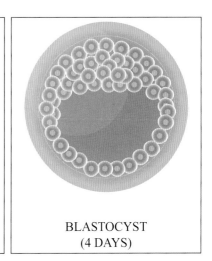

BLASTOCYST
(4 DAYS)

# 2A Arithmetic sequences and series

## ▉ Use of the formula for the $n$th term of an arithmetic sequence

An **arithmetic sequence** is formed by adding or subtracting the same number to get to the next term. For example:

1, 4, 7, 10, 13, ...   has a common difference of $+3$

19, 15, 11, 7, 3, ...   has a common difference of $-4$

In general, if the common difference is $d$ (which can be negative), then

| 1st term | 2nd term | 3rd term | 4th term | 5th term |
|----------|----------|----------|----------|----------|
| $u_1$ | $u_1 + d$ | $u_1 + 2d$ | $u_1 + 3d$ | $u_1 + 4d$ |

From this, we can suggest a formula for the $n$th term.

---

**KEY POINT 2.1**

For an arithmetic sequence with common difference $d$,

$$u_n = u_1 + (n-1)d.$$

---

**CONCEPTS – GENERALIZATION**

This formula has not been formally proved – that would require a method called mathematical induction. However, do you think it is a convincing **generalization** of the pattern? Sometimes systematic listing in the correct way allows easy generalization. Would it have been more or less obvious if it had been written as $u_n = u_1 + nd - d$?

---

**WORKED EXAMPLE 2.1**

An arithmetic sequence has first term 7 and common difference 8.

Find the 10th term.

Use $u_n = u_1 + (n-1)d$ ............................ $u_n = 7 + 8(n-1)$
with $u = 7$ and $d = 8$

So,

Let $n = 10$ ............................ $u_{10} = 7 + 8(10-1) = 79$

**WORKED EXAMPLE 2.2**

The third term of an arithmetic sequence is 10 and the 15th term is 46.

Find the first term and the common difference.

Use $u_n = u_1 + (n-1)d$ to write the first bit of information as an equation ·········· $u_3 = 10$

Therefore, $u_1 + 2d = 10$

And likewise with the second bit of information ·········· $u_{15} = 46$

Therefore, $u_1 + 14d = 46$

Now solve simultaneously with your GDC

From the GDC,

$u_1 = 4, d = 3$

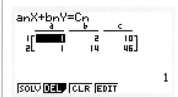

**WORKED EXAMPLE 2.3**

Find the number of terms in the arithmetic sequence 1, 4, 7,... , 100.

This is an arithmetic sequence with first term 1 and common difference 3 ·········· $u_1 = 1, d = 3$

Find the formula for the $n$th term ·········· So, $u_n = 1 + 3(n-1)$

Set this equal to 100 and solve for $n$ ·········· $1 + 3(n-1) = 100$

$$3(n-1) = 99$$

$$n - 1 = 33$$

$$n = 34$$

## ▣ Use of the formula for the sum of the first $n$ terms of an arithmetic sequence

It is also useful to have a formula for the sum of the first $n$ terms of an arithmetic sequence. This sum is sometimes called an **arithmetic series**. There are two versions of this.

**Tip**

Just as $u_1$ is the first term, $u_2$ the second term and so on, $S_2$ is the sum of the first two terms, $S_3$ the sum the first three terms and so on. The sum of the first $n$ terms is therefore $S_n$.

**KEY POINT 2.2**

For an arithmetic sequence with common difference $d$,

● $S_n = \dfrac{n}{2}[2u_1 + (n-1)d]$

or

● $S_n = \dfrac{n}{2}(u_1 + u_n)$

The second formula in Key Point 2.2 follows directly from the first:

$$S_n = \frac{n}{2}\left[2u_1 + (n-1)d\right]$$

$$= \frac{n}{2}\left[u_1 + u_1 + (n-1)d\right]$$

$$= \frac{n}{2}(u_1 + u_n)$$

---

**Proof 2.1**

Prove that for an arithmetic sequence with common difference $d$, $S_n = \frac{n}{2}\left[2u_1 + (n-1)d\right]$.

Write out the first couple of terms and the last couple $\cdots\cdots$ $S_n = a + [a+d] + \ldots + [a + (n-2)d] + [a + (n-1)d]$

Write down the sum again but in the opposite order $\cdots\cdots$ $S_n = [a + (n-1)d] + [a + (n-2)d] + \ldots + [a+d] + a$

Now adding the two expressions gives $n$ identical terms $\cdots\cdots$ $2S_n = [2a + (n-1)d] + [2a + (n-1)d] + \ldots + [2a + (n-1)d] + [2a + (n-1)d]$

$$2S_n = n[2a + (n-1)d]$$

$$S_n = \frac{n}{2}[2a + (n-1)d]$$

---

Carl Friedrich Gauss (1777–1855) was one of the most renowned mathematicians of the nineteenth century. There is a famous story of how, when he was about nine years old, his teacher wanted to keep the class quiet and so asked them to add up all the integers from 1 to 100. Within seconds Gauss had replied with the correct answer, much to the annoyance of his teacher! It is thought he had used a very similar method to the above proof to arrive at the answer.

---

**WORKED EXAMPLE 2.4**

An arithmetic sequence has first term 5 and common difference –2.

Find the sum of the first 10 terms.

$$u_1 = 5,\ d = -2$$

Use $S_n = \frac{n}{2}\left[2u_1 + (n-1)d\right]$ $\cdots\cdots\cdots\cdots\cdots\cdots\cdots$ $S_{10} = \frac{10}{2}\left[2 \times 5 + (10 - 1) \times (-2)\right]$

$$= -40$$

---

**WORKED EXAMPLE 2.5**

An arithmetic sequence has first term 10 and last term 1000. If there are 30 terms, find the sum of all the terms.

$$u_1 = 10, \; u_n = 1000$$

Use $S_n = \dfrac{n}{2}\left[u_1 + u_n\right]$ ·························· $S_{30} = \dfrac{30}{2}\left[10 + 1000\right]$

$$= 15\,150$$

---

## ■ Use of sigma notation for sums of arithmetic sequences

Instead of writing out the terms of a sequence in a sum, you will often see this expressed in a shorthand form using sigma notation.

**Tip**

There is nothing special about the letter $r$ here – any letter could be used.

**KEY POINT 2.3**

● $\displaystyle\sum_{r=1}^{r=n} u_r = u_1 + u_2 + u_3 + \ldots + u_n$

The value of $r$ at the bottom of the sigma (here $r = 1$) shows where the counting starts.

The value of $r$ at the top of the sigma (here $r = n$) shows where the counting stops.

**Tip**

Be careful! In Worked Example 2.6 it is easy to think that there are seven terms in the sequence but there are actually eight ($u_3$, $u_4$, $u_5$, $u_6$, $u_7$, $u_8$, $u_9$ and $u_{10}$).

**WORKED EXAMPLE 2.6**

Evaluate

$$\sum_{r=3}^{10} (5r + 2).$$

Substitute the first few values of $r$ into the formula: $r = 3$, $r = 4$, $r = 5$ ·····

$$\sum_{r=3}^{10} (5r + 2) = (5 \times 3 + 2) + (5 \times 4 + 2) + (5 \times 5 + 2) + \ldots$$

This is the sum of an arithmetic sequence with $u_1 = 17$, $d = 5$ and $n = 8$ ···············

$$= 17 + 22 + 27 + \ldots$$

Use $S_n = \dfrac{n}{2}\left[2u_1 + (n-1)d\right]$ ···············

$$= \dfrac{8}{2}\left[2 \times 17 + (8 - 1)5\right]$$

$$= 276$$

---

Although you will be expected to identify the kind of sum in Worked Example 2.6 as being arithmetic and then show how you have found the value using the formula for $S_n$, it is possible to do this directly on your calculator.

```
Σ(5R+2,R,3,10)
                        276




FMin FMax ΣΓ logₐb        ▷
```

If you know the sum of the terms of an arithmetic sequence you can find the number of terms using technology.

---

**WORKED EXAMPLE 2.7**

The sum of the first $n$ terms of the arithmetic sequence 3, 7, 11, 15, … is 465. Find $n$.

This is an arithmetic sequence with first term 3 and common difference 4 ................. $u_1 = 3, d = 4$

So, $S_n = \dfrac{n}{2}[2 \times 3 + (n-1)4]$

Find and (if you want to) simplify the formula for the sum of the first $n$ terms .........

$= \dfrac{n}{2}[6 + 4n - 4]$

$= \dfrac{n}{2}[2 + 4n]$

$= n + 2n^2$

The sum of the first $n$ terms is 465 ......... $S_n = 465$

Therefore, $2n^2 + n = 465$

Use the table function on your GDC to search for the smallest positive value of $n$ that gives 465 ............

From GDC,
$n = 15$

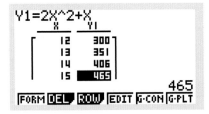

---

 As well as using technology, this problem can be thought of as creating a quadratic equation, which can be solved by methods such as factorizing (see Chapter 15).

 ## ■ Applications of arithmetic sequences

Arithmetic sequences can be used to model any process where something tends to change by a constant amount. In particular, you should know that **simple interest** means that the same amount is added into an account each year.

---

**WORKED EXAMPLE 2.8**

A savings account pays 10% simple interest each year. If Rashid invests €3000 in this account, how much will he have after 5 years?

10% of 3000 = 300, so 300 will be added on each year. It is a good idea to write out the first couple of terms to see what is happening .........

The amount in the account after
1 year: $3000 + 300 = 3300$
2 years: $3300 + 300 = 3600$

This is an arithmetic sequence with $u_1 = 3300$ and $d = 300$.

Use $u_n = u_1 + (n-1)d$ ......... $u_5 = 3300 + 4 \times 300$

$= 4500$

## Be the Examiner 2.1

Ben starts a new job on 1 January 2020 earning £20 000. His salary will go up by £750 on 1 January each year. What will he earn in 2030?

Which is the correct solution? Identify the errors made in the incorrect solutions.

| Solution 1 | Solution 2 | Solution 3 |
|---|---|---|
| $u_{10} = 20\,000 + (10-1)750$ <br> $= 26\,750$ <br> He will earn £26 750. | $S_{10} = \dfrac{10}{2}[2 \times 20\,000 + (10-1)750]$ <br> $= 233\,750$ <br> He will earn £233 750. | $u_{11} = 20\,000 + (11-1)750$ <br> $= 27\,500$ <br> He will earn £27 500. |

## ■ Analysis, interpretation and prediction where a model is not perfectly arithmetic in real life

If the difference between data terms isn't exactly the same, an arithmetic sequence can still be used to model the process.

### WORKED EXAMPLE 2.9

The data below show the number of schools taking the IB diploma in different years:

| 2013 | 2014 | 2015 | 2016 | 2017 |
|---|---|---|---|---|
| 2155 | 2211 | 2310 | 2487 | 2667 |

Modelling the data as an arithmetic sequence, predict the number of schools taking the IB in 2023.

The difference between years is not constant, so all we can do is take the average difference between the first and last given terms ⋯⋯⋯⋯ $d = \dfrac{2667 - 2155}{4} = 128$

Use $u_n = u_1 + (n-1)\,d$, with $u_1 = 2155$ and $d = 128$. ⋯⋯⋯ $u_{11} = 2155 + (11-1)128$

Note that 2023 is the 11th term of the sequence $\qquad\qquad = 3435$

In this example we assumed that the arithmetic sequence had to be fitted to the first (and last) terms in the given data. You will see in Chapter 6 that linear regression provides a better way of modelling the arithmetic sequence that does not require this assumption.

### CONCEPTS – MODELLING

What are the underlying assumptions in applying an arithmetic **model** to this data? How valid do you think these assumptions are? Do you think this model is more useful for predicting future behaviour or analysing past behaviour to identify anomalous years?

## TOK Links

Is all knowledge concerned with the identification and use of patterns? One sequence that you may be interested in researching further is the Fibonacci sequence, in which each term is the sum of the two preceding terms: 1, 1, 2, 3, 5, 8, 13, 21... This sequence of numbers, and the related golden ratio, can be seen occurring naturally in biological contexts, as well as in human endeavours such as art and architecture. You might like to investigate how the photographs on the covers of all four Hodder Education books for IB Diploma Mathematics exhibit this sequence of numbers.

## Exercise 2A

For questions 1 and 2, use the method demonstrated in Worked Example 2.1. to find the required term of the arithmetic sequence.

1   a   $u_1 = 5$, $d = 4$. Find $u_{12}$.
    b   $u_1 = 11$, $d = 3$. Find $u_{20}$.

2   a   $u_1 = 13$, $d = -7$. Find $u_6$.
    b   $u_1 = 11$, $d = -3$. Find $u_{18}$.

For questions 3 to 5, use the method demonstrated in Worked Example 2.2 to find the required feature of the arithmetic sequence.

3   a   $u_1 = 7$, $u_5 = 15$. Find $d$.
    b   $u_1 = 19$, $u_9 = 75$. Find $d$.

4   a   $u_1 = 10$, $u_{21} = 0$. Find $d$.
    b   $u_1 = 100$, $u_9 = 80$. Find $d$.

5   a   $u_8 = 12$, $u_{11} = 33$. Find $u_1$ and $d$.
    b   $u_7 = 11$, $u_{12} = 56$. Find $u_1$ and $d$.

For questions 6 to 8, use the method demonstrated in Worked Example 2.1 and Worked Example 2.2 to find a formula for the $n$th term of the arithmetic sequence described.

6   a   $u_1 = 8$, $d = -3$
    b   $u_1 = 10$, $d = 4$

7   a   $u_1 = 3$, $u_5 = 11$
    b   $u_1 = 10$, $u_{11} = 50$

8   a   $u_2 = 6$, $u_5 = 21$
    b   $u_3 = 14$, $u_6 = 5$

For questions 9 to 10, use the method demonstrated in Worked Example 2.3 to find the number of terms in each of the given arithmetic sequences.

9   a   2, 8, 14,..., 200
    b   5, 9, 13, ..., 97

10   a   13, 7, 1, ..., −59
     b   95, 88, 81, ..., −101

For questions 11 to 13, use the method demonstrated in Worked Example 2.4 (and Worked Example 2.2) to find the given sum.

11   a   $u_1 = 9$, $d = 4$. Find $S_{12}$.
     b   $u_1 = 17$, $d = 2$. Find $S_{20}$.

12   a   $u_1 = 34$, $d = -7$. Find $S_{11}$.
     b   $u_1 = 9$, $d = -11$. Find $S_8$.

13   a   $u_1 = 7$, $u_6 = 32$. Find $S_{11}$.
     b   $u_1 = 9$, $u_5 = 41$. Find $S_{12}$.

For questions 14 and 15, use the method demonstrated in Worked Example 2.5 to find the given sum.

14   a   $u_1 = 9$, $u_8 = 37$ Find $S_8$.
     b   $u_1 = 4$, $u_{12} = 61$ Find $S_{12}$.

15   a   $u_1 = 27$, $u_{10} = -36$ Find $S_{10}$.
     b   $u_1 = 50$, $u_{200} = -50$ Find $S_{200}$.

For questions 16 to 18, use the method demonstrated in Worked Example 2.6 to evaluate the given sum.

16   a   $\displaystyle\sum_{r=1}^{8} 4$

     b   $\displaystyle\sum_{r=1}^{11} 9$

17   a   $\displaystyle\sum_{r=1}^{7} (2r + 7)$

     b   $\displaystyle\sum_{r=1}^{10} (3r + 4)$

18   a   $\displaystyle\sum_{r=5}^{17} (6r - 4)$

     b   $\displaystyle\sum_{r=3}^{8} (12r - 7)$

For questions 19 and 20, use the method demonstrated in Worked Example 2.7 to find the number of terms in the arithmetic sequence.

19  a  4, 9, 14...  $S_n = 442$        20  a  $u_1 = 20, d = -1, S_n = 200$

    b  8, 9, 10...  $S_n = 125$             b  $u_1 = 100, d = -5, S_n = 1000$

21  An arithmetic sequence has first term 7 and common difference 11.

    a  Find the 20th term of the sequence.

    b  Find the sum of the first 20 terms.

22  The first term of an arithmetic sequence is 3 and the second term is 7.

    a  Write down the common difference.

    b  Find the eighth term of the sequence.

    c  Find the sum of the first 15 terms.

23  In an arithmetic sequence, the second term is 13 and the common difference is 5.

    a  Write down the first term.

    b  Find the sum of the first 10 terms.

24  The first term of an arithmetic sequence is −8 and the 16th term is 67.

    a  Find the common difference.

    b  Find the 25th term.

25  Sam invests £300 at 4% simple interest.  She does not withdraw any money.

    a  How much does she have in her account at the end of the first year?

    b  How much does she have in her account at the end of the 10th year?

26  On Daniel's first birthday, his grandparents started a saving account with £100. On each of his subsequent birthdays they put £10 in the account. What is the total amount they have saved for Daniel after they have deposited money on his 21st birthday? You may ignore any interest accrued.

27  The height of each step in a stairway follows an arithmetic sequence. The first step is 10 cm off the ground and each subsequent step is 20 cm higher. If the staircase is 270 cm high, how many steps does it have?

28  An arithmetic sequence has first term 11 and last term 75.

    a  Given the common difference is 8, find the sum of the terms.

    b  Given the common difference is 4, find the sum of the terms.

29  In an arithmetic sequence, the 10th term is 26 and the 30th term is 83. Find the 50th term.

30  In an arithmetic sequence, the first term is 8, the common difference is 3 and the last term is 68.

    a  How many terms are there in this sequence?

    b  Find the sum of all the terms of the sequence.

31  The 7th term of an arithmetic sequence is 35 and the 18th term is 112.

    a  Find the common difference and the first term of the sequence.

    b  Find the sum of the first 18 terms.

32  An arithmetic sequence has $u_{10} = 16$ and $u_{30} = 156$.

    a  Find the value of $u_{50}$.

    b  Find $\displaystyle\sum_{r=1}^{20} u_r$.

33  Find $\displaystyle\sum_{r=1}^{16} (5r - 3)$.

34  Three consecutive terms of an arithmetic sequence are $3x + 1$, $2x + 1$, $4x - 5$. Find the value of $x$.

35  The fifth term of an arithmetic sequence is double the second term, and the seventh term of the sequence is 28. Find the 11th term.

36  The sum of the first three terms of an arithmetic sequence equals the tenth term, and the seventh term of the sequence is 27. Find the 12th term.

**37** A sequence is defined by $u_{n+1} = u_n + 4$. If $u_1 = 10$ find $\displaystyle\sum_{r=1}^{20} u_r$.

**38** a The arithmetic sequence $U$ has first term 7, common difference 12 and last term 139. Find the sum of terms in $U$.

  b The arithmetic sequence $V$ has the same first and last terms as sequence $U$ but has common difference 6. Find the sum of terms in $V$.

  c The arithmetic sequence $W$ has the same first term and common difference as $V$, and the same number of terms as $U$. Find the sum of terms in $W$.

**39** Joe plays a game five times and scores 53, 94, 126, 170 and finally 211.

Modelling the data as an arithmetic sequence, predict the score he might get on his 10th attempt, if his pattern continues consistently.

**40** A survey of the number of ducks seen on a lake on the first four days of the inward migration period records

3, 12, 23 and then 33 ducks on the lake.

Modelling the data as an arithmetic sequence, predict the number of ducks seen on the sixth day, if the pattern continues consistently.

**41** The first three terms of an arithmetic sequence are 1, $a$, $3a + 5$. Evaluate the fourth term of the sequence.

**42** In an effort to reduce his screen time, Stewart reduced his consumption by 5 minutes each day. On day one of the program he looks at a screen for 200 minutes.

  a Which is the first day on which his screen time is reduced to zero?

  b For how many minutes in total has Stewart looked at a screen during this program?

**43** As part of a health program Theo tries to increase the number of steps he takes each day by 500 steps.

On the first day he walks 1000 steps. Assume that he sticks to this plan.

  a How long will it take him to get to 10 000 steps in a day?

  b How long will it take him to complete a total of 540 000 steps.

**44** The sum of the first $n$ terms of an arithmetic sequence is denoted by $S_n$. Given that $S_n = 2n^2 + n$,

  a find the first two terms of the sequence

  b hence find the 50th term.

**45** The sum of the first $n$ terms of a sequence is given by $S_n = n^2 + 4n$. Find an expression for the $n$th term of the sequence.

**46** The ninth term of an arithmetic sequence is four times larger than the third term. Find the value of the first term.

**47** What is the sum of all three-digit multiples of 7?

**48** What is the sum of all whole numbers from 0 up to (and including) 100 which are multiples of 5 but not multiples of 3.

**49** A rope of length 4 m is cut into 10 sections whose lengths form an arithmetic sequence. The largest resulting part is 3 times larger than the smallest part. Find the size of the smallest part.

**50** The first four terms of an arithmetic sequence are 4, $a$, $b$, $a - b$. Find the sixth term.

**51** a A sequence has formula $u_n = a + nd$. Prove that there is a constant difference between consecutive terms of the sequence.

  b A sequence has formula $u_n = an^2 + bn$. Prove that the differences between consecutive terms forms an arithmetic sequence.

**52** Alessia writes down a list of the first $n$ positive integers in order.

  a Show that by the time she has written down the number 19 she has written down 29 digits.

  b Her final list contains a total of 342 digits. What was the largest number in her list?

# 2B Geometric sequences and series

## ◼ Use of the formula for the $n$th term of a geometric sequence

A **geometric sequence** has a common ratio between each term. For example,

| | |
|---|---|
| 3, 6, 12, 24, 48, ... | has a common ratio of 2 |
| 80, 40, 20, 10, 5, ... | has a common ratio of $\frac{1}{2}$ |
| 2, –6, 18, –54, 162, ... | has a common ratio of –3 |

In general, if the common ratio is $r$ (which can be negative), then

| 1st term | 2nd term | 3rd term | 4th term | 5th term |
|---|---|---|---|---|
| $u_1$ | $u_1 r$ | $u_1 r^2$ | $u_1 r^3$ | $u_1 r^4$ |

From this, we can suggest a formula for the $n$th term.

---

**KEY POINT 2.4**

For a geometric sequence with common ratio $r$, $u_n = u_1 r^{n-1}$.

---

**WORKED EXAMPLE 2.10**

A geometric sequence has first term 3 and common ratio –2. Find the 10th term.

Use $u_n = u_1 r^{n-1}$ with $u_1 = 3$ and $r = -2$ .......................... $u_n = 3(-2)^{n-1}$

So, $u_{10} = 3(-2)^9 = -1536$

---

**WORKED EXAMPLE 2.11**

The second term of a geometric sequence is 6 and the fourth term is 96. Find the possible values of the first term and the common ratio.

Use $u_n = u_1 r^{n-1}$ to write the first bit ·········································· $u_2 = 6$
of information as an equation

Therefore, $u_1 r^1 = 6$

Do likewise with the second bit of information ···························· $u_4 = 96$
Therefore, $u_1 r^3 = 96$

These are non-linear simultaneous equations ····················· Therefore $\dfrac{u_1 r^3}{u_1 r} = \dfrac{96}{6}$
We can solve them by dividing them

So $\qquad r^2 = 16$

There are two numbers which ·············································· $r = \pm 4$
square to make 16

If $r = 4$ then $u_1 = 1.5$

Substitute each of these values into $u_1 r = 6$ ·····················
If $r = -4$ then $u_1 = -1.5$

You will see in Chapter 3 that these equations can also be solved directly on your GDC.

---

**WORKED EXAMPLE 2.12**

Find the number of terms in the geometric sequence 1, 2, 4, 8, ..., 512.

This is a geometric sequence with ·································· $u_1 = 1, r = 2$
first term 1 and common ratio 2

Find the formula for the $n$th term ·································· So, $u_n = 1 \times 2^{n-1}$

$u_n = 512$

The $n$th term is 512 ·································· Therefore, $2^{n-1} = 512$

Use the table function on
your GDC to search for the ·································· From GDC, $n = 10$
value of $n$ that gives 512

| ◀ 1.1 | 1.2 ▷ | *Unsaved ▼ | | ⬛❎ |
|---|---|---|---|---|
| | ᴬn | ᴮu_n | ᶜ | ᴰ |
| ◆ | | =2^('n[]-1) | | |
| 7 | 7 | 64 | | |
| 8 | 8 | 128 | | |
| 9 | 9 | 256 | | |
| 10 | 10 | 512 | | |
| 11 | 11 | 1024 | | |
| B10 | =512 | | | ◀▶ |

Note that you could also solve the equation $2^{n-1} = 512$ using logs with the methods of Section 1C.

## ■ Use of the formula for the sum of the first $n$ terms of a geometric sequence

Just as for arithmetic sequences, there is a formula for finding the sum of the first $n$ terms of a geometric sequence. This sum is sometimes called a **geometric series**.

**Tip**

The second formula in Key Point 2.5 follows from the first on multiplication of the numerator and denominator by −1.

**KEY POINT 2.5**

For a geometric sequence with common ratio $r$:

- $S_n = \dfrac{u_1(1 - r^n)}{1 - r}, \quad r \neq 1$

  or

- $S_n = \dfrac{u_1(r^n - 1)}{r - 1}, \quad r \neq 1$

### Proof 2.2

Prove that for a geometric sequence with common ratio $r$,

$$S_n = \frac{u_1(1-r^n)}{1-r}.$$

Write out the first few terms and the last few $\quad\cdots\cdots\cdots\cdots$ $S_n = u_1 + u_1 r + u_1 r^2 + \ldots + u_1 r^{n-2} + u_1 r^{n-1}$

Multiply through by $r$ $\quad\cdots\cdots\cdots\cdots$ $rS_n = u_1 r + u_1 r^2 + \ldots + u_1 r^{n-2} + u_1 r^{n-1} + u_1 r^n$

Subtract to remove all the terms in common $\quad\cdots\cdots\cdots$ $S_n - rS_n = u_1 - u_1 r^n$

Factorize both sides $\quad\cdots\cdots$ $S_n(1-r) = u_1(1-r^n)$

$$S_n = \frac{u_1(1-r^n)}{1-r}$$

At what point in the proof above do you use the fact that $r \neq 1$? Can you find a formula for the sum of the first $n$ terms when $r = 1$?

### WORKED EXAMPLE 2.13

A geometric sequence has first term 16 and common ratio $-0.5$.

Find the sum of the first eight terms to three significant figures.

$$u_1 = 16, \ r = -0.5$$

Use $S_n = \dfrac{u_1(1-r^n)}{1-r}$ $\quad\cdots\cdots\cdots\cdots\cdots\cdots\cdots$ $S_8 = \dfrac{16\left(1-(-0.5)^8\right)}{1-(-0.5)}$

$$= 10.6 \ (3 \text{ s.f.})$$

## ▌ Use of sigma notation for sums of geometric sequences

### WORKED EXAMPLE 2.14

Evaluate

$$\sum_{r=2}^{8} 2 \times 4^r.$$

Substitute the first few values of $r$ into the formula: $r = 2, r = 3, r = 4$ $\quad\cdots\cdots\cdots$ $\displaystyle\sum_{r=2}^{8} 2 \times 4^r = (2 \times 4^2) + (2 \times 4^3) + (2 \times 4^4) + \ldots$

This is the sum of a geometric sequence with $u_1 = 32, r = 4$ and $n = 7$ $\quad\cdots\cdots$ $= 32 + 128 + 512 + \ldots$

$$= \frac{32(1-4^7)}{1-4}$$

Use $S_n = \dfrac{u_1(1-r^n)}{1-r}$ $\quad\cdots\cdots\cdots$ $= 174752$

As with sums of arithmetic sequences in Section 2A, you can check the calculation in Worked Example 2.14 on your GDC:

```
Σ(2×4^R,R,2,8)
                                      174752



FMin FMax Σ( log_ab            ▷
```

## Applications of geometric sequences

Geometric sequences can be used to model any process where something tends to change by a constant factor.

Often this will be expressed in terms of a percentage change.

---

**KEY POINT 2.6**

An increase of $r\%$ is equivalent to multiplying by the factor $1+\dfrac{r}{100}$.

---

**EXAMPLE 2.15**

A model predicts that the number of students taking the IB in a region will increase by 5% each year. If the number of IB students in the region is currently 12 000, predict the number taking the IB in 5 years' time.

A 5% increase corresponds to multiplication by 1.05. It is a good idea to write out the first couple of terms to see what is happening

Number of IB students in

1 year's time: $12\,000 \times 1.05$

2 years' time: $12\,000 \times 1.05^2$

This is a geometric sequence with

$u_1 = 12\,000 \times 1.05$ and $r = 1.05$

Use $u_n = u_1 r^{n-1}$

$u_5 = u_1 r^4$

$= (12\,000 \times 1.05) \times 1.05^4$

$\approx 15\,315$

---

**You are the Researcher**

A Farey sequence of order $n$ is a list of reduced fractions in order of size between 0 and 1 which have a denominator of at most $n$. They have many algebraic and geometric properties and applications which you might like to research.

## Exercise 2B

For questions 1 to 3, use the method demonstrated in Worked Example 2.10 to find the required term of the geometric sequence.

1　a　$u_1 = 5$, $r = 4$. Find $u_{12}$.　　　　　2　a　$u_1 = 5$, $r = -5$. Find $u_6$.　　　　　3　a　$u_1 = 32$, $r = -\frac{1}{2}$. Find $u_{11}$.

　　b　$u_1 = 2$, $r = 3$. Find $u_7$.　　　　　　b　$u_1 = 11$, $r = -2$. Find $u_{18}$.

　　　　　　　　　　　　　　　　　　　　　　　　　　　　　　　　　　　　b　$u_1 = 54$, $r = -\frac{1}{3}$. Find $u_{10}$.

For questions 4 to 6, use the method demonstrated in Worked Example 2.11 to find all possible values of the first term and common ratio for the following geometric sequences.

4　a　$u_3 = 28$, $u_8 = 896$　　　　　5　a　$u_3 = 12$, $u_9 = 768$　　　　　6　a　$u_2 = -6$, $u_6 = -96$

　　b　$u_5 = 108$, $u_8 = 2916$　　　　　b　$u_5 = 45$, $u_7 = 405$　　　　　b　$u_8 = 56$, $u_{12} = 3.5$

For questions 7 and 8, use the method demonstrated in Worked Example 2.12 to find the number of terms in each of the geometric sequences described.

7　a　$u_1 = 3$, $r = 5$, $u_n = 9375$　　　　　8　a　$3, 6, 12, \ldots, 12\,288$

　　b　$u_1 = 17$, $r = 2$, $u_n = 2176$　　　　　b　$5, 15, 45, \ldots, 10\,935$

For questions 9 to 11, use the method demonstrated in Worked Example 2.13 to find the required sum of the geometric sequence described.

9　a　$u_1 = 5$, $r = 3$. Find $S_7$.　　　　10　a　$u_1 = 96$, $r = 0.5$. Find $S_7$.　　　　11　a　$u_1 = 192$, $r = -\frac{1}{4}$. Find $S_5$.

　　b　$u_1 = 3$, $r = 4$. Find $S_6$.　　　　　b　$u_1 = 162$, $r = \frac{1}{3}$. Find $S_5$.　　　　　b　$u_1 = 216$, $r = -\frac{1}{3}$. Find $S_6$.

For questions 12 to 15, use the formula for the sum of terms from a geometric sequence and the method demonstrated in Worked Example 2.14 to evaluate the following expressions.

12　a　$\displaystyle\sum_{r=1}^{5} 3^r$　　　　　　　　　　13　a　$\displaystyle\sum_{r=1}^{8} 2 \times 5^r$

　　b　$\displaystyle\sum_{r=1}^{4} 7^r$　　　　　　　　　　　b　$\displaystyle\sum_{r=1}^{9} 11 \times 3^r$

14　a　$\displaystyle\sum_{r=3}^{8} 7 \times 3^r$　　　　　　　15　a　$\displaystyle\sum_{r=2}^{7} 72 \times \left(-\frac{1}{2}\right)^r$

　　b　$\displaystyle\sum_{r=4}^{11} 8 \times 5^r$　　　　　　　　b　$\displaystyle\sum_{r=6}^{10} 24\,057 \times \left(-\frac{1}{3}\right)^r$

16　A geometric sequence has first term 128 and common ratio 0.5.

　　a　Find the eighth term of the sequence.

　　b　Find the sum of the first eight terms.

17　The first term of a geometric sequence is 3 and the second term is 6.

　　a　Write down the common ratio.

　　b　Find the sixth term.

　　c　Find the sum of the first 10 terms.

18　In a geometric sequence the second term is 24 and the fifth term is 81.

　　a　Find the common ratio.

　　b　Find the value of the seventh term.

19　A student studies the amount of algae in a pond over time. She estimates the area of algae on the pond on day 1 of her study is $15\,\text{cm}^2$, and she predicts that the area will double every eight days.

　　According to her projection, what area will the algae cover at the start of the ninth week?

20 A yeast culture is grown in a sugar solution. The concentration of sugar in the solution at the start of the culture is $1.2\,mg\,ml^{-1}$, and halves every 2 days. What is the concentration of the sugar solution after 12 days?

21 The first term of a geometric sequence is 8 and the sum of the first two terms is 12. Find the sum of the first five terms.

22 The time taken for a computer learning program to identify a face decreases by 20% every time it sees that face. The first time the face is observed it takes 5 seconds to identify it. How long does it take to identify a face on the 10th attempt?

23 During a drought, the volume of water in a reservoir decreases by 8% every day. At the beginning of day one of the drought it contains $5000\,m^3$:

   a  Find the volume at the start of the fifth day.

   b  How many days does it take to use up $3000\,m^3$ of water?

24 The fifth term of a geometric sequence is eight times larger than the second term. Find the ratio $\dfrac{S_8}{u_1}$.

25 🌐  According to a common legend, the game of chess was invented by Sissa ben Dahir of the court of King Shiram in India. The King was so impressed he asked Sissa to name his reward. Sissa asked for one grain of wheat on the first square of the chessboard, 2 on the second, 4 on the third and so on. The King thought that Sissa was being very foolish to ask for so little.

   a  How many grains would be on the 64th square of the chessboard?

   b  How many grains would be on the chessboard in total?

   c  The annual worldwide production of wheat is approximately $7.5 \times 10^{14}$ g. If one grain of wheat has a mass of $0.1$ g, how many years would it take to produce enough wheat to satisfy Sissa's reward? Give your answer to the nearest 50 years.

## TOK Links

How useful is your intuition as a way of knowing for dealing with very large numbers?
Does relating new knowledge to established knowledge to put big numbers in some more meaningful context (such as in part **c** of this question) improve your understanding?

26 Find an expression for the $n$th term of the geometric sequence $xy^2$, $y^3$, $x^{-1}y^4$ …

27 If $u_1 = 3$ and $u_{n+1} = 2u_n$ find $\displaystyle\sum_{r=1}^{10} u_r$.

28 The first three terms of a geometric sequence are $1$, $x$, $2x^2 + x$.

   Find the value of the 10th term.

29 Evaluate $\displaystyle\sum_{r=1}^{10} \dfrac{6^r}{2^r}$.

30 A basketball is dropped vertically. The first bounce reaches a height of $0.6$ m and subsequent bounces are 80% of the height of the previous bounce.

   a  Find the height of the fifth bounce.

   b  Find the total distance travelled by the ball from top of the first bounce to the top of the fifth bounce.

   c  Suggest why this model is no longer accurate by the 20th bounce.

# 2C Financial applications of geometric sequences and series

## ▉ Compound interest

One common application of geometric sequences is **compound interest**. This is calculated by applying a percentage increase to an initial sum, then applying the same percentage increase to the new sum and so on.

Interest rates are often quoted annually but then compounded over different periods – for example, monthly, quarterly or half yearly. The quoted annual interest rate is split equally amongst each of these periods.

............................

**Tip**

Be warned – this means that the actual annual interest rate is not always the same as the quoted one when it is compounded over shorter time intervals.

............................

---

**WORKED EXAMPLE 2.16**

A savings account has an annual interest rate of 3% compounded monthly.

Find the amount in the account 18 months after $5000 is invested.

The monthly interest rate will be $\dfrac{3}{12} = 0.25\%$

0.25% is equivalent to

multiplying by $1 + \dfrac{0.25}{100} = 1.0025$ .............. After 18 months the amount in the account will be $5000 \times 1.0025^{18} = \$5229.85$

---

 This calculation can also be done on your GDC using the TVM (Time Value of Money) package.

The variables you need to enter are:

- n – the **number** of time periods (here 18 months)
- I% – the **annual** interest rate
- PV – the present value. Conventionally this is negative for an investment.

```
Compound Interest:End
n   =18
I%  =3
PV  =-5000
PMT=0
FV  =5229.845602
P/Y=12                  ↓
 n  I% PV PMT FV AMT
```

- FV is the future value, which is the answer in Worked Example 2.16. Additionally, if you enter this quantity, the calculator can be used to find one of the other values.
- P/Y – the number of payments per year (here 12 as each period is 1 month). Some calculators might allow you to specify separately P/Y as the number of payments made each year and C/Y as the number of compounding periods per year. You should always make both of these equal for all problems in the IB.

**Tip**

In the TVM package, depreciation corresponds to a negative interest rate.

## ■ Annual depreciation and inflation

An asset, such as a car, will tend to lose value or **depreciate**. The method of calculating the value of an asset after depreciating is exactly the same as for compound interest, except that the percentage change will be negative. For example, an annual depreciation rate of 15% would correspond to multiplication by $1 - \frac{15}{100} = 0.85$.

Over time, prices tend to rise. The average percentage increase in prices over a year is known as the **inflation rate**.

The effect of inflation is that the value of money decreases or depreciates over time. The value of money after being adjusted for inflation is known as the **value in real terms**. If nothing else is said then you should assume that this is relative to the initial time.

For example, if the inflation rate is 2% in a given year, then £100 at the beginning of the year will be worth, in real terms, £98 at the end of the year.

### KEY POINT 2.7

The percentage change in real terms, $r\%$, is given by $r = c - i$, where $c\%$ is the given percentage change and $i\%$ is the inflation rate.

### CONCEPTS – APPROXIMATION

In fact, the method in Key Point 2.7 for finding the value in real terms is just an **approximation** – which is very good for small percentage changes. The true value would be found by

multiplication by $\dfrac{1 + \dfrac{c}{100}}{1 + \dfrac{i}{100}}$.

### WORKED EXAMPLE 2.17

An investment account offers 2% annual interest. Jane invests £1000 in this account for 5 years. During those 5 years there is inflation of 2.5% per annum.

What is the value in real terms of Jane's investment after those 5 years.

| | |
|---|---|
| Use $r = c - i$ to find the annual real terms percentage change | The annual real terms percentage change is $2 - 2.5 = -0.5\%$ |
| A decrease of $0.5\%$ means multiplication by $1 - \dfrac{0.5}{100} = 0.995$ | So, after 5 years the value in real terms is $1000 \times 0.995^5 \approx £975$ |

**Tip**

This can also be done using the TVM package with an annual interest rate of −0.5%.

### CONCEPTS – APPROXIMATION

In Worked Example 2.17, which quantities are exact and which are uncertain? Would it be appropriate to give the answer to two decimal places, as is often done when doing financial arithmetic? Compare the confidence you have in your model in Worked Examples 2.16 and 2.17.

## Exercise 2C

For questions 1 to 3, use the TVM package on your calculator or the method demonstrated in Worked Example 2.16 to find the final value of the investment.

1　a　$2000 invested at 6% compounded annually for 3 years.

　　b　$3000 invested at 4% compounded annually for 18 years.

2　a　$500 invested at 2.5% compounded quarterly for 6 years.

　　b　$100 invested at 3.5% compounded twice-yearly for 10 years.

3　a　$5000 invested at 5% compounded monthly for 60 months.

　　b　$800 invested at 4% compounded monthly for 100 months.

For questions 4 and 5, use the TVM package on your calculator to find how long is required to achieve the given final value.

4　a　$100 invested at 5% compounded annually. £1000 required.

　　b　$500 invested at 4% compounded annually. £600 required.

5　a　$300 invested at 2% compounded monthly. £400 required.

　　b　$1000 invested at 1.5% compounded monthly. £1100 required.

For questions 6 and 7, use the TVM package on your calculator to find the annual interest rate required to achieve the given final value. Assume that interest is compounded annually.

6　a　€500 invested for 10 years. €1000 required.

　　b　€500 invested for 10 years. €800 required.

7　a　€100 invested for 5 years. €200 required.

　　b　€100 invested for 10 years. €200 required.

For questions 8 and 9, use the TVM package on your calculator or the method demonstrated in Worked Example 2.16 to find the final value of an asset after the given depreciation. Assume this depreciation is compounded annually.

8　a　Initial value $1200, annual depreciation 10% for 10 years.

　　b　Initial value $1200, annual depreciation 20% for 10 years.

9　a　Initial value £30 000, annual depreciation 15% for 5 years.

　　b　Initial value £30 000, annual depreciation 25% for 5 years.

For questions 10 and 11, use the TVM package on your calculator or the method demonstrated in Worked Example 2.17 to find the real value of cash (that is, no interest paid) subject to the inflation described, relative to the initial value.

10　a　$100 after 1 year of inflation at 3%.

　　b　$300 after 1 year of inflation at 2%.

11　a　£1000 after 10 years of inflation at 3%.

　　b　£5000 after 10 years of inflation at 5%.

For questions 12 and 15, use the TVM package on your calculator or the method demonstrated in Worked Example 2.17 to find the real value of an investment subject to the inflation described. All interest is compounded annually.

12　a　$100 after 1 year of interest at 5% and inflation at 2%.

　　b　$5000 after 1 year of interest at 4% and inflation at 1.4%.

13　a　£100 after 1 year of interest at 4% and inflation at 10%.

　　b　£2500 after 1 year of interest at 3% and inflation at 5%.

14　a　€1000 after 10 years of interest at 3.5% and inflation at 3%.

　　b　€500 after 5 years of interest at 5% and inflation at 2%.

15　a　$1000 after 10 years of interest at 5% and inflation at 10%.

　　b　$5000 after 8 years of interest at 3% and inflation at 5%.

**16**　A savings account pays interest at 3% per year (compounded annually). If £800 is deposited in the account, what will the balance show at the end of 4 years?

**17** A bank charges 5% annual interest on loans. If a seven-year loan of £10000 is taken, with no repayments until the seven years end, how much will need to be paid at the end of the loan term?

**18** A savings account has an annual interest rate of 4% compounded monthly. Find the account balance 30 months after $8000 is invested.

**19** A bank loan has a published annual interest rate of 5.8%, compounded monthly. What is the balance 18 months after a loan of €15000 is taken out, if no payments are made?

**20** A car is bought for £20000. Each year there is 15% depreciation. Find the expected value of the car after 5 years.

**21** £1000 is invested at an annual interest rate of 6%.

   a   Is the return better if the interest is compounded annually or monthly?

   b   Over 10 years, what is the difference in the return between these two compounding methods?

**22** A car is bought for £15000. It depreciates by 20% in the first year and 10% in subsequent years. There is an average of 2.5% inflation each year. What is the real terms value of the car after 5 years? Give your answer to the nearest £10.

**23** A company buys a manufacturing machine for $20000, with an expected lifetime of 8 years and scrap value $1500. The depreciation rate is set at 30%. Complete the table of asset values for the 8 years. Give all values to the nearest dollar.

| Year | Start-year value ($) | Depreciation expense ($) | End-year value($) |
|------|---------------------|--------------------------|-------------------|
| 1 | 20000 | | |
| 2 | | | |
| 3 | | | |
| 4 | | | |
| 5 | | | |
| 6 | | | |
| 7 | | | |
| 8 | | | |

**24** £1000 is invested for 5 years in an account paying 6.2% annual interest. Over the same period, inflation is judged to be 3.2% annually. What is the real terms percentage increase in value of the investment at the end of the 5 years?

**25** A bank advertises a loan charging 12% annual interest compounded monthly. What is the equivalent annual interest rate if compounded annually?

**26** A government pledges to increase spending on education by 5% each year.

   a   If the inflation rate is predicted to be 2.5% each year, what is the real terms increase per year?

   b   If this is pledged for a 5-year period, what is the overall real terms percentage increase in spending on education?

**27**  In 1923 Germany suffered from hyperinflation. The annual inflation rate was approximately $10^{11}$%.

Use the **exact** formula for inflation given in the Concepts box on page 40 to answer the following:

   a   At the beginning of 1923 a loaf of bread costs approximately 250 marks. How much did it cost at the end of 1923?

   b   Siegfried had 2000000 marks in savings at the beginning of 1923. He invested it in a savings account paying 20% interest. How much was it worth in real terms at the end of 1923?

   c   Anna had a mortgage of 25 million marks at the beginning of 1923. Interest of 15% was added to this mortgage during 1923. Anna made no payments. What is the debt in real terms by the end of 1923?

**TOOLKIT:** Problem Solving

Pick any natural number. If it is even, halve it. If it is odd, multiply by three, then add one, then continually repeat this process with the result. What number will you end up with? Use a spreadsheet to come up with a conjecture. For further practice using spreadsheets to investigate sequences, refer back to the introductory toolkit chapter.

This result has been suspected for some time, but it is still unproven. You might like to see if you can prove it for any special cases, or investigate how long it takes the sequence to first reach its endpoint for different starting values.

## Checklist

■ You should know the formula for the $n$th term of an arithmetic sequence:

   □ $u_n = u_1 + (n-1)d$

■ You should know the two versions of the formula for the sum of the first $n$ terms of an arithmetic sequence:

   □ $S_n = \dfrac{n}{2}\left[2u_1 + (n-1)d\right]$ or

   □ $S_n = \dfrac{n}{2}(u_1 + u_n)$

■ You should know the formula for the $n$th term of a geometric sequence:

   □ $u_n = u_1 r^{n-1}$

■ You should know the two versions of the formula for the sum of the first $n$ terms of a geometric sequence:

   □ $S_n = \dfrac{u_1(1-r^n)}{1-r}, \quad r \neq 1$ or

   □ $S_n = \dfrac{u_1(r^n-1)}{r-1}, \quad r \neq 1$

■ You should be able to work with arithmetic and geometric sequences given in sigma notation:

   □ $\displaystyle\sum_{r=1}^{r=n} u_r = u_1 + u_2 + u_3 + \ldots + u_n$

      — The value of $r$ at the bottom of the sigma (here $r = 1$) shows where the counting starts.
      — The value of $r$ at the top of the sigma (here $r = n$) shows where the counting stops.

■ You should be able to apply arithmetic and geometric sequences to real-world problems. This will often involve percentage change:

   □ A change of $r\%$ is equivalent to multiplying by the factor $1 + \dfrac{r}{100}$

   □ An annual inflation rate of $i\%$ is equivalent to multiplying by a factor $\dfrac{1}{1 + \dfrac{i}{100}}$ each year to find the value in real terms.

## ■ Mixed Practice

**1** Pierre invests 5000 euros in a fixed deposit that pays a nominal annual interest rate of 4.5%, compounded *monthly*, for 7 years.
  **a** Calculate the value of Pierre's investment at the end of this time. Give your answer correct to two decimal places.

Carla has 7000 dollars to invest in a fixed deposit which is compounded *annually*. She aims to double her money after 10 years.
  **b** Calculate the minimum annual interest rate needed for Carla to achieve her aim.

Mathematical Studies SL May 2015 TZ1 Paper 1 Q10

**2** Only one of the following four sequences is arithmetic and only one of them is geometric.

$$a_n = 1, 2, 3, 5, \ldots$$
$$b_n = 1, \frac{3}{2}, \frac{9}{4}, \frac{27}{8}, \ldots$$
$$c_n = 1, \frac{1}{2}, \frac{1}{3}, \frac{1}{4}, \ldots$$
$$d_n = 1, 0.95, 0.90, 0.85, \ldots$$

  **a** State which sequence is
    **i** arithmetic,
    **ii** geometric.
  **b** For **another** geometric sequence $e_n = -6, -3, -\frac{3}{2}, -\frac{3}{4}, \ldots$
    **i** write down the common ratio;
    **ii** find the **exact** value of the 10th term. Give your answer as a fraction.

Mathematical Studies SL May 2015 TZ2 Paper 1 Q9

**3** In an arithmetic sequence $u_8 = 10$, $u_9 = 12$.
  **a** Write down the value of the common difference.
  **b** Find the first term.
  **c** Find the sum of the first 20 terms.

**4** In a geometric sequence the first term is 2 and the second term is 8.
  **a** Find the common ratio.
  **b** Find the fifth term.
  **c** Find the sum of the first eight terms.

**5** A company projects a loss of $100 000 in its first year of trading, but each year it will make $15 000 more than the previous year. In which year does it first expect to make a profit?

**6** A car has initial value $25 000. It falls in value by $1500 each year. How many years does it take for the value to reach $10 000?

**7** One week after being planted, a sunflower is 20 cm tall and it subsequently grows by 25% each week.
  **a** How tall is it 5 weeks after being planted?
  **b** In which week will it first exceed 100 cm?

**8** Kunal deposits 500 euros in a bank account. The bank pays a nominal annual interest rate of 3% compounded quarterly.
  **a** Find the amount in Kunal's account after 4 years, assuming no further money is deposited. Give your answer to two decimal places.
  **b** How long will it take until there has been a total of 100 euros paid in interest?

**9** At the end of 2018 the world's population was 7.7 billion. The annual growth rate is 1.1%. If this growth rate continues
   **a** estimate the world's population at the end of 2022
   **b** what is the first year in which the population is predicted to exceed 9 billion?

**10** The second term of an arithmetic sequence is 30. The fifth term is 90.
   **a** Calculate
      **i** the common difference of the sequence;
      **ii** the first term of the sequence.

   The first, second and fifth terms of this arithmetic sequence are the first three terms of a geometric sequence.
   **b** Calculate the seventh term of the **geometric** sequence.

*Mathematical Studies SL May 2015 TZ1 Paper 1 Q7*

**11** The sum of the first $n$ terms of an arithmetic sequence is given by $S_n = 6n + n^2$.
   **a** Write down the value of
      **i** $S_1$;
      **ii** $S_2$.

   The $n$th term of the arithmetic sequence is given by $u_n$.
   **b** Show that $u_2 = 9$.
   **c** Find the common difference of the sequence.
   **d** Find $u_{10}$.
   **e** Find the lowest value of $n$ for which $u_n$ is greater than 1000.
   **f** There is a value of $n$ for which $u_1 + u_2 + \ldots + u_n = 1512$.
      Find the value of $n$.

*Mathematical Studies SL May 2015 TZ2 Paper 2 Q3*

**12** The first three terms of an arithmetic sequence are $x$, $2x + 4$, $5x$. Find the value of $x$.

**13** The audience members at the first five showings of a new play are: 24, 34, 46, 55, 64.
   **a** Justify that the sequence is approximately arithmetic.
   **b** Assuming that the arithmetic sequence model still holds, predict the number of audience members at the sixth showing.

**14** Evaluate $\displaystyle\sum_{r=1}^{12} \frac{6^r}{3^r}$.

**15** According to a business plan, a company thinks it will sell 100 widgets in its first month trading, and 20 more widgets each month than the previous month. How long will it take to sell a total of 4000 widgets?

**16** Find an expression for the $n$th term of the geometric sequence $a^2b^2$, $a^4b$, $a^6$.

**17** When digging a rail tunnel there is a cost of $10 000 for the first metre. Each additional metre costs $500 dollars more per metre (so the second metre costs $10 500). How much does it cost to dig a tunnel 200 m long?

**18** When Elsa was born, her grandparents deposited $100 in her savings account and then on subsequent birthdays they deposit $150 then $200 then $250 and so on in an arithmetic progression. How much have they deposited in total just after her 18th birthday?

**19** £5000 is invested for 3 years in an account paying 5.8% annual interest. Over the same period, inflation is judged to be 2.92% annually. What is the real terms percentage increase in value of the investment at the end of the 3 years?

**20** A company purchases a computer for $2000. It assumes that it will depreciate in value at a rate of 10% annually. If inflation is predicted to be 2% annually, what is the real terms value of the computer after 4 years?

**21** Cameron invests $1000. He has a choice of two schemes:

Scheme A offers $25 every year.

Scheme B offers 2% interest compounded annually.

Over what periods of investment (in whole years) is scheme A better than scheme B?

**22** In a game, $n$ small pumpkins are placed 1 metre apart in a straight line. Players start 3 metres before the first pumpkin.

Start

Each player **collects** a single pumpkin by picking it up and bringing it back to the start. The nearest pumpkin is collected first. The player then collects the next nearest pumpkin and the game continues in this way until the signal is given for the end.

Sirma runs to get each pumpkin and brings it back to the start.
**a** Write down the distance, $a_1$, in metres that she has to run in order to **collect** the first pumpkin.
**b** The distances she runs to **collect** each pumpkin form a sequence $a_1, a_2, a_3, \ldots$ .
   **i** Find $a_2$.    **ii** Find $a_3$.
**c** Write down the common difference, $d$, of the sequence.

The final pumpkin Sirma **collected** was 24 metres from the start.
**d** **i** Find the total number of pumpkins that Sirma **collected**.
   **ii** Find the total distance that Sirma ran to **collect** these pumpkins.

Peter also plays the game. When the signal is given for the end of the game he has run 940 metres.
**e** Calculate the total number of pumpkins that Peter **collected**.
**f** Calculate Peter's distance from the start when the signal is given.

Mathematical Studies SL November 2014 Paper 2 Q5

**23** The seventh, third and first terms of an arithmetic sequence form the first three terms of a geometric sequence.

The arithmetic sequence has first term $a$ and non-zero common difference $d$.
**a** Show that $d = \frac{a}{2}$.

The seventh term of the arithmetic sequence is 3. The sum of the first $n$ terms in the arithmetic sequence exceeds the sum of the first $n$ terms in the geometric sequence by at least 200.
**b** Find the least value of $n$ for which this occurs.

Mathematics HL November 2014 Paper 2 Q7

**24** An athlete is training for a marathon. She considers two different programs. In both programs on day 1 she runs 10 km.

In program A she runs an additional 2 km each day compared to the previous day.

In program B she runs an additional 15% each day compared to the previous day.

  **a** In which program will she first reach 42 km in a day. On what day of the program does this occur?

  **b** In which program will she first reach a total of 90 km run. On what day of the program does this occur?

**25** A teacher starts on a salary of £25 000. Each year the teacher gets a pay rise of £1500. The teacher is employed for 30 years.

  **a** Find their salary in their final year.

  **b** Find the total they have earned during their teaching career.

  **c** If the inflation rate is on average 1.5% each year, find the real value of their final salary at the end of their final year in terms of the value at the beginning of their career, giving your answer to the nearest £100.

**26** The 10th term of an arithmetic sequence is two times larger than the fourth term.

Find the ratio: $\dfrac{u_1}{d}$.

**27** If $a$, $b$, $c$, $d$ are four consecutive terms of an arithmetic sequence, prove that $2(b - c)^2 = bc - ad$.

# 3 Core: Functions

## ESSENTIAL UNDERSTANDINGS

- Models are depictions of real-life events using expressions, equations or graphs while a function is defined as a relation or expression involving one or more variables.
- Creating different representations of functions to model the relationships between variables, visually and symbolically as graphs, equations and/or tables represents different ways to communicate mathematical ideas.

## In this chapter you will learn…

- about function notation
- how to determine the domain and range of a function
- about the use of functions in modelling real-life events
- about the idea of an inverse function
- how to draw graphs of inverse functions
- how to sketch a graph from given information
- how to find key features of graphs using a GDC
- how to find intersections of graphs using a GDC
- how to solve equations using a GDC.

### CONCEPTS

The following concepts will be addressed in this chapter:
- Different **representations** of functions – symbolically and visually as graphs, equations and tables – provide different ways to communicate mathematical **relationships**.
- Moving between different forms to **represent** functions allows for deeper understanding and provides different approaches to problem solving.

■ **Figure 3.1** Can we quantify the relationships between inputs and outputs in the examples shown?

## PRIOR KNOWLEDGE

Before starting this chapter, you should already be able to complete the following:

1  Use technology to sketch the graph of $y = x^2 + x - 12$.
2  Find the values of $x$ for which $2x - 3 \geqslant 0$.
3  Evaluate on your calculator $\ln 1$.
4  Solve $0.1 = e^x$.
5  Sketch $y = |x - 1|$ using technology.

In mathematics, a process that takes inputs and generates outputs is called a function. These functions can be represented in many different forms, such as a graph, algebraically or in a table. Sometimes it will be necessary to use one of these for a particular reason, but often you can choose the one that is most appropriate to work with for the task at hand.

Functions have a wide range of uses in mathematics. They are studied as objects in their own right, but knowing the properties of a wide range of functions also helps you to model many real-world situations.

## Starter Activity

Look at Figure 3.1. In small groups identify any inputs and outputs you can think of that are related to each of the images.

**Now look at this problem:**

A particular mathematical rule is defined to be the sum of a number and all its digits.

If the answer is 19, what was the number?

# 3A Concept of a function

## ■ Function notation

A function is a rule that maps each input value, $x$, to a single output value, $f(x)$.

For example, the rule 'square the input value and then add 3' would be written in function notation as:

■  $f(x) = x^2 + 3$

or alternatively as:

■  $f : x \mapsto x^2 + 3$

A function can be

■  one-to-one: when each output value comes from a single input value, for example, $y = x^3$

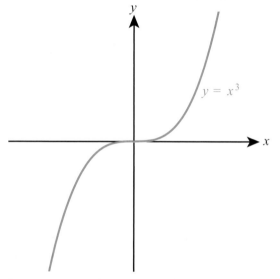

■  many-to-one: when an output value can come from more than one different input value, for example, $y = x^2$.

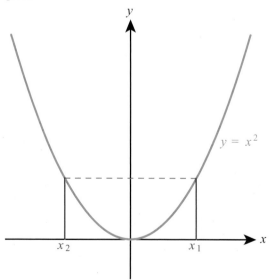

## CONCEPTS – REPRESENTATION

Does the graphical **representation** or the algebraic representation of the function make it more obvious whether the function is one-to-one? Which one would be more useful in finding $y$ when $x = 13$? One key skill is deciding which representation of a function is most useful in solving different types of problems.

The notation $f(x)$ was first used by the Swiss mathematician Leonhard Euler in 1750.

### WORKED EXAMPLE 3.1

If $f(x) = 2x + 7$ find $f(3)$.

Substitute $x = 3$ into $2x + 7$ ································· $f(3) = 2(3) + 7$
$$= 13$$

### WORKED EXAMPLE 3.2

Determine, with reasons, if either of the following graphs shows a function of the form $y = f(x)$.

a

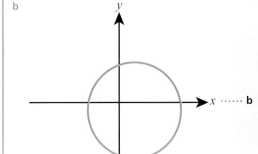

a   There are no values of $x$ that map to more than one output, so this is a function.

b

b   There are values of $x$ that map to two different outputs, so this is not a function.

## Domain and range

As well as knowing what the function does to the input value, you also need to know what inputs are allowed.

### Tip

If no domain is specified, you can assume that the domain is over all real numbers, sometimes written as $x \in \mathbb{R}$.

**KEY POINT 3.1**

The domain of a function is the set of all allowed input values.

There are three main things which you should consider when determining the largest possible domain of a function.

- You can never square root a negative number.
- You can never divide by zero.
- You can never take the logarithm of a negative number or zero.

**WORKED EXAMPLE 3.3**

Find the largest possible domain of $f(x) = \sqrt{2 - x}$.

The expression under the square root must be greater than or equal to zero

Largest possible domain is given by

$$2 - x \geqslant 0$$

$$x \leqslant 2$$

### Be the Examiner 3.1

Find the largest possible domain of $f(x) = \frac{1}{x-2}$.

Which is the correct solution? Identify the errors made in the incorrect solutions.

| Solution 1 | Solution 2 | Solution 3 |
|---|---|---|
| Largest possible domain is given by: $x - 2 > 0$ $x > 2$ | Largest possible domain is given by: $x - 2 \neq 0$ $x \neq 2$ | Largest possible domain is given by: $x \neq -2$ |

Once you know the domain of a function, you can find the possible output values.

**KEY POINT 3.2**

The range of a function is the set of all possible outputs.

To find the range it is best to start with a graph of the function.

**WORKED EXAMPLE 3.4**

Find the range of $f(x) = \sqrt{2 - x}$, $x \leqslant 0$.

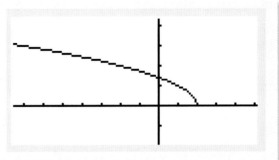

Sketch the graph on your GDC ···················

The calculator shows the graph over the largest possible domain, but we have restricted it to only the left of the $y$-axis. We need to find its value there, in this case $f(0)$. We could do this using the calculator, or substitute numbers into the formula ························································· $f(0) = \sqrt{2}$

From the graph we can see that the restricted function is always greater than its value at $x = 0$ ···················· Range is $f(x) \geqslant \sqrt{2}$

## ⚙ ■ The concept of a function as a mathematical model

Functions can often be created and used to predict the results observed in real life from a particular process – this is known as producing a mathematical model of a process. By understanding the underlying theory of the process involved, the particular function needed can be derived.

For example, Zipf's Law relates the population of a city to the rank of that city's population (largest, second largest, etc).

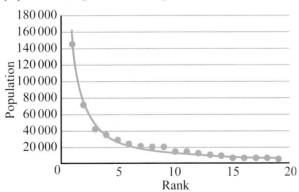

This model does not give exactly the results observed in real life, as you can see from the observed data points and the predictions made by the graph of the function. Just because a model is not perfect does not mean that it is useless. You need to understand what the limitations of any given model might be. Things to consider:

■ Check at large and small values that you get a sensible result.

■ Check whether the range is sensible in the given context.

■ Think about how accurate the model needs to be, for example, the dosage of drugs for a patient compared to the volume of water in a swimming pool.

■ Think about whether there is some theoretical underpinning for the model, such as, does the function used follow from a mathematical derivation, or is the function just chosen in order to fit the observed data.

---

**WORKED EXAMPLE 3.5**

The population of pigeons, P(t) on an island at a time t years after being introduced is modelled by $P(t) = 100t^2$.

State four reasons why this model is unlikely to be perfectly accurate.

See if there are any particular values of $t$ that do not give sensible values for the number of pigeons

Think about the biological situation – you are not expected to have any special knowledge here, just to use common sense

In reality the population would increase more quickly in the spring/summer, for example

1  $P(0) = 0$, which says that there are no pigeons at the start of the time period. This is not consistent with pigeons being introduced to the island.

2  $P\left(\dfrac{1}{3}\right) = \dfrac{100}{9}$, so the model predicts a non-integer number of pigeons after 4 months.

3  As $t$ gets large the population grows without limit. This is not realistic as eventually space or resources such as food will run out.

4  The model does not allow for any randomness in the growth rate of the population, or any differences due to seasonality.

---

**TOK Links**

Even though the model is not perfect, this does not mean it is not useful. It might give reasonable estimates of the approximate population over the first 10 years. In more advanced work, models do not predict single numbers, but a range of possible values the output can take. Which is better – a precise answer unlikely to be perfectly correct or an imprecise answer which is likely to include the correct answer?

---

**Tip**

Note that $f^{-1}(x)$ *does not* mean $\dfrac{1}{f(x)}$, as you might be tempted to think from the laws of exponents for numbers.

## ■ Informal concept of an inverse function

The inverse function, $f^{-1}$, of a function f reverses the effect of f. For example, if $f(2) = 4$, then $f^{-1}(4) = 2$.

It is important to note that only one-to-one functions have inverses. For example, the many-to-one function $f(x) = x^2$ does not have an inverse because if you know the output is 9, it is impossible to determine whether the input was −3 or 3.

---

**WORKED EXAMPLE 3.6**

If $f(x) = 3x + 5$ find $f^{-1}(11)$.

To find $f^{-1}(11)$, solve $f(x) = 11$

$$3x + 5 = 11$$
$$3x = 6$$
$$x = 2$$
So, $f^{-1}(11) = 2$

## Graphical interpretation of an inverse function

When you find the inverse of a function, you are making the input of f the output of $f^{-1}$ and the output of f the input of $f^{-1}$.

Graphically you are swapping the $x$ and $y$ coordinates, which is achieved by reflecting the graph of $y = f(x)$ in the line $y = x$.

**KEY POINT 3.3**

The graph of $y = f^{-1}(x)$ is a reflection of the graph $y = f(x)$ in the line $y = x$.

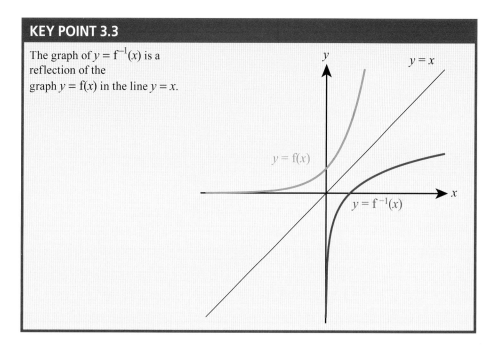

**WORKED EXAMPLE 3.7**

Sketch the inverse function of the following graph.

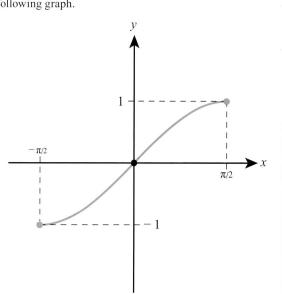

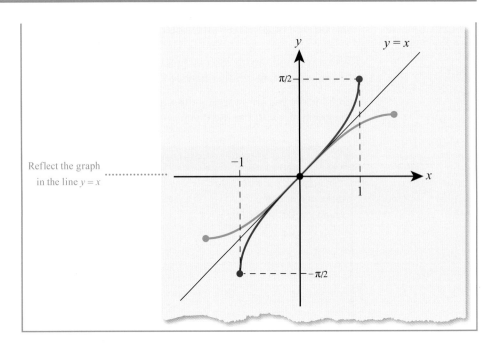

Reflect the graph
in the line $y = x$

## CONCEPTS – REPRESENTATION

Functions can be **represented** by rules, tables, graphs or mapping diagrams.
Different representations are appropriate in different problems and choosing which
representation to use can be a key tactical decision.

## Exercise 3A

For questions 1 to 4, use the method shown in Worked Example 3.1.

1   If f($x$) = $x$ + 5, find
    a   f(4)
    b   f(9)

2   If f($x$) = 5$x$ − 8, find
    a   f(3)
    b   f(11)

3   If f($x$) = 7 − 3$x$, find
    a   f(8)
    b   f(−4)

4   If f($x$) = $x^3$ − 2$x^2$ + 1, find
    a   f(−2)
    b   f(−3)

For questions 5 to 7, use the method shown in Worked Example 3.2 to determine whether the following graphs represent functions.

5   a

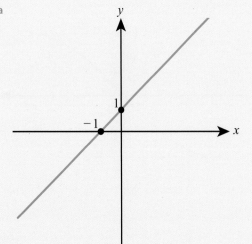

    b

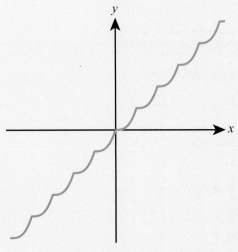

6  a

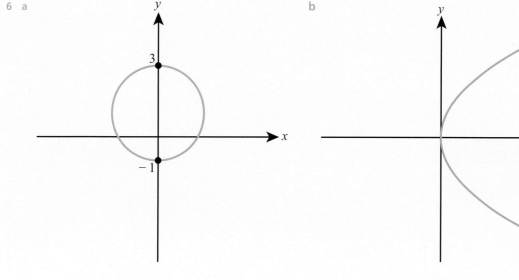

b

7  a

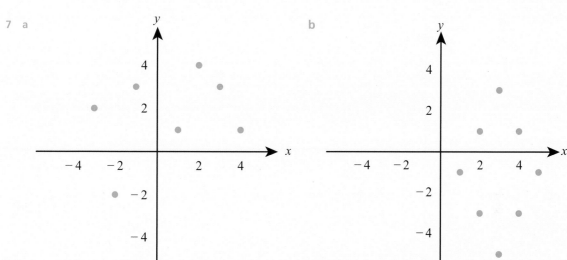

b

For questions 8 and 9, use technology to sketch the graph and then the method shown in Worked Example 3.2 to determine whether the following equations for $y$ describe a function of $x$.

8  a  $y = \pm\sqrt{x}$, $x \geqslant 0$

   b  $y = \sqrt{x^2}$, $x \in \mathbb{R}$

9  a  $y = x^{\frac{1}{3}}$, $x \in \mathbb{R}$

   b  $y = \dfrac{2}{x^2}$, $x \neq 0$

For questions 10 to 14, use the method shown in Worked Example 3.3 to find the largest possible domain of the given real functions.

10  a  $f(x) = 3x^2 - 2$

    b  $g(x) = 7x^3 - 2x$

11  a  $f(x) = \dfrac{1}{x}$

    b  $f(x) = -\dfrac{3}{x}$

12  a  $f(x) = 2\sqrt{x + 5}$

    b  $g(x) = 3\sqrt{2x - 7}$

13  a  $f(x) = \ln(5x + 3)$

    b  $g(x) = \ln(2x - 8)$

14  a  $f(x) = \dfrac{3}{5 - 2x}$

    b  $g(x) = \dfrac{2x}{3 + x}$

For questions 15 to 18, use the method shown in Worked Example 3.4 to find the range of each function.

15  a  $f(x) = x^2 - 2, x \in \mathbb{R}$                              b  $g(x) = 3x^2 + 7, x \in \mathbb{R}$

16  a  $f(x) = 2x + 8, x \leqslant 5$                              b  $g(x) = 7 - 3x, x > 1$

17  a  $f(x) = \sqrt{4 + x}, x \geqslant -4$                      b  $g(x) = 3\sqrt{5 - 2x}, x < 2.5$

18  a  $f(x) = 2 + \ln(x - 7), x \geqslant 8$                   b  $g(x) = 3 - \ln(2x + 1), x > 0$

For questions 19 to 21, use the method shown in Worked Example 3.6.

19  a  If $f(x) = 5x$, find $f^{-1}(200)$.                       b  If $g(x) = 7x$, find $g^{-1}(-49)$.

20  a  If $f(x) = 3x^5$, find $f^{-1}(3)$.                         b  If $g(x) = 2x^3$, find $g^{-1}(16)$.

21  a  If $f(x) = 6x^2 + 1, x \leqslant 0$, find $f^{-1}(55)$.   b  If $g(x) = \sqrt{3 + x^2}, x \leqslant 0$, find $g^{-1}(2)$.

For questions 22 to 25, each graph represents the function f(x). Use the method shown in Worked Example 3.7 to sketch the inverse function $f^{-1}(x)$ or state that there is no inverse function.

22  a

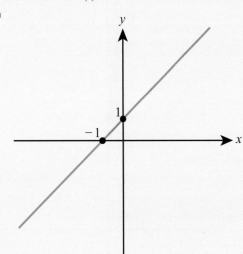

b

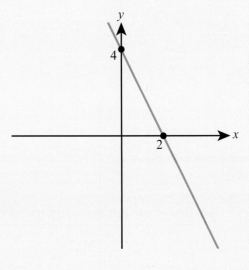

23  a

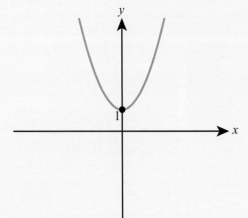

b

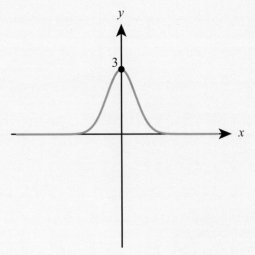

24  a

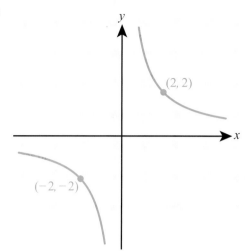

b

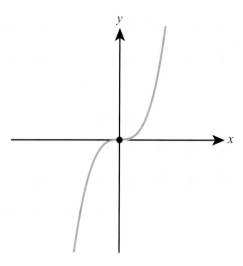

25  a

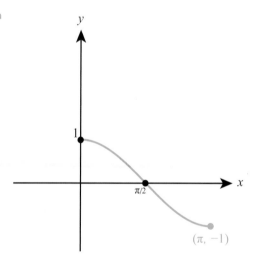

b

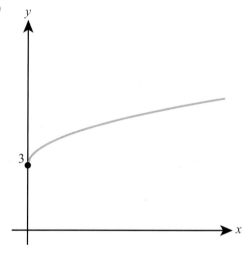

26  Given a function $g(x) = 4x - 5$,

a  evaluate $g(-2)$

b  solve the equation $g(x) = 7$.

27  Given the function $h(x) = \dfrac{x-5}{3}$, solve the equation $h(x) = 12$.

28  The speed, $v \, \mathrm{m\,s^{-1}}$, of a car at the time $t$ seconds is modelled by the function $v(t) = 3.8t$.

a  Find the speed of the car after 1.5 seconds.

b  Comment on whether this model should be used to predict the speed of the car after 30 seconds.

29  a  Write down the largest possible domain of the function $f(x) = \dfrac{3}{(x-5)^2}$.

b  Evaluate $f(2)$.

30  A function is defined for all real numbers by $q(x) = 3x^2 - 2$.

a  Evaluate $q\left(\dfrac{1}{2}\right)$, giving your answer as an exact fraction.

b  Find the range of the function.

c  Solve the equation $q(x) = 46$.

**31** A technology company wants to model the number of smartphones in the world. They propose the following model: $N = 2.3e^{0.098t} + 1.2$

where $N$ billions is the number of smartphones at the time $t$ years from now.

 a According to this model, how many smartphones will there be in the world in 5 years' time?

 b Give one reason why this model might not give a correct prediction for the number of smartphones 50 years from now.

**32** One pound sterling can be converted into \$1.30.

 a If the amount in pounds sterling is $x$ find the function $f(x)$ which tells you the amount this is worth in dollars.

 b Explain the meaning of the function $f^{-1}(x)$.

**33** A function is defined by the following table:

| $x$ | 0 | 1 | 2 | 3 | 4 | 5 | 6 |
|---|---|---|---|---|---|---|---|
| $f(x)$ | 3 | 6 | 4 | 2 | 0 | 1 | 5 |

 a Find $f^{-1}(6)$.

 b Solve $f(x) = x + 2$.

**34** a Find the largest possible domain of the function $f(x) = \sqrt{2x - 5}$.

 b Find the range of the function for the domain from part **a**.

 c Solve the equation $f(x) = 3$.

**35** The size ($N$) of a population of fish in a lake, $t$ months after they are first introduced, is modelled by the equation $N = 150 - 90e^{-0.1t}$.

 a How many fish were initially introduced into the lake?

 b Find the value of $N$ when $t = 15$. According to this model, how many fish will there be in the lake after 15 months?

 c Give one reason why this model is not perfectly accurate.

**36** A function is given by $f(x) = \dfrac{x}{2} + 5$.

 a Evaluate $f(18)$.

 b Find $f^{-1}(7)$.

**37** The diagram shows the graph of $y = g(t)$. Copy the graph and on the same axes sketch the graph of $y = g^{-1}(t)$.

**38** A function is defined by $g(x) = \log_3 x$.

 a State the largest possible domain of g.

 b Evaluate $g(81)$.

 c Find $g^{-1}(-2)$.

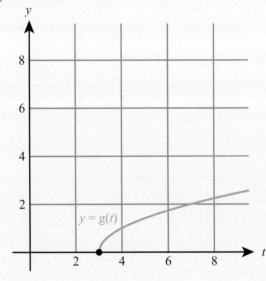

**39** Find the largest possible domain of the function $n(x) = \log_5(3x - 15)$.

**40** a Find the largest possible domain of the function $h(x) = \log(7 - 3x)$.

b Find $h^{-1}(2)$.

**41** For a function $f(x) = 10 - 3x$ defined on the domain $x \leqslant 2$:

a Evaluate $f(-3)$.

b Find the range of the function.

c Explain why the equation $f(x) = 1$ has no solutions.

**42** The diagram shows the graph of $y = f(x)$.

a Find $f(4)$.

b Find $f^{-1}(4)$.

c Copy the graph and on the same axes sketch the graph of $y = f^{-1}(x)$.

**43** A function is defined by $N(t) = 3e^{-0.4t}$.

a Evaluate $N(7)$.

b Find $N^{-1}(2.1)$.

**44** a Find the largest domain of the function

$$f(x) = \frac{3}{4 - \sqrt{x - 1}}.$$

b For the domain found in part a find the range of the function.

**45** A function is defined by $f(x) = x^3 + x - 8, x \geqslant 0$.

a Use technology to sketch $y = f(x)$.

b Sketch on the same axis $y = f^{-1}(x)$.

c Solve $f(x) = f^{-1}(x)$.

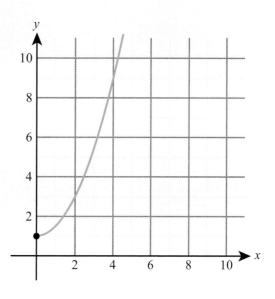

# 3B Sketching graphs

## ◼ Key features of graphs

There are several important features of the graph $y = f(x)$ that you need to be able to identify.

■ **Intercepts** – these are the points where the graph meets the axes. Where it meets the $y$-axis is called the $y$-intercept. The points where it meets the $x$-axis are called the $x$-intercepts of the graph or the zeros of $f(x)$ or the roots of $f(x) = 0$.

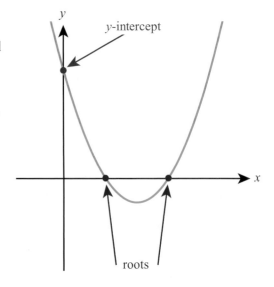

■ Vertices (singular: vertex) – these are points where the graph reaches a maximum or minimum point and changes direction. For example, points $A$ and $B$ are both **vertices of the graph**:

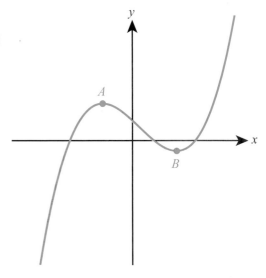

■ **Asymptotes** – these are lines to which a graph tends but never reaches. For example, the two dotted lines are both asymptotes. The red one is a vertical asymptote and the green one a horizontal asymptote:

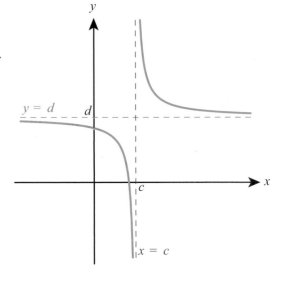

## Tip

For most functions you will meet, any lines of symmetry will pass through either the vertices or the vertical asymptotes of the graph. This is the easiest way to find them.

■ Symmetries – in this course you might be asked to find vertical lines of symmetry.

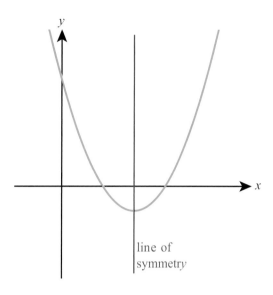

line of symmetry

### You are the Researcher

There are many other types of symmetry in graphs, such as rotational symmetry and translational symmetry. You might want to think about how these can be described using function notation. A particularly important type of symmetry is enlargement symmetry, when zooming in on a graph produces a similar graph. This creates a type of picture called a fractal, that as well as being beautiful has applications in fields from compressing computer files to analysing animals' pigmentation.

 You can use your GDC to help you draw graphs and identify these key features.

### WORKED EXAMPLE 3.8

Use technology to sketch $y = \dfrac{x-3}{x+4}$, labelling all **axis intercepts** and asymptotes.

Draw the graph and find the $y$-intercept:

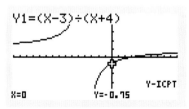

$y$-intercept is $(0, -0.75)$

The $x$-intercept can be found using the appropriate function on your calculator, such as, 'Root' or 'G-solve'

$x$-intercept is $(3, 0)$

Zoom out to get a better view of the shape of the graph. There looks to be one vertical asymptote and one horizontal asymptote. Use 'Trace' or the equivalent function on your GDC to find these ($x = -4$, $y =$ ERROR means the vertical asymptote is at $x = -4$)

Vertical asymptote at $x = -4$
Horizontal asymptote at $y = 1$

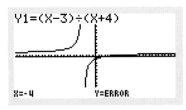

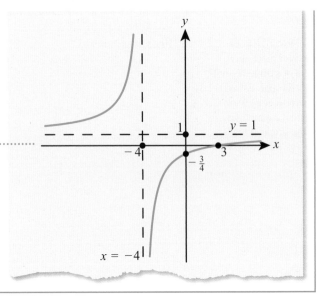

Now sketch the graph, labelling
all the features found ··············

---

## WORKED EXAMPLE 3.9

Find the minimum value of f($x$) = $x\mathrm{e}^x$. Hence find the range of f($x$).

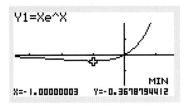

The minimum value of f($x$) is $-0.368$

Hence the range is f($x$) $\geqslant -0.368$

---

## WORKED EXAMPLE 3.10

a   Find the coordinates of the vertices on the graph of f($x$) = $x^4 - 4x^3 + 2x^2 + 4x - 3$.

b   Given that the curve $y$ = f($x$) has a line of symmetry, find its equation.

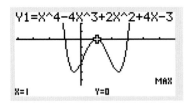

·············· The vertices are $(1, 0)$, $(-0.414, -4)$, $(2.414, -4)$

From the graph you can see that there
is a line of symmetry passing vertically ·············· The line of symmetry is $x = 1$
through the maximum point

## ■ Creating a sketch from information given

**WORKED EXAMPLE 3.11**

Sketch a graph with the following properties:
■ The range is $f(x) \leq 0$ or $f(x) > 1$
■ There are vertical asymptotes at $x = \pm 2$
■ The only axis intercept is the origin, which is also a vertex.

**Step 1**: Range $f(x) \leq 0$ or $f(x) > 1$ gives
a horizontal asymptote at $y = 1$

There are also two vertical asymptotes at $x = \pm 2$

Draw the asymptotes and identify the
regions corresponding to the range

**Step 2**: Vertex at $O$ (together with lower part
of range) means the curve must lie in the
lower part of the graph for $-2 < x < 2$

Fill in the graph for $-2 < x < 2$ as it must tend to
$-\infty$ as $x$ approaches either vertical asymptote

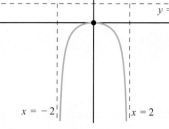

A possible graph is:

**Step 3**: One or both of the parts of the graph
given by the intervals $x < -2$ and $x > 2$ ....................
must fill in the upper part of the range

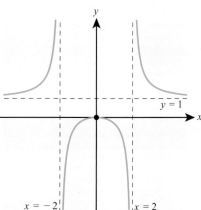

## ■ Creating a sketch from context

**WORKED EXAMPLE 3.12**

The function $N(d)$ gives the total number of seeds distributed within a distance $d$ of the centre of the tree. The tree is modelled as a cylinder of radius 0.2 m. The tree releases a total of 300 seeds. There are only 50 seeds within 1 m of the tree, and half are distributed more than 3 m from the tree.

The graph starts at 0.2 m and has a horizontal asymptote at $N(d) = 300$. In context, it must be increasing as $d$ is increasing. The points $(1, 50)$ and $(3, 150)$ should also be labelled

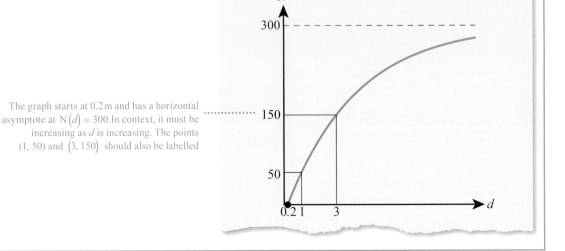

## ■ Finding the intersection of graphs using technology

**WORKED EXAMPLE 3.13**

Find the points of intersection of $y = x^3 - x^2$ and $y = 2x - 1$.

Draw both graphs and find the coordinates of their points of intersection

The points of intersection are $(-1.25, -3.49)$, $(0.445, -0.110)$ and $(1.80, 2.60)$.

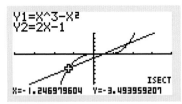

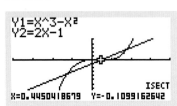

Move the cursor right to find the other two

The ideas in this section will be developed and extended in Chapter 14.

## ■ Solving equations using graphs

Instead of just being asked to solve an equation, you might instead be asked to find:

■ The **roots of an equation** – these are exactly the same thing as the solutions of the equation.

■ The **zeros of a function** – these are the values of $x$ for which $f(x) = 0$.

### WORKED EXAMPLE 3.14

Find the zeros of the function $f(x) = \dfrac{1}{x} + x - \sqrt{x} - 2$.

Graph the function and find any roots ················· The zeros are $x = 0.450$ and $x = 3.63$.

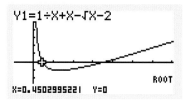

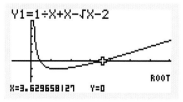

Move the cursor right to find the second root. You should zoom out enough to make sure that you have not missed any other roots

A report by Charles Godfrey from the Fourth International Congress of Mathematicians lamented that the reliance of students on modern technology to solve equations meant that 'there was a danger that the boys would become helpless in dealing with the most straight-forward algebraic expressions'. You might not find this surprising until you find out that this was a conference from 1908 and the modern technology that was causing this concern was graph paper! This new method of solving equations had only recently become feasible with a massive reduction in the price of graph paper and there were serious concerns that there were not enough teachers who were able to use this revolutionary and innovative method.

**WORKED EXAMPLE 3.15**

Find the root(s) of the equation $x^2 = e^x$.

Rearrange as a single function equal to zero ............................................ $x^2 = e^x$

$$x^2 - e^x = 0$$

Now graph the function $f(x) = x^2 - e^x$ and find any roots ............. The root is $x = -0.703$

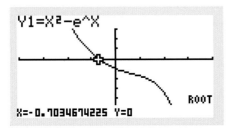

**CONCEPTS – REPRESENTATION AND RELATIONSHIPS**

An alternative to the approach in Worked Example 3.15 would be to draw the graphs $y = x^2$ and $y = e^x$ and find their point of intersection. You need to use the fact that the solutions of $x^2 - e^x = 0$ and $x^1 = e^x$ are equivalent and both have a **relationship** to their graphical **representation**.

All equations can be solved graphically, but not all equations can be solved algebraically. For example, in Worked Example 3.15 there is no way of exactly solving this equation and representing the answer using functions from school mathematics. However, such equations do exist in real-world situations, so choosing an appropriate representation of both the problem and the solution is vital.

## Exercise 3B

For questions 1 to 3, show the graph of the given function on your calculator and then use the method shown in Worked Example 3.8 to sketch the curve in the interval $-5 \leqslant x < 5$ and $-5 \leqslant y < 5$, labelling all axis intercepts and asymptotes.

1  a  $f(x) = \dfrac{1}{x+2}$

   b  $g(x) = \dfrac{x-1}{2x}$

2  a  $f(x) = \dfrac{x+3}{3x-3}$

   b  $g(x) = \dfrac{x-1}{x+2}$

3  a  $f(x) = \dfrac{3x^2 - 12}{x^2}$

   b  $g(x) = 5xe^{-x}$

For questions 4 to 6, use the method demonstrated in Worked Example 3.9 to find the range of each function, using technology to determine maximum and minimum values where needed.

4  a  $f(x) = 3x^2 - 4x + 5$

   b  $g(x) = 7 - 5x - 2x^2$

5  a  $f(x) = \ln(x^2 + 3x + 3)$

   b  $g(x) = \ln(2x^2 + 2x + 5)$

6  a  $f(x) = xe^{1-x^2}$

   b  $g(x) = e^{1-x^2}\cos 4x$

For questions 7 and 8, use technology and the method demonstrated in Worked Example 3.10 to find the coordinates of any vertices on the graph of each curve. Where there is a vertical line of symmetry, state its equation.

7  a  $f(x) = 5x^2 - 7x + 3$

   b  $g(x) = 5 - 4x - 7x^2$

8  a  $f(x) = x^4 + 4x^3 + 5x^2 + 2x + 2$

   b  $g(x) = 4x^4 - 8x^3 + 4x^2 - 1$

For questions 9 to 11, use the method demonstrated in Worked Example 3.11 to sketch a graph with the given properties.

9   a   The graph of f($x$) has a zero at 2 and no vertices. It has a single vertical asymptote at $x = 3$ and a horizontal asymptote at $y = 1$. The range of f($x$) is f($x$) $\neq 1$.

    b   The graph of g($x$) has a zero at $-3$ and no vertices. It has a single vertical asymptote at $x = 1$ and a horizontal asymptote at $y = -2$. The range of g($x$) is g($x$) $\neq -2$.

10   a   The range of f($x$) is $0 < f(x) < 3$. The graph has no vertices and has $y$-axis intercept 2. The function is always increasing, so f($a$) $<$ f($b$) whenever $a < b$.

    b   The graph of g($x$) has zeros at 1 and $-2$ and no vertices. It has a single vertical asymptote at $x = 0$ and for large positive and negative values of $x$, g($x$) $\approx x + 1$ The range of g($x$) is $\mathbb{R}$.

11   a   The graph of f($x$) has zeros at 3 and at 1 and no vertices. It has a single vertical asymptote at $x = 2$ and a horizontal asymptote at $y = 2$. The range of f($x$) is f($x$) $< 2$.

    b   The graph of g($x$) has zeros at 1 and 3 and a vertex at $(2, -2)$. It has vertical asymptotes at $x = 0$ and $x = 4$ and range g($x$) $\geqslant -2$ or g($x$) $< -8$.

For questions 12 to 14, use technology to find the points of intersection of the following pairs of equations; refer to the method shown in Worked Example 3.13. Give your answers to three significant figures.

12   a   $y = x^2 - 7$ and $y = x^3 + x$

    b   $y = 2x^3 + 1$ and $y = x^2 - 3x - 2$

13   a   $y = x^3 + 3x^2 + 3$ and $y = x^2 + x + 5$

    b   $y = x^4 - 4x^3 + 11x - 3$ and $y = x^2 - 5x + 9$

14   a   $y = x^4 + x^3 - 2x^2 - 2$ and $y = x^3 - x^2$

    b   $y = x^3 - 3x^2 + 2x + 7$ and $y = x^3 + x^2 - 6x + 3$

For questions 15 and 16, use your calculator to find the zeros of the following functions, drawing on the method shown in Worked Example 3.14. Give your answers correct to three significant figures.

15   a   f($x$) $= x^3 - 5\sqrt{x} + 1$             16   a   f($x$) $= 3\ln x - x + 1$

    b   g($x$) $= 5x\sqrt{x} - 7x + 1$              b   g($x$) $= e^x - 2x - 7$

For questions 17 and 18, use your calculator to find the roots of the following equations, drawing on the method shown in Worked Example 3.15. Give your answers correct to three significant figures.

17   a   $x^3 = 4\sqrt{x} - 1$                   18   a   $x^2 + e^{3x} = xe^x + 5$

    b   $7x\sqrt{x} = x^2 + 1$                b   $\ln(2x^2 + 1) = 3 - x - x^2$

**19**   Sketch the graph of $y = \ln(x^2 + 3x + 5)$ and write down the coordinates of its vertex.

**20**   Sketch the graph of $y = 7 - 3e^{-0.5x}$. Label any axis intercepts and find the equation of the horizontal asymptote.

**21**   a   Find the largest possible domain of the function f($x$) $= \dfrac{3x - 1}{x + 2}$.

    b   Sketch the graph of $y = $ f($x$), stating the equations of any asymptotes.

**22**   Sketch a graph with a $y$-intercept 3, no vertical asymptotes and horizontal asymptote $y = 1$.

**23**   Find the coordinates of the point of intersection of the graph $y = 5 - x$ and $y = \dfrac{1}{2}e^x$.

**24**   Find the coordinates of the maximum point on the graph of $y = \dfrac{10}{x^2 + 2x + 5}$.

**25**   A manufacturing firm uses the following model for their monthly profit: $P = 9.4q - 0.02q^2 - 420$

    where $P$ is the profit and $q$ is the number of items produced that month.

    How many items a month should the firm produce in order to maximize the profit?

**26**   Find the equation of the line of symmetry of the graph of $y = \dfrac{1}{x^2 + 2x + 3}$.

**27**   Sketch the graph of $y = \dfrac{1}{x^2 - 4}$. State the equations of all the asymptotes.

**28**   Sketch a graph with the following properties:

    ■   It has a vertical asymptote $x = 1$ and a horizontal asymptote $y = 3$.

    ■   Its only axis intercepts are $(-1, 0)$ and $(0, -3)$.

**29** A graph is used to show the number of students, $N$, who scored fewer than $m$ marks in a test. 80 students took the test, which was marked out of 50 marks. The lowest mark was 15 and the highest mark was 45. Exactly half the students scored fewer than 30 marks. Sketch one possible graph showing this information.

**30** A car is moving at a speed of $26\ \mathrm{m\,s^{-1}}$ when it starts to brake. After 0.5 seconds its speed has reduced to $15\ \mathrm{m\,s^{-1}}$ and after another 0.5 seconds it has reduced to $10\ \mathrm{m\,s^{-1}}$. Sketch a possible graph showing how the speed of the car changes with time.

**31** A small business wants to model how the money spent on advertising affects their monthly profit. After collecting some data, they found that:

- with no money spent on advertising in a month, they made a monthly loss of $200
- when they spent $120 or $400 on advertising in a month, their monthly profit was $350
- they made the largest profit, $600, when they spent $180 on advertising in a month.

Sketch one possible graph showing how the monthly profit varies with money spent on advertising.

**32** Solve the equation $x^2 = \dfrac{1}{x+1}$.

**33** Find all the roots of the equation $4 - x^2 = \dfrac{1}{x+1}$.

**34** Find the zeros of the function $f(x) = x^4 - 2x^3 - 5x^2 + 12x - 4$.

**35** Find all the roots of the equation $\left|4 - x^2\right| = x^2$.

**36** Find the maximum value of the function $f(x) = xe^{-0.4x}$.

**37** Sketch the graph of $y = \dfrac{\ln(x-2)}{x}$.

**38** Solve the equation $\ln\left|x-1\right| = \left|\ln(x-1)\right|$.

---

**TOOLKIT: Problem Solving**

Are Cartesian graphs the best way of representing functions? The website https://imaginary.org/ challenges this concept by using 3D images, animations and even music to convey information about functions and equations. There is often a link between simple equations and beautiful representations. See if you can explore the Surfer program to model mathematical sculptures of architecture, the natural world or abstract ideas.

---

## Checklist

- You should understand function notation. A function can be written as, for example,
  - $f(x) = 3x - 2$ or
  - $f : x \mapsto 3x - 2$

- You should know how to find the domain and range of a function.
  - The domain of a function is the set of all allowed input values.
  - The range of a function is the set of all possible outputs.

- You should understand that functions can be used in modelling real-life events, and that there are often limitations of such models.

- You should understand that an inverse function reverses the effect of the original function.
  - The graph of $y = f^{-1}(x)$ is a reflection of the graph $y = f(x)$ in the line $y = x$.

- You should be able to use your GDC to:
  - sketch the shape of graphs
  - find key features of graphs such as axis intercepts, vertices, asymptotes
  - find intersections of graphs
  - solve equations.

## ■ Mixed Practice

**1** Let $G(x) = 95e^{(-0.02x)} + 40$, for $20 \leqslant x \leqslant 200$.
   **a** Sketch the graph of G.
   **b** Robin and Pat are planning a wedding banquet. The cost per guest, G dollars, is modelled by the function $G(n) = 95e^{(-0.02n)} + 40$, for $20 \leqslant n \leqslant 200$, where $n$ is the number of guests.

   Calculate the **total** cost for 45 guests.

Mathematics SL May 2015 TZ1 Paper 2 Q5

**2** Find the zeros of the function $f(x) = x - 3\ln x$.

**3** **a** Find the largest possible domain of the function $g(x) = \sqrt{x + 5}$.
   **b** Solve the equation $g(x) = \dfrac{1}{x + 3}$.

**4** Find the coordinates of the points of intersection of the graph $y = 3 - x^2$ and $y = 2e^x$.

**5** Sketch the graph of $y = |x^2 - 4x - 5|$, showing the coordinates of all axis intercepts and any maximum and minimum points.

**6** **a** Sketch the graph of $y = \dfrac{3x - 1}{x - 3}$.
   **a** Hence find the domain and range of the function $f(x) = \dfrac{3x - 1}{x - 3}$.

**7** A speed of a car, $v\,\text{m s}^{-1}$, at time $t$ seconds, is modelled by the equation $v = 18te^{-0.2t}$.
   **a** Find the speed of the car after 1.5 seconds.
   **b** Find two times when the speed of the car is $10\,\text{m s}^{-1}$.
   **c** How long does it take for the car to reach the maximum speed?

**8** The speeds of two runners are modelled by the equations: $v_1 = 8 - 6e^{-0.5t}$, $v_2 = 2 + 3t^2 - t^3$

   where $v$ is in $\text{m s}^{-1}$ and $t$ is the time in seconds, with $0 \leqslant t \leqslant 2$.
   **a** Show that both runners start at the same speed.
   **b** After how long will they be running at the same speed again?

**9** Water has a lower boiling point at higher altitudes. The relationship between the boiling point of water ($T$) and the height above sea level ($h$) can be described by the model $T = -0.0034h + 100$ where $T$ is measured in degrees Celsius (°C) and $h$ is measured in **metres** from sea level.
   **a** Write down the boiling point of water at sea level.
   **b** Use the model to calculate the boiling point of water at a height of 1.37 km above sea level.

   Water boils at the top of Mt. Everest at 70 °C.
   **c** Use the model to calculate the height above sea level of Mt. Everest.

Mathematical Studies SL May 2012 TZ2 Paper 1 Q6

**10** To convert temperature in degrees Celsius, $x$ to degrees Fahrenheit, $f(x)$ a website says to multiply by 1.8 then add 32.
   **a** Find an expression for $f(x)$ in terms of $x$.
   **b** What is the interpretation of $f^{-1}(x)$ in this context?

**11** **a** $f : x \rightarrow 3x - 5$ is a mapping from the set $S$ to the set $T$ as shown on the right. Find the values of $p$ and $q$.

   **b** A function g is such that $g(x) = \dfrac{3}{(x - 2)^2}$.

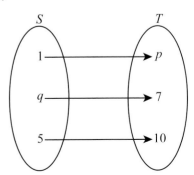

    **i**   State the domain of the function g($x$).

    **ii**  State the range of the function g($x$).

    **iii** Write down the equation of the vertical asymptote.

Mathematical Studies SL November 2007 Paper 1 Q12

**12** A function is defined on the domain $x \geq 7$ by the equation $f(x) = 3x - 1$.

    **a**  Find the range of f.

    **b**  Find $f^{-1}(35)$.

**13** Solve the equation $\dfrac{2x - 1}{x + 4} = 4 - x$.

**14** Find all the roots of the equation $\left| x - 2 \right| = \dfrac{1}{x}$.

**15** Find the domain and range of the function $h(x) = \dfrac{18}{x^2 - 9}$.

**16** Find the range of the function $f(x) = x^2 - 7x + 3$ defined on the domain $0 < x < 6$.

**17** A function is defined by $f(x) = 5e^x - 4x$.

    **a**  Find the smallest possible value of $f(x)$.

    **b**  Solve the equation $f^{-1}(x) = 2$.

**18** The function $g(x) = 8 - 3x^2$ is defined on the domain $-3 < x < 2$.

    **a**  Find the range of the function.

    **b**  Solve the equation $g(x) = 5$.

    **c**  Explain why the equation $g(x) = -20$ has no solutions.

**19** Sketch a possible graph of a function which is defined for all real values of $x$, has a horizontal asymptote $y = 5$, and crosses the coordinate axes at (0, 3) and (−1, 0).

**20** A cup of tea initially has temperature $90°\,C$ and is left to cool in a room of temperature $20°\,C$. Sketch the graph showing how the temperature of the tea changes with time.

**21** In a biology experiment, there are initially 600 bacteria. The population of bacteria increases but, due to space constraints, it can never exceed 2000. Sketch a possible graph showing how the number of bacteria changes with time.

**22** A child pushes a toy car in the garden, starting from rest, and then lets it go. The speed of the car, $v\,\mathrm{ms}^{-1}$, after $t$ seconds is modelled by the equation $v = 3te^{-t}$.

    **a**  Find the maximum speed of the car.

    **b**  Suggest why this is not a good model for the speed of the car after 20 seconds.

**23** Solve the equation $\left| e^{2x} - \dfrac{1}{x + 2} \right| = 2$.

Mathematics HL May 2005 TZ2 Paper 2 Q2

**24** **a**  Sketch the graph of $y = \dfrac{x - 12}{\sqrt{x^2 - 4}}$.

    **b**  Write down

       **i**  the $x$-intercept;

       **ii**  the equations of all asymptotes.

Mathematics HL May 2005 TZ1 Paper 1 Q15

**25** Solve the equation $e^x = x^3 - 2$.

**26** A function is defined by $f(x) = \dfrac{x^2 - 9x}{5x + 1}$ for $x > 1$. Find the minimum value of this function.

**27** A function is defined by $f(x) = \dfrac{10x^2 + 7}{x^2 - 4x}$.

   **a** Write down the largest possible domain of the function.

   **b** Find the range of the function for the domain from part **a**.

**28 a** Find the largest domain of the function $g(x) = \dfrac{2x}{3 + \ln x}$.

   **b** For the domain found in part **a** find the range of the function.

**29** A function is defined by $g(x) = 2x + \ln(x - 2)$.

   **a** State the largest possible domain and corresponding range of g.

   **b** Sketch the graph of $y = g(x)$ and $y = g^{-1}(x)$ on the same set of axes.

   **c** Solve the equation $g(x) = g^{-1}(x)$.

# Core: Coordinate geometry

## ESSENTIAL UNDERSTANDINGS

■ Geometry allows us to quantify the physical world, enhancing our spatial awareness in two and three dimensions.

### In this chapter you will learn...

■ how to find the gradient and intercepts of straight lines
■ how to find the equation of a straight line in different forms
■ about the gradients of parallel and perpendicular straight lines
■ how to find the point of intersection of two straight lines
■ how to find the distance between two points in three dimensions
■ the midpoint of two points in three dimensions.

## CONCEPTS

The following concepts will be addressed in this chapter:

■ The properties of shapes are highly dependent on the dimension they occupy in **space**.
■ The **relationships** between the length of the sides and the size of the angles in a triangle can be used to solve many problems involving position, distance, angles and area.

## PRIOR KNOWLEDGE

Before starting this chapter, you should already be able to complete the following:

1 Find the length $AB$ in the following triangle:

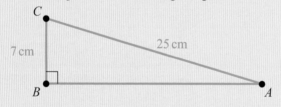

2 The points $P$ and $Q$ have coordinates $(2, -3)$ and $(-4, 5)$ respectively. Find
  a  the distance between $P$ and $Q$
  b  the midpoint of $PQ$.

■ **Figure 4.1** Can we use mathematics to model the gradients of these slopes?

The trajectory of a plane taking off or the spokes on a wheel are approximately straight lines, and it can be useful to describe these in an accurate way. For example, an air traffic controller would need to know whether the straight-line trajectory of any particular plane misses all the other planes that are nearby. But a straight-line model won't be perfectly accurate – does this mean the model isn't useful?

As well as physical situations, there are many other relationships that can be modelled by straight lines, for example, the distance travelled and fare paid for a taxi journey.

## Starter Activity

Look at the images in Figure 4.1 and discuss the following questions: What is the best measure of steepness? Can we always tell whether a slope is going up or down?

**Now look at this problem:**

Consider the following data:

| $x$ | 0 | 1 | 2 | 3 | 4 |
|-----|-----|-----|-----|-----|-----|
| $y$ | 18 | 22 | 26 | 30 | 34 |

Can you predict the value of $y$ when $x = 1.5$ if

a   $y$ represents the temperature of water in a kettle in °C and $x$ the time in seconds after turning it on?

b   $y$ represents the number of bees in a hive in thousands and $x$ represents time in years after the hive is created?

---

**LEARNER PROFILE – Open-minded**
Is mathematics a universal language? Why is mathematical notation so similar in different countries when other aspects of language differ so widely? Why are there still differences? What does 1,245 mean in your country? Does it mean the same everywhere? Is one convention better than another?

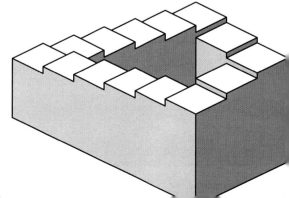

# 4A Equations of straight lines in two dimensions

## ■ Gradient and intercepts

The **gradient** of a straight line is a measure of how steep the line is – the larger the size of the number, the steeper the line.

The gradient can be positive, negative or zero:

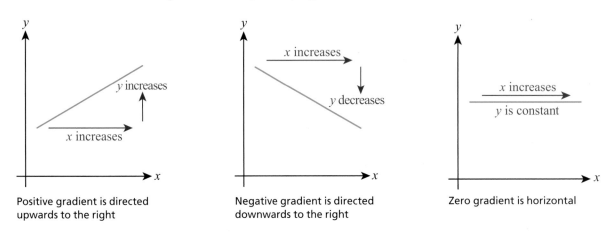

Positive gradient is directed
upwards to the right

Negative gradient is directed
downwards to the right

Zero gradient is horizontal

More precisely, the gradient is the change in the $y$-coordinate for every 1 unit of increase in the $x$-coordinate, or equivalently, the change in $y$ divided by the change in $x$.

Given two points on the line, $(x_1, y_1)$ and $(x_2, y_2)$, the change in $y$ is $(y_2 - y_1)$ and the change in $x$ is $(x_2 - x_1)$.

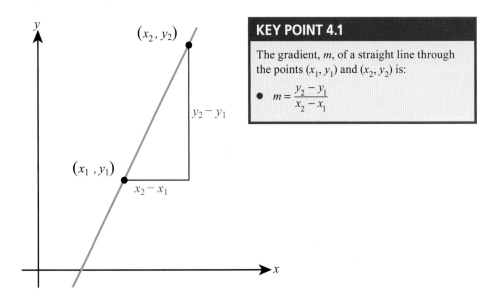

**KEY POINT 4.1**

The gradient, $m$, of a straight line through the points $(x_1, y_1)$ and $(x_2, y_2)$ is:

$$\bullet \quad m = \frac{y_2 - y_1}{x_2 - x_1}$$

---

**WORKED EXAMPLE 4.1**

Find the gradient of the line connecting (3, 2) and (5, −4).

Use $m = \dfrac{y_2 - y_1}{x_2 - x_1}$ with $(x_1, y_1) = (3, 2)$ and $(x_2, y_2) = (5, -4)$ $\cdots\cdots\cdots\cdots\cdots$ $m = \dfrac{-4 - 2}{5 - 3}$

$$= -3$$

---

# Be the Examiner 4.1

Find the gradient of the line through the points (3, −2) and (−4, 7).

Which is the correct solution? Identify the errors made in the incorrect solutions.

| Solution 1 | Solution 2 | Solution 3 |
|---|---|---|
| $m = \dfrac{-2 - 7}{3 - (-4)}$ | $m = \dfrac{7 - 2}{-4 - 3}$ | $m = \dfrac{-2 - 3}{7 - (-4)}$ |
| $= -\dfrac{9}{7}$ | $= -\dfrac{5}{7}$ | $= -\dfrac{5}{11}$ |

---

 Intercepts were first mentioned in Chapter 3.

**WORKED EXAMPLE 4.2**

Find the $x$ and $y$ intercepts of the line $2x + 3y - 6 = 0$.

The $x$-intercept will occur where $y = 0$ $\cdots\cdots\cdots\cdots$

When $y = 0$:
$$2x - 6 = 0$$
$$x = 3$$
So $x$-intercept is (3, 0)

The $y$-intercept will occur where $x = 0$ $\cdots\cdots\cdots\cdots$

When $x = 0$:
$$3y - 6 = 0$$
$$y = 2$$
So $y$-intercept is (0, 2)

## ▮ Different forms of the equation of a straight line

There are various ways of expressing the equation of a straight line.

### KEY POINT 4.2

The straight line with gradient $m$ and $y$-intercept $c$ has equation $y = mx + c$.

Is maths really an 'international language'? Different countries historically have slightly different preferences for the letters they use to refer to the gradient and intercept of a straight line:

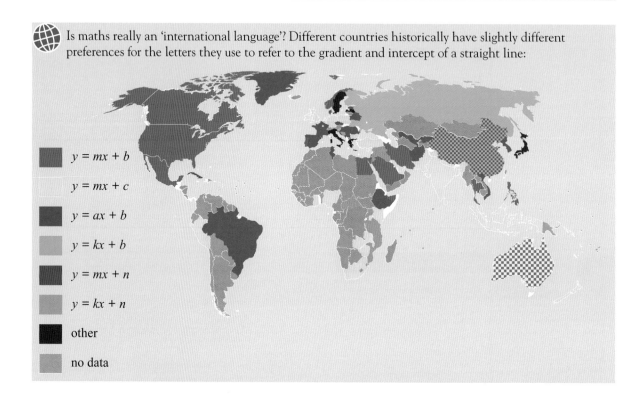

- $y = mx + b$
- $y = mx + c$
- $y = ax + b$
- $y = kx + b$
- $y = mx + n$
- $y = kx + n$
- other
- no data

### CONCEPTS – SPACE

Key Point 4.2 actually only gives the equation of a line in two-dimensional **space**. In 3D it is much harder to describe a line using Cartesian coordinates. The number of dimensions the shape that we are trying to describe has, changes the rules we can use.

### WORKED EXAMPLE 4.3

Find the equation of the line with gradient 4 and $y$-intercept −3 in the form $ax + by + d = 0$.

Since we have information ............................ $m = 4$ and $c = -3$
about gradient and $y$-intercept,       Therefore, equation is $y = 4x - 3$
we should use $y = mx + c$

Rearrange into the required form ............................ $-4x + y + 3 = 0$

## Tip

The form
$ax + by + c = 0$ is
sometimes referred to
as the 'general form'.

> **WORKED EXAMPLE 4.4**
>
> a   Find the equation of the line $3x + 4y + 6 = 0$ in the form $y = mx + c$.
>
> b   Hence write down the gradient and $y$-intercept of the line.
>
> Rearrange to make $y$ the subject  ·········· **a**  $3x + 4y + 6 = 0$
> $$4y = -3x - 6$$
> $$y = -\frac{3}{4}x - \frac{3}{2}$$
>
> Once in the form $y = mx + c$, the  ·········· **b**  Therefore,
> gradient is $m$ and the $y$-intercept is $c$
> $$\text{gradient} = -\frac{3}{4}$$
> $$y\text{-intercept} = -\frac{3}{2}$$

If you know a point on the line and the gradient, there is another, more direct form of the equation of the line to use instead of $y = mx + c$.

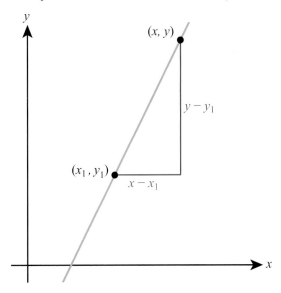

The point $(x, y)$ is any general point on the line. Using this and the known point $(x_1, y_1)$ you can write the gradient as

$$m = \frac{y - y_1}{x - x_1}$$

Rearranging this gives an alternative form of the equation of a straight line.

## Tip

This is sometimes
referred to as 'point–
gradient form'.

> **KEY POINT 4.3**
>
> The straight line with gradient $m$, passing through the point $(x_1, y_1)$, has equation
> $y - y_1 = m(x - x_1)$.

**WORKED EXAMPLE 4.5**

Find the equation of the line with gradient 2 through the point (3, 5).

The question does not ask for the
answer in any particular form,
so it can be left like this

$$y - y_1 = m(x - x_1)$$
$$y - 5 = 2(x - 3)$$

**WORKED EXAMPLE 4.6**

Find the equation of the line connecting (−1, 1) and (3, −7).

First find the gradient using $m = \dfrac{y_2 - y_1}{x_2 - x_1}$

$$m = \frac{-7 - 1}{3 - (-1)}$$
$$= -2$$

Then find the equation of the line using
$y - y_1 = m(x - x_1)$. You can choose
either of the points to be $(x_1, y_1)$

$$y - y_1 = m(x - x_1)$$
$$y - 1 = -2(x - (-1))$$
$$y - 1 = -2(x + 1)$$

## ■ Parallel lines

**KEY POINT 4.4**

Parallel lines have the same gradient.

**WORKED EXAMPLE 4.7**

Find the equation of the line parallel to $3x + y + 7 = 0$ through the point (2, 5).

First rearrange into the form
$y = mx + c$

$$3x + y + 7 = 0$$
$$y = -3x - 7$$

Any line parallel to $y = -3x - 7$
will have gradient −3

Equation of line parallel to this and
passing through (2, 5) is
$$y - 5 = -3(x - 2)$$

## ■ Perpendicular lines

The gradients of perpendicular lines are also related.

**KEY POINT 4.5**

If two lines with gradients $m_1$ and $m_2$ are perpendicular, then $m_1 m_2 = -1$.

**Tip**

Key Point 4.5 is often
used in the form
$m_1 = -\dfrac{1}{m_2}$.

## Proof 4.1

Prove that if a line $l_1$ has gradient $m$, then a perpendicular line $l_2$ has gradient $-\dfrac{1}{m}$.

If line $l_1$ has gradient $m$, then a right-angled triangle can be drawn as shown with horizontal side of length 1 and vertical side of length $m$ ...... The line $l_1$ has gradient $m = \dfrac{m}{1}$:

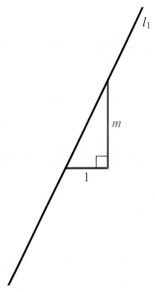

The right-angled triangle is also rotated by 90° making the horizontal side length $m$ and the vertical side length −1 (since it is going down) ...... Rotating $l_1$ through 90° gives a perpendicular line $l_2$:

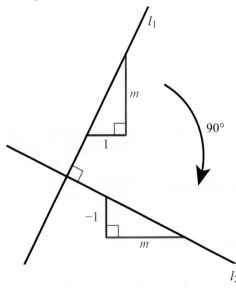

The gradient is $\dfrac{\text{change in } y}{\text{change in } x}$ ...... $l_2$ has gradient $\dfrac{-1}{m}$

---

**WORKED EXAMPLE 4.8**

Find the equation of the line perpendicular to $y = 5x + 2$ through the point $(1, 6)$. Give your answer in the form $ax + by + c = 0$, where $a, b, c \in \mathbb{Z}$.

Use $m_1 = -\dfrac{1}{m_2}$ ............ Gradient of any line perpendicular to $y = 5x + 2$ is $-\dfrac{1}{5}$.
Equation of line through $(1, 6)$ is

$$y - 6 = -\frac{1}{5}(x - 1)$$

Multiply through by 5 ··············· $5y - 30 = -(x - 1)$

$$5y - 30 = -x + 1$$

$$x + 5y - 31 = 0$$

---

## ■ Intersection of two lines

Two lines will intersect at one point as long as the lines are not parallel or identical. To find the point of intersection, you need to solve the equations of the lines simultaneously.

**Tip**

Although you should be able to do this algebraically, as demonstrated in Worked Example 4.9, you will most often just need to be able to use your GDC to solve simultaneous equations.

**WORKED EXAMPLE 4.9**

Find the point of intersection of $y = 3x + 2$ and $y = x + 4$.

Where the lines intersect, the $y$-coordinates must be equal, so replace $y$ in the second equation with the expression for $y$ from the first ·············· $3x + 2 = x + 4$

Solve for $x$ ·············· $2x = 2$

$$x = 1$$

Find the corresponding ·············· When $x = 1$, $y = 3(1) + 2 = 5$
$y$ value by substituting $x = 1$ into either of the equations

So the coordinates of the intersection point are $(1, 5)$

---

**You are the Researcher**

Using coordinate geometry to represent straight lines is a relatively recent innovation, suggested by René Descartes (1596–1650). You might like to research how the ancient Greek, Babylonian and Chinese mathematicians all described and worked with straight lines.

---

**TOOLKIT: Modelling**

Find a roadmap of your country and model the major roads using straight lines. Use this model to calculate the shortest routes between two points. What assumptions are you making in your model? How could you improve the model?

## Exercise 4A

For questions 1 to 3, use the method demonstrated in Worked Example 4.1 to find the gradient of the line connecting points $A$ and $B$.

1   a   $A(1, 5)$ and $B(3, 7)$                        2   a   $A(4, 1)$ and $B(2, 7)$                        3   a   $A(3, 3)$ and $B(7, 5)$

    b   $A(2, 2)$ and $B(5, 8)$                          b   $A(7, 3)$ and $B(5, 5)$                          b   $A(-1, 8)$ and $B(5, 5)$

For questions 4 to 6, use the method demonstrated in Worked Example 4.2 to find the coordinates of the points where the line crosses the $x$- and $y$-axes.

4   a   $y = 3x - 6$                        5   a   $2y = 5 - x$                        6   a   $4x + 3y - 12 = 0$

    b   $y = 8 - 2x$                          b   $3y = 2x + 6$                          b   $2x - 3y - 9 = 0$

For questions 7 to 9, use the method demonstrated in Worked Example 4.3 to find the equation of the straight line with given gradient and $y$-intercept. Give your answer in the form $ax + by + d = 0$.

7   a   Gradient 2, $y$-intercept 3                        8   a   Gradient $-2$, $y$-intercept 4                        9   a   Gradient $\frac{1}{2}$, $y$-intercept $\frac{7}{2}$

    b   Gradient 5, $y$-intercept 1                          b   Gradient $-3$, $y$-intercept 7                          b   Gradient $-\frac{1}{3}$, $y$-intercept 3

For questions 10 to 12, use the method demonstrated in Worked Example 4.4 to rewrite the line equation in the form $y = mx + c$ and state the gradient and $y$-intercept of the line.

10   a   $2x - y + 5 = 0$                        11   a   $2x + 3y + 6 = 0$                        12   a   $3x + 5y - 7 = 0$

     b   $3x - y + 4 = 0$                           b   $5x + 2y - 10 = 0$                           b   $11x + 2y + 5 = 0$

For questions 13 to 15, use the method demonstrated in Worked Example 4.5 to give the equation of the line.

13   a   Gradient 2 through $(1, 4)$                        14   a   Gradient $-5$ through $(-1, 3)$                        15   a   Gradient $\frac{2}{3}$ through $(1, -1)$

     b   Gradient 3 through $(5, 2)$                           b   Gradient $-2$ through $(2, -1)$                           b   Gradient $-\frac{3}{4}$ through $(3, 1)$

For questions 16 to 19, use the method demonstrated in Worked Example 4.6 to find the equation of the line connecting points $A$ and $B$ in the form $ax + by + c = 0$.

16   a   $A(3, 1)$, $B(5, 5)$                        17   a   $A(3, 7)$, $B(5, 5)$                        18   a   $A(5, 1)$, $B(1, 3)$

     b   $A(-2, 3)$, $B(1, 9)$                           b   $A(-2, 4)$, $B(-1, 1)$                           b   $A(-2, 3)$, $B(6, -1)$

19   a   $A(3, 1)$, $B(7, -2)$

     b   $A(-2, 7)$, $B(3, -1)$

For questions 20 to 24, use the method demonstrated in Worked Example 4.7 to find the equations of the required lines.

20   a   Parallel to $y = 3x + 2$ through $(0, 4)$                        21   a   Parallel to $y = 1.5x + 4$ through $(1, 5)$

     b   Parallel to $y = -x + 2$ through $(0, 9)$                           b   Parallel to $y = 0.5x - 3$ through $(-1, -3)$

22   a   Parallel to $x - y - 3 = 0$ through $(2, -4)$                        23   a   Parallel to $x + 3y - 1 = 0$ through $(1, 1)$

     b   Parallel to $2x - y - 5 = 0$ through $(3, -1)$                           b   Parallel to $x + 5y + 4 = 0$ through $(-3, -1)$

24   a   Parallel to $2x - 5y - 3 = 0$ through $(5, 2)$

     b   Parallel to $3x + 4y - 7 = 0$ through $(-2, 1)$

For questions 25 to 29, use the method demonstrated in Worked Example 4.8 to find the equations of the required lines.

25   a   Perpendicular to $y = x + 2$ through $(0, 5)$                        26   a   Perpendicular to $y = \frac{1}{4}x + 2$ through $(1, 5)$

     b   Perpendicular to $y = 3x + 2$ through $(0, 8)$                           b   Perpendicular to $y = -\frac{1}{3}x$ through $(2, 4)$

27   a   Perpendicular to $x - 5y - 3 = 0$ through $(-1, 2)$                        28   a   Perpendicular to $2x + y - 7 = 0$ through $(1, 3)$

     b   Perpendicular to $x - 2y - 3 = 0$ through $(2, 3)$                           b   Perpendicular to $3x + y + 1 = 0$ through $(-2, -4)$

29   a   Perpendicular to $5x + 2y = 0$ through $(-1, 7)$

     b   Perpendicular to $3x - 2y = 0$ through $(3, -3)$

For questions 30 to 33, use the method demonstrated in Worked Example 4.9 or technology to find the point of intersection of the given lines.

30   a   $x + 2y = 5$ and $x - 2y = -3$          31   a   $y = x + 2$ and $y = 4 - x$

     b   $2x + 3y = 13$ and $x - y = 4$                b   $y = x + 3$ and $y = 7 - x$

32   a   $y = 2x + 1$ and $y = x + 3$           33   a   $y = 3 - 2x$ and $y = 3x + 1$

     b   $y = 3x + 5$ and $y = x - 1$                 b   $y = 7 - 3x$ and $y = 2x - 4$

**34**   a   Calculate the gradient of the line passing through the points $A(1, -7)$ and $B(5, 0)$.

     b   Find the equation of the line through the point $C(8, 3)$ which is perpendicular to the line $AB$. Give your answer in the form $y = mx + c$.

**35**

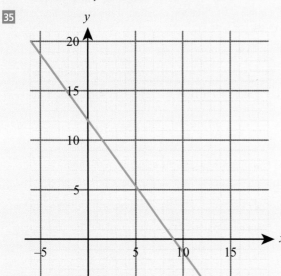

     a   Calculate the gradient of the line shown in the diagram.

     b   Find the equation of the line in the form $ax + by = c$, where $a$, $b$ and $c$ are positive integers.

**36**   A line has equation $5x + 7y = 17$.

     a   Find the gradient of the line.

     b   Find the $x$-intercept of the line.

**37**   A straight line connects points $A(-3, 1)$ and $B(7, 3)$.

     a   Find the gradient of the line.

     b   Find the equation of the line in the form $ax + by = c$, where $a$, $b$ and $c$ are integers.

**38**   A straight line $l_1$ has equation $7x + 2y = 42$.

     a   Find the gradient of the line $l_1$.

     b   Determine whether the point $P(8, -5)$ lies on the line $l_1$.

     c   Find the equation of the line $l_2$ which passes through $P$ and is perpendicular to $l_1$. Give your answer in the form $y = mx + c$.

**39**   A mountain is modelled as a cone of height 8800 m and base width 24 000 m. Based on this model, what is the gradient of the mountain?

**40** Point $P$ has coordinates $(0, 11)$. Point $Q$ is 8 units right of $P$. Point $R$ has coordinates $(k, -3)$ and is directly below $Q$.

a  Write down the coordinates of $Q$.

b  Write down the value of $k$.

c  Write down the equation of the line $QR$.

d  Find the area of the triangle $PQR$.

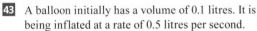

**41** Line $l_1$ has gradient $-\dfrac{3}{2}$ and crosses the $y$-axis at the point $(0, -6)$.

Line $l_2$ passes through the points $(-5, 1)$ and $(7, -2)$.

a  Write down the equation of $l_1$.

b  Find the equation of $l_2$, giving your answer in the form $ax + by = c$.

c  Find the coordinates of the point of intersection of $l_1$ and $l_2$.

**42** A drawbridge consists of a straight plank pivoted at one end.

When the other end of the bridge is 3.5 m above the horizontal, the plank is at a gradient of 0.7. Find the length of the plank.

**43** A balloon initially has a volume of 0.1 litres. It is being inflated at a rate of 0.5 litres per second.

a  Find an expression for the volume of the balloon after $t$ seconds.

b  The balloon pops when its volume reaches 5 litres. How long would this take?

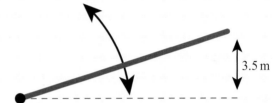

3.5 m

**44** Hooke's law states that the force ($F$ Newtons) required to stretch a spring $x$ metres is given by $F = kx$.

a  State the units of $k$.

b  If the numerical value of $k$ is 0.3, find the force required to extend the spring by 0.06 m.

c  Would a stiffer spring have a larger, smaller or the same value of $k$?

d  The spring breaks when a force of 0.14 N is applied. How much has it extended at the point of breaking?

**45** A mobile phone contract has a fixed charge of \$5 per month plus \$1 per 100 minutes spent talking.

The monthly cost is \$$C$ when Joanna talks for $m$ minutes.

a  Find a relationship of the form $C = am + b$

b  Find the cost if Joanna talks for a total of 3 hours in a month.

c  An alternative contract has no fixed charge, but charges \$1 per 50 minutes spent talking. How many minutes should Joanna expect to talk each month for the first contract to be better value?

**46** A company receives £10 per item it sells. It has fixed costs of £2000 per month.

a  Find an expression for the profit $P$ if the company sells $n$ items.

b  How many items must the company sell to make a profit of £1500.

c  The company refines the model it is using. It believes that its fixed costs are actually £1200 but it has a variable cost of £2 per item sold. Find a new expression for the profit in terms of $n$.

d  At what value of $n$ do the two models predict the same profit? Based on the new model, will the company have to sell more or fewer items to make a profit of £1500?

**47** Consider a quadrilateral with vertices $A(-4, 3)$, $B(3, 8)$, $C(5, -1)$ and $D(-9, -11)$.

a  Show that the sides $AB$ and $CD$ are parallel.

b  Show that $ABCD$ is not a parallelogram.

**48** A car can drive up a road with maximum incline 0.3. Find the minimum length of road which is required for a car to go up a mountain road from sea level to a point 400 m above sea level.

# 4B Three-dimensional coordinate geometry

## ■ The distance between two points

The idea of using Pythagoras' theorem to find the distance between two points in two dimensions can be extended to three dimensions.

---

**KEY POINT 4.6**

The distance $d$ between the points $(x_1, y_1, z_1)$, and $(x_2, y_2, z_2)$ is:

- $d = \sqrt{(x_2 - x_1)^2 + (y_2 - y_1)^2 + (z_2 - z_1)^2}$

---

**Proof 4.2**

Prove that the distance, $d$, between $(x_1, y_1, z_1)$ and $(x_2, y_2, z_2)$ is

$d = \sqrt{(x_2 - x_1)^2 + (y_2 - y_1)^2 + (z_2 - z_1)^2}$.

Consider the two points $(x_1, y_1, z_1)$ and $(x_2, y_2, z_2)$ as being at opposite ends of a diagonal of a cuboid:

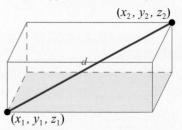

Use Pythagoras in 2D for a right-angled triangle lying in the base of the cuboid to find the length of the diagonal ......... The length of the diagonal of the base of the cuboid, $l$, is related to the sides of the base by

$$l^2 = (x_2 - x_1)^2 + (y_2 - y_1)^2$$

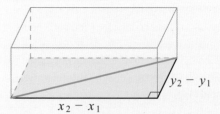

Then use Pythagoras in 2D again for the right-angled triangle with base $l$ and hypotenuse $d$ ......... Then, $\quad d^2 = l^2 + (z_2 - z_1)^2$

$$= (x_2 - x_1)^2 + (y_2 - y_1)^2 + (z_2 - z_1)^2$$

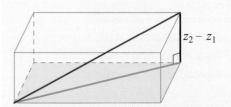

Therefore, $d = \sqrt{(x_2 - x_1)^2 + (y_2 - y_1)^2 + (z_2 - z_1)^2}$

**WORKED EXAMPLE 4.10**

Find the distance between $(1, 2, 3)$ and $(3, -2, -4)$.

Use $d = \sqrt{(x_2 - x_1)^2 + (y_2 - y_1)^2 + (z_2 - z_1)^2}$ ............

$d = \sqrt{(3 - 1)^2 + (-2 - 2)^2 + (-4 - 3)^2}$

$= \sqrt{4 + 16 + 49}$

$= \sqrt{69}$

$= 8.31$ (3 s.f.)

## ■ The midpoint

The midpoint of the line connecting two points can be thought of as the average of the coordinates of the two points.

**KEY POINT 4.7**

The midpoint, $M$, of the points $(x_1, y_1, z_1)$ and $(x_2, y_2, z_2)$ is:

- $M = \left( \dfrac{x_1 + x_2}{2}, \dfrac{y_1 + y_2}{2}, \dfrac{z_1 + z_2}{2} \right)$

**WORKED EXAMPLE 4.11**

Find the midpoint of $(3, 4, 8)$ and $(5, 5, -5)$.

Use $M = \left( \dfrac{x_1 + x_2}{2}, \dfrac{y_1 + y_2}{2}, \dfrac{z_1 + z_2}{2} \right)$ ............

$M = \left( \dfrac{3 + 5}{2}, \dfrac{4 + 5}{2}, \dfrac{8 + (-5)}{2} \right)$

$= (4, 4.5, 1.5)$

## Exercise 4B

For questions 1 to 3, use techniques from your prior learning to find the exact distance between the two points given.

1  a  $(0, 0)$ and $(4, 3)$

   b  $(0, 0)$ and $(5, 12)$

2  a  $(7, 5)$ and $(2, -7)$

   b  $(4, -1)$ and $(-2, 7)$

3  a  $(3, 5)$ and $(-2, 3)$

   b  $(-2, 5)$ and $(4, -2)$

For questions 4 to 6, use the method demonstrated in Worked Example 4.10 to find the exact distance between the two points given.

4  a  $(0, 0, 0)$ and $(1, 2, 2)$

   b  $(0, 0, 0)$ and $(2, 6, 9)$

5  a  $(4, 1, 3)$ and $(6, 5, 7)$

   b  $(-3, 2, -1)$ and $(3, -1, 5)$

6  a  $(1, 5, -3)$ and $(4, -3, 2)$

   b  $(-4, -3, 7)$ and $(2, 2, 9)$

For questions 7 to 9, use the method demonstrated in Worked Example 4.11 to find the midpoint of line segment $AB$.

7  a  $A(1, 1)$ and $B(7, -3)$

   b  $A(5, 2)$ and $B(3, 10)$

8  a  $A(1, -3, 7)$ and $B(5, -5, -1)$

   b  $A(2, 4, -8)$ and $B(-4, 0, 6)$

9  a  $A(4, 3, 2)$ and $B(7, 1, -8)$

   b  $A(-2, -2, 11)$ and $B(-9, 9, 3)$

10  a  Find the coordinates of the midpoint of the line segment connecting points $A(-4, 1, 9)$ and $B(7, 0, 2)$.

   b  Find the length of the segment $AB$.

11  Point $A$ has coordinates $(4, -1, 2)$. The midpoint of $AB$ has coordinates $(5, 1, -3)$. Find the coordinates of $B$.

12  The midpoint of the line segment connecting the points $P(-4, a, 1)$ and $Q(b, 1, 8)$ is $M(8, 2, c)$. Find the values of $a$, $b$ and $c$.

**13** Points $A$ and $B$ have coordinates $(3, -18, 8)$ and $(2, -2, 11)$. Find the distance of the midpoint of $AB$ from the origin.

**14** A birdhouse of height 1.6 m is located at the origin. A bird flies in a straight line from the birdhouse to the top of a 8.3 m tall tree located at the point with coordinates $(14, 3)$. The flight takes 3.5 seconds. Find the average speed of the bird.

**15** The distance of the point $(k, 2k, 5k)$ from the origin is 30. Find the positive value of $k$.

**16** The distance between points $(k, k, 0)$ and $(1, -1, 3k)$ is $\sqrt{46}$. Find the two possible values of $k$.

**17** The point $(2a, a, 5a)$ is twice as far from the origin as the point $(-4, 1, 7)$. Find the positive value of $a$.

**18** The midpoint of the line segment connecting points $P\,(3a + 1, 2a)$ and $Q\,(5 - b, b + 3)$ has coordinates $(4, -5)$.

   a  Find the values of $a$ and $b$.

   b  Find the equation of the line connecting $P$ and $Q$.

**19** Line $l_1$ has equation $4x - 7y = 35$.

   a  Find the gradient of the line $l_1$.

   b  Line $l_2$ passes through the point $N(-4, 2)$ and is perpendicular to $l_1$. Find the equation of $l_2$ in the form $ax + by = c$, where $a$, $b$ and $c$ are integers.

   c  Find the coordinates of $P$, the point of intersection of the lines $l_1$ and $l_2$.

   d  Hence find the shortest distance from point $N$ to the line $l_1$.

**20** A tent has vertices $A(0, 0, 0)$, $B(3, 0, 0)$, $C(3, 2, 0)$, $D(0, 2, 0)$, $E(0.2, 1, 1.8)$ and $F(2.8, 1, 1.8)$, where $A$, $B$, $C$, $D$ form the base. All lengths are in metres.

   a  State the height of the tent.

   b  An extra support needs to be added from the midpoint of $BC$ to the corner $E$. Find the length of this support.

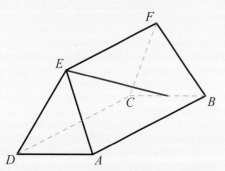

**21** $A(1, 4)$ and $C(5, 10)$ are opposite vertices of a square $ABCD$.

   a  Find the midpoint of $AC$.

   b  Find the coordinates of $B$ and $D$.

**22** The points $A(1, 1)$ and $C(8, 8)$ form the opposite ends of the diagonal of rhombus $ABCD$.

   a  Using the fact that each diagonal of a rhombus perpendicularly bisects the other, find the equation of the line $BD$.

   b  If the gradient of line segment $AB$ is $\dfrac{4}{3}$, find the coordinates of $B$ and $D$.

   c  Find the length of each side of the rhombus.

**23** An empty room is 3 m by 4 m by 5 m. A spider and a fly start in one corner of the room.

   a  The fly travels to the furthest corner of the room by flying and stops there. What is the shortest distance it can travel to get there?

   b  The spider then chases after the fly by crawling along any wall. What is the shortest distance it can travel to get to the fly's final position?

## Checklist

▪ You should be able to find the gradient of a straight line given two points on the line: $m = \dfrac{y_2 - y_1}{x_2 - x_1}$

▪ You should be able to work with the equation of a straight line in several different forms:
  □ Gradient–intercept form: $y = mx + c$
  □ General form: $ax + by + c = 0$
  □ Point–gradient form: $y - y_1 = m(x - x_1)$

▪ You should know that parallel lines have the same gradient.

▪ You should know that if two lines with gradients $m_1$ and $m_2$ are perpendicular, then $m_1 m_2 = -1$.

▪ You should be able to find the intersection of two straight lines by solving simultaneous equations.

▪ You should be able to find the distance between two points in three dimensions:
  □ $d = \sqrt{(x_2 - x_1)^2 + (y_2 - y_1)^2 + (z_2 - z_1)^2}$

▪ You should be able to find the midpoint of two points in three dimensions:
  □ $M = \left( \dfrac{x_1 + x_2}{2}, \dfrac{y_1 + y_2}{2}, \dfrac{z_1 + z_2}{2} \right)$

---

## ■ Mixed Practice

**1** $P(4, 1)$ and $Q(0, -5)$ are points on the coordinate plane.
  **a** Determine the
    **i** coordinates of $M$, the midpoint of $P$ and $Q$
    **ii** gradient of the line drawn through $P$ and $Q$
    **iii** gradient of the line drawn through $M$, perpendicular to $PQ$.

  The perpendicular line drawn through $M$ meets the $y$-axis at $R(0, k)$.
  **b** Find $k$.

*Mathematical Studies SL May 2007 Paper 1 Q10*

**2** The midpoint, $M$, of the line joining $A(s, 8)$ to $B(-2, t)$ has coordinates $M(2, 3)$.
  **a** Calculate the values of $s$ and $t$.
  **b** Find the equation of the straight line perpendicular to $AB$, passing through the point $M$.

*Mathematical Studies SL November 2007 Paper 1 Q13*

**3** The straight line, $L_1$, has equation $2y - 3x = 11$. The point $A$ has coordinates $(6, 0)$.
  **a** Give a reason why $L_1$ **does not** pass through $A$.
  **b** Find the gradient of $L_1$.
  $L_2$ is a line perpendicular to $L_1$. The equation of $L_2$ is $y = mx + c$.
  **c** Write down the value of $m$.
  $L_2$ **does** pass through $A$.
  **d** Find the value of $c$.

*Mathematical Studies SL May 2013 Paper 1 TZ1 Q10*

**4** Two points have coordinates $A(3, 6)$ and $B(-1, 10)$.
  **a** Find the gradient of the line $AB$.

  Line $l_1$ passes through $A$ and is perpendicular to $AB$.
  **b** Find the equation of $l_1$.
  **c** Line $l_1$ crosses the coordinate axes at the points $P$ and $Q$. Calculate the area of the triangle $OPQ$, where $O$ is the origin.

**5** Two points have coordinates $P(-1, 2)$ and $Q(6, -4)$.
  **a** Find the coordinates of the midpoint, $M$, of $PQ$.
  **b** Calculate the length of $PQ$.
  **c** Find the equation of a straight line which is perpendicular to $PQ$ and passes through $M$.

**6** Two points have coordinates $P(-1, 2, 5)$ and $A(6, -4, 3)$.
  **a** Find the coordinates of the midpoint, $M$, of $PQ$.
  **b** Calculate the length of $PQ$.

**7** The line $l_1$ with equation $x + 2y = 6$ crosses the $y$-axis at $P$ and the $x$-axis at $Q$.
  **a** Find the coordinates of $P$ and $Q$.
  **b** Find the exact distance between $P$ and $Q$.
  **c** Find the point where the line $y = x$ meets $l_1$.

**8** The line connecting points $M(3, -5)$ and $N(-1, k)$ has equation $4y + 7x = d$.
  **a** Find the gradient of the line.
  **b** Find the value of $k$.
  **c** Find the value of $d$.

**9** The vertices of a quadrilateral have coordinates $A(-3, 8)$, $B(2, 5)$, $C(1, 6)$ and $D(-4, 9)$.
  **a** Show that $ABCD$ is a parallelogram.
  **b** Show that $ABCD$ is not a rectangle.

**10** Show that the triangle with vertices $(-2, 5)$, $(1, 3)$ and $(5, 9)$ is right angled.

**11** The distance of the point $(-4, a, 3a)$ from the origin is $\sqrt{416}$. Find two possible values of $a$.

**12** The midpoint of the line joining points $A(2, p, 8)$ and $B(-6, 5, q)$ has coordinates $(-2, 3, -5)$.
  **a** Find the values of $p$ and $q$.
  **b** Calculate the length of the line segment $AB$.

**13** The cross section of a roof is modelled as an isosceles triangle. The width of the house is 8 m and the height of the roof is 6 m. Find the gradient of the side of the roof.

**14** The lines with equations $y = \dfrac{1}{2}x - 3$ and $y = 2 - \dfrac{2}{3}x$ intersect at the point $P$. Find the distance of $P$ from the origin.

**15** Triangle $ABC$ has vertices $A(-4, 3)$, $B(5, 0)$ and $C(4, 7)$.
  **a** Show that the line $l_1$ with equation $y = 3x$ is perpendicular to $AB$ and passes through its midpoint.
  **b** Find the equation of the line $l_2$ which is perpendicular to $AC$ and passes through its midpoint.
  **c** Let $S$ be the intersection of the lines $l_1$ and $l_2$. Find the coordinates of $S$ and show that it is the same distance from all three vertices.

**16** Health and safety rules require that ramps for disabled access have a maximum gradient of 0.2. A straight ramp is required for accessing a platform 2 m above ground level. What is the closest distance from the platform the ramp can start?

**17** The point $C(2, 0, 2)$ is plotted on the diagram on the right.
  **a** Copy the diagram and plot the points $A(5, 2, 0)$ and $B(0, 3, 4)$.
  **b** Calculate the coordinates of $M$, the midpoint of $AB$.
  **c** Calculate the length of $AB$.

  Mathematical Studies SL November 2005 Paper 1 Q15

**18** Two points have coordinates $A(-7, 2)$ and $B(5, 8)$.
  **a** Calculate the gradient of $AB$.
  **b** Find the coordinates of $M$, the midpoint of $AB$.
  **c** Find the equation of the line $l_1$ which passes through $M$ and is perpendicular to $AB$. Give your answer in the form $ax + by = c$, where $a$, $b$ and $c$ are integers.
  **d** Show that the point $N(1, 1)$ lies on $l_1$.
  **e** Hence find the perpendicular distance from $N$ to the line $AB$.

**19** A straight line has equation $3x - 7y = 42$ and intersects the coordinate axes at points $A$ and $B$.
  **a** Find the area of the triangle $AOB$, where $O$ is the origin.
  **b** Find the length of $AB$.
  **c** Hence find the perpendicular distance of the line from the origin.

**20** Line $l_1$ has equation $y = 5 - \dfrac{1}{2}x$ and crosses the $x$-axis at the point $P$. Line $l_2$ has equation $2x - 3y = 9$ and intersects the $x$-axis at the point $Q$. Let $R$ be the intersection of the lines $l_1$ and $l_2$. Find the area of the triangle $PQR$.

**21** $A(8, 1)$ and $C(2, 3)$ are opposite vertices of a square $ABCD$.
  **a** Find the equation of the diagonal $BD$.
  **b** Find the coordinates of $B$ and $D$.

**22** Quadrilateral $ABCD$ has vertices with coordinates $(-3, 2)$, $(4, 3)$, $(9, -2)$ and $(2, -3)$. Prove that $ABCD$ is a rhombus but not a square.

**23** A car drives up a straight road with gradient 0.15. How far has the car travelled when it has climbed a vertical distance of $20\,\mathrm{m}$?

**24** A bridge consists of two straight sections, each pivoted at one end. When both sections of the bridge are raised 6 m above the horizontal, each section has gradient 0.75. Find the distance, $d$, between the two closest end points of the two sections.

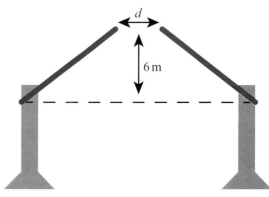

# 5 Core: Geometry and trigonometry

## ESSENTIAL UNDERSTANDINGS

- Geometry and trigonometry allow us to quantify the physical world, enhancing our spatial awareness in two and three dimensions. This topic provides us with the tools for analysis, measurement and transformation of quantities, movements and relationships.

### In this chapter you will learn...

- how to find the volume and surface area of three-dimensional solids
- how to work with trigonometric ratios and inverse trigonometric ratios
- how to find the angle between two intersecting lines in two dimensions
- how to use the sine rule to find sides and angles in non-right-angled triangles
- how to use the cosine rule to find sides and angles in non-right-angled triangles
- how to find the area of a triangle when you do not know the perpendicular height
- how to find the angle between two intersecting lines in three-dimensional shapes
- how to find the angle between a line and a plane in three-dimensional shapes
- how to construct diagrams from given information
- how to use trigonometry in questions involving bearings
- how to use trigonometry in questions involving angles of elevation and depression.

### CONCEPTS

The following concepts will be addressed in this chapter:
- Volume and surface area of shapes are determined by formulae, or general mathematical **relationships** or rules expressed using symbols or variables.
- The **relationships** between the length of the sides and the size of the angles in a triangle can be used to solve many problems involving position, distance, angles and area.

■ **Figure 5.1** How else are triangles used in the real world?

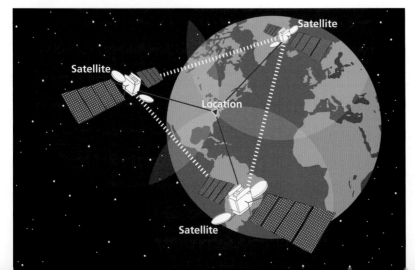

### PRIOR KNOWLEDGE

Before starting this chapter, you should already be able to complete the following:

**1** Find the angle $\theta$:

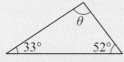

**2** Find the length $AC$:

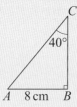

**3** Find the area of the following triangle:

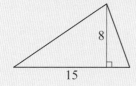

**4** Find the angles $\alpha$ and $\beta$:

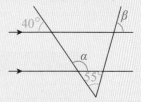

**5** Point $A$ is on a bearing of 230° from point $B$. Find the bearing of $B$ from $A$.

**6** Find the volume and surface area of the solid cylinder shown:

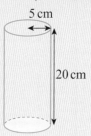

Geometry was initially all about measuring the real world, with applications from architecture to astronomy. But now mathematicians consider geometry to be the study of what stays the same when other things change. There are many different branches of geometry that seemingly have little to do with the real world that classical geometry models.

## Starter Activity

Look at Figure 5.1. In small groups, discuss why triangles are of such fundamental importance in the real world.

**Now look at this problem:**

Draw any right-angled triangle containing an angle of 50°.

Label $a$ as the shortest side of your triangle and $b$ as the longest (the hypotenuse).

With a ruler, measure the lengths of the sides $a$ and $b$ and then work out:

a $\quad a + b$        b $\quad a - b$        c $\quad ab$        d $\quad \dfrac{a}{b}$

Compare your answers with other people's answers. What do you notice?

# 5A Volumes and surface areas of three-dimensional solids

In addition to the volume and surface area of cuboids, cylinders and prisms that you already know, you also need the formulae for the volume and surface area of other common three-dimensional shapes such as spheres, cones and pyramids.

**KEY POINT 5.1**

| Shape | Volume | Surface area |
|---|---|---|
| Sphere of radius $r$ 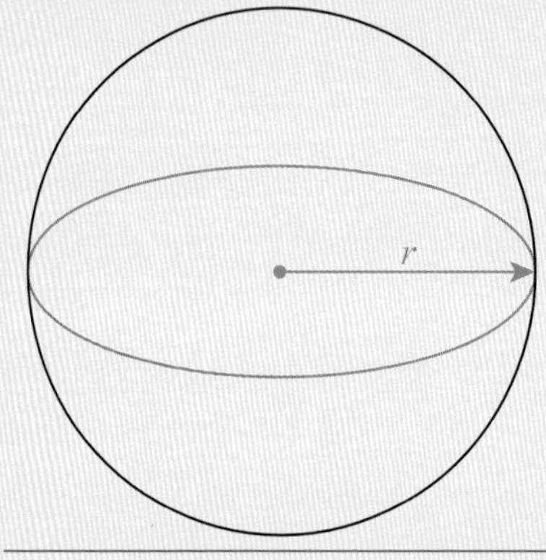 | $\frac{4}{3}\pi r^3$ | $4\pi r^2$ |
| Cone of base radius $r$, height $h$ and slope length $l$ 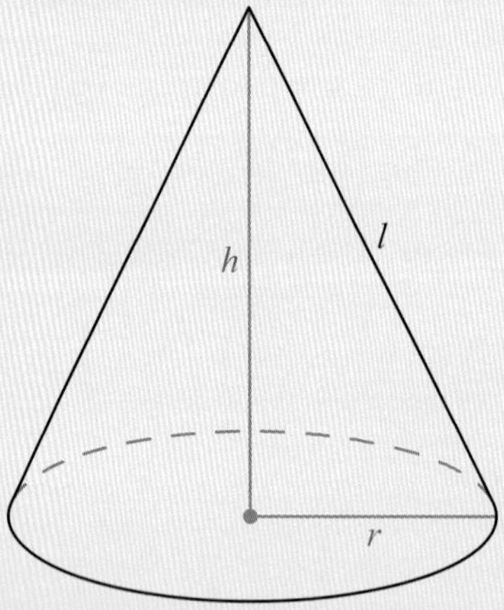 | $\frac{1}{3}\pi r^2 h$ | $\pi r l + \pi r^2$ |
| Pyramid of base area $B$ and height $h$ 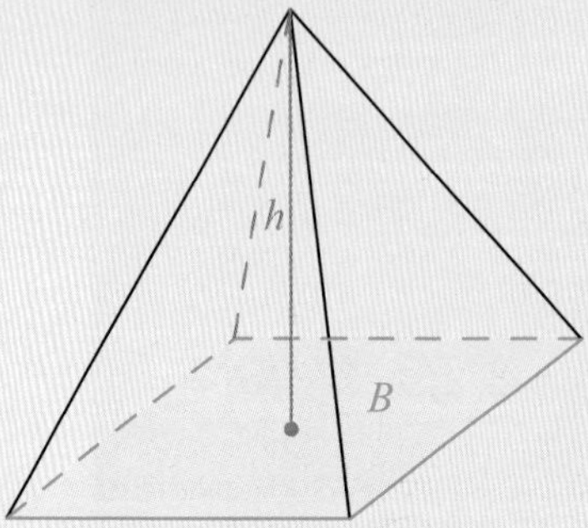 | $\frac{1}{3}Bh$ | Area of triangular sides + $B$ |

**Tip**

The surface area of a cone is in two parts. The curved surface area is $\pi r l$ and the base area is $\pi r^2$.

**WORKED EXAMPLE 5.1**

Find the surface area of the hemisphere shown.

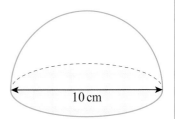

The surface area is made up
of half a sphere and a circle

Surface area $= \dfrac{4\pi r^2}{2} + \pi r^2$

The diameter is 10 cm
so the radius is 5 cm

$= \dfrac{4\pi(5)^2}{2} + \pi(5)^2$

$= 75\pi$

$\approx 236\,\text{cm}^2$

**WORKED EXAMPLE 5.2**

The following shape models an ice cream as a hemisphere
attached to a cone.

Find the volume of this shape.

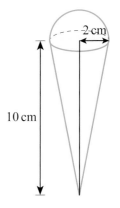

Add together the volume
of the cone and the volume
of the hemisphere

Volume $= \dfrac{1}{3}\pi r^2 h + \dfrac{2}{3}\pi r^3$

$= \dfrac{1}{3}\pi(2)^2(10) + \dfrac{2}{3}\pi(2)^3$

The radius of the cone and
the hemisphere are both 2

$= \dfrac{40}{3}\pi + \dfrac{16}{3}\pi$

$= \dfrac{56}{3}\pi$

$\approx 58.6\,\text{cm}^3$

You might have to first use your knowledge of Pythagoras' theorem to find unknown lengths.

**WORKED EXAMPLE 5.3**

Find the surface area and volume of this solid pyramid.

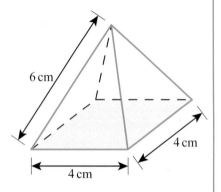

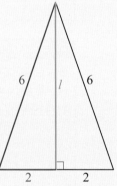

To find the area of each triangular side, we need their perpendicular height. For this, use Pythagoras $\cdots\cdots\cdots\cdots\cdots\cdots$

$$l^2 = 6^2 - 2^2$$
$$\text{So, } l = \sqrt{32} = 4\sqrt{2}$$

Use Area $= \frac{1}{2}bh$ to find the $\cdots\cdots$ area of each triangle

Area of triangular side $= \frac{1}{2} \times 4 \times 4\sqrt{2}$
$$= 8\sqrt{2}$$

There are four triangular sides and a square base $\cdots\cdots\cdots$

So area of pyramid $= 4 \times 8\sqrt{2} + 4^2$
$$= 32\sqrt{2} + 16$$
$$\approx 61.3 \text{cm}^2$$

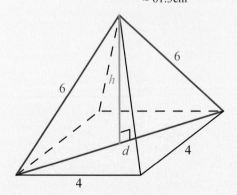

For the volume, we need the perpendicular height of the pyramid. First find the length of the diagonal of the base $\cdots\cdots\cdots\cdots\cdots$

$$d^2 = 4^2 + 4^2$$
$$\text{So, } d = \sqrt{32} = 4\sqrt{2}$$

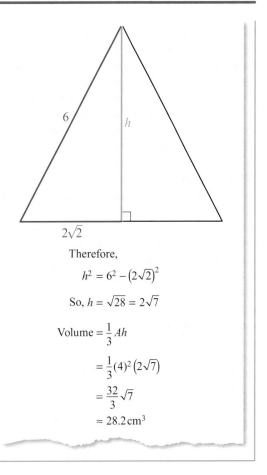

The right-angled triangle
including the perpendicular
height will have a base that is half
of the diagonal, that is, $2\sqrt{2}$

Therefore,

$$h^2 = 6^2 - \left(2\sqrt{2}\right)^2$$

So, $h = \sqrt{28} = 2\sqrt{7}$

$$\text{Volume} = \frac{1}{3}Ah$$

$$= \frac{1}{3}(4)^2\left(2\sqrt{7}\right)$$

$$= \frac{32}{3}\sqrt{7}$$

$$\approx 28.2\,\text{cm}^3$$

## Exercise 5A

For questions 1 to 4, use Key Point 5.1 and the method demonstrated in Worked Example 5.1 to find the curved surface area and volume of the following solids.

1   a   A sphere with radius 1.8 cm

    b   A sphere with radius 11.2 m

3   a   A hemisphere with radius 5.2 m

    b   A hemisphere with radius 4 km

2   a   A sphere with diameter 1.6 m

    b   A sphere with diameter 0.2 m

4   a   A cone with base radius 3 cm and height 4,1 cm

    b   A cone with base radius 0.2 mm and height 7 mm

For questions 5 to 8, use Key Point 5.1 and the method demonstrated in Worked Example 5.1 to find the exact total surface area and volume of the following solids.

5   a   A sphere with radius 2 cm

    b   A sphere with radius 3 m

7   a   A hemisphere with radius $\frac{1}{2}$ m

    b   A hemisphere with radius $\frac{2}{3}$ mm

6   a   A sphere with diameter 8 m

    b   A sphere with diameter 12 m

8   a   A cone with base radius 5 cm and height 12 cm

    b   A cone with base radius 4 cm, height 3 cm

For questions 9 to 11, use Key Point 5.1 to calculate the volume of the following shapes.

9   a   A triangular-based pyramid with base area 6 cm$^2$ and height 3 cm

    b   A triangular-based pyramid with base area 8 cm$^2$ and height 6 cm

10   a   A square-based pyramid with base area 5 cm$^2$ and height 12 cm

    b   A square-based pyramid with base area 9 cm$^2$ and height 6 cm

11   a   A cone with base area 8 cm$^2$ and height 10 cm

    b   A cone with base area 2 cm$^2$ and height 4 cm

For questions 12 to 15, use the method demonstrated in Worked Example 5.3 to calculate the volume and total surface area of the following right pyramids and cones.

12  a

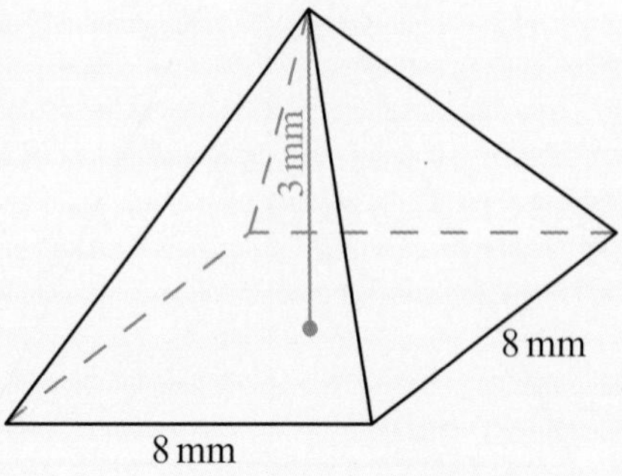

b

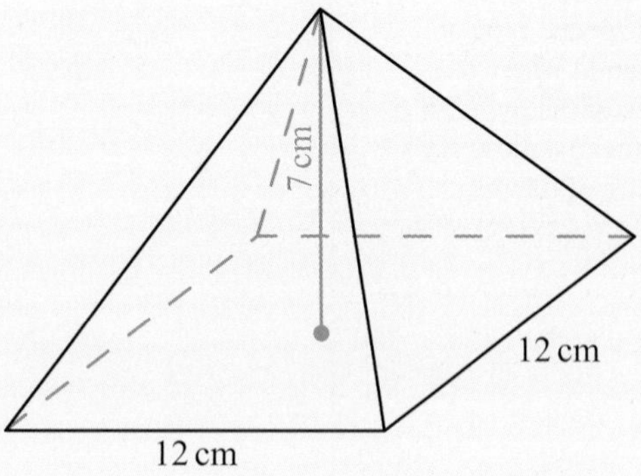

13  a

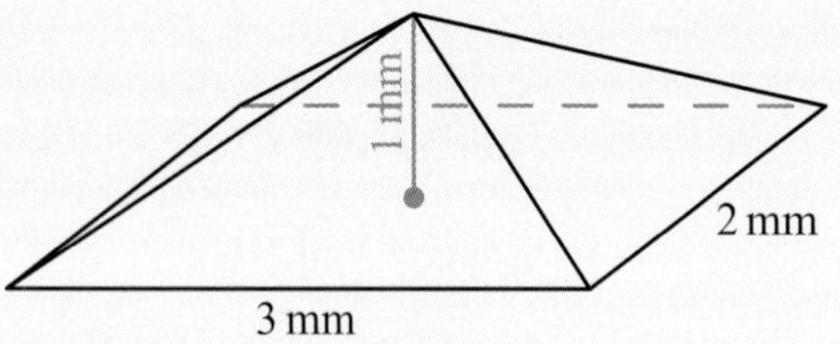

b

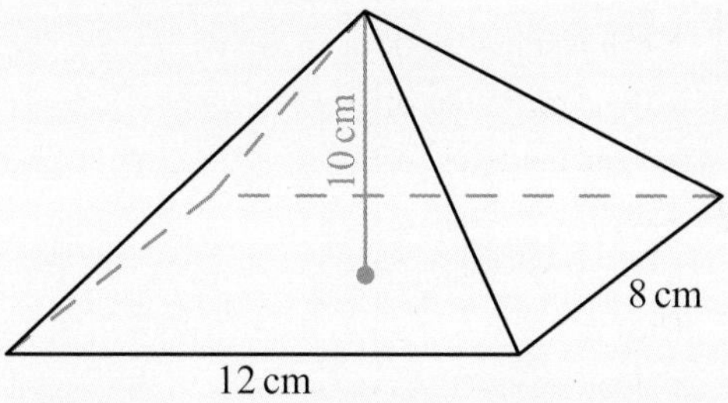

14  a

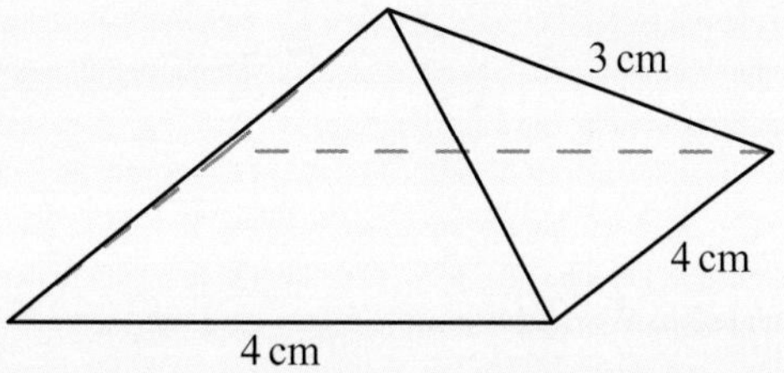

b

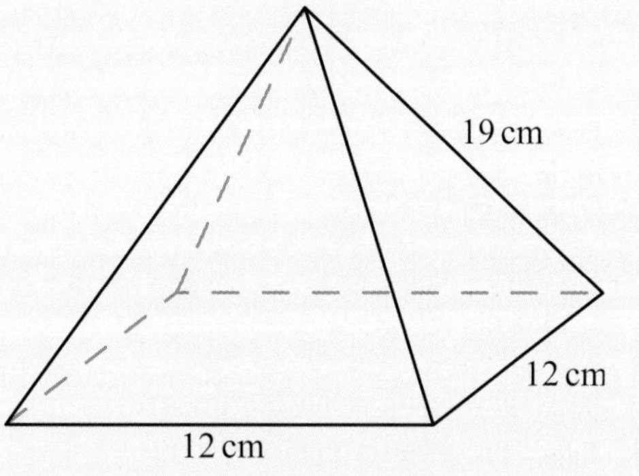

15  a

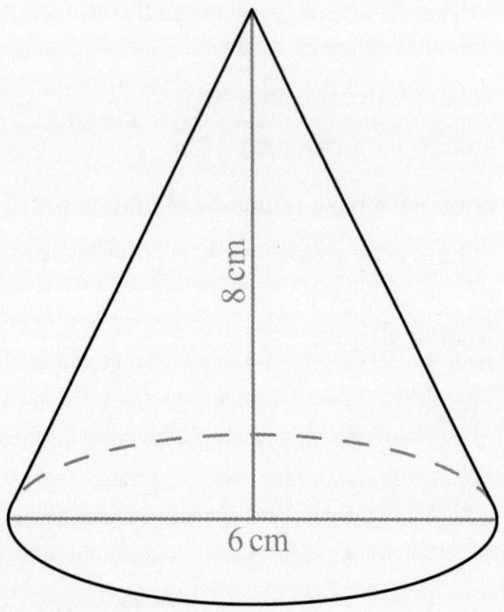

b

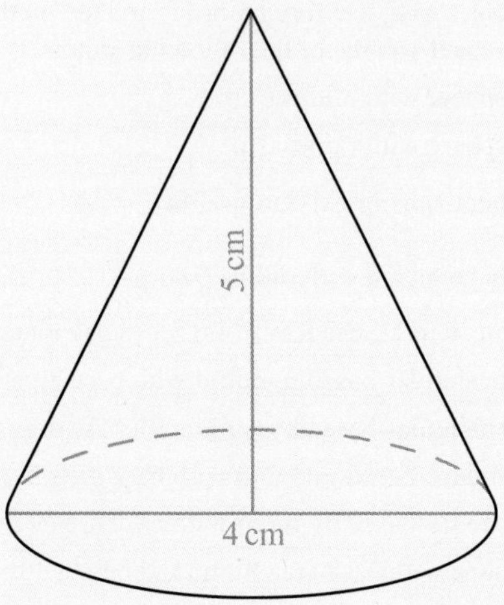

**16** Find the surface area of a sphere with diameter 15 cm.

**17** A solid hemisphere has radius 3.2 cm. Find its total surface area.

**18** The base of a conical sculpture is a circle of diameter 3.2 m. The height of the sculpture is 3.1 m. Find the volume of the sculpture.

**19** Find the volume of a solid hemisphere of radius 4 cm.

**20** Find the volume of a sphere with diameter 18.3 cm.

**21** A cylinder with length 1 m and radius 5 cm has a cone of the same radius and length 10 cm attached to one end. Find the surface area and volume of this compound shape.

**22** A fence post can be modelled as a cylinder with a hemisphere on top. The cylinder and the hemisphere have the same diameter of 28 cm. The total height of the post is 73 cm. Find the volume of the whole post, giving your answer in standard form.

**23** A cone has vertical height of 12 cm and slant height of 17 cm.

   a  Find the radius of the base of the cone.

   b  Find the total surface area of the cone.

**24** A cylinder has height 12.3 cm and volume 503.7 cm³.

   a  Find the radius of the base of the cylinder.

   b  Find the height of a cone which has the same radius and same volume as the cylinder.

**25** A metal ornament is made in the shape of a cone with radius 4.7 cm and height 8.3 cm.

   a  Find the volume of the ornament.

The ornament is melted down and made into a ball.

   b  Find the radius of the ball.

**26** A toy rocket can be modelled as a cone on top of a cylinder. The cone and the cylinder both have the diameter of 18 cm. The height of the cylinder is 23 cm and the height of the whole toy is 35 cm.

   a  Find the slant height of the cone.

   b  Find the total surface area of the toy rocket.

**27** The base of a pyramid is a square of side 12 cm. Each side face is an isosceles triangle with sides 12 cm, 15 cm and 15 cm.

   a  Find the area of each side face.

   b  Hence find the total surface area of the pyramid.

   c  Find the volume of the pyramid.

**28** A cone with height 6 cm and radius 2 cm has a hemispherical hole with radius 1 cm bored into the centre of its base. Find the surface area and volume of this compound shape.

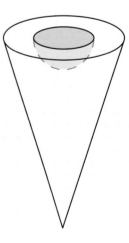

**29** Two hemispheres, with $A$ the centre of each, are joined at their flat surfaces. One has radius 8 mm and the other radius 10 mm. Find the total surface area and volume of this compound shape.

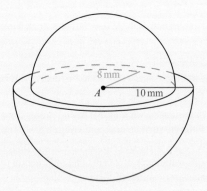

**30** A conic frustum is formed by slicing the cone tip from a larger cone, with the slice parallel to the cone base. Find the volume and total surface area of a frustum formed by taking a right cone with radius 8 mm and height 30 mm and removing the cone tip with height 12 mm.

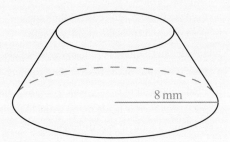

**31** A square-based pyramid has height 24 cm and volume 1352 m$^3$

   a Find the length of one side of the base.

   b Find the total surface area of the pyramid.

**32** Find the surface area of the sphere whose volume is 354 m$^3$.

**33** A pharmaceutical company intends to make a new tablet, where each tablet takes the form of a cylinder capped at each end with a hemisphere. The initial tablet design has radius 2 mm and cylinder length 8 mm.

   a Find the surface area and volume of this shape.

   b After trialing the tablet, it is found that the radius of the tablet must be decreased by 10% for ease of swallowing, and a safer dose occurs when the volume is decreased by 10%. If it takes the same general shape as before, find the total length of the new tablet.

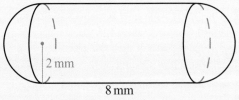

**34** A solid metal fencepost has the shape shown below, consisting of a cylinder with radius 4 cm and length 1.8 m, with a conical spike at one end (length 0.1 m) and a hemisphere at the other. Both have the same radius as the main post.

   a Find the volume of metal needed to manufacture a batch of 1000 such posts.

   b Find the approximate volume of paint needed for the batch, allowing that the paint is applied to produce a 0.4 mm thick coating. Justify any assumptions you make.

# 5B Rules of trigonometry

## ■ Revision of right-angled trigonometry

You are already familiar with using sine, cosine and tangent in right-angled triangles to find lengths and angles.

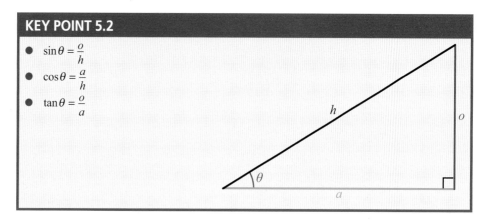

**KEY POINT 5.2**

- $\sin\theta = \dfrac{o}{h}$
- $\cos\theta = \dfrac{a}{h}$
- $\tan\theta = \dfrac{o}{a}$

**CONCEPTS – RELATIONSHIPS**

One of the key ideas of geometry is that it is the study of what stays the same when things change. The **relationships** here reflect the fact that, for similar shapes, ratios of side lengths remain constant.

**Tip**

Remember that you need to use $\sin^{-1}$, $\cos^{-1}$, $\tan^{-1}$ when finding angles.

**WORKED EXAMPLE 5.4**

For the following triangle, find the size of angle $A$.

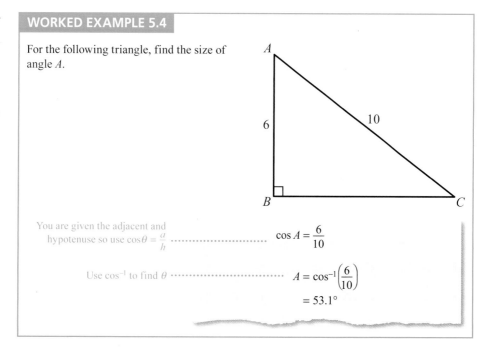

You are given the adjacent and hypotenuse so use $\cos\theta = \dfrac{a}{h}$ ·········· $\cos A = \dfrac{6}{10}$

Use $\cos^{-1}$ to find $\theta$ ·········· $A = \cos^{-1}\left(\dfrac{6}{10}\right)$

$= 53.1°$

## ■ The size of an angle between intersecting lines

Given a straight line, you can always draw a right-angled triangle by taking a vertical from the line to the *x*-axis, and then use trigonometry to find the angle the line makes with the *x*-axis.

### WORKED EXAMPLE 5.5

Find the angle between the line $y = 2x$ and the positive $x$-axis.

Sketch the line and form a right-angled triangle with horizontal distance 1 and vertical height 2 (since the gradient of the line is 2)

You know the opposite and adjacent sides to the angle $A$, so use $\tan \theta = \dfrac{o}{a}$

$$\tan \theta = \frac{2}{1}$$

$$\theta = \tan^{-1} 2 = 63.4°$$

Once you know how to find the angle between a line and the $x$-axis, you can use it to find the angle between two intersecting lines.

### WORKED EXAMPLE 5.6

Find the acute angle between $y = 3x + 1$ and $y = 2 - x$.

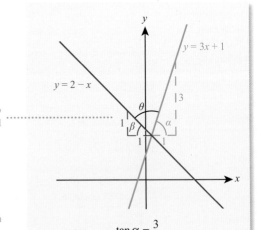

Start by sketching the two lines so you can see which angle you need

Find the angle each line makes with the horizontal. For $y = 3x + 1$, the gradient is 3, so a right-angled triangle with base 1 will have height 3

$$\tan \alpha = \frac{3}{1}$$

$$\alpha = \tan^{-1} 3 = 71.6°$$

For $y = 2 - x$, although the gradient is negative ($-1$), you are only interested in the lengths of the sides of the right-angled triangle, which must be positive. So for a base of 1, the height will be 1

$$\tan \beta = \frac{1}{1}$$

$$\beta = \tan^{-1} 1 = 45°$$

Use the fact that the angles on a line sum to 180°

$$\text{So, } \theta = 180 - (71.6 + 45)$$

$$= 63.4°$$

## ▇ The sine rule

If a triangle does not contain a right angle you cannot directly use your previous knowledge of trigonometry to find lengths or angles. However, there are some new rules which can be applied. The first of these is called the sine rule. The sine rule is useful if you know the length of a side and the size of the angle opposite that side (as well as either one other side or angle).

**KEY POINT 5.3**

The sine rule:

- $$\frac{a}{\sin A} = \frac{b}{\sin B} = \frac{c}{\sin C}$$

or, equivalently,

- $$\frac{\sin A}{a} = \frac{\sin B}{b} = \frac{\sin C}{c}$$

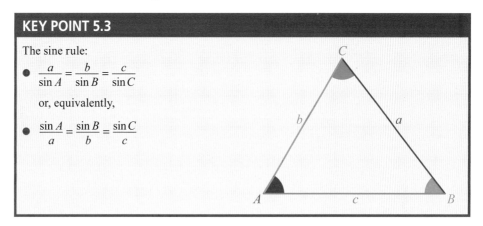

The diagram in Key Point 5.3 shows the convention that angles are labelled using capital letters and the opposite side is labelled using the equivalent lowercase letter.

**Proof 5.1**

Prove that in any triangle $\dfrac{a}{\sin A} = \dfrac{b}{\sin B}$.

Creating right-angled triangles allows you to use right-angled trigonometry ········ Divide the triangle $ABC$ into two right-angled triangles:

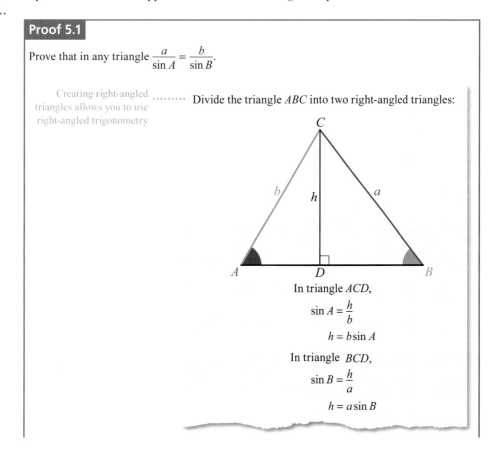

In triangle $ACD$,

$$\sin A = \frac{h}{b}$$

$$h = b \sin A$$

In triangle $BCD$,

$$\sin B = \frac{h}{a}$$

$$h = a \sin B$$

But $h$ is common to both triangles so the two expressions can be equated ············· So, $a \sin B = b \sin A$

$$\frac{a}{\sin A} = \frac{b}{\sin B}$$

## WORKED EXAMPLE 5.7

Find length $b$ in the following triangle.

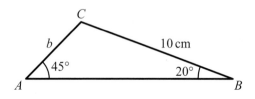

Use $\dfrac{a}{\sin A} = \dfrac{b}{\sin B}$ ················· By the sine rule,

$$\frac{10}{\sin 45} = \frac{b}{\sin 20}$$

$$b = \frac{10}{\sin 45} \times \sin 20 = 4.84 \text{ cm}$$

## WORKED EXAMPLE 5.8

In triangle $ABC$, angle $B = 70°$, $c = 2.6$ and $b = 3.2$. Find angle $A$.

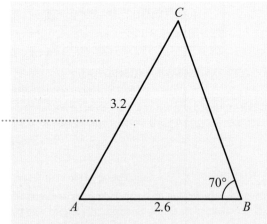

The first thing to do is to draw a diagram. It does not have to be perfectly to scale, but it is often a good idea to make it look roughly correct. In this example the lengths were not given units, so they can be labelled without units

You cannot find $A$ directly as you do not know the length $a$, so start by using $\dfrac{\sin B}{b} = \dfrac{\sin C}{c}$ to find $C$

By the sine rule,

$$\frac{\sin 70}{3.2} = \frac{\sin C}{2.6}$$

$$\sin C = \frac{\sin 70}{3.2} \times 2.6$$

$$C = \sin^{-1}\left(\frac{\sin 70}{3.2} \times 2.6\right) = 49.8°$$

The angles in a triangle sum to $180°$ ················· So,

$$A = 180 - 70 - 49.8$$

$$= 60.2°$$

.........................

**Tip**

The second version is just a rearrangement of the first to make it easier to find a missing angle.

.........................

.........................

**Tip**

You can use the cosine rule with any letter as the subject on the left-hand side of the formula, as long as you make sure the angle matches this. For example, it could be written as $b^2 = a^2 + c^2 - 2ac\cos B$ or $c^2 = a^2 + b^2 - 2ab\cos C$.

.........................

# ■ The cosine rule

The cosine rule is needed if you are given two sides and the angle between them, or all three sides but no angle. In these cases, you cannot use the sine rule.

**KEY POINT 5.4**

The cosine rule:
- $a^2 = b^2 + c^2 - 2bc\cos A$
  or, equivalently,
- $\cos A = \dfrac{b^2 + c^2 - a^2}{2bc}$

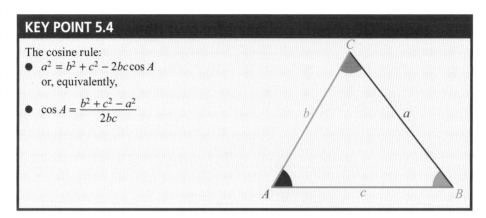

**Proof 5.2**

Prove that $a^2 = b^2 + c^2 - 2bc\cos A$.

Again, start by creating two right-angled triangles ......... Divide the triangle *ABC* into two right-angled triangles:

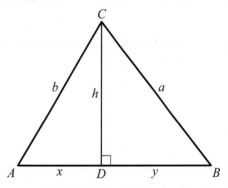

Start by writing down an expression for $a^2$ .................... In triangle *BCD*,
$$a^2 = h^2 + (c-x)^2$$

There are two variables we do not want ( $h$ and $x$ ) so we need two equations from the other triangle to eliminate these variables .... In triangle *ACD*,
$$h^2 = b^2 - x^2$$
and
$$\cos A = \frac{x}{b}$$
$$x = b\cos A$$

So,

Expand the brackets in the ............ $a^2 = h^2 + (c-x)^2$
expression from *BCD*
$$= h^2 + c^2 - 2cx + x^2$$

Substitute in the two expressions .................. $= b^2 - x^2 + c^2 - 2cb\cos A + x^2$
from the second triangle
$$= b^2 + c^2 - 2bc\cos A$$

**WORKED EXAMPLE 5.9**

Find length $c$ in the following diagram.

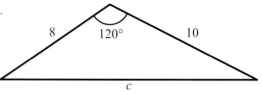

Use $c^2 = a^2 + b^2 - 2ab \cos C$ (remember the angle must match the side you choose as the subject on the LHS) $\cdots\cdots\cdots$

Remember to square root to find $c$ $\cdots\cdots\cdots$

By the cosine rule,

$$c^2 = 10^2 + 8^2 - 2(10)(8)\cos 120$$
$$= 244$$

So, $c = 15.6$

**WORKED EXAMPLE 5.10**

Find the angle $B$ in the following triangle.

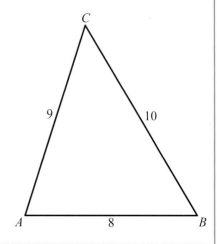

Use $\cos B = \dfrac{a^2 + c^2 - b^2}{2ac}$ $\cdots\cdots\cdots\cdots$

By the cosine rule,

$$\cos B = \frac{10^2 + 8^2 - 9^2}{2(10)(8)}$$

$$B = \cos^{-1}\left(\frac{10^2 + 8^2 - 9^2}{2(10)(8)}\right) = 58.8°$$

### You are the Researcher

There is also a rule called the tan rule which has fallen out of favour because all problems solved using it can be solved using a combination of the sine and cosine rules. However, it does have several interesting applications and proofs that you could research.

## Be the Examiner 5.1

In triangle $ABC$, $AB = 5.9\,\text{cm}$, $AC = 12\,\text{cm}$, $A\hat{B}C = 60°$ and $A\hat{C}B = 25°$.

Find the length of side $BC$.

Which is the correct solution? Identify the errors made in the incorrect solutions.

| Solution 1 | Solution 2 | Solution 3 |
|---|---|---|
| By cosine rule, $BC^2 = 5.9^2 + 12^2 - 2(5.9)(12)\cos 60$ $= 108.1$ So, $BC = 10.4\,\text{cm}$ | By cosine rule, $BC^2 = 5.9^2 + 12^2 - 2(5.9)(12)\cos 25$ $= 50.4768$ So, $BC = 7.10\,\text{cm}$ | By cosine rule, $BC^2 = 5.9^2 + 12^2 - 2(5.9)(12)\cos 95$ $= 191.151$ So, $BC = 13.8\,\text{cm}$ |

## ■ Area of a triangle

You already know you can use Area $= \frac{1}{2}bh$ to find the area of a triangle if you know the base and the perpendicular height.

Often, though, you will not know the perpendicular height. In that case there is an alternative formula.

### Tip

In this formula, the angle you are interested in ($C$) is always between the two side lengths you are interested in ($a$ and $b$).

**KEY POINT 5.5**

● Area $= \frac{1}{2}ab\sin C$

**Proof 5.3**

Prove that for a triangle

Area $= \frac{1}{2}ab\sin C$.

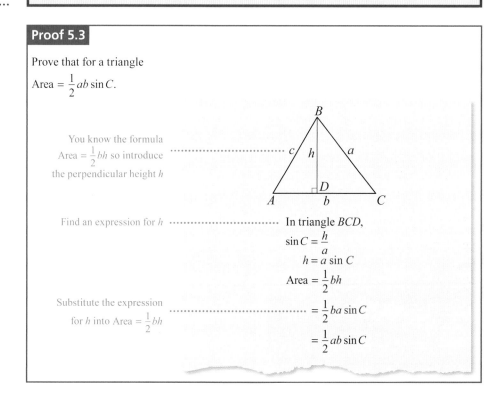

You know the formula Area $= \frac{1}{2}bh$ so introduce the perpendicular height $h$ ·······

Find an expression for $h$ ·······   In triangle $BCD$,

$$\sin C = \frac{h}{a}$$
$$h = a\sin C$$
$$\text{Area} = \frac{1}{2}bh$$

Substitute the expression for $h$ into Area $= \frac{1}{2}bh$ ·······   $= \frac{1}{2}ba\sin C$

$$= \frac{1}{2}ab\sin C$$

**WORKED EXAMPLE 5.11**

Find the area of the triangle on the right.

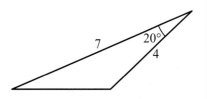

Use Area $= \frac{1}{2} ab \sin C$

Remember that the angle $C$ will be between the side lengths $a$ and $b$

Area $= \frac{1}{2}(7)(4) \sin 20$

$= 4.79$

**WORKED EXAMPLE 5.12**

The triangle on the right has area $15 \, \text{cm}^2$.

Find the acute angle marked $x$.

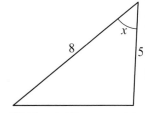

Use Area $= \frac{1}{2} ab \sin C$, with Area $= 15$

$15 = \frac{1}{2} \times 5 \times 8 \sin x$

Solve for $x$

$15 = 20 \sin x$

$\sin x = \frac{3}{4}$

$x = 48.6°$

## Exercise 5B

For questions 1 to 3, use the method demonstrated in Worked Example 5.4 to find the required angle in right-angled triangle $ABC$, where angle $A$ is 90°.

1  a  $AB = 7 \, \text{cm}$ and $AC = 4 \, \text{cm}$. Find angle $ABC$.

   b  $AB = 8 \, \text{mm}$ and $AC = 5 \, \text{mm}$. Find angle $ACB$.

2  a  $BC = 5 \, \text{cm}$ and $AB = 4 \, \text{cm}$. Find angle $ACB$.

   b  $AB = 7 \, \text{mm}$ and $BC = 11 \, \text{mm}$. Find angle $ACB$.

3  a  $AB = 13 \, \text{cm}$ and $BC = 17 \, \text{cm}$. Find angle $ABC$.

   b  $AC = 9 \, \text{mm}$ and $BC = 15 \, \text{mm}$. Find angle $ACB$.

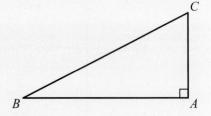

For questions 4 to 9, use the techniques from your prior learning to find the required side in right-angled triangle $ABC$, where angle $A$ is 90° as above.

4  a  Angle $ABC = 30°$ and $BC = 5$. Find $AC$.

   b  Angle $ABC = 40°$ and $BC = 6$. Find $AC$.

5  a  Angle $ABC = 50°$ and $AC = 6$. Find $BC$.

   b  Angle $ABC = 60°$ and $AC = 10$. Find $BC$.

6  a  Angle $ABC = 20°$ and $BC = 10$. Find $BA$.

   b  Angle $ABC = 45°$ and $BC = 20$. Find $BA$.

7  a  Angle $ABC = 55°$ and $AB = 10$. Find $BC$.

   b  Angle $ABC = 70°$ and $AB = 9$. Find $BC$.

8    a    Angle $ABC = 15°$ and $AB = 10$. Find $AC$.

     b    Angle $ABC = 65°$ and $AB = 12$. Find $AC$.

9    a    Angle $ABC = 2°$ and $AC = 12$. Find $AB$.

     b    Angle $ABC = 80°$ and $AC = 24$. Find $AB$.

For questions 10 to 12, use the method demonstrated in Worked Example 5.5 to find the angle between the two lines.

10    a    Find the angle between the positive $x$-axis and $y = 3x$.

     b    Find the angle between the positive $x$-axis and $y = 5x$.

11    a    Find the angle between the positive $y$-axis and $y = 3x$.

     b    Find the angle between the positive $y$-axis and $y = \frac{7}{2}x$.

12    a    Find the angle between the positive $y$-axis and $y = -\frac{5}{2}x$.

     b    Find the angle between the positive $y$-axis and $y = -\frac{9}{2}x$.

For questions 13 to 15, use the method demonstrated in Worked Example 5.6 to find the angle between the two lines.

13    a    Find the acute angle between the lines $y = 5x$ and $y = 3x$.

     b    Find the acute angle between the lines $y = 2x$ and $y = \frac{7}{2}x$.

14    a    Find the acute angle between the lines $y = 5x$ and $y = -3x$.

     b    Find the acute angle between the lines $y = 2x$ and $y = -3x$.

15    a    Find the acute angle between the lines $y = 5x - 3$ and $y = \frac{5}{2}x + 1$.

     b    Find the acute angle between the lines $y = 9x - 6$ and $y = 4 - 7x$.

For questions 16 to 18, use the method demonstrated in Worked Example 5.7, applying the sine rule to find the required side length.

16    a    In triangle $ABC$, angle $A = 45°$, angle $B = 30°$ and $a = 12\,$cm. Find $b$.

     b    In triangle $ABC$, angle $A = 60°$, angle $B = 45°$ and $a = 8\,$mm. Find $b$.

17    a    In triangle $PQR$, angle $P = 70°$, angle $Q = 40°$ and $QR = 5\,$cm. Find $PR$.

     b    In triangle $PQR$, angle $P = 60°$, angle $Q = 15°$ and $QR = 11\,$mm. Find $PR$.

18    a    In triangle $ABC$, angle $A = 23°$, angle $C = 72°$ and $a = 1.3\,$cm. Find $b$.

     b    In triangle $ABC$, angle $A = 39°$, angle $C = 74°$ and $a = 2.8\,$mm. Find $b$.

For questions 19 to 21, use the method demonstrated in Worked Example 5.8, applying the sine rule to find the required angle.

19    a    In triangle $ABC$, angle $A = 75°$. $a = 11\,$cm and $b = 7\,$cm. Find angle $B$.

     b    In triangle $ABC$, angle $A = 82°$. $a = 9\,$cm and $b = 7\,$cm. Find angle $B$.

20    a    In triangle $PQR$, angle $P = 104°$. $QR = 2.8\,$cm and $PQ = 1.7\,$cm. Find angle $Q$.

     b    In triangle $PQR$, angle $P = 119°$. $QR = 13\,$cm and $PQ = 7\,$cm. Find angle $Q$.

21    a    In triangle $ABC$, angle $A = 84°$. $a = 7.3\,$cm and $c = 7.1\,$cm. Find angle $B$.

     b    In triangle $ABC$, angle $A = 70°$. $a = 89\,$m and $c = 81\,$m. Find angle $B$.

For questions 22 to 24, use the method demonstrated in Worked Example 5.9, applying the cosine rule to find the unknown side length $x$.

22    a    In triangle $ABC$, angle $A = 60°$. $b = 4\,$cm and $c = 7\,$cm. Find side length $a$.

     b    In triangle $ABC$, angle $A = 60°$. $b = 5\,$mm and $c = 8\,$mm. Find side length $a$.

23    a    In triangle $PQR$, angle $P = 45°$. $PR = 7\,$cm and $PQ = 6\,$cm. Find side length $QR$.

     b    In triangle $PQR$, angle $P = 50°$. $PR = 8\,$mm and $PQ = 7\,$mm. Find side length $QR$.

24    a    In triangle $ABC$, angle $B = 70°$. $a = 5\,$cm and $c = 4\,$cm. Find side length $b$.

     b    In triangle $ABC$, angle $C = 65°$. $a = 4\,$mm and $b = 7\,$mm. Find side length $c$.

For questions 25 to 27, use the method demonstrated in Worked Example 5.10, applying the cosine rule to find the unknown angle.

25  a   In triangle $ABC$, $a = 7$ cm, $b = 8$ cm and $c = 11$ cm. Find angle $A$.

    b   In triangle $ABC$, $a = 10$ cm, $b = 12$ cm and $c = 15$ cm. Find angle $A$.

26  a   In triangle $PQR$, $PQ = 14.2$ cm, $QR = 5.1$ cm and $PR = 11$ cm. Find angle $Q$.

    b   In triangle $PQR$, $PQ = 16$ cm, $QR = 13$ cm and $PR = 17$ cm. Find angle $R$.

27  a   In triangle $ABC$, $a = 5.7$ cm, $b = 8.1$ cm and $c = 6.6$ cm. Find angle $C$.

    b   In triangle $ABC$, $a = 1.4$ cm, $b = 2.5$ cm and $c = 1.9$ cm. Find angle $B$.

For questions 28 to 30, use the method demonstrated in Worked Example 5.11 to find the area of each triangle.

28  a   In triangle $ABC$, $a = 12$ cm, $b = 5$ cm and angle $C = 72°$.

    b   In triangle $ABC$, $a = 6$ mm, $b = 7$ mm and angle $C = 30°$.

29  a   In triangle $PQR$, $PR = 19$ cm, $QR = 17$ cm and angle $R = 52°$.

    b   In triangle $PQR$, $PR = 3.5$ mm, $QR = 2.1$ mm and angle $R = 28°$.

30  a   In triangle $ABC$, $a = 37$ cm, $c = 51$ cm and angle $B = 42°$.

    b   In triangle $ABC$, $b = 61$ mm, $c = 71$ mm and angle $A = 52°$.

For questions 31 to 33, use the method demonstrated in Worked Example 5.12 to find the unknown value.

31  a   In triangle $ABC$, $a = 15.5$ cm, $b = 14.7$ cm and the area is $90$ cm$^2$. Find acute angle $C$.

    b   In triangle $ABC$, $a = 4.8$ mm, $b = 5.2$ mm and the area is $11$ mm$^2$. Find acute angle $C$.

32  a   In triangle $PQR$, $PR = 12$ cm, angle $R = 40°$ and the area is $32$ cm$^2$. Find length $QR$.

    b   In triangle $PQR$, $QR = 8$ mm, angle $R = 35°$ and the area is $12$ mm$^2$. Find length $PR$.

33  a   In triangle $ABC$, $a = 5.2$ cm, angle $B = 55°$ and the area is $15$ cm$^2$. Find length $c$.

    b   In triangle $ABC$, $a = 6.8$ mm, angle $B = 100°$ and the area is $28$ mm$^2$. Find length $c$.

**34**  Find the length marked $x$ in the diagram.

**35**  Find the angle marked $\theta$ in the diagram.

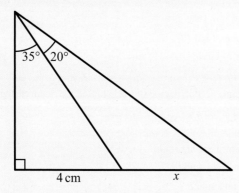

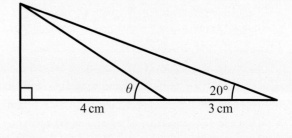

**36**  The diagram shows the line with equation $y = \dfrac{1}{2}x + 3$.

    a   Write down the coordinates of both axis intercepts.

    b   Find the size of the angle between the line and the $x$-axis.

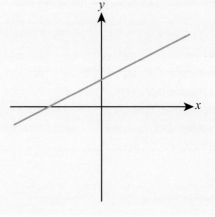

**37** a  Draw a line with equation $y = 3x - 1$, labelling both axis intercepts.

b  Find the size of the angle that the line makes with the *x*-axis.

**38** Find the length marked *a*.

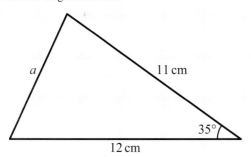

**39** Find the missing angles in this triangle.

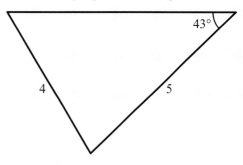

**40** Find the size of the angle $C\hat{A}B$.

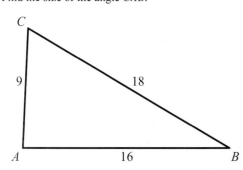

**41** For the triangle shown in the diagram:

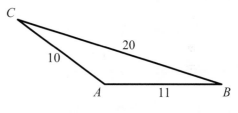

a  Find the size of angle $\hat{C}$.

b  Find the area of the triangle.

**42** The area of this triangle is 241. Find the length of *AB*.

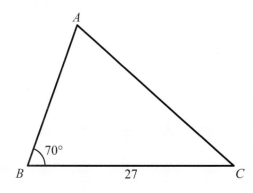

**43** Find the length marked *x* and the angle marked $\theta$.

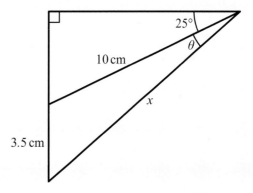

**44** Find the acute angle that the line with equation $4x + 5y = 40$ makes with the *x*-axis.

**45** a  Show that the lines $y = 2x - 8$ and $y = \frac{1}{4}x - 1$ intersect on the *x*-axis.

b  Find the angle between the two lines.

**46** Find the acute angle between the lines $2x - 5y = 7$ and $4x + y = 8$.

**47** In triangle $ABC$, $A = 40°$, $B = 60°$ and $a = 12\,\text{cm}$. Find the length of side *b*.

**48** In triangle $ABC$, $A = 45°$, $b = 5\,\text{cm}$ and $c = 8\,\text{cm}$. Find the length of side *a*.

**49** In triangle $ABC$, the sides are $a = 4\,\text{cm}$, $b = 6\,\text{cm}$, $c = 8\,\text{cm}$. Find angle *A*.

**50** Triangle $XYZ$ has $X = 66°$, $x = 10\,\text{cm}$ and $y = 8\,\text{cm}$. Find angle *Y*.

**51** Triangle $PQR$ has $P = 102°$, $p = 7\,\text{cm}$ and $q = 6\,\text{cm}$. Find angle *R*.

**52** In triangle $ABC$, $B = 32°$, $C = 64°$ and $b = 3\,\text{cm}$. Find side *a*.

**53** Find the length of the side *BC*.

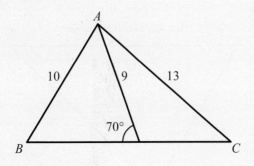

**54** The area of this triangle is 26. *θ* is acute. Find the value of *θ* and the length of *AB*.

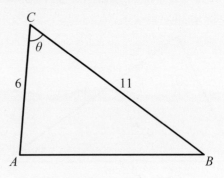

**55** Find the value of *x*.

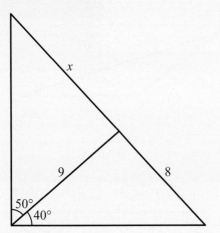

**56** Express *h* in terms of *d*.

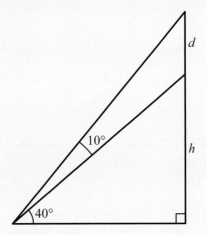

**57** In the diagram below, *AC* = 8.

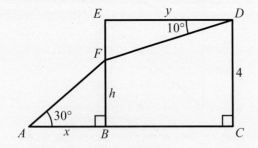

a  Express *x* and *y* in terms of *h*.

b  Hence find the value of *h*.

# 5C Applications of trigonometry

There are many different situations that can arise in which the rules of trigonometry can be applied. The best approach is to always draw a good diagram and look for appropriate triangles, especially right-angled triangles.

## ■ Angles between two intersecting lines in 3D shapes

The key idea when finding angles in 3D shapes is to look for useful triangles.

**WORKED EXAMPLE 5.13**

For the following cuboid,

a . find the acute angle $HAG$

b  find the acute angle between $AG$ and $EC$.

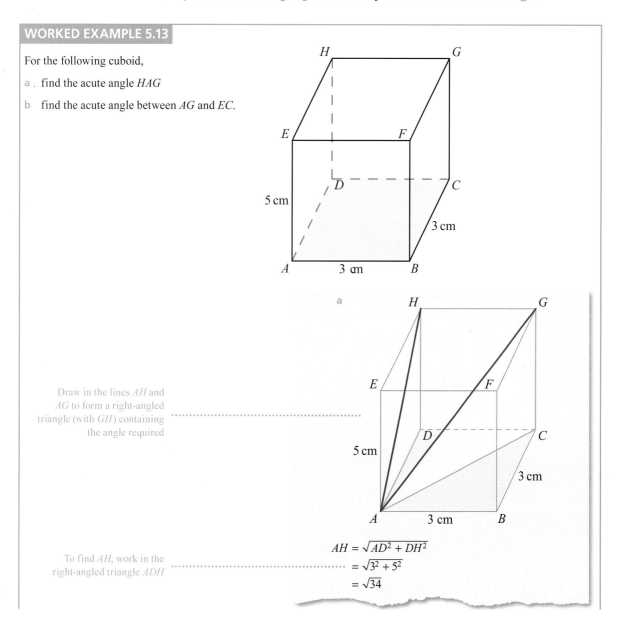

a

Draw in the lines $AH$ and $AG$ to form a right-angled triangle (with $GH$) containing the angle required

To find $AH$, work in the right-angled triangle $ADH$

$$AH = \sqrt{AD^2 + DH^2}$$
$$= \sqrt{3^2 + 5^2}$$
$$= \sqrt{34}$$

Now work in the right-
angled triangle *AGH*

In triangle *AGH*,

$$\tan A = \frac{3}{\sqrt{34}}$$

$$A = \tan^{-1}\left(\frac{3}{\sqrt{34}}\right)$$

$$= 27.2°$$

$$\therefore \text{ angle } HAG = 27.2°$$

Draw in the diagonals
*AG* and *EC*

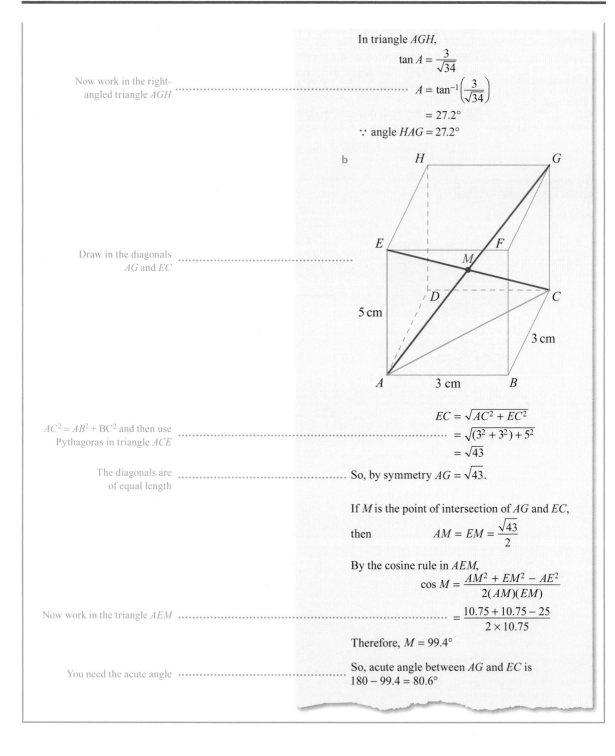

b

$AC^2 = AB^2 + BC^2$ and then use
Pythagoras in triangle *ACE*

$$EC = \sqrt{AC^2 + EC^2}$$

$$= \sqrt{(3^2 + 3^2) + 5^2}$$

$$= \sqrt{43}$$

The diagonals are
of equal length

So, by symmetry $AG = \sqrt{43}$.

If *M* is the point of intersection of *AG* and *EC*,

then

$$AM = EM = \frac{\sqrt{43}}{2}$$

By the cosine rule in *AEM*,

$$\cos M = \frac{AM^2 + EM^2 - AE^2}{2(AM)(EM)}$$

Now work in the triangle *AEM*

$$= \frac{10.75 + 10.75 - 25}{2 \times 10.75}$$

Therefore, $M = 99.4°$

You need the acute angle

So, acute angle between *AG* and *EC* is
$180 - 99.4 = 80.6°$

**WORKED EXAMPLE 5.14**

The diagram below on the right a square-based right pyramid.

a   Find angle *EBC*.

b   Find angle *EBD*.

c   Find the height, *h*.

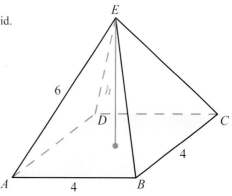

**a**

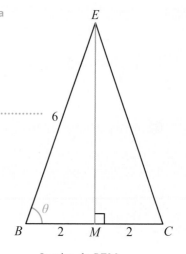

You could use the cosine rule in triangle *BCE*, but it is easier to put in the perpendicular from *E* and work in the right-angled triangle *BEM*

In triangle *BEM*,

$$\cos\theta = \frac{BM}{BE}$$

$$\theta = \cos^{-1}\left(\frac{2}{6}\right)$$

$$= 70.5°$$

∴ angle EBC = 70.5°

**b**   Let *N* be the midpoint of *DB*, perpendicular from *E*.

Work first in the base plane and then in the triangle *BEN*

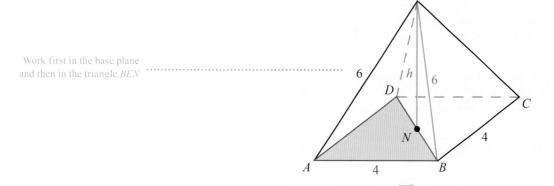

$$BD = \sqrt{AB^2 + AD^2}$$
$$= \sqrt{4^2 + 4^2}$$
$$= 4\sqrt{2}$$

$BN = \frac{1}{2}BD$ ..................................................... So, $BN = 2\sqrt{2}$

In triangle $BEN$,
$$\cos\alpha = \frac{BN}{BE}$$
$$\alpha = \cos^{-1}\left(\frac{2\sqrt{2}}{6}\right)$$
$$= 61.9°$$

c $\quad h = \sqrt{BE^2 - BN^2}$

Use Pythagoras' theorem
in triangle *BEN* ..................................................... $= \sqrt{36 - 8}$
$$= 2\sqrt{7}$$

## ■ Angles between a line and a plane in 3D shapes

You can form a right-angled triangle by
projecting the line onto the plane. The
angle you need will be in this triangle at
the point of intersection between the line
and the plane.

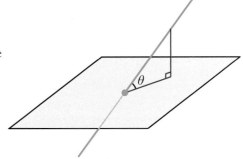

Find the angle between the line *AG* and the plane *ABCD* in the diagram below.

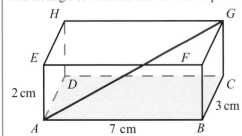

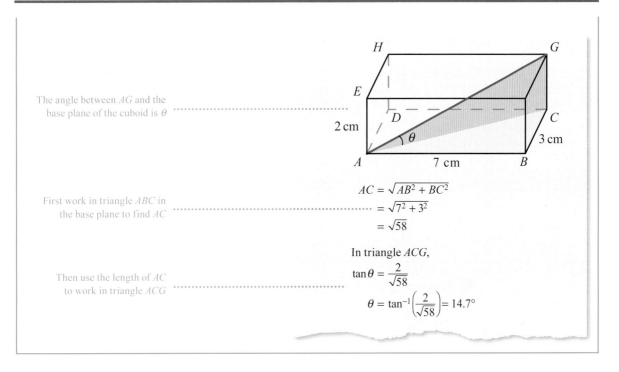

The angle between $AG$ and the base plane of the cuboid is $\theta$

First work in triangle $ABC$ in the base plane to find $AC$

$$AC = \sqrt{AB^2 + BC^2}$$
$$= \sqrt{7^2 + 3^2}$$
$$= \sqrt{58}$$

Then use the length of $AC$ to work in triangle $ACG$

In triangle $ACG$,
$$\tan\theta = \frac{2}{\sqrt{58}}$$
$$\theta = \tan^{-1}\left(\frac{2}{\sqrt{58}}\right) = 14.7°$$

**WORKED EXAMPLE 5.16**

Find the angle between the line $AD$ and the plane $ABC$ in the triangular-based right pyramid shown on the right.

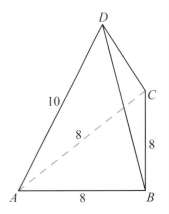

Let the perpendicular from $D$ intersect the base plane $ABC$ at $M$ and let $N$ be the midpoint of $AB$:

In the base plane, $AM$ will bisect the angle at $A$ (which is 60° as triangle $ABC$ is equilateral)

From the base plane, find $AM$ .................................................

In triangle $AMN$,

$$\cos 30 = \frac{4}{AM}$$

$$AM = \frac{8\sqrt{3}}{3}$$

... and then use this in the triangle $ADM$ .................................................

In triangle $ADM$,

$$\cos\theta = \frac{AM}{AD} = \frac{4\sqrt{3}}{15}$$

$$\theta = \cos^{-1}\left(\frac{4\sqrt{3}}{15}\right)$$

$$= 62.5°$$

## ■ Bearings and constructing diagrams from information

Bearings are a common way of describing the direction an object is travelling or the position of two objects relative to each other. They are angles measured clockwise from north.

Bearings are one of a large number of mathematical ideas which originated from navigation. The word 'geometry' comes from the ancient Greek word for 'measuring the Earth'. How important is it for all ships and aeroplanes to use the same conventions when describing journeys?

### WORKED EXAMPLE 5.17

A ship leaves a port on a bearing of 020° and travels 100 km. It unloads some of its cargo then travels on a bearing of 150° for 200 km to unload the rest of its cargo. Find

a   the distance it must now travel to return to the original port

b   the bearing it must travel on to return to the original port.

Start by drawing the situation described ·····················

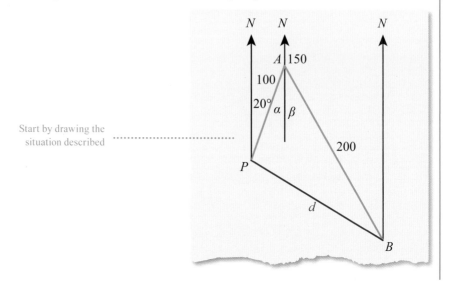

Split the angle *PAB* using a vertical through *A* so you can relate α and β to the information you have .................... **a** α = 20° by alternate angles.

β = 180 − 150 = 30°

So, $P\hat{A}B = 20 + 30 = 50°$

You can now work in the triangle *PAB* .................... By cosine rule in triangle *PAB*,

$d^2 = 100^2 + 200^2 - 2 \times 100 \times 200 \cos 50$

$= 24\,288.495\,61$

So,

$d = \sqrt{24\,288.495\,61}$

$= 156\,\text{m}$

**b** By sine rule in *PAB*,

$$\frac{\sin APB}{200} = \frac{\sin 50}{155.848}$$

$$APB = \sin^{-1}\left(\frac{\sin 50}{155.848} \times 200\right)$$

$$= 79.4°$$

## ■ Angles of elevation and depression

An angle of elevation is an angle above the horizontal and an angle of depression is an angle below the horizontal. These are often used to describe the objects from a viewer's perspective.

### WORKED EXAMPLE 5.18

When a man stands at a certain distance from a building, the angle of elevation of the top of the building is 21.8°. When the man walks a further 50 m away in the same direction, the new angle of elevation is 11.3°.

If the measurements are being taken at the man's eye level, which is 1.8 m above the ground, find the height of the building.

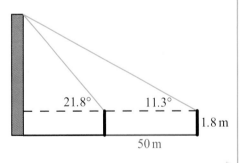

Work first in the non-right-angled triangle in order to find the length that is shared with the right-angled triangle .........................

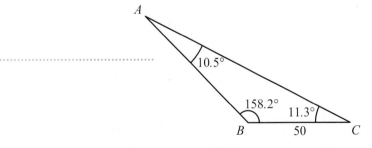

Angles on a straight line ............................................... $A\hat{B}C = 180 - 21.8 = 158.2°$

Sum of angles in a triangle ............................................... $B\hat{A}C = 180 - 158.2 - 11.3 = 10.5°$

By sine rule,

$$\frac{AB}{\sin 11.3} = \frac{50}{\sin 10.5}$$

You will need the sine rule ...........................................

$$AB = \frac{50}{\sin 10.5} \times \sin 11.3$$

$$= 53.8 \, m$$

Now work in the right-angled triangle ...........................................

$$\sin 21.8 = \frac{AD}{53.7618}$$

$$AD = 20.0 \, m$$

Add on the man's height to get the height of the building ...........................................

Therefore,

height of building $= 20.0 + 1.8 = 21.8 \, m$

## Exercise 5C

Questions 1 to 5 refer to the cuboid shown on the right.

Use the method demonstrated in Worked Example 5.13 to find the required angles.

1  a  $AC = 7$ cm, $AE = 5$ cm. Find angle $ACE$.

    b  $FH = 6$ cm, $HD = 10$ cm. Find angle $HFD$.

2  a  $BD = 10$ cm, $AE = 3$ cm. Find angle $ACE$.

    b  $HC = 4$ cm, $CB = 5$ cm. Find angle $CEB$.

3  a  $AE = 3$ cm, $AB = 4$ cm, $AD = 5$ cm. Find angle $HBA$.

    b  $AD = 4$ cm, $DC = 10$ cm, $DH = 6$ cm. Find angle $GDF$.

4  a  $AE = 3$ cm, $AB = 4$ cm, $AD = 5$ cm. Find angle $HFC$.

    b  $AE = 6$ cm, $AB = 3$ cm, $AD = 10$ cm. Find angle $GEB$.

5  a  $AE = 3$ cm, $AB = 4$ cm, $AD = 5$ cm. Find the acute angle between $AG$ and $HB$.

    b  $AE = 6$ cm, $AB = 3$ cm, $AD = 10$ cm. Find the acute angle between $CE$ and $AG$.

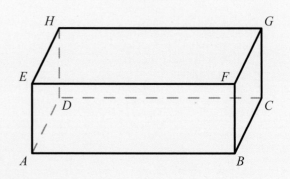

Questions 6 to 9 refer to a right pyramid with rectangular base *ABCD* and vertex *M*. *O* is the centre of the base so that *OM* is the height of the pyramid.

Use the method demonstrated in Worked Example 5.14 to find the required lengths and angles.

6  a  $AC = 5$ cm, $OM = 6$ cm. Find angle *ACM*.

  b  $AC = 9$ cm, $OM = 4$ cm. Find angle *ACM*.

7  a  $AB = 5$ cm, $BC = 7$ cm, $OM = 5$ cm. Find angle *ACM*.

  b  $AB = 4$ mm, $BC = 8$ mm, $OM = 5$ mm. Find angle *ACM*.

8  a  $AB = BC = 12$ cm, $AM = 17$ cm. Find angle *ABM*.

  b  $AB = BC = 11$ cm, $AM = 8$ cm. Find angle *ABM*.

9  a  $AB = 6$ cm, $BC = 5$ cm, angle $MAO = 35°$. Find pyramid height *OM*.

  b  $AB = 9$ m, $BC = 7$ m, $AM = 10$ m. Find pyramid height *OM*.

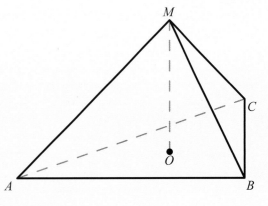

Questions 10 to 12 refer to a cuboid as labelled in questions 1–5.

Use the method demonstrated in Worked Example 5.15 to find the smallest angle between the diagonal *AG* and the base plane *ABCD*, given the following measurements.

10  a  $BH = 15$, $DH = 9$

  b  $BH = 25$, $DH = 17$

11  a  $AB = 5$, $AD = 8$, $AE = 7$

  b  $AB = 8$, $AD = 3$, $AE = 9$

12  a  $AE = 6$, $AF = 7$, $AH = 11$

  b  $AE = 1.2$, $AF = 2.1$, $CF = 4.3$

Questions 13 to 15 refer to a right pyramid as labelled in questions 6–9. Use the method demonstrated in Worked Example 5.15 to find the angle between the edge *AM* and plane *ABCD* in each case.

13  a  $AB = BC = 7$ cm, $OM = 5$ cm

  b  $AB = BC = 7$ mm, $OM = 5$ mm

14  a  $AB = 9$ cm, $BC = 4$ cm, $OM = 6$ cm

  b  $AB = 11$ mm, $BC = 12$ mm, $OM = 15$ mm

15  a  $AB = 3$ cm, $BC = 7$ cm, $AM = 4$ cm

  b  $AB = 6$ mm, $BC = 8$ mm, $AM = 13$ mm

Question 16 refers to a right pyramid with triangular base *ABC* and vertex *M*. *O* is the centre of the base so that *OM* is the height of the pyramid.

Use the method demonstrated in Worked Example 5.16 to find the angle between the edge *AM* and plane *ABC* in each case.

16  a  *ABC* is an equilateral triangle with side length 5 cm, $OM = 4$ cm.

  b  *ABC* is an equilateral triangle with side length 11 mm, $OM = 6$ mm.

In each of questions 17 to 19, a drone is sent on two legs of a journey. Use the method demonstrated in Worked Example 5.17 to calculate:

  i  the distance

  ii  the bearing on which the drone must travel to return to its original position.

17  a  2 km north and then 1.8 km on a bearing 145°.

  b  3.3 km south and then 2.1 km on a bearing 055°.

18  a  1.7 km east and then 2.3 km on a bearing 085°.

  b  6.8 km west and then 9.1 km on a bearing 035°.

19  a  2.2 km on a bearing 130° and then 1.8 km on a bearing 145°.

b  13 km on a bearing 220° and then 17 km on a bearing 105°.

20  Viewed from 40 m away, a building has an angle of elevation of 55°. Find the height of the building.

21  For the cuboid shown in the diagram:

a  Find the length of *HB*.

b  Find the angle between *HB* and *BD*.

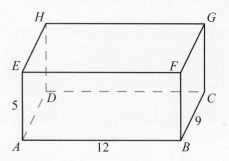

22  The diagram shows a square – based pyramid.

a  Draw a sketch of triangle *ACE*, labelling the lengths of all the sides.

b  Find the height of the pyramid.

c  Find the angle between *AE* and *EC*.

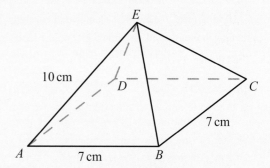

23  A cone has diameter 12 cm and vertical height 9 cm. Find the angle $\theta$ between the sloping edge and the base.

24  From his tent, Mario can see a tree 120 m away on a bearing of 056°. He can also see a rock that is due east of his tent and due south of the tree.

a  Sketch and label a diagram showing this information.

b  Hence find the distance from the rock to the tree.

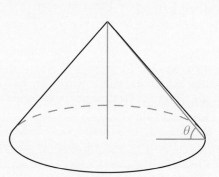

25  A ship sails from a port due north for 1.2 km. It then changes direction and sails another 0.8 km on a bearing of 037°.

a  Sketch and label a diagram showing this information.

b  Find the final distance of the ship from the port.

26  Alec has an eye-level 160 cm above the ground. He stands 6.5 m from a tree. He can see the top of the tree at an angle of elevation of 62°.

Find the height of the tree.

27  From her window, 9 m above ground, Julia observes a car at an angle of depression of 12°. Find the distance of the car from the bottom of Julia's building.

28  After recording the angle of elevation of the top of a statue at an unknown distance from the statue's base, a student walks exactly 5 m directly away from the statue along horizontal ground and records a second angle of elevation. The two angles recorded are 17.7° and 12.0°. Find the height of the statue.

29  From the top of his lighthouse, the keeper observes two buoys, the nearest of which is directly to the east of the lighthouse and the second buoy 18 m south of the first. The surface of the water is still.

If the angle of depression to the first buoy is 42.5° and the angle of depression to the second buoy is 41.3°, find the height of the lighthouse above sea level, to the nearest metre.

**30** A cuboid *ABCDEFGH* with sides 4 cm, 7 cm and 9 cm is shown in the diagram.

   a Find the length of the diagonal *AG*.

   b Find the angle between the diagonal *AG* and the base *ABCD*.

   c Find the angle between *AG* and the side *AB*.

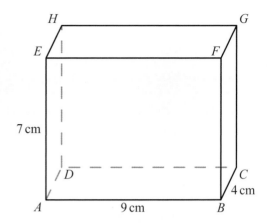

**31** In a cuboid with base *ABCD* and upper surface *EFGH*, the three face diagonals have lengths *AC* = 13 m, *AF* = 7 m, *CF* = 11 m.

   a Find the length of the body diagonal *AG*, giving your answer to four significant figures.

   b Find the smallest angle between line *AG* and the base of the cuboid.

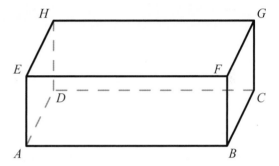

**32** The cuboid *ABCDEFGH* is shown in the diagram on the right.

   a Find the lengths of *AG* and *CE*.

   b Find the acute angle between *AG* and *CE*.

**33** A dog runs 220 m on a bearing of 042° and then a further 180 m on a bearing of 166°. Find the distance and the bearing on which the dog must run to return to the starting position.

**34** A lighthouse is 2.5 km from the port on a bearing of 035°. An island is 1.3 km from the port on a bearing of 130°. Find the distance and the bearing of the lighthouse from the island.

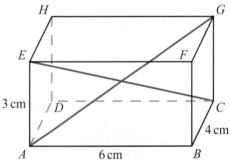

**35** The Great Pyramid of Giza has a square base of side 230 m. Each of the sloping edges makes an angle of 42° with the base. Find the height of the pyramid.

**36** Ramiz stands at the point *R*, 19.5 m from the base *B* of a vertical tree. He can see the top of the tree, *T*, at an angle of elevation of 26°.

   a Find the height of the tree.

   Mia can see the top of the tree at an angle of elevation of 41°.

   b Find Mia's distance from the bottom of the tree.

   c Given that the distance between Mia and Ramiz is 14.7 m, find the size of the angle $R\hat{B}M$ correct to the nearest degree.

**37** A visitor at an art gallery sits on the floor, 2.4 m from the wall, and looks up at a painting. He can see the bottom of the painting at an angle of elevation of 55° and the top of the painting at an angle of elevation of 72°. Find the height of the painting.

**38** The Louvre pyramid in Paris is a square-based right pyramid made mainly of glass. The square base has sides of 34 m and the height of the pyramid is 21.6 m.

a An air conditioning company recommend one unit per 1000 m$^3$ of air volume. How many units are needed to air-condition the Louvre pyramid? What assumptions are you making?

On one day, the external temperature is 20 degrees below the required internal temperature. In these conditions, the rate at which energy is lost through the glass is 192 Watts per m$^2$.

b What is the total power required to heat the pyramid to offset the energy lost through the glass?

c Health and safety regulations say that scaffolding must be used to clean any glass building with a maximum angle of elevation greater than 50°. Do the cleaners need to use scaffolding? Justify your answer.

**39** Building regulations in a city say that the maximum angle of elevation of a roof is 35°. A building has a footprint of 7 m by 5 m. The roof must be an isosceles triangle-based prism with a vertical line of symmetry.

a Find the maximum height of the roof.

b Find the maximum volume.

Only parts of the roof above 0.6 m are classed as usable.

c What percentage of the floor area is usable?

d What percentage of the volume is usable?

**40** A cuboid *ABCDEFGH* with sides 4 cm, 7 cm and 9 cm is shown in the diagram. A triangle is formed by the diagonals *BG*, *GE* and *EB* of three neighbouring faces.

Find the area of the triangle *BGE*.

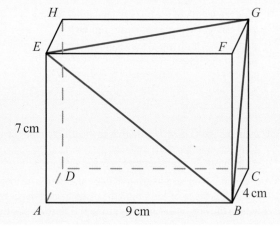

**41** The base of a pyramid *VABCD* is a square *ABCD* with side length 20 cm. The sloping edges have length 23 cm. *M* is the midpoint of the edge *AB* and *N* is the midpoint of the edge *BC*. Find the size of the angle *MVN*.

**42** Amy cycles around the park. She cycles 120 m on a bearing of 070°, then 90 m on a bearing of 150° and finally 110 m on a bearing of 250°. How far is she from her starting point?

---

**TOOLKIT: Modelling**

Find a local building and model it using some of the shapes you have met. Use trigonometry to estimate the dimensions of the shapes. Calculate the volume and surface area of the building, trying to keep track of the estimated size of any errors in your values.

Use these values to answer real world questions about the buildings such as:
- How much paint would be required to paint it?
- How much does it cost to maintain the building's temperature?
- How long would it take to clean the building?
- How energy efficient is the building?

See if you can compare your answers to available data on these values.

## Checklist

▪ You should be able to find the volume and surface area of three-dimensional solids:

| Shape | Volume | Surface area |
|---|---|---|
| Sphere of radius $r$ | $\dfrac{4}{3}\pi r^3$ | $4\pi r^2$ |
| Cone of base radius $r$, height $h$ and slant height $l$ | $\dfrac{1}{3}\pi r^2 h$ | $\pi r l + \pi r^2$ |
| Pyramid of base area $B$ and height $h$ | $\dfrac{1}{3}Bh$ | Area of triangular sides $+B$ |

▪ You should be able to find the angle between two intersecting lines in two dimensions.

▪ You should be able to use the sine rule to find side lengths and angles in non-right-angled triangles:

☐ $\dfrac{a}{\sin A} = \dfrac{b}{\sin B} = \dfrac{c}{\sin C}$     or, equivalently,     ☐ $\dfrac{\sin A}{a} = \dfrac{\sin B}{b} = \dfrac{\sin C}{c}$

▪ You should be able to use the sine rule to find side lengths and angles in non-right-angled triangles:

☐ $a^2 = b^2 + c^2 - 2bc \cos A$     or, equivalently,     ☐ $\cos A = \dfrac{b^2 + c^2 - a^2}{2bc}$

▪ You should be able to find the area of a triangle when you do not know the perpendicular height:

☐ $\text{Area} = \dfrac{1}{2}ab \sin C$

▪ You should be able to find the angle between two intersecting lines in three-dimensional shapes.

▪ You should be able to find the angle between a line and a plane in three-dimensional shapes.

▪ You should be able to construct diagrams from given information.

▪ You should be able to use trigonometry in questions involving bearings.

▪ You should be able to use trigonometry in questions involving angles of elevation and depression.

---

## ■ Mixed Practice

**1** Viewed from 50 m away, a building has an angle of elevation of 35°. Find the height of the building.

**2** The cube in the diagram has side 16 cm.
  **a** Find the lengths of $AC$ and $AG$.
  **b** Draw a sketch of triangle $ACG$, labelling the lengths of all the sides.
  **c** Find the angle between $AC$ and $AG$.

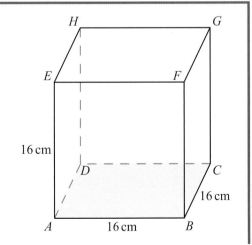

**3** The base of a right pyramid is a square of side 23 cm.
The angle between $AC$ and $AE$ is 56°.
   **a** Find the length of $AC$.
   **b** Find the height of the pyramid.
   **c** Find the length of $AE$.

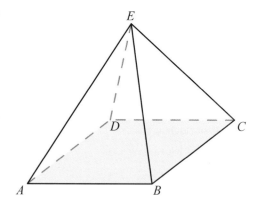

**4** A cone has radius 5 cm and vertical height 12 cm.
Find the size of angle $\theta$.

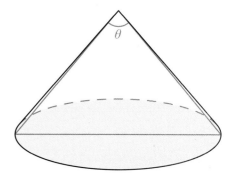

**5** $A$ and $B$ are points on a straight line as shown
on the graph on the right.
   **a** Write down the $y$-intercept of the line $AB$.
   **b** Calculate the gradient of the line $AB$.
The acute angle between the line $AB$ and the
$x$-axis is $\theta$.
   **c** Calculate the size of $\theta$.

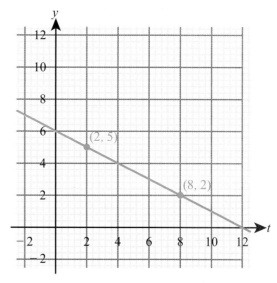

Mathematical Studies SL May 2009 Paper 1 TZ1 Q4

**6** The quadrilateral $ABCD$ has $AB = 10$ cm. $AD = 12$ cm and $CD = 7$ cm.
The size of angle $ABC$ is $100°$ and the size of angle $ACB$ is $50°$.
  **a** Find the length of $AC$ in centimetres.
  **b** Find the size of angle $ADC$.

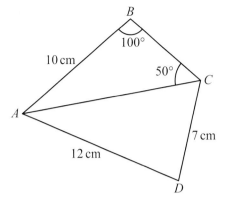

Mathematical Studies SL May 2010 Paper 1 TZ2 Q5

**7** In triangle $ABC$, $A = 50°$, $B = 70°$ and $a = 10$ cm. Find the length of side $b$.

**8** In triangle $ABC$, $A = 15°$, $b = 8$ cm and $c = 10$ cm. Find the length of side $a$.

**9** In triangle $ABC$, the sides are $a = 3$ cm, $b = 5$ cm and $c = 7$ cm. Find angle $A$.

**10** Triangle $XYZ$ has $X = 42°$, $x = 15$ cm and $y = 12$ cm. Find angle $Y$.

**11** Triangle $PQR$ has $P = 120°$, $p = 9$ cm and $q = 4$ cm. Find angle $R$.

**12** In triangle $ABC$, $B = 32°$, $C = 72°$ and $b = 10$ cm. Find side $a$.

**13** After recording the angle of elevation of the top of a tower at an unknown distance from the tower's base, a student walks exactly $20$ m directly away from the tower along horizontal ground and records a second angle of elevation. The two angles recorded are $47.7°$ and $38.2°$. Find the height of the tower.

**14** All that remains intact of an ancient castle is part of the keep wall and a single stone pillar some distance away. The base of the wall and the foot of the pillar are at equal elevations.

From the top of the keep wall, the tip of the pillar is at an angle of depression of $23.5°$ and the base of the pillar is at an angle of depression of $37.7°$.

The wall is known to have a height of $41$ m. Find the height of the pillar, to the nearest metre.

**15 a** Sketch the lines with equations $y = \frac{1}{3}x + 5$ and $y = 10 - x$, showing all the axis intercepts.

  **b** Find the coordinates of the point of intersection between the two lines.
  **c** Find the size of the acute angle between the two lines.

**16** A square-based right pyramid has height $26$ cm. The angle between the height and one of the sloping edges is $35°$. Find the volume of the pyramid.

**17** The base of a cuboid $ABCDEFGH$ is a square of side $6$ cm. The height of the cuboid is $15$ cm. $M$ is the midpoint of the edge $BC$.
  **a** Find the angle between $ME$ and the base $ABCD$.
  **b** Find the size of the angle $HME$.

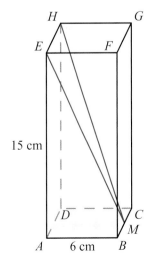

**18** In triangle $ABC$, $AB = x$, $AC = 2x$, $BC = x + 4$ and $B\hat{A}C = 60°$. Find the value of $x$.

**19** The area of this triangle is 84 units$^2$. Find the value of $x$.

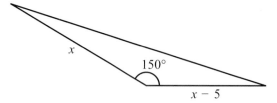

**20** A 30 m tall tower and a vertical tree both stand on horizontal ground. From the top of the tower, the angle of depression of the bottom of the tree is 50°. From the bottom of the tower, the angle of elevation of the top of the tree is 35°. Find the height of the tree.

**21** Tennis balls are sold in cylindrical tubes that contain four balls. The radius of each tennis ball is 3.15 cm and the radius of the tube is 3.2 cm. The length of the tube is 26 cm.
   **a** Find the volume of one tennis ball.
   **b** Calculate the volume of the empty space in the tube when four tennis balls have been placed in it.

Mathematical Studies SL May 2009 Paper 1 TZ1 Q13

**22** The diagram shows a right triangular prism, $ABCDEF$, in which the face $ABCD$ is a square.

$AF = 8$ cm, $BF = 9.5$ cm, and angle $BAF$ is 90°.
   **a** Calculate the length of $AB$.
   M is the midpoint of $EF$.
   **b** Calculate the length of $BM$.
   **c** Find the size of the angle between $BM$ and the face $ADEF$.

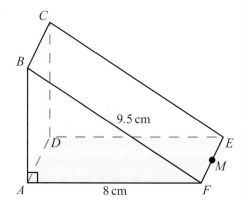

Mathematical Studies SL November 2012 Paper 1 Q12

**23** **Part A**
   The diagram on the right shows a square-based right pyramid. $ABCD$ is a square of side 10 cm. $VX$ is the perpendicular height of 8 cm. $M$ is the midpoint of $BC$.
   **a** Write down the length of $XM$.
   **b** Calculate the length of $VM$.
   **c** Calculate the angle between $VM$ and $ABCD$.

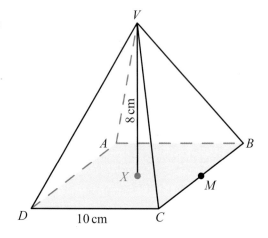

**Part B**

A path goes around a forest so that it forms the
three sides of a triangle. The lengths of two sides
are 550 m and 290 m. These two sides meet at an
angle of 115°. A diagram is shown on the right.

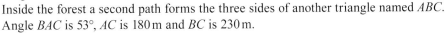

a Calculate the length of the third side of the
   triangle. Give your answer correct to the
   nearest 10 m.
b Calculate the area enclosed by the path that
   goes around the forest.

Inside the forest a second path forms the three sides of another triangle named *ABC*.
Angle *BAC* is 53°, *AC* is 180 m and *BC* is 230 m.

c Calculate the size of angle *ACB*.

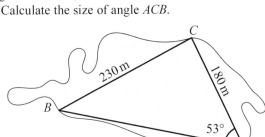

Mathematical Studies SL November 2009 Paper 2 Q1

**24** The diagram shows an office tower of total height 126 metres.
It consists of a square-based pyramid *VABCD* on top of a cuboid
*ABCDPQRS*.

*V* is directly above the centre of the base of the office tower.

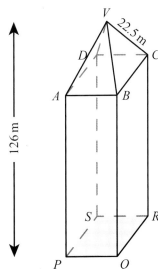

The length of the sloping edge *VC* is 22.5 metres and the angle
that *VC* makes with the base *ABCD* (angle *VCA*) is 53.1°.

a  i  Write down the length of *VA* in metres.
   ii Sketch the triangle *VCA* showing clearly the length of *VC*
      and the size of angle *VCA*.
b Show that the height of the pyramid is 18.0 metres correct to 3
  significant figures.
c Calculate the length of *AC* in metres.
d Show that the length of *BC* is 19.1 metres correct to 3
  significant figures.
e Calculate the volume of the tower.

To calculate the cost of air conditioning, engineers must estimate
the weight of air in the tower. They estimate that 90% of the volume
of the tower is occupied by air and they know that $1\,m^3$ of air weighs 1.2 kg.

f Calculate the weight of air in the tower.

Mathematical Studies SL May 2010 Paper 2 TZ1 Q4

**25** Find the area of the triangle formed by the lines $y = 8 - x$, $2x - y = 10$ and $11x + 2y = 25$.

# ESSENTIAL UNDERSTANDINGS

■ Statistics is concerned with the collection, analysis and interpretation of data.
■ Statistical representations and measures allow us to represent data in many different forms to aid interpretation.

## In this chapter you will learn...

■ about the concepts of population and sample
■ about discrete and continuous data
■ about potential bias in sampling
■ about a range of sampling techniques and their effectiveness
■ how to identify and interpret outliers
■ about frequency distributions
■ how to estimate the mean for a grouped frequency table
■ how to find the modal class
■ how to use your GDC to find the mean, median and mode
■ how to use your GDC to find the quartiles of discrete data
■ how to use your GDC to find standard deviation
■ about the effect of constant changes on a data set
■ how to construct and use statistical diagrams such as histograms, cumulative frequency graphs and box-and-whisker plots
■ about scatter graphs and how to add a line of best fit
■ how to calculate (using your GDC) and interpret a numerical measure of linear correlation
■ how to use your GDC to find the equation of the line of best fit (the regression line)
■ how to use the regression line to predict values not in the data set and how to interpret the coefficients of the regression line
■ about piecewise linear models.

■ **Figure 6.1** How can statistics be used to mislead?

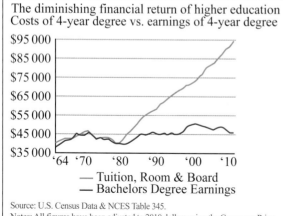

The diminishing financial return of higher education
Costs of 4-year degree vs. earnings of 4-year degree

— Tuition, Room & Board
— Bachelors Degree Earnings

Source: U.S. Census Data & NCES Table 345.
Notes: All figures have been adjusted to 2010 dollars using the Consumer Price Index from the BLS.

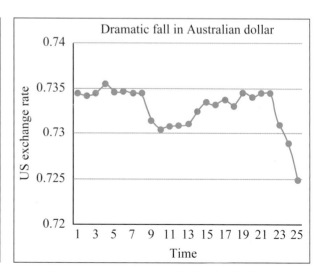

## PRIOR KNOWLEDGE

Before starting this chapter, you should already be able to complete the following:

1   For the data 1, 2, 2, 5, 6, 6, 6, 10, find:
   a   the mean
   b   the median
   c   the mode
   d   the range.

2   For the straight line with equation $y = 1.5x + 7$, state
   a   the gradient
   b   the coordinates of the point where the line crosses the $y$ axis.

## Starter Activity

Look at the graphs in Figure 6.1 and discuss what message they are trying to convey. What techniques do they use to persuade you?

**Now look at this problem:**

The annual salaries of people working in a small business (in thousands of £s) are:

10, 10, 10, 15, 15, 20, 20, 25, 30, 100.

a   Work out the mean, median and mode.
b   Which average would a union representative wish to use?
c   Which would the owner wish to use?
d   Which is the most meaningful?

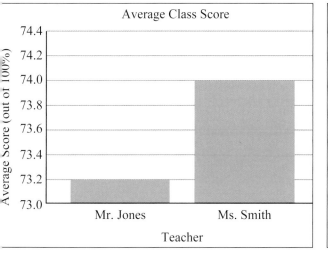

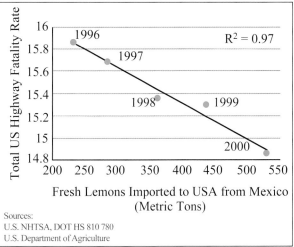

# 6A Sampling

## ▓ Sampling concepts

If you want to find the average height of an adult female in the UK, one approach would be to measure the height of every person in this **population**, that is, every adult female in the UK. Of course, in reality, this is not practical, so instead you could estimate the population average by taking a **sample** of the adult female population. The average height of the sample will most likely differ from the population average, but if the sample is well chosen it should provide a reasonable estimate.

Data such as height, which can take any value, limited only by the accuracy of the instrument you are measuring with, are known as **continuous**.

Data that can take only distinct values, such as the number of people who are taller than 2 metres, are known as **discrete**.

---

**WORKED EXAMPLE 6.1**

Below is an extract from a report on workplace productivity:

'492 employees from a single company participated in the study. Those who were on flexible hours showed a 5.3% increase in productivity. This suggests that all workers in the UK should have access to flexible hours.'

Identify the sample and the population in the extract. From the information you are given, do you think it is reasonable to generalize from the sample to the population in this case?

> The sample is the employees from the company who participated.
>
> The population is all workers in the UK.
>
> There is no indication that the sample of employees from this company is representative of the population of all workers in the UK, so it is not reasonable to generalize.

---

## ▓ Reliability of data sources and bias in sampling

If we want a sample to provide a good estimate of a population value such as the mean, then the sample needs to be representative of the population. This means that the distribution of the values in the sample is roughly the same as in the whole population.

This will not be the case if the sampling procedure is **biased**. This does not necessarily mean an intent to get the wrong answer. It means using a sample which does not represent the population of interest. For example, if we wanted to find out about people's political views and decided to conduct a phone poll in the middle of a week day, our results might be skewed by getting a disproportionate number of retired people's responses.

## CONCEPTS – VALIDITY

In 1948 the Chicago Tribune ran a telephone survey that suggested an overwhelming victory for Thomas Dewey in the US presidential election. They were so confident that they ran a newspaper with a headline announcing Dewey's win. The picture shows the actual winner, Harry Truman, holding this paper. Why did the poll get it so wrong? It was because in 1948 those people who had a phone were significantly better off than the average US citizen, so their sample was entirely unrepresentative! It is very important that when assessing any statistics you consider their **validity**. This means the extent to which you are answering the question you really want to answer. In this example, finding out what proportion of telephone owners would vote for Truman does not answer the question about what proportion of the electorate would vote for Truman.

**TOOLKIT:** Problem Solving

For each of the following situations, explain how the sampling procedure was biased. You may like to research these situations further.

■ In 2013, Google used the number of people searching for flu related terms on the internet to predict the number of people who would go to the doctor to seek flu treatment. They overestimated the true number by 140%.

■ In 1936, the Literary Digest used a poll of 10 million people with 2.4 million responses to predict the outcome of the presidential election in the USA. They got the result spectacularly wrong, but George Gallup predicted the correct result by asking 50 000 people.

■ In 2014, an app used in Boston, USA, for reporting pot holes led to the least damaged roads being repaired.

In Section 6B, you will see precisely what 'very large' and 'very small' mean in this context.

## ■ Interpretation of outliers

Any extreme value – either very large or very small – compared to the rest of the data set is said to be an **outlier**.

You should be aware of any outliers in a data set. Sometimes these will be perfectly valid data values, but at other times these might be errors and therefore should be discarded.

### WORKED EXAMPLE 6.2

Alana was analysing data from questionnaires asking schools for the time in hours they spend each year teaching mathematics. Alana's statistical package flagged up one item as an outlier. All the rest of her data were between 120 and 200 hours. What would you suggest Alana should do if the outlier has the value:

a   −175

b   4

c   240?

The number of hours of teaching cannot be negative so this must be an error ......... **a** This is likely to be a data entry error. She should check the original questionnaire and update to the correct value.

This is far too low to be a plausible response ......... **b** This is likely to be a school giving the number of hours per week rather than per year, but it could be a data entry error again. She should check the original questionnaire but if found not to be a data entry error then discard this value from the analysis.

This seems rather large but is not obviously an error ......... **c** This seems like an extreme value rather than an error, so it should not be discarded from the analysis.

## Links to: Environmental Systems and Societies

The NASA Nimbus satellites collected data on the ozone layer from the early 1970s. Unfortunately, the data were processed by a program that automatically filtered outliers. That meant evidence of the huge 'hole' in the ozone layer above the Antarctic was effectively discarded. The hole was first reported by the British Antarctic Survey in 1985. The historic data were rerun without the outlier-filter and evidence of the hole was seen as far back as 1976. It is not always a good idea to remove outliers – sometimes they are the most important part of the data!

## ■ Sampling techniques and their effectiveness

There are several different methods for selecting samples, each of which has strengths and weaknesses.

### Simple random sample

This is the type of sample most people have in mind when they talk about random samples.

**KEY POINT 6.1**

With simple random sampling, every possible sample (of a given size) has an equal chance of being selected.

This is good in theory and will produce an unbiased sample, but in practice it is difficult to do as you need a list of the entire population from which to select the sample, and then you need to actually obtain data from everyone in the chosen sample.

### Convenience sampling

This approach avoids the difficulties of simple random sampling.

**KEY POINT 6.2**

With convenience sampling, respondents are chosen based upon their availability.

## Tip

Don't think that just because a sampling method is unbiased the chosen sample will necessarily be representative of the population. It might just so happen that by chance the chosen sample contains only extreme values.

This does not produce a random sample, but if the sample size is large enough it can still provide useful information. However, it can introduce bias if the group consists of very similar members and, as such, results may not be generalizable to the population.

## Systematic sampling

This requires a list of all participants ordered in some way.

> **KEY POINT 6.3**
>
> With systematic sampling, participants are taken at regular intervals from a list of the population.

This may be more practical than using, say, a random number generator to select a sample, but you still need a list of the entire population and it is less random than simple random sampling due to the fact that selections are no longer independent.

## Stratified sampling

Another random sampling method is stratified sampling.

> **KEY POINT 6.4**
>
> With stratified sampling the population is split into groups based on factors relevant to the research, then a random sample from each group is taken in proportion to the size of that group.

This produces a sample representative of the population over the factors identified, but again you need a list of the entire population, this time with additional information about each member so as to identify those with particular characteristics.

## Quota sampling

This is a common alternative to stratified sampling.

> **KEY POINT 6.5**
>
> With quota sampling, the population is split into groups based on factors relevant to the research, then convenience sampling from each group is used until a required number of participants are found.

This produces a sample representative of the population for the factors identified, but the convenience sampling element means it can introduce bias if the group consists of very similar members and as such results may not be generalizable to the population.

---

**WORKED EXAMPLE 6.3**

In a survey, researchers questioned shoppers in a shopping mall until they had responses from 100 males and 100 females in the 18–35 age range. Name the sampling technique and state one advantage and one disadvantage compared to a simple random sample.

> Quota sampling.
>
> One advantage over simple random sampling is that quota sampling is practical. To take a simple random sample, researchers would have to know in advance who was going to the shopping mall that day.
>
> One disadvantage is that the shoppers who are prepared to stop and talk to the researchers may not be representative of all shoppers.

---

**WORKED EXAMPLE 6.4**

A school has 60% boys and 40% girls. Describe how a stratified sample of 50 students could be formed, reflecting the gender balance of the school.

60% of the sample of 50 need to be boys ············· Number of boys in sample $= \dfrac{60}{100} \times 50 = 30$

The remainder of the ············· So, number of girls in sample = 20
sample of 50 must be girls

---

**CONCEPTS – APPROXIMATION**

Most advanced statistics is about trying to make some inference about a population based on a sample. For example, if the mean IQ of a sample of 50 students in a school is 120 we might think this is a good estimate of the mean IQ of all students in that school. Although this is not perfect, if the sample was representative then this is a good **approximation** to the true value. With further study of statistics you might even be able to suggest how far from the true value this might be.

## Exercise 6A

For questions 1 to 4, use the method demonstrated in Worked Example 6.4 to determine the number of each type of participant or item required for a stratified sample.

1   a   In a wildlife park, 30% of big cats are lions and 70% are tigers. Select 30 big cats.

    b   Of ice creams sold in a café, 45% are strawberry and 55% chocolate. Select 20 ice creams.

2   a   A school is attended by 120 boys and 80 girls. Select 40 pupils.

    b   There are 84 pupils studying Maths HL and 126 pupils studying Maths SL. Select 45 pupils.

3   a   25% of pupils at school play football, 35% play hockey and 40% play basketball. Select 40 pupils.

    b   30% of fish caught are cod, 45% are haddock and 25% are mackerel. Select a sample of 20 fish.

4   a   A manufacturer produced 240 chairs, 90 tables and 40 beds. Select a sample of 37 pieces of furniture.

    b   A park has 64 oak trees, 56 willows trees and 32 chestnut trees. Select 19 trees.

5   Anke wants to find out the proportion of households in Germany who have a pet. For her investigation, she decides to ask her friends from school, which is located in the centre of a large city, whether their family owns a pet.

    a   What is the relevant population for Anke's investigation?

    b   Name the sampling method that Anke is using.

    c   State one reason why Anke's sample may not be representative of the population.

6   Leonie wants to collect information on the length of time pupils at her school spend on homework each evening. She thinks that this depends on the school year, so her sample should contain some pupils from each year group.

    a   What information does Leonie need in order to be able to select a stratified sample?

Leonie decides to ask pupils in the lunch queue until she has responses from at least 10 pupils in each year group.

    b   Name this sampling method.

    c   Having collected and analysed the data, Leonie found two outliers. For each value, suggest whether it should be kept or discarded.

      i   10 minutes

      ii   20 hours

7   A student wants to conduct an investigation into attitudes to environmental issues among the residents of his village. He decides to talk to the first 20 people who arrive at his local bus stop.

    a   Name this sampling technique.

    b   State the population relevant to his investigation.

c   Give one reason why a sample obtained in this way may not be representative of the whole population.

d   Explain why it would be difficult to obtain a simple random sample in this situation.

8   Joel obtains a random sample of 20 pupils from a college in order to conduct a survey. He finds that his sample contains the following numbers of students of different ages:

| Age | 16 | 17 | 18 | 19 |
|---|---|---|---|---|
| Number | 6 | 0 | 7 | 7 |

Mingshan says: 'There are no 17-year-olds in the sample, so your sampling procedure must have been biased.'

Comment on Mingshan's statement.

9   A manufacturer wants to test the lifetime, in hours, of its light bulbs.

a   Are the data they need discrete or continuous?

b   Explain why they need to test a sample, rather than the whole population.

c   Each lightbulb produced is given a serial number. Explain how to obtain a systematic sample consisting of 5% of all lightbulbs produced.

10   A shop owner wants to find out what proportion of the scarves she sells are bought by women. She thinks that this may depend on the colour of scarves, so she records the gender of the customers who bought the first 10 of each colour of scarf.

a   Give the name of this type of sampling.

b   Explain why a stratified sample would be more appropriate.

The shop owner knows that of all the scarves she sells, 30% are red, 30% are green, 25% are blue and 15% are white. She has a large set of historical sales records which identify the gender of the purchaser and are sorted by colour of scarf sold. She does not want to look through all of them, so she will take a sample of 40 records.

c   How many scarves of each colour should be included to make this a stratified sample?

11   A park ranger wants to estimate the proportion of adult animals in a wildlife park that are suffering from a particular disease. She believes that of the adults of these species present in the park, 20% are deer, 30% are tigers, 40% are wolves and 10% are zebras. She decides to observe animals until she has recorded 10 deer, 15 tigers, 20 wolves and 5 zebras.

a   State the name of this sampling method.

b   Why might this sampling method be better than a convenience sample of the first 50 animals she encounters?

c   Explain why a simple random sample may be difficult to obtain in this situation.

12   Shakir wants to find out the average height of students in his school. He decides to use his friends from the basketball team as a sample.

a   Is Shakir collecting discrete or continuous data?

b   Identify one possible source of bias in his sample.

Shakir decides to change his sampling technique. He obtains an alphabetical list of all students in the school and selects every 10th student for his sample.

c   Name this sampling technique.

d   Explain why this does not produce a simple random sample.

13   The table shows the number of cats, dogs and fish kept as pets by a group of children.

| Cat | Dog | Fish |
|---|---|---|
| 27 | 43 | 30 |

A sample of 20 pets is required, where the type of pet may be a relevant factor in the investigation.

Find the number of each type of pet that should be included in a stratified sample.

**14** Dan needs to select a sample of 20 children from a school for his investigation into the amount of time they spend playing computer games. He thinks that age and gender are relevant factors and so chooses a stratified sample. The table shows the number of pupils of each relevant age and gender at the school.

| Gender/Age | 12 | 13 | 14 |
|---|---|---|---|
| Boys | 40 | 52 | 50 |
| Girls | 0 | 37 | 21 |

a Create a similar table showing how many pupils from each group should be selected for the sample.

b Do you think that it would be reasonable to generalize the results from Dan's investigation to all 12- to 14-year-olds in the country?

**15** A zoologist wants to investigate the distribution of the number of spots on ladybirds. She prepares the following table to record her data:

| Number of spots | 2 | 7 | 10 |
|---|---|---|---|
| Number of ladybirds | | | |

a Are the data she is collecting discrete or continuous?

b Would it be possible to collect the data for the whole population of ladybirds in a particular field?

c The zoologist collects her data by counting the number of spots on the first 100 ladybirds she catches. State the name for this sampling procedure.

d Give one reason why the results from this sample may not be generalizable to the whole population.

**16** Ayesha selects a sample of six children from a primary school and measures their heights in centimetres. She finds that five of the children are taller than the national average for their age.

a Comment whether each of the following could be a sensible conclusion for Ayesha to draw:

 i The children at this school are taller than average.

 ii The sampling methods must have been biased.

 iii This just happens to be an unusual sample.

b If instead, Ayesha took a sample of 60 children and found that 50 were taller than the national average, how would your answer to a change?

c In a larger sample, Ayesha identified two values as outliers: −32 cm and 155 cm. For each of the values, comment whether it should be kept or discarded from further analysis.

**17** In many situations, researchers want to find out about something which people do not want to admit – for example, criminal activity. Randomized Response Theory (RRT) is one method which allows researchers to estimate the proportion of a population with the trait without ever knowing if an individual has that trait.

a Research subjects take one of three cards at random and are asked to follow the instructions on the card. The cards have the following text:

 Card 1: Say yes.

 Card 2: Say no.

 Card 3: Say yes if you have ever taken illegal drugs, otherwise say no.

 If 24 out of 60 research subjects say 'Yes' estimate the percentage of the population who have taken illegal drugs.

b A sample of students were shown two statements.

 Statement 1: I have cheated on a test.

 Statement 2: I have never cheated on a test

They were asked to secretly roll a dice and if they got a 6 answer honestly true or false to the first statement, otherwise answer honestly true or false to the second statement.

 i If 20% of students have cheated on a test, how many out of 120 students would be expected to answer 'True'?

 ii If 48 out of 120 students say 'True' estimate the percentage of the students who have cheated on a test.

**18** Ecologists wanted to estimate the number of cod in the North Sea. They captured 50 000 cod and humanely labelled them. Six months later they captured a sample of 40 000 cod and 20 are found to be labelled.

a Estimate the number of cod in the North Sea.

b State two assumptions required for your calculation in part **a**.

## TOK Links

If you do an internet search for 'How many adult cod are in the North Sea', you might get a surprising answer – lots of sources say that there are only 100. This is based on a research paper which used the threshold for 'adulthood' for a different species of cod. Can you find any other examples when the definitions of terms in statistical arguments are poorly defined? What criteria do you use when judging the authenticity of information you get on the internet? Is it possible to remove misleading information from common knowledge?

# 6B Summarizing data

## ■ Frequency distributions

A frequency distribution is a table showing all possible values a variable can take and the number of times the variable takes each of those values (the frequency).

Often the data will be presented in groups (or classes). For continuous data, these groups must cover all possible data values in the range, so there can be no gaps between the classes.

---

**WORKED EXAMPLE 6.5**

The following table shows the times, recorded to the nearest minute, taken by students to complete an IQ test.

| Time, $t$ (minutes) | 1 | 2 | 3 | 4 | 5 | 6 | 7 | 8 | 9 | 10 |
|---|---|---|---|---|---|---|---|---|---|---|
| Frequency | 1 | 2 | 5 | 12 | 16 | 14 | 10 | 4 | 2 | 1 |

a How many students took the test?

b Copy and complete the following table to summarize this information:

| Time, $t$ (minutes) | $0.5 \leqslant t < 3.5$ | $3.5 \leqslant t < 6.5$ | $6.5 \leqslant t < 10.5$ |
|---|---|---|---|
| Frequency | | | |

Sum all the frequencie ······· **a** $1 + 2 + 5 + 12 + 16 + 14 + 10 + 4 + 2 + 1 = 67$

The interval $0.5 \leqslant t < 3.5$ ······· **b**
includes anyone who
took 1, 2 or 3 min

| Time, $t$ (minutes) | $0.5 \leqslant t < 3.5$ | $0.5 \leqslant t < 3.5$ | $0.5 \leqslant t < 3.5$ |
|---|---|---|---|
| Frequency | 8 | 42 | 17 |

The interval $3.5 \leqslant t < 6.5$
includes anyone who
took 4, 5, or 6 min

The interval $6.5 \leqslant t < 10.5$
includes anyone who
took 7, 8, 9 or 10 min

■ Measures of central tendency

It is very useful to have one number that represents the whole data set, and so it makes sense that this measures the centre of the data set. Such a value is known as an average.

You already know that there are three commonly used measures for the average: the mean (often given the symbol $\bar{x}$), the median and the mode. Usually you will just find these from your GDC.

WORKED EXAMPLE 6.6

For the data set 1, 1, 2, 6, 8 find:

a  the mean                    b  the median                    c  the mode.

Enter the list of data in the GDC and then read off the mean ($\bar{x}$)... ····· **a** mean = 3.6

```
1-Variable
x̄      =3.6
Σx     =18
Σx²    =106
xσn    =2.87054001
xσn-1  =3.2093613
n      =5                        ↓
```

... the median (med) and the mode (mod) ····· **b** median = 2
                                              **c** mode = 1

```
1-Variable
minX =1                          ↑
Q1   =1
Med  =2
Q3   =7
maxX =8
Mod  =1                          ↓
```

CONCEPTS – REPRESENTATION

Is it useful to **represent** a set of data using a single number? What information is lost when we do this?

You should also be able to use the formula for the mean of $n$ items:

**KEY POINT 6.6**

- $\bar{x} = \dfrac{\sum x}{n}$

**WORKED EXAMPLE 6.7**

If the mean of 1, 2, 4, 5, $a$, $2a$ is 8, find the value of $a$.

Use the formula $\bar{x} = \dfrac{\sum x}{n}$ .............................. $\dfrac{1+2+4+5+a+2a}{6} = 8$

Solve for $x$ ................................................. $12 + 3a = 48$

$3a = 36$

$a = 12$

The formula can also be applied to frequency tables.

**WORKED EXAMPLE 6.8**

Given that the mean of the following data set is 4.2, find $y$.

| Data value | 1 | 2.5 | 6 | $y$ |
|---|---|---|---|---|
| Frequency | 4 | 5 | 8 | 3 |

You can use the formula $\bar{x} = \dfrac{\sum x}{n}$ again. The data ........ $\dfrac{(1 \times 4) + (2.5 \times 5) + (6 \times 8) + 3y}{20} = 4.2$

value 2.5 occurs 5 times, so it will contribute $2.5 \times 5$
to the total when you are summing the data values

The total frequency is $4 + 5 + 8 + 3 = 20$

$\dfrac{64.5 + 3y}{20} = 4.2$

Solve for $x$ ·················· $64.5 + 3y = 84$

$3y = 19.5$

$y = 6.5$

## Estimation of the mean from grouped data

If you are dealing with grouped data you can find an estimate for the mean by replacing each group with its midpoint.

**WORKED EXAMPLE 6.9**

For the following frequency table, estimate the mean value.

| $x$ | $0 \leqslant x < 4$ | $4 \leqslant x < 10$ | $10 \leqslant x < 20$ |
|---|---|---|---|
| Frequency | 15 | 35 | 50 |

Re-write the table replacing each group ·············· 
with the value at its midpoint

| Midpoint | 2 | 7 | 15 |
|---|---|---|---|
| Frequency | 15 | 35 | 50 |

For example, the midpoint of the
interval $0 \leqslant x < 4$ is $\dfrac{0+4}{2} = 2$

Enter the midpoints and the frequencies in
the GDC and then find the mean ···················· Mean ≈ 10.25

| | List 1 | List 2 | List 3 | List 4 |
|---|---|---|---|---|
| SUB | | | | |
| 1 | 2 | 15 | | |
| 2 | 7 | 35 | | |
| 3 | 15 | 50 | | |
| 4 | | | | |

| 1VAR | 2VAR | REG | | SET |

## Modal class

It is not possible to find the mode from grouped data as we do not have any information on individual data values, so the best we can do is identify the modal group – this is simply the group that has the highest frequency.

### You are the Researcher

This definition is fine if all the groups have the same width, but if that is not the case you have to be careful that a group is not labelled as a modal class just because it is wider than other groups. This leads on to an idea called frequency density which is used in some types of histograms and leads naturally on to the ideas of continuous random variables.

### WORKED EXAMPLE 6.10

Find the modal class in the table below:

| $x$ | $10 \leqslant x < 14$ | $14 \leqslant x < 18$ | $18 \leqslant x < 22$ |
|---|---|---|---|
| **Frequency** | 6 | 14 | 22 |

The modal class is the group with ···················· Modal class is $18 \leqslant x < 22$
the highest frequency

There are many websites which tell you statistics about different countries. What are the benefits of sharing and analysing data from different countries? If you know the average wage in the US, or the average life expectancy in India, what does that tell you about that country?

 ## ■ Quartiles of discrete data

Once ordered from lowest to highest, the lower quartile (often abbreviated to $Q_1$) is the value that is one-quarter of the way through the data set and the upper quartile (or $Q_3$) is the value that is three-quarters of the way through the data set.

So, the quartiles (together with the median, which can be written as $Q_2$), divide the data set into four parts.

**Tip**

There are slightly
different methods
for working out the
quartiles, so you may
find that the values
from your GDC differ
from those obtained by
hand. The values from
your GDC will always
be acceptable though.

### WORKED EXAMPLE 6.11

Find the upper and lower quartiles of 3, 3, 5, 5, 12, 15.

The lower quartile is $Q_1$ and the upper quartile $Q_3$ ············ Lower quartile = 3
Upper quartile = 12

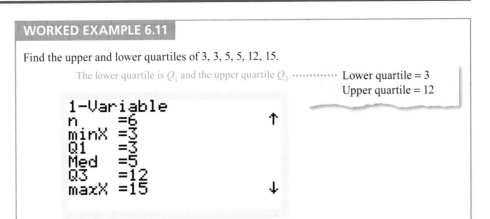

## ◼ Measures of dispersion (interquartile range, standard deviations and variance)

Once you have a measure of the centre of the data set (the average), it is useful to have a measure of how far the rest of the data set is from that central value.

This distance from the average is known as the dispersion (or spread) of the data, and like averages there are several measures of dispersion.

You are already familiar with the **range** as a measure of dispersion. An adjusted version of this is the **interquartile range** (often abbreviated to IQR), which measures the distance between the upper and lower **quartiles** (that is, the width of the central half of the data set).

### KEY POINT 6.7

- $IQR = Q_3 - Q_1$

The **standard deviation** (often denoted by $\sigma$) is another measure of dispersion, which can be thought of as the mean distance of each point from the mean. The **variance** is the square of the standard deviation ($\sigma^2$).

You only need to be able to find the standard deviation using your GDC.

### WORKED EXAMPLE 6.12

In a quality control process, eggs are weighed and the following 10 masses, in grams, are found:
64, 65, 68, 64, 65, 70, 75, 60, 64, 69.
Find the standard deviation, variance and interquartile range of the data.

The standard deviation in this screenshot is $x\sigma_n$, although
it is displayed differently in differently models ············ Standard deviation = 3.98

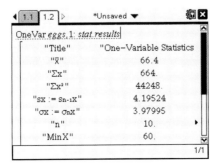

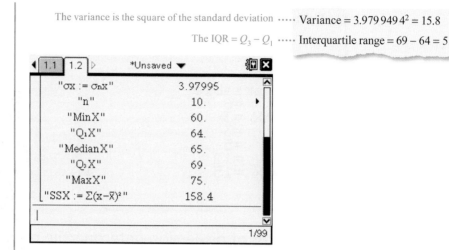

The variance is the square of the standard deviation ····· Variance $= 3.979\,949\,4^2 = 15.8$

The IQR $= Q_3 - Q_1$ ····· Interquartile range $= 69 - 64 = 5$

## CONCEPTS – VALIDITY

Suppose I wanted to compare the spread of lengths of two species of snake. A sample of species A has an IQR of 16 cm and a sample of species B has a standard deviation of 12 cm. Can I say which species has a greater spread? The answer is no, for two reasons:

- I cannot directly compare two different measures of spread.
- I cannot assume that just because the spread of one sample is larger than another that this will also be the case for the two populations.

Both of these points are issues of **validity**. You need to consider the validity of your statistical analysis as well as the sampling process.

## Identifying outliers

One way of deciding when a particularly large or small value qualifies as an outlier is to see whether it is far enough above the upper quartile or below the lower quartile when compared to the general spread of the data values.

The exact definition you need for this course is:

## KEY POINT 6.8

The data value $x$ is an outlier if $x < Q_1 - 1.5(Q_3 - Q_1)$ or $x > Q_3 + 1.5(Q_3 - Q_1)$.

## WORKED EXAMPLE 6.13

A set of data has lower quartile 60 and upper quartile 70. Find the range of values for which data would be flagged as outliers.

$x$ is an outlier if

$x < Q_1 - 1.5(Q_3 - Q_1)$ ... ·····················  $x < 60 - 1.5(70 - 60)$

$x < 45$

... or if

$x > Q_3 + 1.5(Q_3 - Q_1)$ ·····················  $x > 70 + 1.5(70 - 60)$

$x > 85$

$x$ is an outlier if $x < 45$ or $x > 85$

## ■ Effect of constant changes on the original data

If you add a constant to every value in a data set you will cause the average to change by that same value, but this shift will not affect the spread of the data.

> **KEY POINT 6.9**
>
> Adding a constant, $k$, to every data value will:
> - change the mean, median and mode by $k$
> - not change the standard deviation or interquartile range.

> **WORKED EXAMPLE 6.14**
>
> The mean of a set of data is 12 and the standard deviation is 15. If 100 is added to every data value, what would be the new mean and standard deviation?
>
> Adding 100 to every value will increase the mean by 100 ········ New mean $= 12 + 100 = 112$
>
> The standard deviation is unaffected ········ New standard deviation $= 15$

You could also multiply every value in the data set by some factor. Again, this will alter the average by the same factor but this time it will also affect the spread of the data too, since the data set will be stretched.

> **KEY POINT 6.10**
>
> Multiplying every data value by a positive constant, $k$, will:
> - multiply the mean, median and mode by $k$
> - multiply the standard deviation and interquartile range by $k$.

> **WORKED EXAMPLE 6.15**
>
> The median of a set of data is 2.4 and the interquartile range is 3.6. If every data item is halved, find the new median and interquartile range.
>
> Halving every value will halve both ···························· New median $= \dfrac{2.4}{2} = 1.2$
> the median and the interquartile range
>
> New interquartile range $= \dfrac{3.6}{2} = 1.8$

## Exercise 6B

For questions 1 and 2, use the method demonstrated in Worked Example 6.5 to complete the grouped frequency tables summarizing the data in each question. State also the total number of observations.

1   The times taken by members of an athletics team running a sprint race are given below, to the nearest second.

| Time $t$ (sec) | 15 | 16 | 17 | 18 | 19 | 20 | 21 | 22 | 23 |
|---|---|---|---|---|---|---|---|---|---|
| Frequency | 2 | 1 | 7 | 5 | 3 | 0 | 1 | 0 | 1 |

Copy and complete each table.

a

| Time $t$ (sec) | $14.5 \leqslant t < 16.5$ | | $18.5 \leqslant t < 20.5$ | |
|---|---|---|---|---|
| Frequency | | | | |

b

| Time *t* (sec) | | $17.5 \leqslant t < 20.5$ | |
|---|---|---|---|
| Frequency | | | |

2 Some books from two parts of a library are sampled, and the number of pages in each book is recorded, rounding down to the nearest hundred.

| Number of pages *n* | | 200 | 300 | 400 | 500 | 600 | 700 | 800 |
|---|---|---|---|---|---|---|---|---|
| Frequency | Historical fiction | 0 | 2 | 2 | 13 | 12 | 7 | 4 |
| | Romantic fiction | 2 | 7 | 12 | 8 | 5 | 4 | 2 |

Copy and complete the tables, representing the actual number of pages in the sampled books.

a Historical fiction:

| Number of pages *n* | | | |
|---|---|---|---|
| Frequency | 4 | 25 | |

b Romantic fiction:

| Number of pages | | | |
|---|---|---|---|
| Frequency | 9 | | 11 |

For questions 3 and 4, use the method demonstrated in Worked Example 6.6 to find the mean, median and mode for each set of data.

3 a 1, 1, 2, 3, 5, 7, 8

   b 2, 4, 5, 6, 6, 7, 9

4 a 5, 3, 2, 6, 3, 5, 4, 3, 8, 8

   b 12, 24, 25, 24, 33, 36, 24, 55, 55, 24

For questions 5 and 6, use the method demonstrated in Worked Example 6.7 to find the value of *x*.

5 a The mean of 2, 5, 8, $x$, $2x$ is 9

   b The mean of 1, 2, 4, 6, $x$, $x+3$ is 5

6 a The mean of $x$, $x+2$, $3x$, $4x+1$ is 8.5

   b The mean of $x-3$, $2x$, $2x+1$, $3x-2$ is 11.2

For questions 7 and 8, use the method demonstrated in Worked Example 6.8 to find the value of *y*.

7 a Mean = 5.6

| Data value | 2 | 4 | 7 | $y$ |
|---|---|---|---|---|
| Frequency | 6 | 10 | 8 | 6 |

   b Mean = 34.95

| Data value | 27 | 31 | 40 | $y$ |
|---|---|---|---|---|
| Frequency | 5 | 8 | 4 | 3 |

8 a Mean = 5.71

| Data value | $y$ | 5.2 | 6.1 | 7.3 |
|---|---|---|---|---|
| Frequency | 26 | 32 | 18 | 24 |

   b Mean = 2.925

| Data value | $y$ | 2.7 | 3.5 | 3.8 |
|---|---|---|---|---|
| Frequency | 5 | 7 | 6 | 2 |

For questions 9 and 10, use the methods demonstrated in Worked Examples 6.9 and 6.10 to:

i estimate the mean

ii find the modal group.

9 a

| $x$ | $0 \leqslant x < 10$ | $10 \leqslant x < 20$ | $20 \leqslant x < 30$ |
|---|---|---|---|
| Frequency | 15 | 12 | 8 |

b

| $x$ | $5 \leqslant x < 15$ | $15 \leqslant x < 25$ | $25 \leqslant x < 35$ |
|---|---|---|---|
| Frequency | 13 | 14 | 15 |

10 a

| $x$ | $10.5 \leqslant x < 12.5$ | $12.5 \leqslant x < 14.5$ | $14.5 \leqslant x < 16.5$ |
|---|---|---|---|
| Frequency | 17 | 23 | 18 |

b

| $x$ | $3.5 \leqslant x < 7.5$ | $7.5 \leqslant x < 11.5$ | $11.5 \leqslant x < 15.5$ |
|---|---|---|---|
| Frequency | 23 | 31 | 40 |

For questions 11 and 12, use the method demonstrated in Worked Example 6.9 to estimate the mean.

11  a

| $x$ | $0 \leqslant x < 5$ | $5 \leqslant x < 12$ | $12 \leqslant x < 20$ |
|---|---|---|---|
| Frequency | 6 | 8 | 11 |

  b

| $x$ | $2 \leqslant x < 5$ | $5 \leqslant x < 6$ | $6 \leqslant x < 9$ |
|---|---|---|---|
| Frequency | 12 | 9 | 5 |

12  a

| $x$ | $2.5 \leqslant x < 3.5$ | $3.5 \leqslant x < 5.5$ | $5.5 \leqslant x < 10.5$ |
|---|---|---|---|
| Frequency | 23 | 31 | 27 |

  b

| $x$ | $10.5 \leqslant x < 20.5$ | $20.5 \leqslant x < 40.5$ | $40.5 \leqslant x < 45.5$ |
|---|---|---|---|
| Frequency | 14 | 11 | 26 |

For questions 13 and 14, use the method demonstrated in Worked Example 6.13 to identify the range of values which would be flagged as outliers.

13  a  Lower quartile = 50, upper quartile = 80

  b  Lower quartile = 16, upper quartile = 20

14  a  $Q_1 = 6.5$, $Q_3 = 10.5$

  b  $Q_1 = 33.5$, $Q_3 = 45.5$

For questions 15 and 16, use the method demonstrated in Worked Example 6.14 to find the new measures of average and spread.

15  a  The mean of a data set is 34 and the standard deviation is 8. Then 12 is added to every data value.

  b  The mean of a data set is 162 and the standard deviation is 18. Then 100 is subtracted from every data value.

16  a  A data set has median 75 and interquartile range 13. Every data item is decreased by 50.

  b  A data set has median 4.5 and interquartile range 2.1. Every data item is decreased by 4.

For questions 17 and 18, use the method demonstrated in Worked Example 6.15 to find the new measures of average and spread.

17  a  The median of a data set is 36 and the interquartile range is 18. Every data item is multiplied by 10.

  b  The median of a data set is 50 and the interquartile range is 26. Every data item is divided by 10.

18  a  A mean of a data set is 16 and the standard deviation is 6. Every data item is halved.

  b  A mean of a data set is 8.2 and the standard deviation is 0.6. Every data item is multiplied by 5.

19  The times (in minutes) taken by a group of children to complete a puzzle were rounded to the nearest minute and recorded in the frequency table:

| Time (min) | 4 | 5 | 6 | 7 | 8 |
|---|---|---|---|---|---|
| Frequency | 5 | 12 | 12 | 8 | 11 |

  a  How many children were in the group?

  b  How many children took more than 6.5 minutes to complete the puzzle?

  c  Find the mean and standard deviation of the times.

20  The masses of some kittens (in kg) are recorded in the following grouped frequency table:

| Mass (kg) | $0.8 \leqslant m < 1.0$ | $1.0 \leqslant m < 1.2$ | $1.2 \leqslant m < 1.4$ | $1.4 \leqslant m < 1.6$ | $1.6 \leqslant m < 1.8$ |
|---|---|---|---|---|---|
| Frequency | 12 | 19 | 22 | 27 | 19 |

  a  How many kittens have a mass less than 1.2 kg?

  b  State the modal group.

  c  Estimate the mean mass of the kittens.

21  The table shows the times (in seconds) of a group of school pupils in a 100 m race.

| Time (s) | $11.5 \leqslant t < 13.5$ | $13.5 \leqslant t < 15.5$ | $15.5 \leqslant t < 17.5$ | $17.5 \leqslant t < 19.5$ | $19.5 \leqslant t < 21.5$ |
|---|---|---|---|---|---|
| Frequency | 1 | 3 | 12 | 8 | 4 |

  a  How many pupils took part in the race?

  b  State the modal group of the times.

  c  Estimate the mean time. Why is this only an estimate?

**22** The lengths of songs on the latest album by a particular artist, in minutes is recorded as:

3.5, 4, 5, 4.5, 6, 3.5, 4, 5.5, 3, 4.5.

Find:

a the median song length

b the interquartile range of the lengths.

The lengths of songs on an album by another artist have the median of 5.5 and interquartile range of 1.

c Write two comments comparing the lengths of songs by the two artists.

**23** For the set of data:

3, 6, 1, 3, 2, 11, 3, 6, 8, 4, 5

a find the median

b calculate the interquartile range.

c Does the data set contain any outliers? Justify your answer with clear calculations.

**24** The frequency table shows foot lengths of a group of adults, rounded to the nearest cm.

| Length (cm) | 21 | 22 | 23 | 24 | 25 | 26 | 27 | 28 | 29 |
|---|---|---|---|---|---|---|---|---|---|
| Frequency | 4 | 7 | 12 | 8 | 11 | 11 | 8 | 4 | 2 |

a How many adults are in the group?

b What is the modal foot length?

c Find the mean foot length.

d Complete the grouped frequency table:

| Length (cm) | $20.5 \leqslant l < 23.5$ | $23.5 \leqslant l < 26.5$ | |
|---|---|---|---|
| Frequency | | | |

e How many adults have foot length above 23.5 cm?

f Does the modal group include the mode?

**25** The mean of the values 2, 5, $a+2$, $2a$ and $3a+1$ is 17.

a Find the value of $a$.

b Find the standard deviation of the values.

**26** The mean of the values $2x$, $x+1$, $3x$, $4x-3$, $x$ and $x-1$ is 10.5.

a Find the value of $x$.

b Find the variance of the values.

**27** A group of 12 students obtained a mean mark of 67.5 on a test. Another group of 10 students obtained a mean mark of 59.3 on the same test. Find the overall mean mark for all 22 students.

**28** Lucy must sit five papers for an exam. In order to pass her Diploma, she must score an average of at least 60 marks. In the first four papers Lucy scored 72, 55, 63 and 48 marks. How many marks does she need in the final paper in order to pass her Diploma?

**29** A set of data is summarized in a frequency table:

| Data values | 2 | 5 | 7 | $a$ | $a+2$ |
|---|---|---|---|---|---|
| Frequency | 5 | 8 | 13 | 14 | 10 |

a Find the mean of the data in terms of $a$.

b Given that the mean is 8.02, find the value of $a$.

**30** The shoe sizes of a group of children are summarized in a frequency table:

| Shoe size | 4.5 | 5 | 6.5 | 7 | $x$ |
|---|---|---|---|---|---|
| Frequency | 4 | 12 | 8 | 4 | 2 |

   a  Given that the mean shoe size is 5.9, find the value of $x$.

   b  Find the median and quartiles of the shoe sizes.

   c  Hence determine whether the value $x$ is an outlier.

**31** The weekly wages of the employees of a small company are:

£215, £340, £275, £410, £960

   a  Find the mean and standard deviation of the wages.

   b  Would the mean or the median be a better representation of the average wage for this company?

   c  The wages are converted into US dollars, with £1 = \$1.31. Find the new mean and standard deviation.

**32**  a  Find the mean and standard deviation of 1, 3 and 8.

   b  Hence write down the mean and standard deviation of 2003, 2009 and 2024.

**33** The temperature measured in degrees Fahrenheit ($F$) can be converted to degrees Celsius ($C$) using the formula:

$$C = \frac{5}{9}F - \frac{160}{9}$$

The average January temperature in Austin, Texas is 51 °F with the standard deviation of 3.6 °F Find the mean and standard deviation of the temperatures in °C.

**34** The mean distance between stations on a certain trail line is 5.7 miles and the variance of the distances is 4.6 miles$^2$. Given that 1 mile = 1.61 km, find the mean and variance of the distances in kilometres.

**35** A data set has median 25 and interquartile range 14. Every data value is multiplied by –3. Find the new median and interquartile range.

**36** The marks of 11 students on a test are:

35, 35, 37, 39, 42, 42, 42, 43, 45, 46, $m$

where $m > 46$. The marks are all integers.

   a  Find the median and the interquartile range of the marks.

   b  Find the smallest value of $m$ for which this data set has an outlier.

**37** Juan needs to take six different tests as part of his job application. Each test is scored out of the same total number of marks. He needs an average of at least 70% in order to be invited for an interview. After the first four tests his average is 68%. What average score does he need in the final two tests?

**38** The table summarizes the History grades of a group of students.

| Grade | 3 | 4 | 5 | 6 | 7 |
|---|---|---|---|---|---|
| Frequency | 1 | 8 | 15 | $p$ | 4 |

   a  Given that the mean grade was 5.25, find the value of $p$.

   b  Find the standard deviation of the grades.

**39** The frequency table below shows 50 pieces of data with a mean of 5.34. Find the values of $p$ and $q$.

| Value | 3 | 4 | 5 | 6 | 7 | 8 |
|---|---|---|---|---|---|---|
| Frequency | 5 | 10 | 13 | 11 | $p$ | $q$ |

**40** Three positive integers $a$, $b$ and $c$, where $a < b < c$, are such that their median is 26, their mean is 25 and their range is 11. Find the value of $c$.

**41** The numbers $a$, $b$, 1, 2, 3 have got a median of 3 and a mean of 4. Find the largest possible range.

**42** The positive integers 1, 3, 4, 10, 10, 16, $x$, $y$ have a median of 6 and mean of 7. Find all possible values of $x$ and $y$ if $x < y$.

# 6C Presenting data

An alternative to summarizing a data set by numerical values such as an average and a measure of dispersion is to represent the data graphically.

There are several ways this can be done.

## ■ Histograms

A histogram uses the height of each bar to represent the frequency of each group from a frequency table.

While it looks much like a bar chart, the difference is that the horizontal scale is continuous for a histogram, whereas there will be gaps between the bars in a bar chart, which is used for discrete data.

---

**WORKED EXAMPLE 6.16**

Using technology, or otherwise, plot a histogram for the following data.

| $x$ | $10 < x \leqslant 20$ | $20 < x \leqslant 30$ | $30 < x \leqslant 40$ | $40 < x \leqslant 50$ |
|---|---|---|---|---|
| **Frequency** | 16 | 10 | 18 | 14 |

The width of each bar is the width of the corresponding group in the frequency table and the height of each bar is the frequency of that group

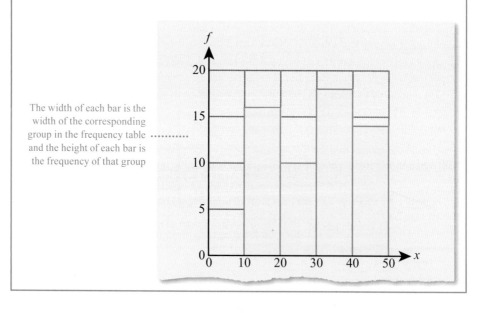

**WORKED EXAMPLE 6.17**

For the histogram on the right estimate the number of items which have $x$ values between 15 and 35.

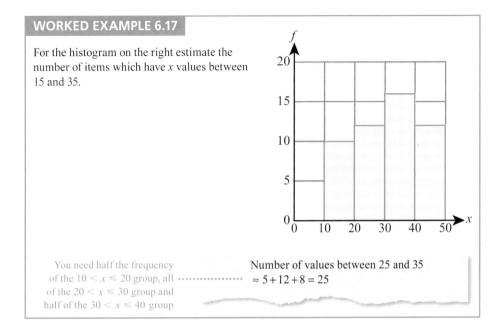

You need half the frequency of the $10 < x \leqslant 20$ group, all of the $20 < x \leqslant 30$ group and half of the $30 < x \leqslant 40$ group

Number of values between 25 and 35
$\approx 5 + 12 + 8 = 25$

## ■ Cumulative frequency graphs

The cumulative frequency is the number of values that are less than or equal to a given point in the data set.

A cumulative frequency graph is a plot of cumulative frequency (on the $y$-axis) against the data values (on the $x$-axis). When working from a grouped frequency table, take the $x$ value at the upper boundary.

**WORKED EXAMPLE 6.18**

Construct a cumulative frequency diagram for the following data.

| $x$ | $10 < x \leqslant 20$ | $20 < x \leqslant 30$ | $30 < x \leqslant 40$ | $40 < x \leqslant 50$ |
|---|---|---|---|---|
| **Frequency** | 10 | 12 | 16 | 12 |

The $x$ values in the cumulative frequency table are the upper boundaries of each group

| $x$ | 20 | 30 | 40 | 50 |
|---|---|---|---|---|
| **Cumulative frequency** | 10 | 22 | 38 | 50 |

Plot the cumulative frequency values against the $x$ values in the cumulative frequency table

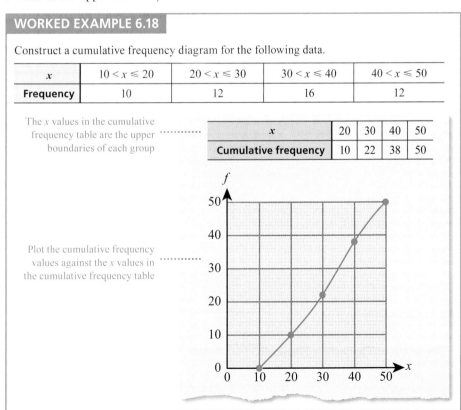

## Using cumulative frequency graphs to find medians, quartiles, percentiles, range and interquartile range

One of the main purposes of a cumulative frequency graph is to estimate the median and interquartile range.

The median and quartiles are particular examples of percentiles, which are points that are a particular percentage of the way through the data set. The median is the 50th percentile, and the lower quartile the 25th percentile.

**WORKED EXAMPLE 6.19**

Use the cumulative frequency graph on the right to estimate:

a　the median

b　the range

c　the interquartile range

d　the 90th percentile.

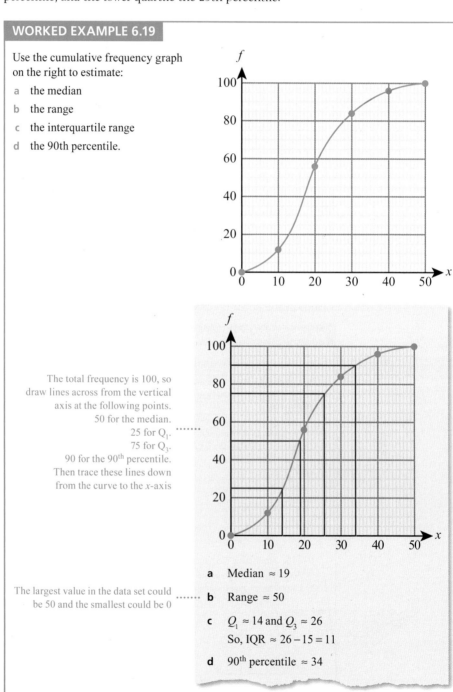

The total frequency is 100, so draw lines across from the vertical axis at the following points.
50 for the median.
25 for $Q_1$.
75 for $Q_3$.
90 for the 90$^{th}$ percentile.
Then trace these lines down from the curve to the *x*-axis

The largest value in the data set could be 50 and the smallest could be 0

a　Median ≈ 19

b　Range ≈ 50

c　$Q_1$ ≈ 14 and $Q_3$ ≈ 26
　　So, IQR ≈ 26 − 15 = 11

d　90$^{th}$ percentile ≈ 34

## ■ Box-and-whisker diagrams

A box-and-whisker diagram represents five key pieces of information about a data set:

- ■ the smallest value
- ■ the lower quartile
- ■ the median
- ■ the upper quartile
- ■ the largest value.

This excludes outliers – they are represented with crosses.

---

**WORKED EXAMPLE 6.20**

Construct a box-and-whisker diagram for the following data:

3, 6, 7, 7, 7, 9, 9, 11, 15, 20.

Use the GDC to find $Q_1$, $Q_3$ and the median

```
1-Variable
n    =10          ↑
minX =3
Q1   =7
Med  =8
Q3   =11
maxX =20          ↓
```

Check to see if there are any outliers ·······································  $x$ is an outlier if

$$x < 7 - 1.5(11 - 7)$$

$$x < 1$$
$$\text{or if}$$

$$x > 11 + 1.5(11 - 7)$$
$$x > 17$$

So, $x = 20$ is an outlier

The minimum value is 3. As 20 is an outlier, you need to mark it with a cross ··············

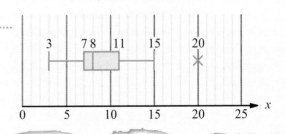

---

  You can use normal box-and-whisker plots to check whether data might plausibly be normally distributed. You will do this in Chapter 8.

## CONCEPTS – REPRESENTATION

Is a picture worth a thousand words? Do you learn more about data by **representing** them as a diagram or summary statistics? Which is easier to manipulate to persuade your intended audience? Are any statistics entirely unbiased?

## Exercise 6C

For questions 1 and 2, use technology or the method demonstrated in Worked Example 16.16 to present the data in a histogram.

1  a

| $x$ | $5 \leqslant x < 10$ | $10 \leqslant x < 15$ | $15 \leqslant x < 20$ | $20 \leqslant x < 25$ |
|---|---|---|---|---|
| **Frequency** | 5 | 10 | 12 | 8 |

b

| $x$ | $0 \leqslant x < 20$ | $20 \leqslant x < 40$ | $40 \leqslant x < 60$ | $60 \leqslant x < 80$ |
|---|---|---|---|---|
| **Frequency** | 18 | 15 | 6 | 13 |

2  a

| $h$ | $1.2 \leqslant h < 2.5$ | $2.5 \leqslant h < 3.8$ | $3.8 \leqslant h < 5.1$ | $5.1 \leqslant h < 6.4$ | $6.4 \leqslant h < 7.7$ |
|---|---|---|---|---|---|
| **Frequency** | 5 | 0 | 12 | 10 | 4 |

b

| $t$ | $30 \leqslant t < 36$ | $36 \leqslant t < 42$ | $42 \leqslant t < 48$ | $48 \leqslant t < 54$ | $54 \leqslant t < 60$ |
|---|---|---|---|---|---|
| **Frequency** | 25 | 31 | 40 | 62 | 35 |

In each of the questions 3 and 4, a data set is presented as a histogram. Use the method demonstrated in Worked Example 6.17 to estimate the number of data items in the given range.

3

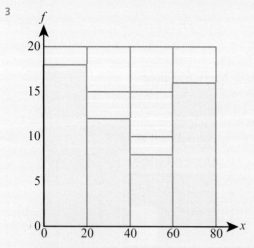

a  $20 \leqslant x < 50$

b  $30 \leqslant x < 80$

4

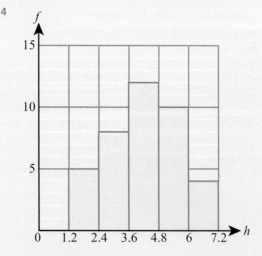

a  $2.7 \leqslant h < 5.4$

b  $4.0 \leqslant h < 4.4$

For questions 5 and 6, use technology or the method demonstrated in Worked Example 6.18 to construct a cumulative frequency diagram for the data in the table.

5  a

| $x$ | $5 \leqslant x \leqslant 10$ | $10 \leqslant x < 15$ | $15 \leqslant x < 20$ | $20 \leqslant x < 25$ |
|---|---|---|---|---|
| **Frequency** | 5 | 10 | 12 | 8 |

b

| $x$ | $0 \leqslant x < 20$ | $20 \leqslant x < 40$ | $40 \leqslant x < 60$ | $60 \leqslant x < 80$ |
|---|---|---|---|---|
| **Frequency** | 18 | 12 | 8 | 16 |

6 a

| h | $1.2 \leqslant h < 2.5$ | $2.5 \leqslant h < 3.8$ | $3.8 \leqslant h < 5.1$ | $5.1 \leqslant h < 6.4$ | $6.4 \leqslant h < 7.7$ |
|---|---|---|---|---|---|
| **Frequency** | 10 | 8 | 4 | 0 | 5 |

b

| t | $30 \leqslant t < 36$ | $36 \leqslant t < 42$ | $42 \leqslant t < 48$ | $48 \leqslant t < 54$ | $54 \leqslant t < 60$ |
|---|---|---|---|---|---|
| **Frequency** | 20 | 0 | 12 | 12 | 25 |

For question 7, use the method demonstrated in Worked Example 6.19 to find:

    i   the median

    iii  the interquartile range

    ii  the range

    iv  the 90th percentile.

7  a

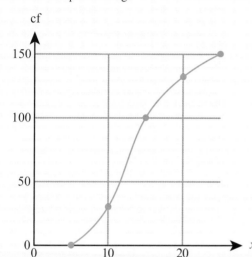

  b

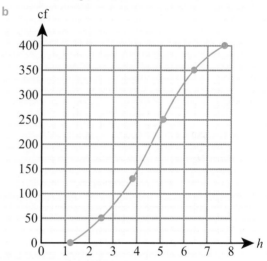

For questions 8 and 9, use the method demonstrated in Worked Example 6.20 to construct a box-and-whisker diagram for the given data.

8  a  3, 5, 5, 6, 8, 9, 9

    b  11, 13, 15, 15, 16, 18, 19, 21, 24, 24

9  a  11, 13, 15, 15, 16, 18, 19, 21, 25, 32

    b  4 , 13, 14, 16, 17, 18, 18, 19, 19

**10** The length of fossils found at a geological dig is summarized in the following table.

| Length (cm) | $0 < l \leqslant 4$ | $4 < l \leqslant 8$ | $8 < l \leqslant 12$ | $12 < l \leqslant 16$ | $16 < l \leqslant 20$ |
|---|---|---|---|---|---|
| **Frequency** | 10 | 15 | 15 | 20 | 20 |

  a  Using technology, draw a histogram to represent this set of data.

  b  Estimate the number of fossils which were longer than 15 cm.

**11** The masses of some apples are summarized in the following table.

| Mass (g) | $80 \leqslant m < 100$ | $100 \leqslant m < 120$ | $120 \leqslant m < 140$ | $140 \leqslant m < 160$ | $160 \leqslant m < 180$ |
|---|---|---|---|---|---|
| **Frequency** | 16 | 14 | 25 | 32 | 18 |

  a  How many apples were weighed?

  b  Using technology, draw a histogram to represent the data.

  c  What percentage of apples had masses between 110 g and 150 g?

**12** The maximum daily temperatures in a particular town, over a period of one year, are summarized in the table.

| Temperature (°C) | $5 \leqslant t < 10$ | $10 \leqslant t < 15$ | $15 \leqslant t < 20$ | $20 \leqslant t < 25$ | $25 \leqslant t < 30$ |
|---|---|---|---|---|---|
| **Frequency** | 43 | 98 | 126 | 67 | 21 |

  a  Use technology to draw a cumulative frequency graph for this data.

  b  Use your graph to estimate:

    i  the median temperature

    ii the interquartile range.

**13** The histogram shows the times taken by a group of pupils to complete a writing task.

a How many pupils completed the task?

b Estimate the percentage of pupils who took more than 22 minutes.

c Complete the grouped frequency table.

| Time (min) | $5 \leqslant t < 10$ | | | |
|---|---|---|---|---|
| Frequency | 7 | | | |

d Hence estimate the mean time taken to complete the task.

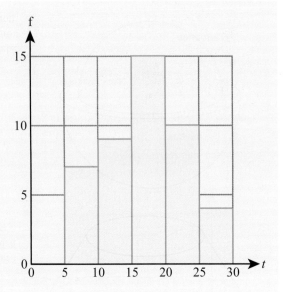

**14** The table summarizes heights of a sample of flowers.

| Height (cm) | $20 \leqslant h < 25$ | $25 \leqslant h < 30$ | $30 \leqslant h < 35$ | $35 \leqslant h < 40$ | $40 \leqslant h < 45$ |
|---|---|---|---|---|---|
| Frequency | 16 | 21 | 34 | 18 | 6 |

a Use technology to produce a cumulative frequency graph for the data.

b Use your graph to estimate the median and the interquartile range of the heights.

c Hence draw a box plot to represent the data.

**15** The number of candidates taking Mathematics SL at a sample of schools is recorded in the grouped frequency table.

| Number of candidates | Number of schools |
|---|---|
| $10 < n \leqslant 30$ | 34 |
| $30 < n \leqslant 50$ | 51 |
| $50 < n \leqslant 70$ | 36 |
| $70 < n \leqslant 90$ | 18 |
| $90 < n \leqslant 110$ | 7 |

a Draw a cumulative frequency graph for the data.

b Estimate the number of schools with more than 60 Mathematics SL candidates.

c Find the median and interquartile range of the data.

d Hence draw a box plot to show the number of Mathematics SL candidates.

The box plot below shows the number of candidates taking History SL at the same group of schools.

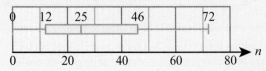

e Write two comments comparing the number of candidates taking Mathematics SL and History SL at this group of schools.

**16**  The histogram summarizes heights of children in a
school.

Use the histogram to estimate the mean height, correct
to the nearest centimetre.

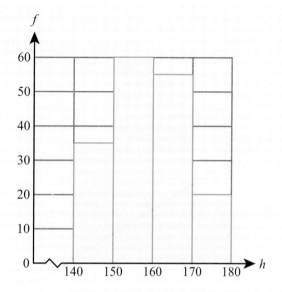

**17**  The cumulative frequency graph summarizes the
examination scores at a particular school.

a  How many students scored below 73?

b  The pass mark for the examination was 55. What
percentage of students passed?

c  Find the 60th percentile of the scores.

d  Estimate the median and quartiles of the scores.
Hence produce a box plot for the data. (You may
assume that there are no outliers.)

e  The box plot below summarizes the scores on the
same exam from a different school. Make two
comments comparing the scores at the two schools.

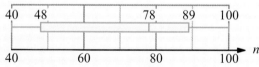

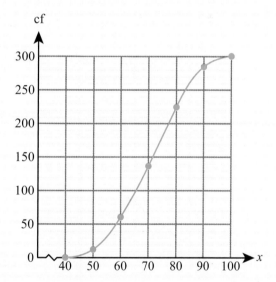

**18**  A group of children recorded the number of siblings
they have:
0, 2, 1, 3, 2, 1, 1, 0, 2, 3, 2, 2, 1, 4, 7.

a  Find the median and quartiles of the data.

b  Calculate the interquartile range.

c  Determine whether this data set contains any outliers.

d  Draw the box plot for the data.

**19** Some information about a data set is summarized in this table.

| Smallest value | 7 |
| --- | --- |
| Second smallest value | 11 |
| Lower quartile | 20 |
| Median | 25 |
| Upper quartile | 28 |
| Second largest value | 32 |
| Largest value | 38 |

a  Determine whether the data set includes any outliers.

b  Draw a box plot to represent the data.

**20** The exam marks of a group of students are recorded in a cumulative frequency table.

| **Mark** | $\leqslant 20$ | $\leqslant 40$ | $\leqslant 60$ | $\leqslant 80$ | $\leqslant 100$ |
| --- | --- | --- | --- | --- | --- |
| **Cumulative frequency** | 6 | 45 | 125 | 196 | 247 |

a  Represent the data in the histogram.

b  Estimate the mean of the marks.

**21** Match each histogram with the cumulative frequency diagram drawn from the same data.

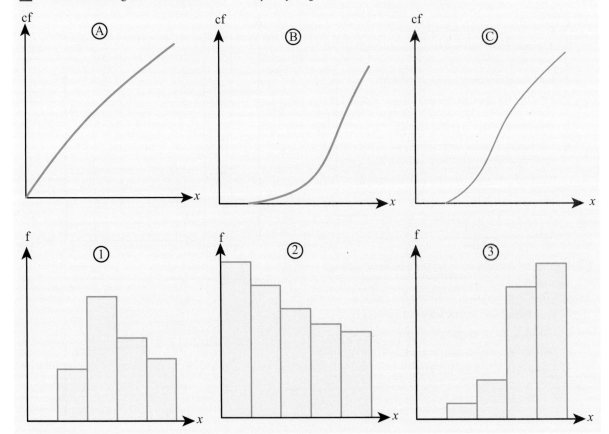

# 6D Correlation and regression

So far, we have only looked at one variable at a time, but sometimes we want to investigate whether there is a relationship between two variables, such as a person's height and mass. These types of data are known as bivariate.

## ◼ Scatter diagrams and lines of best fit

A good starting point for investigating a possible relationship between two variables is to plot one variable against the other on a scatter graph.

If it looks like there is a linear relationship, then we can plot a line of best fit.

---

**KEY POINT 6.11**

The line of best fit will always pass through the mean point $(\bar{x}, \bar{y})$.

---

**WORKED EXAMPLE 6.21**

For the data below:

a find the mean point

b create a scatter diagram, showing the mean point

c draw a line of best fit by eye.

| $x$ | 1 | 2 | 3 | 4 | 4 | 5 | 6 | 6 | 7 | 9 |
|---|---|---|---|---|---|---|---|---|---|---|
| $y$ | 0 | 2 | 4 | 4 | 5 | 6 | 7 | 10 | 9 | 10 |

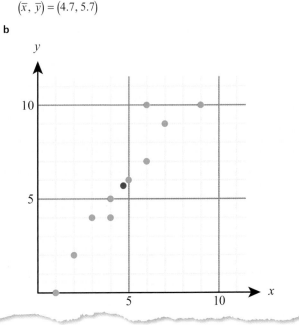

Use the GDC to find $\bar{x}$ and $\bar{y}$: ····· **a** From GDC,

$$(\bar{x}, \bar{y}) = (4.7, 5.7)$$

```
2-Variable
 Σx  =47
 Σx² =273
 xσn =2.28254244
 xσn-1=2.40601099
 n   =10
 ȳ   =5.7
```

**b**

When fitting the line of best fit, ...... **c**
make sure it passes through
$(\overline{x}, \overline{y}) = (4.7, 5.7)$

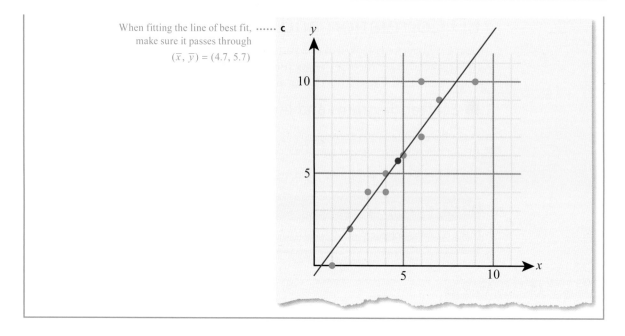

## ◼ Linear correlation of bivariate data

The extent to which two variables are related is called correlation. In this course we will focus on linear correlation:

Positive correlation           Negative correlation           No correlation

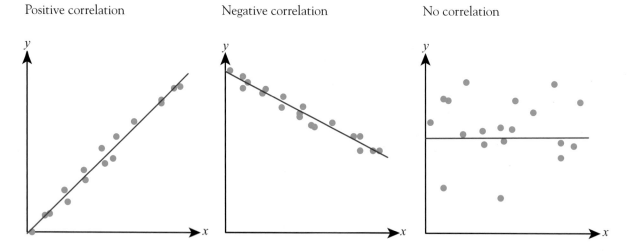

The closer the points are to the line of best fit, the stronger the correlation.

**WORKED EXAMPLE 6.22**

Describe the linear correlation in the scatter graph below.

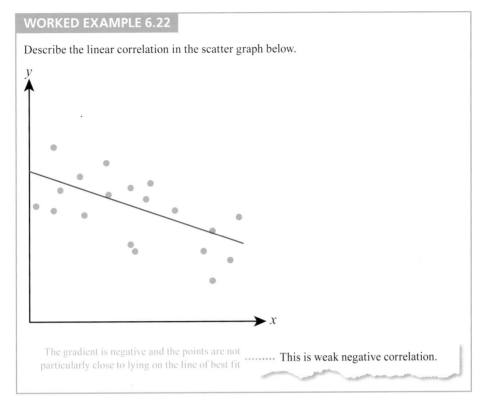

The gradient is negative and the points are not particularly close to lying on the line of best fit ......... This is weak negative correlation.

## Pearson's product-moment correlation coefficient, *r*

Rather than just describing the correlation as 'strong positive' or 'weak negative', etc, there is a numerical value called the Pearson's product-moment correlation coefficient (given the letter *r*) that can be used to represent the strength of linear correlation. You need to be able to find *r* using your calculator.

**Tip**

Just because *r* is close to 1 or to −1, this strong correlation does not necessarily mean that one variable causes a change in the other. The correlation might just be a coincidence or be due to a third hidden variable.

**KEY POINT 6.12**

The Pearson's product-moment correlation coefficient, *r*, is such that $-1 \leqslant r \leqslant 1$.
● $r \approx 1$ means strong positive linear correlation
● $r \approx 0$ means no linear correlation
● $r \approx -1$ means strong negative linear correlation

While it seems clear that a value of *r* such as 0.95 gives good evidence of positive linear correlation, it might not be so clear whether a value such as 0.55 does.

In fact, this depends on how many data points we have – the fewer points there are, the nearer to 1 the value of *r* needs to be to give good evidence of positive linear correlation (and the nearer to −1 for negative correlation).

The 'cut-off' values at which the value of *r* provides significant evidence of positive (or negative) correlation are known as critical values.

**WORKED EXAMPLE 6.23**

a   For the data below, use technology to find the value of Pearson's product-moment correlation coefficient.

| $x$ | 4 | 5 | 4 | 8 | 9 | 10 | 11 | 8 | 6 | 0 |
|---|---|---|---|---|---|---|---|---|---|---|
| $y$ | 9 | 4 | 5 | 9 | 8 | 1 | 2 | 5 | 8 | 11 |

b   Interpret qualitatively what this suggests.

A table suggests that the critical value of $r$ when there are 10 pieces of data is 0.576.

c   What does this mean for your data?

Use your GDC to find $r$: ······ a   From GDC,
$r = -0.636$

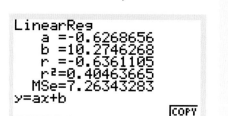

```
LinearReg
  a =-0.6268656
  b =10.2746268
  r =-0.6361105
  r²=0.40463665
  MSe=7.26343283
y=ax+b
                      COPY
```

b   There is weak negative correlation, that is, a general trend that as $x$ increases, $y$ decreases.

The critical value given will always be positive, but this same value can be used ······ c   $r < -0.576$, so there is significant
to check for negative correlation                       evidence of negative correlation.

 ■ Equation of the regression line of $y$ on $x$

Once there is evidence of linear correlation, you would like to know what the precise nature of the linear relationship is between the variables – that is, you would like to know the equation of the line of best fit. This is known as the regression line.

**WORKED EXAMPLE 6.24**

For the data below, find the equation of the regression line using technology.

| $x$ | 1 | 2 | 3 | 4 | 4 | 5 | 6 | 6 | 7 | 9 |
|---|---|---|---|---|---|---|---|---|---|---|
| $y$ | 0 | 2 | 4 | 4 | 5 | 6 | 7 | 10 | 9 | 10 |

The GDC will give you the values of the coefficients $a$ and $b$ in the ·············· straight-line equation $y = ax + b$

Regression line is
$y = 1.33x - 0.534$

```
LinearReg
  a =1.32629558
  b =-0.5335892
  r =0.94742821
  r²=0.89762022
  MSe=1.30662188
y=ax+b
                      COPY
```

**TOOLKIT:** Problem Solving

For each of the following data sets, find **a** the mean of $x$ and $y$, **b** the standard deviation of $x$ and $y$, **c** the correlation coefficient, and **d** the equation of the regression line. What does this suggest about the data sets?

| | I | | II | | III | | IV |
|---|---|---|---|---|---|---|---|
| $x$ | $y$ | $x$ | $y$ | $x$ | $y$ | $x$ | $y$ |
| 10.0 | 8.04 | 10.0 | 9.14 | 10.0 | 7.46 | 8.0 | 6.58 |
| 8.0 | 6.95 | 8.0 | 8.14 | 8.0 | 6.77 | 8.0 | 5.76 |
| 13.0 | 7.58 | 13.0 | 8.74 | 13.0 | 12.74 | 8.0 | 7.71 |
| 9.0 | 8.81 | 9.0 | 8.77 | 9.0 | 7.11 | 8.0 | 8.84 |
| 11.0 | 8.33 | 11.0 | 9.26 | 11.0 | 7.81 | 8.0 | 8.47 |
| 14.0 | 9.96 | 14.0 | 8.10 | 14.0 | 8.84 | 8.0 | 7.04 |
| 6.0 | 7.24 | 6.0 | 6.13 | 6.0 | 6.08 | 8.0 | 5.25 |
| 4.0 | 4.26 | 4.0 | 3.10 | 4.0 | 5.39 | 19.0 | 12.50 |
| 12.0 | 10.84 | 12.0 | 9.13 | 12.0 | 8.15 | 8.0 | 5.56 |
| 7.0 | 4.82 | 7.0 | 7.26 | 7.0 | 6.42 | 8.0 | 7.91 |
| 5.0 | 5.68 | 5.0 | 4.74 | 5.0 | 5.73 | 8.0 | 6.89 |

Now use technology to create scatterplots of each data set.

These data sets are called Anscombe's quartet. They highlight the importance of visualizing your data.

## ■ Use of the regression line for prediction purposes

Once you have found the equation of the regression line, you can use it to predict the $y$ value for a given $x$ value. If the $x$ value you use in the equation is within the range of the data set (interpolation) then the prediction can be considered reliable, but if the $y$ value is beyond the data set (extrapolation), the prediction should be treated with caution as there is no guarantee that the relationship continues beyond the observed values.

**WORKED EXAMPLE 6.25**

Based on data with $x$ values between 10 and 20 and a correlation coefficient of 0.93, a regression line is formed:

$y = 10.2x - 5.3$.

a   Use this line to predict the value of $y$ when:

   i   $x = 3$

   ii   $x = 13$.

b   Comment on the reliability of your answers in **a**.

Substitute each value of $x$ into ......... **a  i**  $y = 10.2(3) - 5.3 = 25.3$
the regression line

   **ii**  $y = 10.2(13) - 5.3 = 127$

**b**  The prediction when $x = 13$ can be considered reliable as 13 is within the range of known $x$ values.

The prediction when $x = 3$, however, cannot be considered reliable, as the relationship has had to be extrapolated significantly beyond the range of given data to make this prediction.

**CONCEPTS – APPROXIMATION AND PATTERNS**

When you are using a regression line you should not assume that your prediction is totally accurate. There will always be natural variation around the line of best fit. However, even though it is only giving us an **approximate** prediction, it can help us to see the underlying **patterns** in the data.

## ■ Interpreting the meaning of regression parameters

The parameters $a$ and $b$ in the regression line $y = ax + b$ are just the gradient and $y$-intercept respectively, which means they can be easily interpreted in context.

**WORKED EXAMPLE 6.26**

Data for the number of flu cases ($f$, in tens of thousands) in a country against the amount spent on promoting a vaccination program ($p$, in millions) are given by:

$f = 12.4 - 2.1p$.

Interpret, in context, the meaning of

a   12.4

b   2.1

in this equation.

12.4 is the $y$-intercept of the line, ····· **a** If no money were spent on promoting a vaccination program
which corresponds to the value of $y$ when $x = 0$.          then the number of flu cases would be
Remember here that $f$ is in                                      $12.4 \times 10^4 = 124\,000$.
tens of thousands

The gradient of the line is –2.1, which ····· **b** For every extra million spent on promoting a vaccination
means that for an increase of 1 unit in $x$          program, the number of flu cases decreases by
there is a decrease of 2.1 units in $y$              $2.1 \times 10^5 = 210\,000$.

**CONCEPTS – PATTERNS**

Especially when there are large amounts of data and many variables, it can be difficult to spot **patterns** visually. Correlation provides an objective way to assess whether there are any underlying linear trends. One of the most useful parts of regression turns out to be deciding whether the underlying regression has a non-zero gradient, as this tells you whether the variables you are considering are actually related in some way.

## ■ Piecewise linear models

In some cases, a scatter graph might show that the same regression line is not a good fit for all the data, but that two (or more) different regression lines do fit very well with different parts of the data set. A model like this is known as piecewise linear.

**WORKED EXAMPLE 6.27**

a   Use technology to plot a scatter diagram of the data below.

| x | 0 | 1 | 3 | 3 | 4 | 5 | 6 | 6 | 7 | 8 | 9 | 9 |
|---|---|---|---|---|---|---|---|---|---|---|---|---|
| y | 1 | 2 | 4 | 3 | 4.5 | 5 | 10 | 11 | 12 | 15 | 17 | 18 |

b   Use your scatter diagram to separate the data into two distinct linear regions. Form a piecewise linear model to fit to the data.

c   Use your model to estimate the value of *y* when *x* is 7.5.

Rather than fit a single line of best fit to the data, ········ **a**
it is clear that there are two distinct regions, so
add a line of best fit separately for each of these

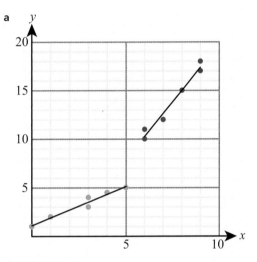

Use your GDC to find the equation ········ **b** For $0 \leqslant x \leqslant 5$,
of each regression line:   $y = 0.808x + 1.10$

```
LinearReg
  a =0.8076923
  b =1.09615384
  r =0.97582104
  r²=0.95222672
 MSe=0.14182692
y=ax+b
              [COPY]
```

```
LinearReg
  a =2.36842105
  b =-3.9298245
  r =0.98582174
  r²=0.9718445
 MSe=0.38596491
y=ax+b
              [COPY]
```

For $6 \leqslant x \leqslant 9$,
$y = 2.37x - 3.93$

For $x = 7.5$, use the second regression line. ········ **c** When $x = 7.5$,
$$y = 2.37(7.5) - 3.93 = 13.8$$

## Exercise 6D

For questions 1 to 3, use the method demonstrated in Worked Example 6.21 to create a scatter diagram, plot the mean point and add a line of best fit by eye.

1 a

| $x$ | 3 | 4 | 6 | 7 | 9 | 10 | 12 | 13 |
|---|---|---|---|---|---|---|---|---|
| $y$ | 2 | 3 | 5 | 4 | 7 | 6 | 5 | 7 |

b

| $x$ | 1 | 2 | 2 | 3 | 5 | 7 | 8 | 8 |
|---|---|---|---|---|---|---|---|---|
| $y$ | −3 | 0 | 1 | −2 | −1 | 2 | 4 | 3 |

2 a

| $x$ | 1 | 1 | 2 | 3 | 5 | 6 | 6 | 8 | 9 |
|---|---|---|---|---|---|---|---|---|---|
| $y$ | 6 | 7 | 5 | 4 | 5 | 4 | 2 | 1 | 2 |

b

| $x$ | 14 | 13 | 15 | 19 | 19 | 22 | 23 | 24 | 22 |
|---|---|---|---|---|---|---|---|---|---|
| $y$ | 14 | 12 | 12 | 11 | 9 | 9 | 8 | 6 | 5 |

3 a

| $x$ | 10 | 12 | 12 | 15 | 16 | 20 | 22 | 25 | 26 | 28 | 31 | 34 |
|---|---|---|---|---|---|---|---|---|---|---|---|---|
| $y$ | 10 | 10 | 13 | 10 | 16 | 12 | 19 | 15 | 20 | 17 | 21 | 21 |

b

| $x$ | 10 | 12 | 12 | 15 | 16 | 20 | 22 | 25 | 26 | 28 | 31 | 34 |
|---|---|---|---|---|---|---|---|---|---|---|---|---|
| $y$ | 21 | 21 | 17 | 20 | 15 | 19 | 12 | 16 | 10 | 13 | 10 | 10 |

For questions 4 to 7, use the method demonstrated in Worked Example 6.22 to describe the linear correlation in the scatter graph.

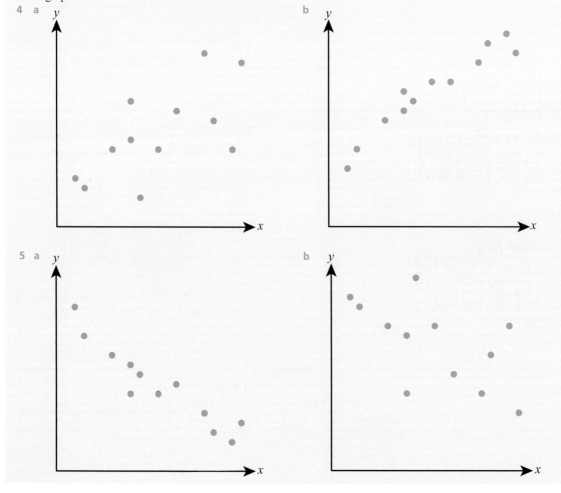

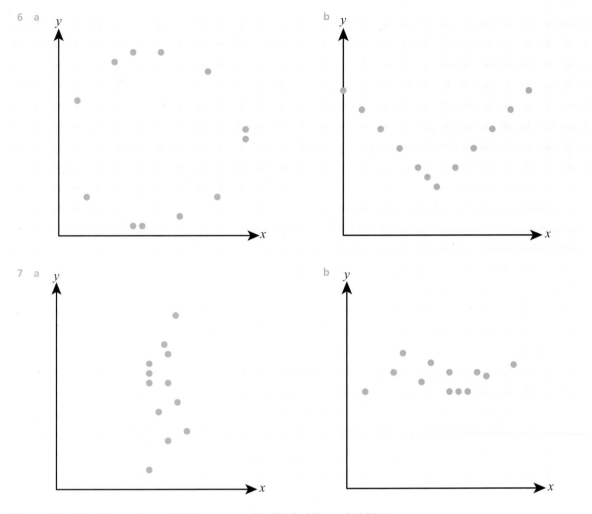

For questions 8 and 9, use the method demonstrated in Worked Example 6.23 to:

    i   find the value of the Pearson's product-moment correlation coefficient

    ii  interpret qualitatively what the correlation coefficient suggests

    iii use the appropriate critical value from this table to decide whether the correlation is significant.

| Number of data pairs | 6 | 7 | 8 | 9 | 10 |
|---|---|---|---|---|---|
| Critical value | 0.621 | 0.584 | 0.549 | 0.521 | 0.497 |

8  a

| $x$ | 3 | 4 | 6 | 7 | 9 | 10 | 12 | 13 |
|---|---|---|---|---|---|---|---|---|
| $y$ | 2 | 3 | 5 | 4 | 7 | 6 | 5 | 7 |

    b

| $x$ | 1 | 1 | 2 | 3 | 5 | 6 | 6 | 8 | 9 |
|---|---|---|---|---|---|---|---|---|---|
| $y$ | 6 | 7 | 5 | 4 | 5 | 4 | 2 | 1 | 2 |

9  a

| $x$ | 3 | 7 | 2 | 5 | 4 | 9 |
|---|---|---|---|---|---|---|
| $y$ | 6 | 10 | 7 | 8 | 9 | 8 |

    b

| $x$ | 14 | 13 | 15 | 19 | 19 | 22 | 23 |
|---|---|---|---|---|---|---|---|
| $y$ | 14 | 10 | 12 | 11 | 9 | 9 | 10 |

For questions 10 and 11, use the method demonstrated in Worked Example 6.24 to find the equation of the regression line.

10  a

| $x$ | 3 | 4 | 6 | 7 | 9 | 10 | 12 | 13 |
|---|---|---|---|---|---|---|---|---|
| $y$ | 2 | 3 | 5 | 4 | 7 | 6 | 5 | 7 |

   b

| $x$ | 1 | 2 | 2 | 3 | 5 | 7 | 8 | 8 |
|---|---|---|---|---|---|---|---|---|
| $y$ | −3 | 0 | 1 | −2 | −1 | 2 | 4 | 3 |

11  a

| $x$ | 1 | 1 | 2 | 3 | 5 | 6 | 6 | 8 | 9 |
|---|---|---|---|---|---|---|---|---|---|
| $y$ | 6 | 7 | 5 | 4 | 5 | 4 | 2 | 1 | 2 |

   b

| $x$ | 14 | 13 | 15 | 19 | 19 | 22 | 23 | 24 | 22 |
|---|---|---|---|---|---|---|---|---|---|
| $y$ | 14 | 12 | 12 | 11 | 9 | 9 | 8 | 6 | 5 |

For questions 12 and 13, you are given the minimum and maximum data values, the correlation coefficient and the equation of the regression line. Use the method demonstrated in Worked Example 6.25 to:

   i    predict the value of $y$ for the given value of $x$

   ii   comment on the reliability of your prediction.

12

| | Minimum value | Maximum value | $r$ | Regression line | $x$ |
|---|---|---|---|---|---|
| a | 10 | 20 | 0.973 | $y = 1.62x − 7.31$ | 18 |
| b | 1 | 7 | −0.875 | $y = −0.625x + 1.37$ | 9 |

13

| | Minimum value | Maximum value | $r$ | Regression line | $x$ |
|---|---|---|---|---|---|
| a | 20 | 50 | 0.154 | $y = 2.71x + 0.325$ | 27 |
| b | 12 | 27 | −0.054 | $y = 4.12x − 2.75$ | 22 |

14  The table shows the data for height and arm length for a sample of 10 15-year-olds.

| Height (cm) | 154 | 148 | 151 | 165 | 154 | 147 | 172 | 156 | 168 | 152 |
|---|---|---|---|---|---|---|---|---|---|---|
| Arm length (cm) | 65 | 63 | 58 | 71 | 59 | 65 | 75 | 62 | 61 | 61 |

   a  Plot the data on a scatter graph.

   b  Describe the correlation between the height and arm length.

   c  Find the mean height and mean arm length. Add the corresponding point to your graph.

   d  Draw the line of best fit by eye.

   e  Use your line of best fit to predict the arm length of a 15-year-old whose height is 150 cm.

   f  Comment on whether or not it would be appropriate to use your line to predict the arm length for

      i    a 15-year-old whose height is 150 cm

      ii   a 15-year-old whose height is 192 cm

      iii  a 72-year-old whose height is 150 cm.

15  The table shows the distance from the nearest train station and the average house price for seven villages.

| Distance (km) | 0.8 | 1.2 | 2.5 | 3.7 | 4.1 | 5.5 | 7.4 |
|---|---|---|---|---|---|---|---|
| Average house price (000s $) | 240 | 185 | 220 | 196 | 187 | 156 | 162 |

   a  Plot the data on a scatter graph.

   b  Describe the correlation between the distance and the average house price.

   c  Find the mean distance and the mean house price.

   d  Draw a line of best fit on your graph.

   e  A new village is to be developed 6.7 km from a train station. Predict the average house price in the new village.

**16** Match the scatter diagrams with the following values of $r$:

A: $r = 0.98$  B: $r = -0.34$  C: $r = -0.93$  D: $r = 0.58$

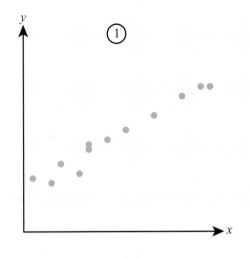

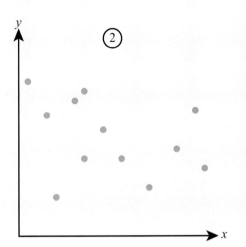

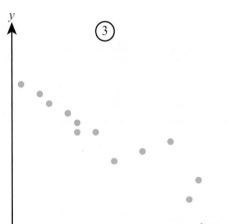

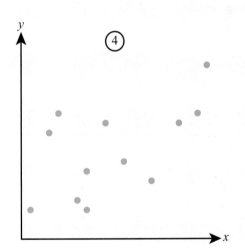

**17** A group of 11 students took a Mathematics test and a chemistry test. Their marks on the two tests are shown in the table.

| Mathematics test mark | 63 | 81 | 57 | 72 | 93 | 47 | 61 | 82 | 65 | 83 | 71 |
|---|---|---|---|---|---|---|---|---|---|---|---|
| Chemistry test mark | 40 | 57 | 46 | 51 | 60 | 37 | 52 | 48 | 33 | 48 | 51 |

a  Find Pearson's product-moment correlation coefficient between the marks on the two tests.

b  Find the equation of the regression line of $y$ on $x$, where $x$ is the mark on the mathematics test and $y$ is the mark on the chemistry test.

c  Another student scored 68 marks on the mathematics test but missed the chemistry test. Use your regression line to estimate the mark they would have got on the chemistry test.

d  Asher says: 'The strong correlation proves that by getting better at mathematics you would also get better at chemistry'. Do you agree with this statement?

**18** A small company records the amount spent on advertising and the profit they made the following month. The results are summarized in the table:

| Amount ($x$) | 120 | 90 | 65 | 150 | 80 | 95 |
|---|---|---|---|---|---|---|
| Profit ($y$) | 1600 | 1300 | 450 | 1650 | 1480 | 1150 |

a Calculate Pearson's product-moment correlation coefficient for the data.

b Interpret your value from part a in context.

c Find the equation of the regression line of $y$ on $x$.

d Use your regression line to predict the profit in the months after the company spent the following sums on advertising:

i $100

ii $200.

e Which of the two estimates in part d would you consider to be more reliable? Explain your answer.

**19** A class of eight students took a History test and a French test. The table shows the marks on the two tests.

| History mark ($x$) | 72 | 47 | 82 | 65 | 71 | 83 | 81 | 57 |
|---|---|---|---|---|---|---|---|---|
| French mark ($y$) | 51 | 60 | 46 | 37 | 52 | 48 | 57 | 41 |

a Find Pearson's product-moment correlation coefficient.

b Find the equation of the regression line of $y$ on $x$.

c The critical value for the correlation coefficient with eight pieces of data is 0.549. Does this suggest that it would be reasonable to use the regression line to predict the mark in the French test for a student who scored 75 marks in the History test? Explain your answer.

**20** The table shows the age (in years) and value (in thousands of dollars) of seven cars.

| Age | 3 | 5 | 12 | 7 | 9 | 7 | 4 |
|---|---|---|---|---|---|---|---|
| Value | 3.2 | 1.1 | 0.6 | 3.1 | 1.8 | 2.4 | 3.4 |

a Use technology to plot the data on a scatter graph.

b Calculate Pearson's product-moment correlation coefficient.

c The critical value of the correlation coefficient for seven pieces of data is 0.584. What does this suggest about your data?

**21** The values of two variables, measured for nine different items, are shown in the table.

| $x$ | 6 | 3 | 5 | 8 | 7 | 5 | 3 | 6 | 9 |
|---|---|---|---|---|---|---|---|---|---|
| $y$ | 31 | 35 | 52 | 22 | 26 | 26 | 41 | 42 | 31 |

a Use technology to plot the data on a scatter graph.

b Describe the correlation suggested by the graph.

c Calculate Pearson's product-moment correlation coefficient.

d The critical value of the correlation coefficient for nine items is 0.521. Do the data suggest that there is significant correlation between the two variables?

**22** Daniel wants to investigate whether there is any relationship between head circumference ($x$ cm) and arm length ($y$ cm). He collects data from six of his friends and finds that the value of Pearson's product-moment correlation coefficient is 0.741 and the equation of the regression line is $y = 1.13x - 17.6$. The head circumferences ranged from 48.5 cm to 53.4 cm and the mean head circumference was 51.7 cm.

a Describe the correlation between the head circumference and arm length.

b Daniel uses his regression line to estimate the arm length of a friend whose head circumference is 49.2 cm. Should this estimate be considered reliable? Explain your answer.

c Find the mean arm length for this sample.

**23** Theo investigates whether spending more time practising his spellings leads to better marks in weekly spelling tests. Each week he records the length of time spent practising and his spelling test mark (out of 20).

| Time (*t* min) | 17 | 5 | 10 | 7 | 25 | 14 | 20 |
|---|---|---|---|---|---|---|---|
| Mark (*m*/20) | 20 | 8 | 6 | 12 | 19 | 16 | 18 |

  a  Find Pearson's product-moment correlation coefficient and describe the correlation between the time spent practising and the test mark.

  b  Find the equation of the regression line of *m* on *t*.

  c  Interpret the gradient and the intercept of your regression line.

**24** A company records its advertising budget, *x*-thousand euros, and its profit, *y*-thousand euros, over a number of years. They find that the correlation coefficient between the two variables is 0.825 and that the equation of the regression line is $y = 3.25x + 13.8$.

  a  Describe, in context, the relationship between the advertising budget and the profit.

  b  Does the value of the correlation coefficient show that increasing the advertising budget will lead to an increase in profit?

  c  Interpret the values

  i  3.25 and

  ii  13.8

  in the equation of the regression line.

**25** The data below show the sales of ice creams at different maximum daily temperatures.

| Temp (°C) | Sales | | Temp (°C) | Sales |
|---|---|---|---|---|
| 9 | 12 | | 25 | 36 |
| 11 | 9 | | 26 | 34 |
| 11 | 14 | | 26 | 37 |
| 13 | 11 | | 27 | 41 |
| 14 | 9 | | 29 | 40 |
| 14 | 13 | | 30 | 41 |
| 15 | 10 | | 31 | 44 |
| 16 | 12 | | 32 | 42 |
| 21 | 32 | | 32 | 45 |
| 23 | 32 | | 34 | 50 |

  a  Using technology, or otherwise, create a scatter diagram illustrating the data.

  b  Suggest what the two regions in your graph might represent.

  c  Describe the correlation between the temperature and the number of ice creams for each of the two regions.

  d  Use an appropriate regression line to estimate the ice cream sales on a day when the maximum temperature is 28°C.

**26** Alessia collected height and mass measurements for a sample of children and adults in the school playground.

| *h* (cm) | 125 | 119 | 152 | 131 | 175 | 121 | 135 | 164 | 158 | 127 |
|---|---|---|---|---|---|---|---|---|---|---|
| *m* (kg) | 23 | 18 | 52 | 29 | 65 | 20 | 31 | 58 | 52 | 25 |

  a  Plot the data on a scatter graph.

  b  Explain why the data contain two linear regions.

  c  Find the midpoint for each of the two regions and add two regression lines to your scatter graph.

  d  Use the appropriate regression line to predict the mass of a child whose height is 120 cm.

**27** The table shows pairs of data values $(x, y)$.

| $x$ | 20 | 10 | 12 | 15 | 7 | 19 | 29 | 24 | 25 | 16 |
|---|---|---|---|---|---|---|---|---|---|---|
| $y$ | 17 | 11 | 12 | 6 | 9 | 16 | 22 | 19 | 20 | 15 |

a   Find the upper and lower quartiles for each of $x$ and $y$.

b   Hence show that there are no outliers in the $x$ data values or the $y$ data values.

c   Using technology, or otherwise, plot the data on a scatter graph.

d   Which point on the scatter graph would you describe as an outlier?

e   Add a line of best fit to your graph

   i   excluding the outlier,

   ii  with the outlier included.

**TOOLKIT: Modelling**

Take a population where you can find the 'true' population statistics – for example, data on the age of pupils in your school, or the incomes in your country. Then try to collect data using different sampling methods. How close do your sample statistics come to the 'true' values. Is there a difference in the accuracy of the mean, median and standard deviation?

## Checklist

- You should be able to understand and work with the concepts of population and sample.

- You should understand when data are discrete or continuous.

- You should be able to identify potential sources of bias in sampling.

- You should be able to understand and evaluate a range of sampling techniques:
  - With simple random sampling, every possible sample (of a given size) has an equal chance of being selected.
  - With convenience sampling, respondents are chosen based upon their availability.
  - With systematic sampling, participants are taken at regular intervals from a list of the population.
  - With stratified sampling, the population is split into groups based on factors relevant to the research, then a random sample from each group is taken in proportion to the size of that group.
  - With quota sampling, the population is split into groups based on factors relevant to the research, then convenience sampling from each group is used until a required number of participants are found.

- You should be able to use your GDC to find the mean, median, mode and range.

- You should be able to use the formula for the mean of data:
  - $\bar{x} = \dfrac{\sum x}{n}$

- You should be able to use your GDC to find the quartiles of discrete data, and know that the interquartile (IQR) is $IQR = Q_3 - Q_1$.

- You should be able to use your calculator to find the standard deviation.

- You should understand the effect of constant changes to a data set.

- Adding a constant, $k$, to every data value will:
  - change the mean, median and mode by $k$
  - not change the standard deviation or interquartile range.

- Multiplying every data value by a constant, $k$, will:
  - multiply the mean, median and mode by $k$
  - multiply the standard deviation and interquartile range by $k$.

- You should know that a data value $x$ is an outlier if $x < Q_1 - 1.5(Q_3 - Q_1)$ or $x < Q_3 + 1.5(Q_3 - Q_1)$.

- You should be able to construct and use statistical diagrams such as histograms, cumulative frequency graphs and box-and-whisker plots.

- You should be able to draw scatter graphs and add a line of best fit.
  - The line of best fit will always pass through the mean point $(\bar{x}, \bar{y})$.

- You should be able to use your GDC to find Pearson's product-moment correlation coefficient and interpret the value:
  - the Pearson's product-moment correlation coefficient, $r$, is such that $-1 \leqslant r \leqslant 1$
  - $r = 1$ means strong positive linear correlation
  - $r = 0$ means no linear correlation
  - $r = -1$ means strong negative linear correlation.

- You should be able to use your GDC to find the regression line, use the regression line to predict values not in the given data set, and interpret the regression coefficients.

- You should be able to fit a piecewise linear model when relevant.

# ■ Mixed Practice

**1** A librarian is investigating the number of books borrowed from the school library over a period of 10 weeks. She decided to select a sample of 10 days and record the number of books borrowed on that day.
- **a** She first suggests selecting a day at random and then selecting every seventh day after that.
  - **i** State the name of this sampling technique.
  - **ii** Identify one possible source of bias in this sample.
- **b** The librarian changes her mind and selects a simple random sample of 10 days instead.
  - **i** Explain what is meant by a simple random sample in this context.
  - **ii** State one advantage of a simple random sample compared to the sampling method from part **a**.
- **c** For the days in the sample, the numbers of books borrowed were:
  17, 16, 21, 16, 19, 20, 18, 11, 22, 14
  Find
  - **i** the range of the data
  - **ii** the mean number of books borrowed per day
  - **iii** the standard deviation of the data.

**2** The table shows the maximum temperature ($T°C$) and the number of cold drinks ($n$) sold by a small shop on a random sample of nine summer days.

| $T$ | 21 | 28 | 19 | 21 | 32 | 22 | 27 | 18 | 30 |
|-----|----|----|----|----|----|----|----|----|----|
| $n$ | 20 | 37 | 21 | 18 | 35 | 25 | 31 | 17 | 38 |

- **a** Using technology, or otherwise, plot the data on a scatter graph.
- **b** Describe the relationship between the temperature and the sales of cold drinks.
- **c** Find the equation of the regression line of $n$ on $T$.
- **d** Use your regression line to estimate the number of cold drinks sold on the day when the maximum temperature is $26°C$.

**3** The masses of 50 cats are summarized in the grouped frequency table:

| Mass (kg) | $1.2 \leqslant m < 1.6$ | $1.6 \leqslant m < 2.0$ | $2.0 \leqslant m < 2.4$ | $2.4 \leqslant m < 2.8$ | $2.8 \leqslant m < 3.2$ |
|-----------|------|------|------|------|------|
| Frequency | 4 | 10 | 8 | 16 | 12 |

- **a** Use this table to estimate the mean mass of a cat in this sample. Explain why your answer is only an estimate.
- **b** Use technology to create a cumulative frequency graph.
- **c** Use your graph to find the median and the interquartile range of the masses.
- **d** Create a box plot to represent the data. You may assume that there are no outliers.

**4** A survey was carried out on a road to determine the number of passengers in each car (excluding the driver). The table shows the results of the survey.

| Number of passengers | 0 | 1 | 2 | 3 | 4 |
|----------------------|----|----|----|----|----|
| Number of cars | 37 | 23 | 36 | 15 | 9 |

- **a** State whether the data are discrete or continuous.
- **b** Write down the mode.
- **c** Use your GDC to find
  - **i** the mean number of passengers per car
  - **ii** the median number of passengers per car
  - **iii** the standard deviation.

Mathematical Studies SL November 2013 Paper 1 Q1

**5** Two groups of 40 students were asked how many books they have read in the last two months. The results for **the first group** are shown in the following table.

| Number of books read | Frequency |
|:---:|:---:|
| 2 | 5 |
| 3 | 8 |
| 4 | 13 |
| 5 | 7 |
| 6 | 4 |
| 7 | 2 |
| 8 | 1 |

The quartiles for these results are 3 and 5.

**a** Write down the value of the median for these results.

**b** Draw a box-and-whisker diagram for these results.

The results for **the second group** of 40 students are shown in the following box-and-whisker diagram.

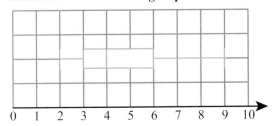

Number of books read

**c** Estimate the number of students **in the second group** who have read at least 6 books.

Mathematical Studies SL May 2015 Paper 1 TZ2 Q4

**6** The following table shows the Diploma score $x$ and university entrance mark $y$ for seven IB Diploma students.

| Diploma score ($x$) | 28 | 30 | 27 | 31 | 32 | 25 | 27 |
|:---|:---:|:---:|:---:|:---:|:---:|:---:|:---:|
| University entrance mark ($y$) | 73.9 | 78.1 | 70.2 | 82.2 | 85.5 | 62.7 | 69.4 |

**a** Find the correlation coefficient.

The relationship can be modelled by the regression line with equation $y = ax + b$.

**b** Write down the value of $a$ and $b$.

Rita scored a total of 26 in her IB Diploma.

**c** Use your regression line to estimate Rita's university entrance mark.

Mathematics SL November 2014 Paper 2 Q2

**7** A student recorded, over a period of several months, the amount of time he waited in the queue for lunch at the college canteen. He summarized the results in this cumulative frequency table.

| Time (minutes) | ⩽ 2 | ⩽ 4 | ⩽ 6 | ⩽ 8 | ⩽ 10 | ⩽ 12 |
|---|---|---|---|---|---|---|
| Cumulative frequency | 4 | 9 | 16 | 37 | 45 | 48 |

  **a** Draw a cumulative frequency graph for the data.
  **b** Use your graph to estimate
    **i** the median
    **ii** the interquartile range of the times
    **iii** the 90th percentile.
  **c** Complete the grouped frequency table:

| Time (min) | $0 < t \leqslant 2$ | $2 < t \leqslant 4$ | | | | |
|---|---|---|---|---|---|---|
| Frequency | | | | | | |

  **d** Estimate the mean waiting time.

**8** The number of customers visiting a shop is recorded over a period of 12 days:
26, 33, 28, 47, 52, 45, 93, 61, 37, 55, 57, 34
  **a** Find the median and the quartiles.
  **b** Determine whether there are any outliers.
  **c** Draw a box-and-whisker diagram for the data.

**9** The heights of a group of 7 children were recorded to the nearest centimetre:
127, 119, 112, 123, 122, 126, 118
  **a** Find the mean and the variance of the heights.
  **b** Each child stands on a 35-centimetre-high stool. Find the mean and variance of their heights.

**10** Theo is keeping a record of his travel expenses. The cost of each journey is $15 plus $3.45 per kilometre. The mean length of Theo's journeys is 11.6 km and the standard deviation of the lengths is 12.5 km. Find the mean and standard deviation of his cost per journey.

**11** The frequency table summarizes data from a sample with mean 1.6. Find the value of $x$.

| $x$ | 0 | 1 | 2 | 3 |
|---|---|---|---|---|
| Frequency | 5 | 6 | 8 | $x$ |

**12** A scientist measured a sample of 12 adult crabs found on a beach, measuring their shell length (*s*) and mass (*m*).

| Shell length (cm) | Mass (g) | | Shell length (cm) | Mass (g) |
|---|---|---|---|---|
| 7.1 | 165 | | 5.9 | 143 |
| 8.1 | 256 | | 8.4 | 190 |
| 8.5 | 194 | | 9.2 | 208 |
| 6.0 | 150 | | 5.1 | 194 |
| 9.0 | 275 | | 6.3 | 217 |
| 5.3 | 204 | | 9.1 | 268 |

**a** Using technology or otherwise plot a scatter diagram to illustrate the scientist's results.

**b** The scientist later realized that the beach contains two species of crab – the Lesser European Crab and the Giant European Crab. Her research suggests that the Giant European crab tends to be heavier than similar-sized Lesser European Crabs. Find the equation for a regression line for the mass of the Giant European crab if its shell length is known.

**c** Find the correlation coefficient for the data for the Giant European crab and comment on your result.

**d** Estimate the mass of a Giant European Crab with shell length 8 cm.

**e** Juvenile Giant European Crabs have a shell length of between 2 and 4 cm. Estimate the possible masses of these crabs and comment on the reliability of your results.

**13** The one hour distances, in miles, covered by runners before (*x*) and after (*y*) going on a new training program are recorded.

The correlation between these two distances is found to be 0.84. The regression line is $y = 1.2x + 2$.

**a** Describe the significance of

**i** the intercept of the regression line being positive

**ii** the gradient of the regression line being greater than 1.

**b** If the previous mean distance is 8 miles, find the new mean.

Their trainer wants to have the data in km. To convert miles to km all the distances are multiplied by 1.6. The new variables in km are *X* and *Y*.

**c** Find

**i** the correlation between *X* and *Y*

**ii** the regression line connecting *X* and *Y*.

# ESSENTIAL UNDERSTANDINGS

- Probability enables us to quantify the likelihood of events occurring and so evaluate risk.
- Both statistics and probability provide important representations which enable us to make predictions, valid comparisons and informed decisions.
- These fields have power and limitations and should be applied with care and critically questioned, in detail, to differentiate between the theoretical and the empirical/observed.

## In this chapter you will learn…

- how to find experimental and theoretical probabilities
- how to find probabilities of complementary events
- how to find the expected number of occurrences
- how to use Venn diagrams to find probabilities
- how to use tree diagrams to find probabilities
- how to use sample space diagrams to find probabilities
- how to calculate probabilities from tables of events
- a formula for finding probabilities of combined events
- how to find conditional probabilities
- about mutually exclusive events
- about independent events.

## CONCEPTS

The following concept will be addressed in this chapter:
- **Modelling** and finding structure in seemingly random events facilitates prediction.

■ **Figure 7.1** What are the chances of a particular outcome in each case?

## PRIOR KNOWLEDGE

Before starting this chapter, you should already be able to complete the following:

1 For the sets $A = \{2, 3, 5, 6, 8\}$ and $B = \{4, 5, 7, 8, 9\}$, find the sets

   a $A \cup B$                                     b $A \cap B$

2 Out of a class of 20 students, 8 play the piano, 5 play the violin and 2 play both.

   Draw a Venn diagram to represent this information.

3 A bag contains six red balls and four blue balls. Two balls are removed at random without replacement.

   Draw a tree diagram to represent this information.

Probability is a branch of mathematics that deals with events that depend on chance. There are many circumstances in which you cannot be certain of the outcome, but just because you do not know for sure what will happen, does not mean you cannot say anything useful. For example, probability theory allows predictions to be made about the likelihood of an earthquake occurring or a disease spreading. Understanding and evaluating risk is crucial in a range of disciplines from science to insurance.

## Starter Activity

Look at the images in Figure 7.1. Discuss what you can say about the outcomes in each case.

**Now look at this problem:**

Flip a coin 10 times and record the number of heads and the number of tails. Could you predict what the outcome would be?

Do you have to get five heads and five tails to conclude that the coin is fair?

**LEARNER PROFILE – Principled**
Are mathematicians responsible for the applications of their work? Some major mathematical breakthroughs have been inspired by and used in warfare, code-breaking and gambling. Should such areas of research be banned?

# 7A Introduction to probability

## ■ Concepts in probability

If you roll a standard fair dice, there are six possible equally likely **outcomes** – 1, 2, 3, 4, 5 and 6 – each having probability $\frac{1}{6}$ of occurring.

You might be interested in a particular combination of these outcomes, such as rolling an even number on the dice. A combination of outcomes is known as an **event**. The probability of an event, $A$, happening is denoted by **P(A)**. This is a number between 0 and 1 inclusive which measures how likely the event is.

You might wish to repeat the roll of the dice several times. Each repetition of the process is known as a **trial**. In practice you often won't know for sure what the probability of any given outcome occurring is, so you have to use experimental data to estimate the probability. One way of estimating the probability of an event is to use the **relative frequency**. This is the fraction of trials which are favourable to an event.

> **You are the Researcher**
>
> Probability is the most common way of representing uncertain events, but there are others you might want to investigate, such as likelihood, odds or possibility theory. They all have different uses and different rules.

**WORKED EXAMPLE 7.1**

A random sample of 80 collectable card packets was opened; 23 were found to contain a special rare card.

Another card packet was purchased. Estimate the probability that it contains a special rare card.

**TOK Links**

How reliable is using information about the past when predicting the future?

You know from the sample data that 23 out of 80 packets contained the rare card so this relative frequency is the best estimate you have of the probability of getting a rare card ·················· P(pack contains rare card) $= \frac{23}{80}$

## ■ Theoretical approach to probability

If you know that all outcomes are equally likely then you can work out probabilities theoretically, rather than needing to approximate them from experimental evidence. To do this you need to list all possible outcomes.

The set of all possible outcomes is called the **sample space**, which is denoted by $U$. For example, the sample space for rolling a standard dice is 1, 2, 3, 4, 5, 6.

Using the notation $n(A)$ to mean the number of outcomes in the event $A$ we have the following result:

**KEY POINT 7.1**

If all outcomes are equally likely, then $P(A) = \dfrac{n(A)}{n(U)}$.

**WORKED EXAMPLE 7.2**

A fair standard six-sided dice is rolled.

Find the probability that the outcome is a prime number.

Use $P(A) = \dfrac{n(A)}{n(U)}$ ............................................ 2, 3 and 5 are prime

$P(\text{prime}) = \dfrac{3}{6}$

There are six outcomes in the sample space and three of these are prime ............................................ $= \dfrac{1}{2}$

## ■ Complementary events

The complement of the event $A$ (denoted by $A'$) occurs when the event $A$ does not happen. Since an event either happens or doesn't, the sum of an event and its complement must be 1.

**KEY POINT 7.2**

● $P(A') = 1 - P(A)$

**WORKED EXAMPLE 7.3**

The probability of no buses arriving in a 10-minute interval is 0.4.

Find the probability of at least one bus arriving in that 10-minute interval.

At least one means every possible outcome except no buses ............................................ $P(\text{at least 1 bus}) = 1 - 0.4$

$= 0.6$

## ■ Expected number of occurrences

If you know the probability of an event happening, then you can use this to estimate how many of the total number of trials will result in that outcome.

**KEY POINT 7.3**

Expected number of occurrences of $A = P(A) \times n$, where $n$ is the number of trials.

**WORKED EXAMPLE 7.4**

An archer has a probability of 0.3 of hitting the bullseye.

If he takes 12 shots in a competition, find the expected number of bullseyes.

Use Expected number $= P(A) \times n$ ............................................ Expected number of bullseyes $= 0.3 \times 12$

$= 3.6$

**Tip**

The expected number does not need to be a whole number.

**TOOLKIT:** Problem Solving

An actuarial study tracked 100 000 people born in the UK in 1940. By 2010, 84 210 were still alive.

a   Use the data to estimate the probability of a person in the UK living to age 70.

b   An insurance company uses this study to price the cost of a life insurance policy. Give two reasons why the answer to **a** might be an underestimate of the probability of a new client living to 70.

c   A 21-year-old takes out a policy which pays out £100 000 on death up to the age of 70, but nothing thereafter. The expected total amount that the 21-year-old will pay is £22 000. Estimate the expected profit of the company on this policy.

d   Why might the 21-year-old take out a policy on which they expect to make a loss?

## Exercise 7A

For questions 1 and 2, use the method demonstrated in Worked Example 7.1 to estimate the probability from given information.

1   a   Out of a random sample of 40 children at a school, 12 were found to have a pet. Estimate the probability that a child at this school has a pet.

    b   A random sample of 60 flowers was collected from a meadow and 8 were found to have six petals. Find the probability that a flower from this meadow has six petals.

2   a   A biased dice is rolled 90 times and 18 sixes were obtained. Find the probability of rolling a six on this dice.

    b   Daniel took 25 shots at a target and hit it 15 times. Find the probability that he hits the target next time.

For questions 3 to 5, use the method demonstrated in Worked Example 7.2 to find the probabilities.

3   A standard six-sided dice is rolled. Find the probability that the outcome is

    a   an even number                                                    b   a multiple of 3.

4   A card is picked at random from a standard pack of 52 cards. Find the probability that the selected card is

    a   a diamond                                                         b   an ace.

5   A card is picked at random from a standard pack of 52 cards. Find the probability that the selected card is

    a   a red 10                                                          b   a black picture card (jack, queen or king).

For questions 6 to 11, use the method demonstrated in Worked Example 7.3.

6   a   The probability that Elsa does not meet any of her friends on the way to school is 0.06. Find the probability that she meets at least one friend on the way to school.

    b   The probability that Asher does not score any goals in a football match is 0.45. Find the probability that he scores at least one goal.

7   a   The probability that a biased coin comes up heads is $\frac{9}{20}$. Find the probability that it comes up tails.

    b   The probability of rolling an even number on a biased dice is $\frac{17}{40}$. Find the probability of rolling an odd number.

8   a   The probability that Daniel is late for school on at least one day during a week is 0.15. Find the probability that he is on time every day.

    b   Theo either walks or cycles to school. The probability that he cycles at least one day a week is 0.87. Find the probability that he walks every day during a particular week.

9   a   When a biased coin is flipped 10 times, the probability of getting more than six tails is $\frac{73}{120}$. What is the probability of getting at most six tails?

    b   When a biased coin is flipped six times, the probability of getting fewer than three heads is $\frac{7}{48}$. What is the probability of getting at least three heads?

10   a   The probability that it rains on at least four days in a week is 0.27. Find the probability that it rains on no more than three days.

     b   The probability that it is sunny on at most two days in a week is 0.34. Find the probability that it is sunny on no fewer than three days.

11   a   The probability that an arrow lands within 20 cm of the target is 0.56. Find the probability that it lands more than 20 cm away.

     b   The probability that a parachutist lands within 20 m of the landing spot is 0.89. Find the probability that she lands more than 20 m away.

For questions 12 to 15, use the method demonstrated in Worked Example 7.4.

12   a   The probability that a packet of crisps contains a toy is 0.2. Find the expected number of toys in 20 packets of crisps.

     b   The probability that a footballer scores a penalty is 0.9. If he takes 30 penalties find the expected number of goals scored.

13   a   The probability that it rains on any particular day is $\frac{2}{5}$. Find the expected number of rainy days in a 30-day month.

     b   The probability that a pack of collectable cards contains a particular rare card is $\frac{3}{70}$. If I buy 140 packs of cards, how many copies of this rare card should I expect to find?

14   a   The probability that a biased coin comes up heads is 0.12. Find the expected number of heads when the coin is tossed 40 times.

     b   The probability that a biased dice lands on a 4 is 0.15. Find the expected number of 4s when the dice is rolled 50 times.

15   a   The probability that Adam is late for work is 0.08. Find the number of days he is expected to be late in a 20-day period.

     b   The probability that Sasha forgets her homework is 0.15. Find the number of days she is expected to forget her homework in a 10-day period.

16   A clinical trial involved 350 participants. 26 of them experienced some negative side-effects from the drug.

     a   Estimate the probability that the drug produces a negative side-effect.

     b   A larger trial is conducted with 900 participants. How many of them are expected to experience negative side-effects?

17   A fair eight-sided dice has sides numbered 1, 1, 1, 2, 2, 3, 4, 5. The dice is rolled 30 times. Estimate the number of times it will land on a 2.

18   A card is selected from a standard pack of 52 cards and returned to the pack. This is repeated 19 more times. Find the expected number of selected cards which are not a diamond.

19   A standard fair $n$-sided dice numbered 1 to $n$ is rolled 400 times and the expected number of 1s is 50. What is the value of $n$?

20   When rolling a biased dice 100 times, the expected number of odd outcomes is three times the expected number of even outcomes. Find the probability of an odd outcome.

21   An actuarial study finds that in a sample of 2000 18-year-old drivers, 124 are involved in an accident. The average cost of such an accident to the insurance company is $15 000. If the company wants to make a 20% profit on policies, what should they charge 18-year-old drivers?

22   An alternative measure of probability is called odds. If the probability of an event is $p$, then the odds are defined as $\frac{p}{1-p}$.

     a   If an event has a probability of 0.6, find its odds.

     b   If an event has odds of 4, find its probability.

23   An urn contains red balls, green balls and blue balls. The ratio of red balls to green balls is 1:3. The ratio of green balls to blue balls is 1:4. A random ball is drawn from the urn. What is the probability that it is a red ball?

24   A computer scientist uses a method called Monte Carlo sampling to estimate the value of $\pi$. He uses a random number generator to generate 100 points at random inside a 1 by 1 square. Of these points, 78 are inside the largest circle which can be drawn in the square. What would be the estimate of $\pi$ she would form on the basis of this sample?

# 7B Probability techniques

## ■ Venn diagrams

You are already familiar with the idea of using Venn diagrams to represent information about two or more sets. Here you will see how they can be used to calculate probabilities.

When filling in a Venn diagram it is a good idea to start in the middle and work outwards.

**WORKED EXAMPLE 7.5**

In a school year of 60 students, 32 study Geography and 40 study Biology. If 18 study both of these subjects, find the probability of a randomly selected student studying neither Geography nor Biology.

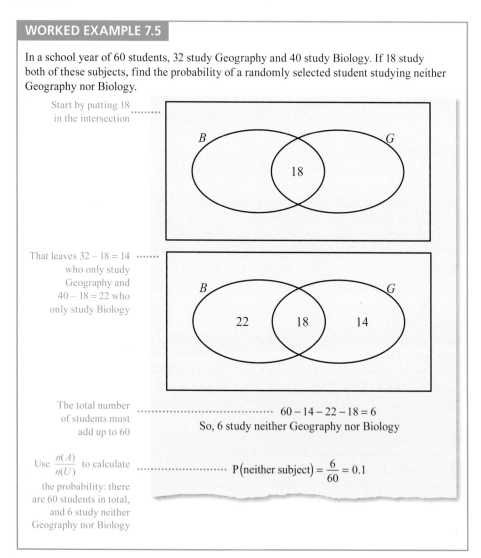

Start by putting 18 in the intersection

That leaves 32 − 18 = 14 who only study Geography and 40 − 18 = 22 who only study Biology

The total number of students must add up to 60

60 − 14 − 22 − 18 = 6
So, 6 study neither Geography nor Biology

Use $\frac{n(A)}{n(U)}$ to calculate the probability: there are 60 students in total, and 6 study neither Geography nor Biology

$P(\text{neither subject}) = \frac{6}{60} = 0.1$

## ■ Tree diagrams

Another technique you are familiar with is tree diagrams. These are used when you have one event followed by another, particularly when the second event depends on the first.

**WORKED EXAMPLE 7.6**

In a city, the probability of it raining is $\frac{1}{4}$. If it rains, then the probability of Anya travelling by car is $\frac{2}{3}$. If it is not raining, then the probability of Anya travelling by car is $\frac{1}{2}$.

Find the overall probability of Anya travelling by car.

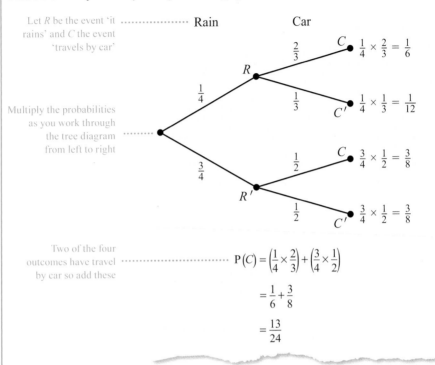

Let $R$ be the event 'it rains' and $C$ the event 'travels by car'

Multiply the probabilities as you work through the tree diagram from left to right

Two of the four outcomes have travel by car so add these

$$P(C) = \left(\frac{1}{4} \times \frac{2}{3}\right) + \left(\frac{3}{4} \times \frac{1}{2}\right)$$

$$= \frac{1}{6} + \frac{3}{8}$$

$$= \frac{13}{24}$$

## ■ Sample space diagrams

Sometimes the easiest way of finding probabilities is to record all possible outcomes in a sample space diagram.

**WORKED EXAMPLE 7.7**

Two fair standard four-sided dice are rolled. Find the probability that the sum is larger than 6.

Draw a sample space diagram consisting of all possible sums of the outcomes on the two dice

|  |  | Dice 1 | | | |
|---|---|---|---|---|---|
|  |  | **1** | **2** | **3** | **4** |
| **Dice 2** | **1** | 2 | 3 | 4 | 5 |
|  | **2** | 3 | 4 | 5 | 6 |
|  | **3** | 4 | 5 | 6 | 7 |
|  | **4** | 5 | 6 | 7 | 8 |

3 outcomes in the sample space are greater than 6 out of 16 in total

$$P(\text{sum} > 6) = \frac{3}{16}$$

## ■ Tables of outcomes

Information may be given in a table. In this case it is a good idea to start by finding the totals of the rows and columns.

---

**WORKED EXAMPLE 7.8**

The following table shows the interaction between type of book and size of book in a random sample of library books.

|  |  | Type of book | |
|---|---|---|---|
|  |  | Fiction | Non-fiction |
| **Number of pages** | 0–200 | 23 | 3 |
|  | 201–500 | 28 | 6 |
|  | 501+ | 8 | 12 |

a Find the probability that a randomly selected book is fiction.

b Alessia randomly selects a non-fiction book. Find the probability that it has 0–200 pages.

Add totals to all columns and rows of the table ............

|  |  | Type of book | | Total |
|---|---|---|---|---|
|  |  | Fiction | Non-fiction |  |
| **Number of pages** | 0–200 | 23 | 3 | 26 |
|  | 201–500 | 28 | 6 | 34 |
|  | 501+ | 8 | 12 | 20 |
| **Total** |  |  | 59 | 21 | 80 |

It is now clear that there are 59 fiction books ...... **a** $P(\text{fiction}) = \dfrac{59}{80}$
out of a total of 80

All the non-fiction books are in the highlighted ...... **b** $P(0\text{–}200 \text{ pages given it is non-fiction}) = \dfrac{3}{21} = \dfrac{1}{7}$
column of the table

---

Part **b** of Worked Example 7.8 is an example of conditional probability which you will meet later in this section.

## ■ Combined events

If you want the probability of either event $A$ or event $B$ occurring, or both occurring, then you can think of this from a Venn diagram as being the probability of being in the union of these two sets $(A \cup B)$.

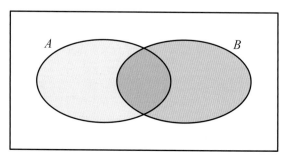

Simply adding the probability of being in $A$ to the probability of being in $B$ will count the intersection $(A \cap B)$ twice so you need to then subtract one lot of the intersection.

**KEY POINT 7.4**

- $P(A \cup B) = P(A) + P(B) - P(A \cap B)$

**WORKED EXAMPLE 7.9**

If $P(A) = 0.4$, $P(B) = 0.3$ and l $P(A \cup B) = 0.5$, find $P(A \cap B)$.

Use $P(A \cup B) = P(A) + P(B) - P(A \cap B)$ ............
and rearrange to find $P(A \cap B)$

$$P(A \cup B) = P(A) + P(B) - P(A \cap B)$$
$$0.5 = 0.4 + 0.3 - P(A \cap B)$$
$$P(A \cap B) = 0.2$$

## ■ Mutually exclusive events

Two events are mutually exclusive if they cannot both happen at the same time, for example, throwing a 1 and throwing a 6 on a dice. In other words, the probability of two mutually exclusive events happening together is zero.

**Tip**

The second formula in Key Point 7.5 follows directly from the formula $P(A \cup B) = P(A) + P(B) - P(A \cap B)$ with $P(A \cap B) = 0$.

**KEY POINT 7.5**

If events $A$ and $B$ are mutually exclusive, then:
- $P(A \cap B) = 0$

or equivalently
- $P(A \cup B) = P(A) + P(B)$

**WORKED EXAMPLE 7.10**

If two events $A$ and $B$ are mutually exclusive and $P(A) = 0.5$ and $P(A \cup B) = 0.8$ find $P(B)$.

Use $P(A \cup B) = P(A) + P(B)$ ............
and rearrange to find $P(B)$

Since $A$ and $B$ are mutually exclusive,
$$P(A \cup B) = P(A) + P(B)$$
$$0.8 = 0.5 + P(B)$$
$$P(B) = 0.3$$

## ■ Conditional probability

If the probability of event $A$ happening is dependent on a previous event, $B$, then the probability is said to be conditional. This can be written as $P(A|B)$, which means the probability of $A$ happening given that $B$ has happened.

**WORKED EXAMPLE 7.11**

A pack of 52 cards contains 13 hearts. If the first card taken from the pack (without replacement) is a heart, what is the probability that the second card will also be a heart.

If the first card is a heart then there are now 12 hearts left out of 51 cards ............
$$P\left(\text{heart 2nd}|\text{heart 1st}\right) = \frac{12}{51} = \frac{4}{17}$$

---

**WORKED EXAMPLE 7.12**

The probability of a randomly chosen student studying only French is 0.4. The probability of a student studying French and Spanish is 0.1.

Find the probability of a student studying Spanish given that they study French.

When the situation is a little more complex, it is a good idea to draw a Venn diagram ·············

Given that the student studies French means that you can ignore anything outside the French set

The remaining probability of speaking Spanish is 0.1 out of a remaining total of 0.5 ············· $P(\text{Spanish}|\text{French}) = \dfrac{0.1}{0.5}$

$$= 0.2$$

---

You can always do conditional probability questions using the Venn diagram method demonstrated in Worked Example 7.12, but this method does suggest a general formula:

$$P(A|B) = \frac{P(A \cap B)}{P(B)}.$$

## ■ Independent events

Two events are independent if one doesn't affect the other; in other words if $P(A|B) = P(A)$. Substituting this into the above formula and rearranging gives the following convenient relationship for independent events:

**KEY POINT 7.6**

If events $A$ and $B$ are independent, then:
● $P(A \cap B) = P(A) \times P(B)$

---

**WORKED EXAMPLE 7.13**

The probability of a randomly chosen individual in a society being male is 0.52. The probability of the individual being of Irish origin is 0.18.

If these two characteristics are independent, find the probability that a randomly chosen individual is a male of Irish origin.

Being a male of Irish origin ················· $P(\text{male} \cap \text{Irish}) = P(\text{male}) \times P(\text{Irish})$
means being male and Irish.
As these two characteristics
are independent you can                                     $= 0.52 \times 0.18$
multiply the probabilities
                                                            $= 0.0936$

> ### CONCEPTS – MODELLING
>
> In reality, there are often links between events which mean that they are not independent. For example, there might be a small difference in genetic or social factors which mean that an individual of Irish origin is more or less likely to be male than other members of the same population. We often make assumptions that such links are small so independence is valid, but you should be aware that you are making these assumptions.

## Be the Examiner 7.1

A bag contains 10 balls labelled uniquely with the numbers 1 to 10. Hanna picks a ball at random.

What is the probability she gets a prime number or an even number?

Which is the correct solution? Identify the errors made in the incorrect solutions.

| Solution 1 | Solution 2 | Solution 3 |
|---|---|---|
| $P(\text{prime}) = \dfrac{4}{10}$ | $P(\text{prime}) = \dfrac{4}{10}$ | $P(\text{prime}) = \dfrac{4}{10}$ |
| $P(\text{even}) = \dfrac{5}{10}$ | $P(\text{even}) = \dfrac{5}{10}$ | $P(\text{even}) = \dfrac{5}{10}$ |
| So, $P(\text{prime or even}) = \dfrac{4}{10} + \dfrac{5}{10}$ $= \dfrac{9}{10}$ | $P(\text{prime and even}) = \dfrac{1}{10}$ So, $P(\text{prime or even}) = \dfrac{4}{10} + \dfrac{5}{10} - \dfrac{1}{10}$ $= \dfrac{8}{10}$ | $P(\text{prime and even}) = \dfrac{4}{10} \times \dfrac{5}{10} = \dfrac{1}{5}$ So, $P(\text{prime or even}) = \dfrac{4}{10} + \dfrac{5}{10} - \dfrac{1}{5}$ $= \dfrac{7}{10}$ |

## Exercise 7B

For questions 1 to 7, use the method demonstrated in Worked Example 7.5 to draw a Venn diagram and then find the required probability.

1   Out of 25 students in a class, 18 speak French, 9 speak German and 5 speak both languages. Find the probability that a randomly chosen student speaks

   a   neither language

   b   German but not French.

2   In a group of 30 children, 12 like apples, 18 like bananas and 10 like both fruits. Find the probability that a randomly chosen child

   a   likes bananas but not apples

   b   likes neither fruit.

3   In a school of 150 pupils, 80 pupils play basketball, 45 play both hockey and basketball and 25 play neither sport. Find the probability that a randomly selected pupil

   a   plays basketball but not hockey

   b   plays hockey but not basketball.

4   In a certain school, 80 pupils study both Biology and History, 35 study neither subject, 45 study only Biology and 60 study only History. Find the probability that a randomly selected pupil

   a   studies at least one of the two subjects

   b   studies History.

5   Out of 35 shops in a town, 28 sell candy, 18 sell ice cream and 12 sell both ice cream and candy. Find the probability that a randomly selected shop sells

   a   ice cream but not candy

   b   candy but not ice cream.

6   At a certain college, the probability that a student studies Geography is 0.4, the probability that they study Spanish is 0.7 and the probability that they study both subjects is 0.2. Find the probability that a randomly selected student

a   studies neither subject                                       b   studies Spanish but not Geography.

7   In a given school, the probability that a child plays the piano is 0.4. The probability that they play both the piano and the violin is 0.08, and the probability that they play neither instrument is 0.3. Find the probability that a randomly selected child

a   plays the violin but not the piano                            b   plays the piano but not the violin.

For questions 8 to 12, use the method demonstrated in Worked Example 7.6 to draw a tree diagram and then find the required probability.

8   Every day, Rahul either drives to work or takes the bus. The probability that he drives is 0.3. If he drives to work the probability that he is late is 0.05. If he takes the bus the probability that he is late is 0.2. Find the probability that Rahul is

a   late for work                                                 b   not late for work.

9   The probability that the school canteen serves pizza for lunch is 0.3. If there is pizza, the probability that they also serve fries is 0.8. If there is no pizza, the probability that they serve fries is 0.5. Find the probability that on a randomly selected day

a   there are no fries                                            b   there are fries.

10  Sophia rolls a fair six-sided dice. If she gets a 6, she tosses a fair coin. If she does not get a 6, she tosses a biased coin which has the probability 0.6 of showing tails. Find the probability that the coin shows

a   heads                                                         b   tails.

11  A bag contains 12 red balls and 18 blue balls. Two balls are taken out at random, one after the other, without replacement. Find the probability that

a   the two balls are the same colour                             b   the second ball is blue.

12  The probability that I revise for a test is 0.8. If I revise, the probability that I pass the test is 0.9. If I do not revise, the probability that I pass is 0.7. Find the probability that

a   I revise and pass the test                                    b   I do not revise but pass the test.

For questions 13 to 17, use the method demonstrated in Worked Example 7.7 to create a sample space diagram showing all possible outcomes, and use it to find the required probability.

13  Two fair six-sided dice are rolled, and the scores are added. Find the probability that the total score is

a   equal to 6                                                    b   greater than 9.

14  Two fair six-sided dice are rolled. Find the probability that

a   the two scores are equal                                      b   the second score is larger than the first.

15  Two fair coins are tossed. Find the probability that

a   both show tails                                               b   the two outcomes are different.

16  A family has two children, born independently. Find the probability that they are

a   a boy and a girl                                              b   two girls.

17  Write out all possible arrangements of the letters T, A, I, L. Hence find the probability that, when those four letters are arranged in a random order,

a   the 'word' ends with two consonants                           b   the two vowels are next to each other.

For questions 18 to 21, use the method demonstrated in Worked Example 7.8 to find the required probability.

18  The table shows the number of students at a school taking different science subjects at Standard Level and Higher Level.

|           | HL | SL |
|-----------|----|----|
| Biology   | 32 | 18 |
| Chemistry | 13 | 25 |
| Physics   | 9  | 12 |

Find the probability that a randomly selected student

a   studies Chemistry                                             b   takes a Standard Level Science subject.

19  The favourite colour and favourite fruit for a group of children are shown in the table.

|  | Apple | Banana | Strawberry |
|---|---|---|---|
| **Red** | 12 | 23 | 18 |
| **Blue** | 16 | 18 | 9 |
| **Green** | 3 | 9 | 6 |

A child is selected at random. Find the probability that

a  their favourite colour is green

b  their favourite fruit is banana.

20  The table shows information on hair colour and eye colour of a group of people.

|  | Blond hair | Brown hair |
|---|---|---|
| **Blue eyes** | 23 | 18 |
| **Brown eyes** | 16 | 20 |

Find the probability that a randomly selected person

a  has blue eyes and brown hair

b  has brown eyes and blond hair.

21  The table shows information on some items of clothing in a shop.

|  | Black | White | Red |
|---|---|---|---|
| **Trousers** | 26 | 8 | 9 |
| **Skirts** | 30 | 12 | 23 |
| **Tops** | 14 | 26 | 15 |

Find the probability that

a  a randomly selected skirt is red

b  a randomly selected white item is a top.

For questions 22 to 24, use the method demonstrated in Worked Example 7.9.

22  a  If $P(A) = 0.3$, $P(B) = 0.8$ and $P(A \cup B) = 0.9$, find $P(A \cap B)$.

  b  If $P(A) = 0.8$, $P(B) = 0.2$ and $P(A \cup B) = 0.7$, find $P(A \cap B)$.

23  a  If $P(A) = 0.6$, $P(B) = 0.4$ and $P(A \cap B) = 0.2$, find $P(A \cup B)$.

  b  If $P(A) = 0.8$, $P(B) = 0.5$ and $P(A \cap B) = 0.4$, find $(A \cup B)$.

24  a  If $P(A) = 0.4$, $P(A \cap B) = 0.3$ and $P(A \cup B) = 0.6$, find $P(B)$.

  b  If $P(A) = 0.7$, $P(A \cap B) = 0.5$ and $P(A \cup B) = 0.9$, find $P(B)$.

For questions 25 and 26, use the method demonstrated in Worked Example 7.10.

25  a  If two events $A$ and $B$ are mutually exclusive and $P(A) = 0.6$ and $P(A \cup B) = 0.8$, find $P(B)$.

  b  If two events $A$ and $B$ are mutually exclusive and $P(A) = 0.4$ and $P(A \cup B) = 0.5$, find $P(B)$.

26  a  Events $A$ and $B$ are mutually exclusive, $P(A) = \dfrac{1}{3}$ and $P(B) = \dfrac{2}{5}$. Find $P(A \cup B)$

  b  Events $A$ and $B$ are mutually exclusive, $P(A) = \dfrac{3}{4}$ and $P(B) = \dfrac{1}{10}$. Find $P(A \cup B)$.

For questions 27 to 30, use the method demonstrated in Worked Example 7.11 to find conditional probabilities.

27  A bag contains 20 red counters and 30 blue counters. Two counters are taken out at random. Given that the first counter is red, find the probability that the second counter is

a  red

b  blue.

28  Two cards are taken out of a standard pack of 52 cards. Given that the first card is a red 10, find the probability that the second card is

a  red

b  a 10.

29   The favourite colour and favourite fruit for a group of children are shown in the table.

|         | Apple | Banana | Strawberry |
|---------|-------|--------|------------|
| Red     | 12    | 23     | 18         |
| Blue    | 16    | 18     | 9          |
| Green   | 3     | 9      | 6          |

Find the probability that

   a   a child's favourite colour is green, given that their favourite fruit is apples

   b   a child's favourite fruit is strawberries, given that their favourite colour is red.

30   The table shows information on hair colour and eye colour of a group of people.

|            | Blond hair | Brown hair |
|------------|-----------|------------|
| Blue eyes  | 23        | 18         |
| Brown eyes | 16        | 20         |

Find the probability that a randomly selected person

   a   has brown eyes, given that they have blond hair

   b   has blond hair, given that they have brown eyes.

For questions 31 to 37, use the method demonstrated in Worked Example 7.12 to draw a Venn diagram and find the required conditional probability.

31   Out of 25 students in a class, 18 speak French, 9 speak German and 5 speak both languages. Find the probability that a randomly chosen student speaks

   a   German, given that they speak French

   b   French, given that they speak German.

32   In a group of 30 children, 12 like apples, 18 like bananas and 10 like both fruits. Find the probability that a randomly chosen child

   a   likes bananas, given that they like apples

   b   likes apples, given that they like bananas.

33   In a school of 150 pupils, 80 pupils play basketball, 45 play both hockey and basketball and 25 play neither sport. Find the probability that a randomly selected pupil

   a   plays basketball, given that they do not play hockey

   b   plays hockey, given that they do not play basketball.

34   In a certain school, 80 pupils study both Biology and History, 35 study neither subject, 45 study only Biology and 60 study only History. Find the probability that a randomly selected pupil

   a   studies Biology, given that they study at least one of the two subjects

   b   studies History, given that they study at least one of the two subjects.

35   Out of 35 shops in a town, 28 sell candy, 18 sell ice cream and 12 sell both ice cream and candy. Find the probability that a randomly selected shop

   a   sells neither, given that it does not sell candy.

   b   sells neither, given that it does not sell ice cream.

36   At a certain college, the probability that a student studies Geography is 0.4, the probability that they study Spanish is 0.7 and the probability that they study both subjects is 0.2. Find the probability that a randomly selected student

   a   studies both subjects, given that they study Geography

   b   studies both subjects, given that they study Spanish.

37   In a school, the probability that a child plays the piano is 0.4. The probability that they play both the piano and the violin is 0.08, and the probability that they play neither instrument is 0.3. Find the probability that a randomly selected child

   a   plays the violin, given that they play at least one instrument

   b   plays the piano, given that they play at least one instrument.

For questions 38 and 39, use the method demonstrated in Worked Example 7.13 to find the combined probability of two independent events.

38   a   The probability that a child has a sibling is 0.7. The probability that a child has a pet is 0.2. If having a sibling and having a pet are independent, find the probability that a child has both a sibling and a pet.

    b   The probability that a child likes bananas is 0.8. The probability that their favourite colour is green is 0.3. If these two characteristics are independent, find the probability that a child's favourite colour is green and they like bananas.

39   a   My laptop and my phone can run out of charge independently of each other. The probability that my laptop is out of charge is 0.2 and the probability that they are both out of charge is 0.06. Find the probability that my phone is out of charge.

    b   The probability that my bus is late is 0.2. The probability that it rains and that my bus is late is 0.08. If the two events are independent, find the probability that it rains.

40   Two fair dice are rolled and the score is the product of the two numbers.

    a   Draw a sample space diagram showing all possible outcomes.

    b   If this were done 180 times, how many times would you expect the score to be at least 20?

41   A fair four-sided dice (numbered 1 to 4) and a fair eight-sided dice (numbered 1 to 8) are rolled and the outcomes are added together. Find the probability that the total is less than 10.

42   A survey of 180 parents in a primary school found that 80 are in favour of building a new library, 100 are in favour of a new dining hall and 40 are in favour of both.

    a   Draw a Venn diagram to represent this information.

    b   Find the probability that a randomly selected parent is in favour of neither a new library, nor a new dining hall.

    c   Find the probability that a randomly selected parent is in favour of a new library, given that they are not in favour of a new dining hall.

43   A bag contains 12 green balls and 18 yellow balls. Two balls are taken out at random. Find the probability that they are different colours.

44   The table summarizes the information on a pupil's punctuality, depending on the weather.

|  | Raining | Not raining |
|---|---|---|
| On time | 35 | 42 |
| Late | 5 | 6 |

    a   Find

        i   P(late|it is raining)                           ii   P(late|it is not raining)

    b   Hence, determine whether being late is independent of the weather. Justify your answer.

45   Asher has 22 toys: 12 are teddies, 5 are green teddies, and 8 are neither green nor teddies.

    a   Draw a Venn diagram showing this information and find the number of toys which are green but not teddies.

    b   Find the probability that a randomly selected toy is green.

    c   Given that a toy is green, find the probability that it is a teddy.

46   Out of 130 students at a college, 50 play tennis, 45 play badminton and 12 play both sports.

    a   How many students play neither sport?

    b   Find the probability that a randomly selected student plays at least one of the two sports.

    c   Given that a student plays exactly one of the two sports, find the probability that it is badminton.

47   A football league consists of 18 teams. My team has a 20% chance of winning against a higher-ranking team and a 70% chance of winning against a lower-ranking team. If my team is currently ranked fifth in the league, what it is the probability that we win the next match against a randomly selected team?

48   An office has two photocopiers. One of them is broken with probability 0.2 and the other with probability 0.3. If a photocopier is selected at random, what it the probability that it is working?

**49** Theo either walks, takes the bus or cycles to school, each with equal probability. The probability that he is late is 0.06, if he walks, it is 0.2 if he takes the bus and 0.1 if he cycles. Using a tree diagram, or otherwise, find the probability that

   a  Theo takes the bus to school and is late

   b  Theo is on time for school.

**50** Three fair six-sided dice are thrown and the score is the sum of the three outcomes. Find the probability that the score is equal to 5.

**51** The table shows information on hair colour and eye colour of a group of adults.

|            | Blue eyes | Brown eyes | Green eyes |
|------------|-----------|------------|------------|
| Blond hair | 26        | 34         | 19         |
| Brown hair | 51        | 25         | 32         |

Find the probability that a randomly selected adult has

   a  blue eyes

   b  blond hair

   c  both blue eyes and blond hair.

**52** Events $A$ and $B$ satisfy: $P(A) = 0.6$, $P(A \cap B) = 0.2$ and $P(A \cup B) = 0.9$. Find $P(B)$.

**53** Events $A$ and $B$ satisfy: $P(A) = 0.7$, $P(A \cup B) = 0.9$ and $P(B') = 0.3$. Find $P(A \cap B)$.

**54** Let $A$ and $B$ be events such that $P(A) = \dfrac{2}{5}$, $P(B|A) = \dfrac{1}{2}$ and $P(A \cup B) = \dfrac{3}{4}$.

   a  Using a Venn diagram, or otherwise, find $P(A \cap B)$.

   b  Find $P(B)$.

**55** A box contains 40 red balls and 30 green balls. Two balls are taken out at random, one after another without replacement.

   a  Given that the first ball is red, find the probability that the second ball is also red.

   b  What is the probability that both balls are red?

   c  Is it more likely that the two balls are the same colour or different colour?

**56** The probability that it rains on any given day is 0.12, independently of any other day. Find the probability that

   a  it rains on at least one of two consecutive days

   b  it rains on three consecutive days.

**57** In a group of 100 people, 27 speak German, 30 speak Mandarin, 40 speak Spanish, 3 speak all three languages. 12 speak only German, 15 speak only Mandarin and 8 speak Spanish and Mandarin but not German.

   a  Draw a Venn diagram showing this information, and fill in the missing regions.

   b  Find the probability that a randomly selected person speaks only Spanish.

   c  Given that a person speaks Spanish, find the probability that they also speak Mandarin.

**58** In a class of 30 pupils, 22 have a phone, 17 have a smart watch and 15 have a tablet. 11 have both a smart watch and a phone, 11 have a phone and a tablet and 10 have a tablet and a smart watch. Everyone has at least one of those items.

   a  Let $x$ be the number of pupils who have all three items. Draw a Venn diagram showing this information.

   b  Use your Venn diagram to show that $x + 22 = 30$.

   c  Hence find the probability that a randomly selected pupil has all three items.

   d  Given that a pupil does not have a tablet, find the probability that they have a phone.

   e  Given that a pupil has exactly two of the items, find the probability that they are a smart watch and a tablet.

**59** a  A mother has two children. If one of them is a boy, what is the probability that they are both boys?

   b  A mother has two children. If the older one is a boy, what is the probability that they are both boys?

**TOOLKIT:** Problem Solving

In a 1950s US TV show, a contestant played the following game. He is shown three doors and told that there is a car behind one of them and goats behind the other two. He is asked to select a door at random. The show host (who knows what is behind each door) then opens one of the other two doors to reveal a goat. The contestant is then given the option to switch to the third door.

a Does your intuition suggest that the contestant should stick, switch or that it does not matter?

b If the car is behind door 1 and the contestant initially selects door 2, explain why they will definitely win if they switch.

c The table shows all possible positions of the car and all possible initial selections. For each combination it shows whether the contestant will win or lose if they switch.

Complete the table.

|  |  | Car behind door | | |
|---|---|---|---|---|
|  |  | 1 | 2 | 3 |
| **Door initially selected** | 1 | L |  |  |
|  | 2 | W |  |  |
|  | 3 |  |  |  |

d Hence decide whether the contestant should switch or stick with the original door.

e How does your intuition change if you exaggerate the problem in the following way?

A contestant is shown 100 doors. He selects door 1, then the host opens all the other doors except door 83, revealing goats behind all of them. Should he stick with door 1 or switch to door 83?

Exaggerating is a very powerful problem-solving technique!

## Checklist

■ You should be able to find experimental and theoretical probabilities. If all outcomes are equally likely then:

$$P(A) = \frac{n(A)}{n(U)}$$

■ You should be able to find probabilities of complementary events: $P(A') = 1 - P(A)$

■ You should be able to find the expected number of occurrences of an event:
  □ Expected number of occurrences of $A$ is $P(A) \times n$, where $n$ is the number of trials.

■ You should be able to use Venn diagrams to find probabilities.

■ You should be able to use tree diagrams to find probabilities.

■ You should be able to use sample space diagrams to find probabilities.

■ You should be able to calculate probabilities from tables of events.

■ You should be able to use the formula for probabilities of combined events:
  □ $P(A \cup B) = P(A) + P(B) - (A \cap B)$

■ You should be able to find conditional probabilities.

■ You should know about mutually exclusive events:
   □ If events $A$ and $B$ are mutually exclusive, then $P(A \cap B) = 0$ or, equivalently,
   □ $P(A \cup B) = P(A) + P(B)$

■ You should know about independent events:
   □ If the events $A$ and $B$ are independent, then $P(A \cap B) = P(A) \times P(B)$

---

## ■ Mixed Practice

**1** In a clinical trial, a drug is found to have a positive effect on 128 out of 200 participants. A second trial is conducted, involving 650 participants. On how many of the participants is the drug expected to have a positive effect?

**2** The table shows the genders and fruit preferences of a group of children.

|  | Boy | Girl |
|---|---|---|
| **Apples** | 16 | 21 |
| **Bananas** | 32 | 14 |
| **Strawberries** | 11 | 21 |

Find the probability that
**a** a randomly selected child is a girl
**b** a randomly selected child is a girl who prefers apples
**c** a randomly selected boy prefers bananas
**d** a randomly selected child is a girl, given that they prefer strawberries.

**3** Every break time at school, Daniel randomly chooses whether to play football or basketball. The probability that he chooses football is $\frac{2}{3}$. If he plays football, the probability that he scores is $\frac{1}{5}$. If he plays basketball, the probability that he scores is $\frac{3}{4}$.

Using a tree diagram, find the probability that Daniel
**a** plays football and scores
**b** does not score.

**4** The Venn diagram shows two sets:
$T = \{$multiples of 3 between 1 and 20$\}$
$F = \{$multiples of 5 between 1 and 20$\}$
**a** Copy the diagram and place the numbers from 1 to 20 (inclusive) in the correct region of the Venn diagram.
**b** Find the probability that a number is
   **i** a multiple of 5
   **ii** a multiple of 5 given that it is not a multiple of 3.

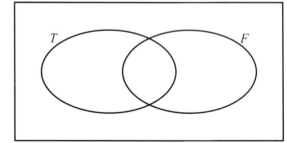

**5** In a certain school, the probability that a student studies Geography is 0.6, the probability that a student studies French is 0.4 and the probability that they study neither subject is 0.2.
**a** Find the probability that a student studies both subjects.
**b** Given that a student does not study French, what is the probability that they study Geography?

**6** Events $A$ and $B$ satisfy: $P(A) = 0.6$, $P(B) = 0.3$, $P(A \cup B) = 0.72$
**a** Find $P(A \cap B)$.
**b** Determine whether $A$ and $B$ are independent.

**7** Events $A$ and $B$ are independent with $P(A) = 0.6$ and $P(B) = 0.8$. Find $P(A \cup B)$.

**8** Events $A$ and $B$ satisfy: $P(A) = 0.7$, $P(B) = 0.8$ and $P(A \cap B) = 0.6$
   **a** Draw a Venn diagram showing events $A$ and $B$.
   **b** Find $P(A \mid B)$.
   **c** Find $P(B \mid A')$.

**9** Elsa has a biased coin with the probability $\frac{1}{3}$ of showing heads, a fair six-sided dice (numbered 1 to 6) and a fair four-sided dice (numbered 1 to 4). She tosses the coin once. If it comes up heads, she rolls the six-sided dice, and if comes up tails she rolls the four-sided dice.
   **a** Find the probability that the dice shows
      **i** a '1'
      **ii** a '6'.
   **b** Find the probability that the number on the dice is a multiple of 3.

**10** An integer is chosen at random from the first 10 000 positive integers. Find the probability that it is
   **a** a multiple of 7
   **b** a multiple of 9
   **c** a multiple of at least one of 7 and 9.

**11** A fair coin is tossed three times.
   **a** Copy and complete the tree diagram showing all possible outcomes.

   Find the probability that
   **b** the coin shows tails all three times
   **c** the coin shows heads at least once
   **d** the coin shows heads exactly twice.

Toss 1   Toss 2   Toss 3

**12** There are six black and eight white counters in a box. Asher takes out two counters without replacement. Separately, Elsa takes out one counter, returns it to the box and then takes another counter. Who has the larger probability of selecting one black and one white counter?

**13** A drawer contains eight red socks, six white socks and five black socks. Two socks are taken out at random. What is the probability that they are the same colour?

**14** A pack of cards in a game contains eight red, six blue and ten green cards. You are dealt two cards at random.
   **a** Given that the first card is blue, write down the probability that the second card is red.
   **b** Find the probability that you get at least one red card.

**15** Alan's laundry basket contains two green, three red and seven black socks.

   He selects one sock from the laundry basket at random.
   **a** Write down the probability that the sock is red.

   Alan returns the sock to the laundry basket and selects two socks at random.
   **b** Find the probability that the first sock he selects is green and the second sock is black.

   Alan returns the socks to the laundry basket and again selects two socks at random.
   **c** Find the probability that he selects two socks of the same colour.

   Mathematical Studies SL May 2013 Paper 1 TZ2 Q10

**16** 100 students at IB College were asked whether they study Music ($M$), Chemistry ($C$) or Economics ($E$) with the following results.

10 study all three

15 study Music and Chemistry

17 study Music and Economics

12 study Chemistry and Economics

11 study Music **only**

6 study Chemistry **only**

**a** Draw a Venn diagram to represent the information above.
**b** Write down the number of students who study Music but not Economics.

There are 22 Economics students **in total**.

**c** **i** Calculate the number of students who study Economics only.
  **ii** Find the number of students who study none of these three subjects.

A student is chosen at random from the 100 that were asked above.
**d** Find the probability that this student
  **i** studies Economics
  **ii** studies Music and Chemistry but not Economics
  **iii** does not study either Music or Economics
  **iv** does not study Music given that the student does not study Economics.

Mathematical Studies SL May 2013 Paper 2 TZ1 Q2

**17** A bag contains 10 red sweets and $n$ yellow sweets. Two sweets are selected at random. The probability that the two sweets are the same colour is $\frac{1}{2}$.
**a** Show that $n^2 - 21n + 90 = 0$.
**b** Hence find the possible values of $n$.

**18** At a large school, students are required to learn at least one language, Spanish or French. It is known that 75% of the students learn Spanish, and 40% learn French.
**a** Find the percentage of students who learn **both** Spanish and French.
**b** Find the percentage of students who learn Spanish, but not French.

At this school, 52% of the students are girls, and 85% of the girls learn Spanish.
**c** A student is chosen at random. Let $G$ be the event that the student is a girl, and let $S$ be the event that the student learns Spanish.
  **i** Find P $(G \cap S)$.
  **ii** Show that $G$ and $S$ are **not** independent.
**d** A boy is chosen at random. Find the probability that he learns Spanish.

Mathematics SL November 2012 Paper 2 Q10

**19** A group of 100 customers in a restaurant are asked which fruits they like from a choice of mangoes, bananas and kiwi fruits. The results are as follows.

15 like all three fruits

22 like mangoes and bananas

33 like mangoes and kiwi fruits

27 like bananas and kiwi fruits

8 like none of these three fruits

*x* like **only** mangoes

**a** **Copy** the following Venn diagram and correctly insert all values from the above information.

The number of customers that like **only** mangoes is equal to the number of customers that like **only** kiwi fruits. This number is half of the number of customers that like **only** bananas.

**b** Complete your Venn diagram from part **a** with this additional information **in terms of *x***.

**c** Find the value of *x*.

**d** Write down the number of customers who like
  **i** mangoes;
  **ii** mangoes or bananas.

**e** A customer is chosen at random from the 100 customers. Find the probability that this customer
  **i** likes none of the three fruits;
  **ii** likes only two of the fruits;
  **iii** likes all three fruits given that the customer likes mangoes and bananas.

**f** Two customers are chosen at random from the 100 customers. Find the probability that the two customers like none of the three fruits.

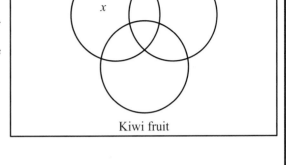

Mathematical Studies SL May 2015 Paper 2 TZ2 Q1

# 8 Core: Probability distributions

## ESSENTIAL UNDERSTANDINGS

- Probability enables us to quantify the likelihood of events occurring and to evaluate risk.
- Both statistics and probability provide important representations which enable us to make predictions, valid comparisons and informed decisions.
- These fields have power and limitations and should be applied with care and critically questioned, in detail, to differentiate between the theoretical and the empirical/observed.
- Probability theory allows us to make informed choices, to evaluate risk, and to make predictions about seemingly random events.

## In this chapter you will learn...

- about discrete random variables and their probability distributions
- how to find the expected value of a discrete random variable and apply this to real-life contexts
- the circumstances under which the binomial distribution is an appropriate model
- how to find the mean and variance of the binomial distribution
- about the key properties of the normal distribution
- how to carry out normal probability calculation and inverse normal calculations on your GDC.

### CONCEPTS

The following concepts will be addressed in this chapter:
- **Modelling** and finding structure in seemingly random events facilitate prediction.
- Different probability distributions provide a **representation** of the relationship between the theory and reality, allowing us to make predictions about what might happen.

■ **Figure 8.1** How are the different variables represented distributed?

## PRIOR KNOWLEDGE

Before starting this chapter, you should already be able to complete the following:

1  A bag contains six red and four blue balls. Two balls are removed at random without replacement. Find:

   a  the probability of getting two red balls

   b  the probability of getting one red ball.

2  Draw a box-and-whisker diagram for the following data: 2, 6, 8, 9, 10, 12.

In some cases, you are equally as likely to see an outcome at the lower or upper end of the possible range as in the middle; for example, getting a 1 or a 6 on a fair dice is just as likely as getting a 3 or 4. However, often the smallest or largest values are far less likely than those around the average; for example, scores in an exam tend to be grouped around the average mark, with scores of 0% or 100% rather less likely!

## Starter Activity

Look at the images in Figure 8.1. Discuss the variables that these extreme cases represent. If you had to draw a histogram representing these variables what would it look like?

**Now look at this problem:**

If you roll a fair dice 10 times, what do you expect the mean score to be?

**LEARNER PROFILE** – Caring

Game theory has combined with ecology to explain the 'mathematics of being nice'. There are some good reasons why some societies are advantaged by working together. Is any action totally altruistic?

# 8A Discrete random variables

## ■ Concept of discrete random variables and their probability distributions

A **discrete random variable** (usually denoted by a capital letter such as $X$) is a variable whose discrete output depends on chance, for example, 'the score on a fair dice'. In this case we might write a statement such as $P(X = 2) = \frac{1}{6}$, which just means 'the probability that the score on the dice is 2 is $\frac{1}{6}$'.

The sample space of a random variable, together with the probabilities of each outcome, is called the probability distribution of the variable. This is often displayed in a table.

---

**WORKED EXAMPLE 8.1**

A drawer has five black socks and three red socks. Two socks are removed without replacement. Find the probability distribution of $X$, the number of red socks removed.

A tree diagram is a useful way of working out the probabilities

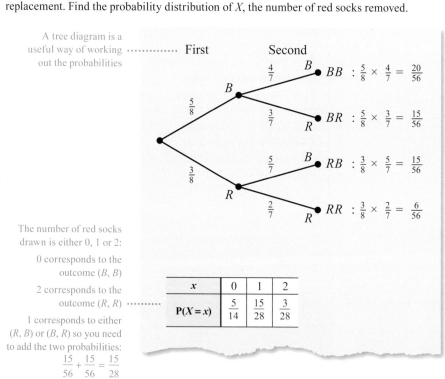

| First | Second | | |
|---|---|---|---|
| | $\frac{4}{7}$ $B$ | $BB$ | $: \frac{5}{8} \times \frac{4}{7} = \frac{20}{56}$ |
| $\frac{5}{8}$ $B$ | $\frac{3}{7}$ $R$ | $BR$ | $: \frac{5}{8} \times \frac{3}{7} = \frac{15}{56}$ |
| $\frac{3}{8}$ $R$ | $\frac{5}{7}$ $B$ | $RB$ | $: \frac{3}{8} \times \frac{5}{7} = \frac{15}{56}$ |
| | $\frac{2}{7}$ $R$ | $RR$ | $: \frac{3}{8} \times \frac{2}{7} = \frac{6}{56}$ |

The number of red socks drawn is either 0, 1 or 2:

0 corresponds to the outcome $(B, B)$

2 corresponds to the outcome $(R, R)$

1 corresponds to either $(R, B)$ or $(B, R)$ so you need to add the two probabilities:
$$\frac{15}{56} + \frac{15}{56} = \frac{15}{28}$$

| $x$ | 0 | 1 | 2 |
|---|---|---|---|
| $P(X = x)$ | $\frac{5}{14}$ | $\frac{15}{28}$ | $\frac{3}{28}$ |

---

Once you had found $P(B, B)$ and $P(R, R)$ from the tree diagram, you didn't actually need to work out $P(B, R) + P(R, B)$ as well because you knew that the sum of all possible outcomes on the right of the tree diagram must be 1.

This leads to a very important property of probability distributions.

**Tip**

Key Point 8.1 just says that the sum of all probabilities must be 1.

**KEY POINT 8.1**

In any discrete probability distribution:

● $\displaystyle\sum_x P(X = x) = 1$

**WORKED EXAMPLE 8.2**

The table below shows the probability distribution of $Y$, the number of cracks observed in a 1 km region of a sewer pipe.

| $y$ | 0 | 1 | 2 | >2 |
|---|---|---|---|---|
| $P(Y = y)$ | 0.45 | 0.3 | 0.15 | $k$ |

a  Find the value of $k$.

b  Find $P(Y \leqslant 1)$.

c  Find $P(Y = 0 | Y \leqslant 1)$.

The sum of the probabilities must be 1 ........ **a**  $0.45 + 0.3 + 0.15 + k = 1$
$$k = 0.1$$

If $Y \leqslant 1$ it can be either 0 or 1 ........ **b**  $P(Y \leqslant 1) = P(Y = 0) + P(Y = 1)$
$$= 0.45 + 0.3$$
$$= 0.75$$

Given $Y \leqslant 1$ means that we only use this part of the table: ........ **c**  $P(Y = 0 | Y \leqslant 1) = \dfrac{0.45}{0.75}$
$$= 0.6$$

| $y$ | 0 | 1 |
|---|---|---|
| $P(Y = y)$ | 0.45 | 0.3 |

## Expected value for discrete data

If you rolled a fair dice enough times you would expect to get a mean score of 3.5 (the mean of 1, 2, 3, 4, 5, 6). This is the theoretical mean or the expected value of the random variable, but in any given number of rolls, you may well get a different outcome for the average.

The expected value, $E(X)$, of a discrete random variable (also referred to as the mean, $\mu$) is the average value you would get if you carried out infinitely-many repetitions.

**KEY POINT 8.2**

For a discrete random variable, $X$:
$$E(X) = \sum_{x} x\, P(X = x)$$

**CONCEPTS – REPRESENTATION**

Using a single value to **represent** the expected outcome will always involve a loss of information. We are not suggesting that this will always be the outcome, or even that it is the most likely outcome. It really only represents what we would predict is the mean outcome if we were allowed an infinite number of observations of the random variable. Whenever we use probability to represent real-life situations, we should be aware of the inherent uncertainty in the predictions. However, if the number of observations is large enough, our predictions can become unerringly accurate. This underpins much of the statistical interpretations of economics and physics. Even though the behaviour of a single molecule of water going down a waterfall or a single customer in a market is difficult to predict, the overall behaviour of groups is remarkably predictable.

---

**WORKED EXAMPLE 8.3**

The random variable $Y$ has the probability distribution $P(Y = y) = 0.125(x^2 - 7x + 14)$ for $y \in \{2, 3, 4\}$

Find the expected value, $E(Y = y)$.

| Evaluate the formula at $y = 1$, 2 and 3 to create a probability distribution table | $y$ | 2 | 3 | 4 |
|---|---|---|---|---|
| | $P(Y = y)$ | 0.5 | 0.25 | 0.25 |

Use $E(Y) = \sum\limits_{y} y\ P(Y = y)$ ............... $E(Y) = (2 \times 0.5) + (3 \times 0.25) + (4 \times 0.25)$

$$= 2.75$$

---

 ## Applications

There are many applications of discrete random variables. They are used to model things from the creation of muon particles in nuclear physics to the number of infected individuals in an epidemic. One particular example you need to know in the IB is the application to games of chance. The definition of a fair game you need to use is one where the expected gain for each participant is zero.

---

**WORKED EXAMPLE 8.4**

Jenny and Hiroki play a game. Hiroki pays Jenny $4 to play, then rolls a fair standard six-sided dice. He gets back the score on the dice in dollars.

Is this a fair game?

Define the random variable so it is clear what you are referring to ...... Let the random variable $X$ be the score on the dice.

Use $E(X) = \sum\limits_{x} x\ P(X = x)$ to find the expected score ...... $E(X) = \left(1 \times \dfrac{1}{6}\right) + \left(2 \times \dfrac{1}{6}\right) + \left(3 \times \dfrac{1}{6}\right) + \left(4 \times \dfrac{1}{6}\right) + \left(5 \times \dfrac{1}{6}\right) + \left(6 \times \dfrac{1}{6}\right)$

$$= 3.5$$

This is the amount Hiroki expects to receive ...... So his expected payout is $3.50

But he pays out $4 to play ...... A player would expect to lose $0.50 when playing this game so it is not fair.

---

### TOK Links

What is meant by a fair game? Within IB maths we use the definition that the expected gain of each player is zero, but there are other possibilities. The expected gain of a player in a lottery is negative, but all the rules are clearly displayed and everybody is free to choose whether or not to play. There are often issues when commonly used words such as 'fair' are used in a precise mathematical context because different people may apply subtly different prior meanings. Language can be both a conduit and a barrier to knowledge.

Is 'expected gain' a useful criterion for deciding on the fairness of a game? What role should mathematics take in judging the ethics of a situation? Should mathematicians be held responsible for unethical applications of their work?

## Exercise 8A

For questions 1 to 6, use the method demonstrated in Worked Example 8.1 to find the probability distribution of $X$.

1  a  A bag contains three blue counters and four yellow counters. Two counters are removed without replacement and $X$ is the number of blue counters removed.

   b  Adam has seven blue shirts and eight purple shirts in the wardrobe. He picks two shirts without looking and $X$ is the number of blue shirts selected.

2  a  A fair coin is tossed twice and $X$ is the number of heads.

   b  A fair coin is tossed twice and $X$ is the number of tails.

3  a  A fair coin is tossed three times, $X$ is the number of tails.

   b  A fair coin is tossed three times, $X$ is the number of heads.

4  a  A fair six-sided dice is rolled once and $X$ is the number shown on the dice.

   b  A fair eight-sided dice is rolled once and $X$ is the number shown on the dice.

5  a  A fair six-sided dice is rolled three times and $X$ is the number of 5s.

   b  A fair eight-sided dice is rolled three times and $X$ is the number of 1s.

6  a  The probability that it rains on any given day is 0.2. $X$ is the number of rainy days in a two-day weekend.

   b  The probability that I go to the gym on any given day is 0.7. $X$ is the number of days I go to the gym during a two-day weekend.

For questions 7 to 9, use the method demonstrated in Worked Example 8.2 to find the value of $k$ and then find the required probabilities.

7  a

| $y$ | 0 | 1 | 2 | >2 |
|---|---|---|---|---|
| $P(Y = y)$ | 0.32 | 0.45 | 0.06 | $k$ |

   i  $P(Y \leqslant 1)$

   ii  $P(Y = 0 | Y \leqslant 1)$

   b

| $y$ | 0 | 1 | 2 | >2 |
|---|---|---|---|---|
| $P(Y = y)$ | 0.12 | 0.37 | 0.25 | $k$ |

   i  $P(Y \leqslant 1)$

   ii  $P(Y = 0 | Y \leqslant 1)$

8  a

| $x$ | 0 | 1 | 2 | 3 | 4 |
|---|---|---|---|---|---|
| $P(X = x)$ | 0.1 | $k$ | 0.3 | 0.3 | 0.1 |

   i  $P(X > 2)$

   ii  $P(X = 3 | X > 2)$

   b

| $x$ | 0 | 1 | 2 | 3 | 4 |
|---|---|---|---|---|---|
| $P(X = x)$ | 0.06 | 0.15 | $k$ | 0.35 | 0.08 |

   i  $P(X > 2)$

   ii  $P(X = 3 | X > 2)$

9  a

| $z$ | 2 | 3 | 5 | 7 | 11 |
|---|---|---|---|---|---|
| $P(Z = z)$ | 0.1 | $k$ | $3k$ | 0.3 | 0.1 |

   i  $P(Z < 7)$

   ii  $P(Z \geqslant 3 \mid Z < 7)$

b

| $z$ | 1 | 4 | 9 | 16 | 25 |
|---|---|---|---|---|---|
| $P(Z = z)$ | 0.1 | $k$ | 0.2 | $2k$ | 0.1 |

   i  $P(Z \leqslant 16)$

   ii  $P(Z > 4 \mid Z \leqslant 16)$

For questions 10 to 12, use the method demonstrated in Worked Example 8.3 to find the expected value of the random variable with the given probability distribution.

10  a

| $y$ | 1 | 2 | 3 |
|---|---|---|---|
| $P(Y = y)$ | 0.4 | 0.1 | 0.5 |

  b

| $y$ | 1 | 2 | 3 |
|---|---|---|---|
| $P(Y = y)$ | 0.4 | 0.3 | 0.3 |

11  a

| $x$ | 0 | 1 | 2 | 3 | 4 |
|---|---|---|---|---|---|
| $P(X = x)$ | 0.1 | 0.2 | 0.3 | 0.3 | 0.1 |

  b

| $x$ | 0 | 1 | 2 | 3 | 4 |
|---|---|---|---|---|---|
| $P(X = x)$ | 0.06 | 0.15 | 0.36 | 0.35 | 0.08 |

12  a

| $z$ | 2 | 3 | 5 | 7 | 11 |
|---|---|---|---|---|---|
| $P(Z = z)$ | 0.1 | 0.2 | 0.3 | 0.3 | 0.1 |

  b

| $z$ | 1 | 4 | 9 | 16 | 25 |
|---|---|---|---|---|---|
| $P(Z = z)$ | 0.1 | 0.4 | 0.2 | 0.2 | 0.1 |

13  A discrete random variable $X$ has the probability distribution given in this table.

| $x$ | 1 | 2 | 3 | 4 |
|---|---|---|---|---|
| $P(X = x)$ | 0.2 | 0.1 | 0.3 | $k$ |

  a  Find the value of $k$.       b  Find $P(X \geqslant 3)$.       c  Find $E(X)$.

14  A probability distribution of Y is given in the table.

| $y$ | 1 | 3 | 6 | 10 |
|---|---|---|---|---|
| $P(Y = y)$ | 0.1 | 0.3 | $k$ | $2k$ |

  a  Find the value of $k$.       b  Find $P(Y < 6)$.       c  Find $E(Y)$.

15  A bag contains eight red sweets and six yellow sweets. Two sweets are selected at random. Let $X$ be the number of yellow sweets.

  a  Find the probability distribution of $X$.

  b  Find the expected number of yellow sweets.

16   Two cards are selected at random from a standard pack of 52 cards.

  a   Find the probability that both cards are hearts.

  b   Find the probability distribution of $H$, the number of hearts drawn.

  c   Find the expected number of hearts.

17   Simon and Olivia play the following game:

  Simon gives Olivia £5. Olivia then tosses a fair coin. If the coin comes up heads she gives Simon £7; if it comes up tails she gives him £2.

  Determine whether this is a fair game.

18   Maria and Shinji play the following game: Shinji selects a card at random from a standard pack of 52 cards. If the card is a diamond, he gives Maria £3; otherwise Maria gives him £$n$.

  Find the value of $n$ which makes this a fair game.

19   A game stall offers the following game: You toss three fair coins. You receive the number of dollars equal to the number of tails.

  How much should the stall charge for one game in order to make the game fair?

20   A fair coin is tossed three times.

  a   Draw a tree diagram showing all possible outcomes.

  b   Find the probability that exactly two heads are rolled.

  Let $X$ be the number of heads.

  c   Find the probability distribution of $X$.

  d   Find the expected number of heads.

21   A fair four-sided dice (with faces numbered 1 to 4) is rolled twice and $X$ is the sum of the two scores.

  a   Find the probability distribution of $X$.

  b   Find the expected total score.

22   A discrete random variable has the probability distribution given by $P(X = x) = \frac{1}{15}(x + 3)$ for $x \in \{1, 2, 3\}$.

  a   Write down $P(X = 3)$.

  b   Find the expected value of $X$.

23   A discrete random variable $Y$ has the probability distribution given by $P(Y = y) = k(y - 1)$ for $y \in \{4, 5, 6\}$.

  a   Find the value of $k$.          b   Find $P(Y \geqslant 5)$.          c   Calculate $E(Y)$.

24   A probability distribution is given by $P(X = x) = \frac{c}{x}$ for $x = 1, 2, 3, 4$.

  a   Find the value of $c$.          b   Find $P(X = 2 | X \leqslant 3)$.          c   Calculate $E(X)$.

25   The discrete random variable $X$ has the probability distribution given in the table.

| $x$ | 1 | 2 | 3 | 4 |
|---|---|---|---|---|
| $P(X = x)$ | $a$ | 0.2 | 0.3 | $b$ |

Given that $E(X) = 2.4$, find the values of $a$ and $b$.

# 8B Binomial distribution

There are certain probability distributions that arise as the result of commonly occurring circumstances. One of these 'standard' distributions is the binomial distribution.

**KEY POINT 8.3**

The binomial distribution occurs when the following conditions are satisfied:
- There is a fixed number of trials.
- Each trial has one of two possible outcomes ('success' or 'failure').
- The trials are independent of each other.
- The probability of success is the same in each trial.

If a random variable $X$ has the binomial distribution with $n$ trials and a constant probability of success $p$, we write $X \sim \text{B}(n, p)$.

## WORKED EXAMPLE 8.5

A drawer contains five black and three red socks. Two socks are pulled out of the drawer without replacement and $X$ is the number of black socks removed.

Does this situation satisfy the standard conditions for $X$ to be modelled by a binomial distribution?

There is a fixed number of trials (2), each trial has two possible ..... outcomes (black or red)

No, since the probability of success is not the same in each trial.

## You are the Researcher

There are many standard probability distributions that you might like to research, all with interesting properties. In Worked Example 8.5, $X$ is actually modelled by a distribution called the hypergeometric distribution.

 You can calculate probabilities from the binomial distribution with your GDC.

## WORKED EXAMPLE 8.6

The number of bullseyes hit by an archer in a competition is modelled by a binomial distribution with 10 trials and a probability 0.3 of success. Find the probability that the archer hits

a exactly three bullseyes

b at least three bullseyes.

Define the random variable ...... Let $X$ be the number of bullseyes. you are going to use.

$X$ is binomial with $n = 10$ and $p = 0.3$ ...... $X \sim \text{B}(10, 0.3)$

Use your GDC, making sure you are ...... **a** $\text{P}(X = 3) = 0.267$ using the option which just finds the probability of a single value of $X$:

```
Binomial P.D
Data      :Variable
x         :3
Numtrial:10
P         :0.3
Save Res:None
Execute
|CALC
```

Use your GDC, making sure you are using ...... **b** $\text{P}(X \geqslant 3) = 1 - \text{P}(X \leqslant 2)$ the option which just finds the probability of a being less than or equal to a value of $X$: $= 1 - 0.383$
$= 0.617$

```
Binomial C.D
Data      :Variable
x         :2
Numtrial:10
P         :0.3
Save Res:None
Execute
|CALC
```

## CONCEPTS – MODELLING

In the example above, it is unlikely that the conditions for the binomial are perfectly met. For example, there might be different wind conditions for the different shots. Missing the previous shot might make the archer lose confidence or perhaps become more focused. However, even though the **model** is not perfect it does not mean it is useless, but the reporting of the results should reflect how certain you are in the assumptions.

## ■ Mean and variance of the binomial distribution

Being a standard distribution, you do not have to work out the expected value (mean) of a binomial distribution as you would for a general discrete probability distribution using the method from Section 8A.

Instead, there is a formula you can use (derived from the general method), and a similar formula for the variance as well.

### KEY POINT 8.4

If $X \sim B(n, p)$ then:

- $E(X) = np$
- $\text{Var}(X) = np(1 - p)$

### WORKED EXAMPLE 8.7

Dima guesses randomly the answers to a multiple-choice quiz. There are 20 questions and each question has five possible answers. If $X$ is the number of questions Dima answers correctly, find

a   the mean of $X$

b   the standard deviation of $X$.

$X$ is binomial with $n = 20$
and $p = \dfrac{1}{5} = 0.2$  ········  $X \sim B(20, 0.2)$

Use $E(X) = np$  ········  **a**  $E(X) = 20 \times 0.2$
$= 4$

Use $\text{Var}(X) = np(1 - p)$  ········  **b**  $\text{Var}(X) = 20 \times 0.2 \times (1 - 0.2)$
$= 3.2$

Standard deviation is the
square root of the variance  ········  So, standard deviation $= 1.79$

# Be the Examiner 8.1

Mia tosses a fair coin 10 times.

What is the probability she gets more than four but fewer than eight heads?

Which is the correct solution? Identify the errors made in the incorrect solutions.

| Solution 1 | Solution 2 | Solution 3 |
|---|---|---|
| $X$ is number of heads | $X$ is number of heads | $X$ is number of heads |
| $X \sim \text{B}(10, 0.5)$ | $X \sim \text{B}(10, 0.5)$ | $X \sim \text{B}(10, 0.5)$ |
| $\text{P}(4 < X < 8) = \text{P}(X \leqslant 7) - \text{P}(X \leqslant 3)$ | $\text{P}(4 < X < 8) = \text{P}(X \leqslant 8) - \text{P}(X \leqslant 3)$ | $\text{P}(4 < X < 8) = \text{P}(X \leqslant 7) - \text{P}(X \leqslant 4)$ |
| $= 0.9453 - 0.1719$ | $= 0.9893 - 0.1719$ | $= 0.9453 - 0.3770$ |
| $= 0.773$ | $= 0.817$ | $= 0.568$ |

# Exercise 8B

For questions 1 to 5, use the method demonstrated in Worked Example 8.5 to decide whether the random variable $X$ can be modelled by a binomial distribution. If not, state which of the conditions are not met. If yes, identify the distribution in the form $X \sim \text{B}(n, p)$.

1  a  A fair coin is tossed 30 times and $X$ is the number of heads.

   b  A fair six-sided dice is rolled 45 times and $X$ is the number of 3s.

2  a  A fair six-sided dice is rolled until it shows a 6 and $X$ is the number of 4s rolled up until that point.

   b  A fair coin is tossed until it shows tails and $X$ is the number of tosses up until that point.

3  a  Dan has eight white shirts and five blue shirts. He selects four shirts at random and $X$ is the number of white shirts.

   b  Amy has seven apples and six oranges. She selects three pieces of fruit at random and $X$ is the number of apples.

4  a  In a large population, it is known that 12% of people carry a particular gene. 50 people are selected at random and $X$ is the number of people who carry the gene.

   b  It is known that 23% of people in a large city have blue eyes. 40 people are selected at random and $X$ is the number of people with blue eyes.

5  a  In a particular town, the probability that a child takes the bus to school is 0.4. A family has five children and $X$ is the number of children in the family who take the bus to school.

   b  The probability that a child goes bowling on a Sunday is 0.15. $X$ is the number of friends, out of a group of 10, who will go bowling next Sunday.

For questions 6 to 12, use the method demonstrated in Worked Example 8.6 to find the required binomial probability.

In questions 6 to 8, $X \sim \text{B}(20, 0.6)$

6  a  $\text{P}(X = 11)$        b  $\text{P}(X = 12)$

7  a  $\text{P}(X \leqslant 12)$       b  $\text{P}(X \leqslant 14)$

8  a  $\text{P}(X > 11)$        b  $\text{P}(X > 13)$

In questions 9 to 12, $X \sim \text{B}\left(16, \frac{1}{3}\right)$

9  a  $\text{P}(X = 4)$         b  $\text{P}(X = 7)$

10  a  $\text{P}(X \geqslant 5)$       b  $\text{P}(X \geqslant 8)$

11  a  $\text{P}(3 < X \leqslant 5)$    b  $\text{P}(2 < X \leqslant 4)$

12  a  $\text{P}(4 \leqslant X < 7)$    b  $\text{P}(3 \leqslant X < 5)$

For questions 13 to 15, an archer has a probability of 0.7 of hitting the target. If they take 12 shots, find the probability that they hit the target:

13   a   exactly six times          b   exactly eight times

14   a   at least seven times       b   at least six times

15   a   fewer than eight times     b   fewer than four times.

For questions 16 to 18, use the method demonstrated in Worked Example 8.7 to find the mean and standard deviation of $X$.

16   a   A multiple-choice quiz has 25 questions and each question has four possible answers. Elsa guesses the answers randomly and $X$ is the number of questions she answers correctly.

     b   A multiple-choice quiz has 30 questions and each question has three possible answers. Asher guesses the answer randomly and $X$ is the number of questions he answers correctly.

17   a   A fair coin is tossed 10 times and $X$ is the number of heads.

     b   A fair coin is tossed 20 times and $X$ is the number of tails.

18   a   A fair dice is rolled 30 times and $X$ is the number of 6s.

     b   A fair dice is rolled 20 times and $X$ is the number of 4s.

19   A fair six-sided dice is rolled 10 times. Let $X$ be the number of 6s.

     a   Write down the probability distribution of $X$.

     b   Find the probability that exactly two 6s are rolled.

     c   Find the probability that more than two 6s are rolled.

20   A footballer takes 12 penalties. The probability that he scores on any shot is 0.85, independently of any other shots.

     a   Find the probability that he will score exactly 10 times.

     b   Find the probability that he will score at least 10 times.

     c   Find the expected number of times he will score.

21   The random variable $Y$ has distribution $B(20, 0.6)$. Find

     a   $P(Y = 11)$            b   $P(Y > 9)$           c   $P(7 \leqslant Y < 10)$       d   the variance of $Y$.

22   A multiple-choice test consists of 25 questions, each with five possible answers. Daniel guesses answers at random.

     a   Find the probability that Daniel gets fewer than 10 correct answers.

     b   Find the expected number of correct answers.

     c   Find the probability that Daniel gets more than the expected number of correct answers.

23   It is known that 1.2% of people in a country suffer from a cold at any time. A company has 80 employees.

     a   What assumptions must be made in order to model the number of employees suffering from a cold by a binomial distribution?

     b   Suggest why some of the assumptions might not be satisfied.

     Now assume that the conditions for a binomial distribution are satisfied. Find the probability that on a particular day,

     c   exactly three employees suffer from a cold

     d   more than three employees suffer from a cold.

24   A fair six-sided dice is rolled 20 times. What is the probability of rolling more than the expected number of sixes?

25   Eggs are sold in boxes of six. The probability that an egg is broken is 0.06, independently of other eggs.

     a   Find the probability that a randomly selected box contains at least one broken egg.

     A customer will return a box to the shop if it contains at least one broken egg. If a customer buys 10 boxes of eggs, find the probability that

     b   she returns at least one box

     c   she returns more than two boxes.

26   An archer has the probability of 0.7 of hitting the target. A round of a competition consists of 10 shots.

     a   Find the probability that the archer hits the target at least seven times in one round.

     b   A competition consists of five rounds. Find the probability that the archer hits the target at least seven times in at least three rounds of the competition.

**27** Daniel has a biased dice with probability $p$ of rolling a 6. He rolls the dice 10 times and records the number of sixes. He repeats this a large number of times and finds that the average number of 6s from 10 rolls is 2.7. Find the probability that in the next set of 10 rolls he gets more than four 6s.

**28** A random variable $X$ follows binomial distribution B $(n, p)$. It is known that the mean of $X$ is 36 and the standard deviation of $X$ is 3. Find the probability that $X$ takes its mean value.

**29** A machine produces electronic components that are packaged into packs of 10. The probability that a component is defective is 0.003, independently of all other components.

a  Find the probability that at least one of the components in the pack is defective.

The manufacturer uses the following quality control procedure to check large batches of boxes:

A pack of 10 is selected at random from the batch. If the pack contains at least one defective component, then another pack is selected from the same batch. If that pack contains at least one defective component, then the whole batch is rejected; otherwise the whole batch is accepted.

b  Find the probability that a batch is rejected.

c  Suggest a reason why the assumption of independence might not hold.

# 8C The normal distribution

## ■ Properties of the normal distribution

So far, we have been dealing with discrete distributions where all the outcomes and probabilities can be listed. With a continuous variable this is not possible, so instead continuous distributions are often represented graphically, with probabilities being given by the area under specific regions of the curve.

One of the most widely occurring distributions is the normal distribution, which is symmetrical with the most likely outcomes around a central value and increasingly unlikely outcomes farther away from the centre:

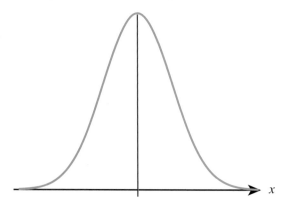

Typical examples of random variables that are modelled by a normal distribution include people's height, mass and scores in an IQ test.

To specify any particular normal distribution, you need to know the central value (its mean, $\mu$) and how spread out the curve is (its standard deviation, $\sigma$). We then write $X \sim N(\mu, \sigma^2)$.

**You are the Researcher**

The normal distribution is not just any curve with a peak in the middle. The basic normal distribution N(0,1) has the formula $\dfrac{1}{\sqrt{2\pi}}e^{-\frac{x^2}{2}}$. You might want to research how this formula was derived by mathematicians such as De Moivre and Gauss.

There are several other similar shaped curves – for example, the Cauchy distribution has formula $\dfrac{1}{\pi\left(1+x^2\right)}$. You might want to research its uses.

It will always be the case that approximately:

68% of the data lie between $\mu \pm \sigma$

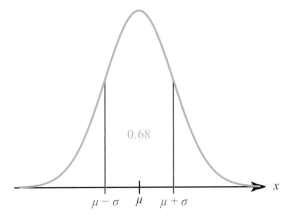

95% of the data lie between $\mu \pm 2\sigma$

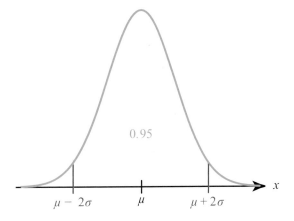

99.7% of the data lie between $\mu \pm 3\sigma$

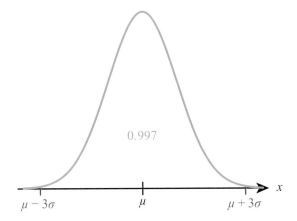

**WORKED EXAMPLE 8.8**

A sample of 20 students was asked about the number of siblings they had. The results are shown below.

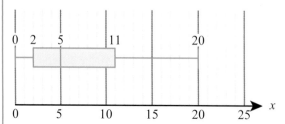

State two reasons why a normal distribution would be a poor model for the number of siblings of students in this population.

The normal distribution is symmetric ...... The sample suggests that the distribution is not symmetric.

The normal distribution is continuous ...... The number of siblings is discrete and not continuous.

**CONCEPTS – MODELLING**

There are many real-world situations where we assume that a normal distribution is an appropriate **model**. There are various more formal tests we can use to see whether data are likely to be taken from a normal distribution – for example, Q–Q plots or the Shapiro–Wilk test. Having ways to check assumptions is a key part of modelling. However, even if the data are not perfectly normal, there are many situations where we say that they are good enough. For example, if discrete data have enough levels, we often say that they are approximately normal.

 ■ Normal probability calculations

As with the binomial distribution, you can calculate probabilities directly from your GDC.

**Tip**

With a continuous variable we usually only use statements such as $P(X < k)$, rather than $P(X \leqslant k)$. In fact, both mean exactly the same – this is not the case for discrete variables!

**WORKED EXAMPLE 8.9**

If $X \sim N(10,25)$ find $P(X > 20)$.

Use the GDC in Normal C.D. mode to find the probability of $X$ taking a range of values. For the upper value just choose a very large number: .................... $P(X > 20) = 0.0228$

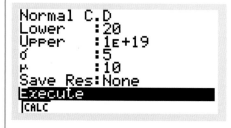

## Inverse normal calculations

You may be given a particular probability and asked to find the corresponding value of $X$. Again, this can be done on your GDC, this time using the Inverse Normal distribution.

---

**WORKED EXAMPLE 8.10**

The waist circumference of adult males is believed to follow a normal distribution with mean 92 cm and standard deviation 4.5 cm. A trouser manufacturer wants to make sure that their smallest size does not exclude more than 5% of the population.

Find the smallest size waist circumference their trousers should fit.

Use the Inverse Normal function on your GDC. ················· $X \sim N\left(92, 4.5^2\right)$
The tail is to the **left** as you want $P(X < x)$

$$P(X < x) = 0.05$$

$$x = 84.6$$

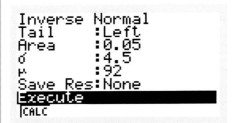

```
Inverse Normal
Tail      :Left
Area      :0.05
σ         :4.5
μ         :92
Save Res:None
Execute
|CALC
```

---

## Exercise 8C

For questions 1 to 8, use the method demonstrated in Worked Example 8.9 to find the required probability.

In questions 1 and 2, $X \sim N(10, 25)$.

  1   a   $P(X > 15)$                              b   $P(X > 12)$

  2   a   $P(X < 13)$                              b   $P(X < 7)$

In questions 3 and 4, $Y \sim N(25, 12)$.

  3   a   $P(18 < Y < 23)$                         b   $P(23 < Y < 28)$

  4   a   $P(Y > 24.8)$                            b   $P(Y > 27.4)$

In questions 5 and 6, $T$ has a normal distribution with mean 12.5 and standard deviation 4.6.

  5   a   $P(T < 11.2)$                            b   $P(T < 13.1)$

  6   a   $P(11.6 < T < 14.5)$                     b   $P(8.7 < T < 12.0)$

In questions 7 and 8, $M$ has a normal distribution with mean 62.4 and variance 8.7.

  7   a   $P(M > 65)$                              b   $P(M > 60)$

  8   a   $P(61.5 < M < 62.5)$                     b   $P(62.5 < M < 63.5)$

For questions 9 to 12, use the method demonstrated in Worked Example 8.10 to find the value of $x$.

In questions 9 and 10, $X \sim N(10, 25)$.

  9   a   $P(X < x) = 0.75$                        b   $P(X < x) = 0.26$

 10   a   $P(X > x) = 0.65$                        b   $P(X > x) = 0.12$

In questions 11 and 12, $X \sim N(38, 18)$.

 11   a   $P(X < x) = 0.05$                        b   $P(X < x) = 0.97$

 12   a   $P(X > x) = 0.99$                        b   $P(X > x) = 0.02$

13  The box plot on the right summarizes the ages of pupils at a school. Does the box plot suggest that the ages are normally-distributed? Give one reason to explain your answer.

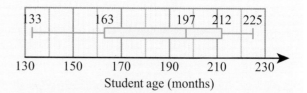

14  The heights of trees in a forest can be modelled by a normal distribution with mean 8.7 m and standard deviation 2.3 m. Elsa wants to find the probability that a tree is between 5 and 10 metres tall.

   a  Sketch a normal distribution curve and shade the area representing the required probability.

   b  Find the probability that a randomly chosen tree is between 5 and 10 metres tall.

15  The battery life of a particular model of a phone can be modelled by a normal distribution with mean 56 hours and standard deviation 8 hours. Find the probability that a randomly selected phone has battery life

   a  between 50 and 60 hours

   b  more than 72 hours.

16  The heights of trees in a forest are modelled by a normal distribution with mean 17.2 m and standard deviation 6.3 m. Find the probability that a randomly selected tree

   a  has height between 15 m and 20 m

   b  is taller than 20 m.

17  Charlotte's times for running a 400 m race are normally distributed with mean 62.3 seconds and standard deviation 4.5 seconds.

   a  What is the probability that Charlotte's time in a randomly selected race is over 65 seconds?

   b  Charlotte ran 38 races this season. What is the expected number of races in which she ran over 65 seconds?

   c  In order to qualify for the national championships, Charlotte needs to run under 59.7 seconds. What is the probability that she will qualify with a single race?

18  The times taken by a group of children to complete a puzzle can be modelled by a normal distribution with mean 4.5 minutes and variance 2.25 minutes$^2$.

   a  What is the probability that a randomly selected child takes more than 7 minutes to complete the puzzle?

   b  Find the length of time such that 90% of the children take less than this time to complete the puzzle.

19  The daily amount of screen time among teenagers can be modelled by a normal distribution with mean 4.2 hours and standard deviation 1.3 hours.

   a  What percentage of teenagers get more than 6 hours of screen time per day?

   b  Let $T$ be the time such that 5% of teenagers get more screen time than this. Find the value of $T$.

   c  In a group of 350 teenagers, how many would you expect to get less than 3 hours of screen time per day?

20  In a long jump competition, the distances achieved by a particular age group can be modelled by a normal distribution with mean 5.2 m and variance 0.6 m$^2$.

   a  In a group of 30 competitors from this age group, how many would you expect to jump further than 6 m?

   b  In a large competition, the top 5% of participants will qualify for the next round. Estimate the qualifying distance.

21  Among 17-year-olds, the times taken to run 100 m are normally distributed with mean 14.3 s and variance 2.2 s$^2$. In a large competition, the top 15% of participants will qualify for the next round. Estimate the required qualifying time.

22  A random variable follows a normal distribution with mean 12 and variance 25. Find

   a  the upper quartile

   b  the interquartile range of the distribution.

23  A random variable follows a normal distribution with mean 3.6 and variance 1.44. Find the 80th percentile of the distribution.

**24** Given that $X \sim N(17, 3.2^2)$, find the value of $k$ such that $P(15 < X < k) = 0.62$.

**25** Given that $Y \sim N(13.2, 5.1^2)$, find the value of $c$ such that $P(c < Y < 17.3) = 0.14$.

**26** Percentage scores on a test, taken by a large number of people, are found to have a mean of 35 with a standard deviation of 20. Explain why this suggests that the scores cannot be modelled by a normal distribution.

**27** The box plot shows the distribution of masses of a large number of children in a primary school.

Student mass (kg)

a   Identify one feature of the box plot which suggests that a normal distribution might be a good model for the masses.

b   The mean mass of the children is 36 kg and the standard deviation is 8.5 kg. Copy and complete the box plot by adding the relevant numbers. You may assume that the end points are the minimum and maximum values which are not outliers.

**28** The reaction time of a sprinter follows a normal distribution with mean 0.2 seconds and standard deviation 0.1 seconds.

a   Find the probability of getting a negative reaction time. Explain in this context why this might be a plausible outcome.

A false start is declared if the reaction time is less than 0.1 seconds.

b   Find the probability of the sprinter getting a false start.

c   In 10 races, find the probability that the sprinter gets a false start more than once.

d   What assumptions did you have to make in part c? How realistic are these assumptions?

**29** A farmer has chickens that produce eggs with masses that are normally distributed with mean 60 g and standard deviation 5 g. Eggs with a mass less than 53 g cannot be sold. Eggs with a mass between 53 g and 63 g are sold for 12 cents and eggs with a mass above 63 g are sold for 16 cents. If the farmer's hens produce 6000 eggs each week, what is the farmer's expected income from the eggs.

## Checklist

■ You should be able to work with discrete random variables and their probability distributions. In any discrete probability distribution:

□  $\sum_x P(X = x)$

■ You should be able to find the expected value of a discrete random variable and apply this to real-life contexts. For a discrete random variable, $X$:

□  $E(X) = \sum_x x \, P(X = x)$

■ You should know the conditions that need to be satisfied for the binomial distribution to be appropriate:

□   There is a fixed number of trials.

□   Each trial has one of two possible outcomes ('success' or 'failure').

□   The trials are independent of each other.

□   The probability of success is the same in each trial.

■ You should be able to find the mean and variance of the binomial distribution. If $X \sim B(n, p)$ then:

□   $E(X) = np$

□   $Var(X) = np(1 - p)$

■ You should know the key properties of the normal distribution.

■ You should be able to carry out normal probability calculations and inverse normal calculations on your GDC.

## ■ Mixed Practice

**1** Random variable $X$ has the probability distribution given in the table.

| $x$ | 1 | 2 | 3 | 4 |
|---|---|---|---|---|
| $P(X = x)$ | 0.2 | 0.2 | 0.1 | $k$ |

   **a** Find the value of $k$.
   **b** Find $P(X \geqslant 3)$.
   **c** Find $E(X)$.

**2** A fair six-sided dice is rolled twelve times. Find the probability of getting
   **a** exactly two 6s
   **b** more than two 1s.

**3** Lengths of films are distributed normally with mean 96 minutes and standard deviation 12 minutes. Find the probability that a randomly selected film is
   **a** between 100 and 120 minutes long
   **b** more than 105 minutes long.

**4** Scores on a test are normally distributed with mean 150 and standard deviation 30. What score is needed to be in the top 1.5% of the population?

**5** A factory making plates knows that, on average, 2.1% of its plates are defective. Find the probability that in a random sample of 20 plates, at least one is defective.

**6** Daniel and Alessia play the following game. They roll a fair six-sided dice. If the dice shows an even number, Daniel gives Alessia $1. If it shows a 1, Alessia gives Daniel $1.50; otherwise Alessia gives Daniel $0.50.
   **a** Complete the table showing possible outcomes of the game for Alessia.

| Outcome | $1 | −$0.50 | −$1.50 |
|---|---|---|---|
| Probability | | | |

   **b** Determine whether the game is fair.

**7** A spinner with four sectors, labelled 1, 3, 6 and $N$ (where $N > 6$) is used in a game. The probabilities of each number are shown in the table.

| Number | 1 | 3 | 6 | $N$ |
|---|---|---|---|---|
| Probability | $\frac{1}{2}$ | $\frac{1}{5}$ | $\frac{1}{5}$ | $\frac{1}{10}$ |

A player pays three counters to play a game, spins the spinner once and receives the number of counters equal to the number shown on the spinner. Find the value of $N$ so that the game is fair.

**8** A random variable $X$ is distributed normally with a mean of 20 and variance 9.
  **a** Find $P(X \leqslant 24.5)$.
  **b** Let $P(X \leqslant k) = 0.85$.
    **i** Represent this information on a copy of the following diagram.

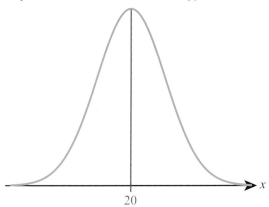

    **ii** Find the value of $k$.

Mathematics SL May 2011 Paper 2 TZ1 Q4

**9** The random variable $X$ has the distribution $B(30, p)$. Given that $E(X) = 10$, find
  **a** the value of $p$;
  **b** $P(X = 10)$;
  **c** $P(X \geqslant 15)$.

Mathematics HL May 2012 Paper 2 TZ2 Q2

**10** The mean mass of apples is 110 g with standard deviation 12.2 g.
  **a** What is the probability that an apple has a mass of less than 100 g?
  **b** Apples are packed in bags of six. Find the probability that more than one apple in a bag has a mass of less than 100 g.

**11** Masses of eggs are normally distributed with mean 63 g and standard deviation 6.8 g. Eggs with a mass over 73 g are classified as 'very large'.
  **a** Find the probability that a randomly selected box of six eggs contains at least one very large egg.
  **b** What is more likely: a box of six eggs contains exactly one very large egg, or that a box of 12 eggs contains exactly two very large eggs?

**12** Bob competes in the long jump. The lengths of his jumps are normally distributed with mean 7.35 m and variance 0.64 m².
  **a** Find the probability that Bob jumps over 7.65 m.
  **b** In a particular competition, a jump of 7.65 m is required to qualify for the final. Bob makes three jumps. Find the probability that he will qualify for the final.

**13** In an athletics club, the 100 m times of all the runners follow the same normal distribution with mean 15.2 s and standard deviation 1.6 s. Eight of the runners have a practice race. Heidi runs the time of 13.8 s. What is the probability that she wins the race?

**14** The amount of paracetamol in a tablet is distributed normally, with mean 500 mg and variance 6400 mg². The minimum dose required for relieving a headache is 380 g. Find the probability that, in a trial of 25 randomly selected participants, more than two receive less than the required dose.

**15** A dice is biased and the probability distribution of the score is given in the following table.

| Score | 1 | 2 | 3 | 4 | 5 | 6 |
|---|---|---|---|---|---|---|
| Probability | 0.15 | 0.25 | 0.08 | 0.17 | 0.15 | 0.20 |

  **a**  Find the expected score for this dice.
  **b**  The dice is rolled twice. Find the probability of getting one 5 and one 6.
  **c**  The dice is rolled 10 times. Find the probability of getting at least two 1s.

**16** A biased four-sided dice is used in a game. The probabilities of different outomes on the dice are given in this table.

| Value | 1 | 2 | 3 | 4 |
|---|---|---|---|---|
| Probability | $\frac{1}{4}$ | $\frac{1}{3}$ | $a$ | $b$ |

A player pays two counters to play the game, rolls the dice once and receives the number of counters equal to the number shown on the dice. Given that the player expects to win one counter when he plays the game three times, find the value of $a$.

**17** The heights of pupils at a school can be modelled by a normal distribution with mean 148 cm and standard deviation 8 cm.
  **a**  Find the interquartile range of the distribution.
  **b**  What percentage of the pupils at the school should be considered outliers?

**18** The mean time taken to complete a test is 10 minutes and the standard deviation is 5 minutes. Explain why a normal distribution would be an inappropriate model for the time taken.

**19** Quality control requires that no more than 2.5% of bottles of water contain less than the labelled volume. If a manufacturer produces bottles containing a volume of water following a normal distribution with mean value of 330 ml and standard deviation 5 ml, what should the labelled volume be, given to the nearest whole number of millilitres?

**20** Jan plays a game where she tosses two fair six-sided dice. She wins a prize if the sum of her scores is 5.
  **a**  Jan tosses the two dice once. Find the probability that she wins a prize.
  **b**  Jan tosses the two dice eight times. Find the probability that she wins three prizes.

Mathematics SL May 2010 Paper 2 TZ2 Q3

**21** The time taken for a student to complete a task is normally distributed with a mean of 20 minutes and a standard deviation of 1.25 minutes.
  **a**  A student is selected at random. Find the probability that the student completes the task in less than 21.8 minutes.
  **b**  The probability that a student takes between $k$ and 21.8 minutes is 0.3. Find the value of $k$.

Mathematics SL November 2013 Paper 2 Q6

**22** The weight, $W$, of bags of rice follows a normal distribution with mean 1000 g and standard deviation 4 g.
  **a**  Find the probability that a bag of rice chosen at random weighs between 990 g and 1004 g.

95% of the bags of rice weigh less than $k$ grams.
  **b**  Find the value of $k$.

For a bag of rice chosen at random, $P(1000 - a < W < 1000 + a) = 0.9$.
  **c**  Find the value of $a$.

Mathematical Studies May 2015 Paper 1 TZ1 Q13

**23** A test has five questions. To pass the test, at least three of the questions must be answered correctly.

The probability that Mark answers a question correctly is $\frac{1}{5}$. Let $X$ be the number of questions that Mark answers correctly.
   **a** **i** Find E($X$).
   **ii** Find the probability that Mark passes the test.

Bill also takes the test. Let $Y$ be the number of questions that Bill answers correctly. The following table is the probability distribution for $Y$.

| $y$ | 0 | 1 | 2 | 3 | 4 | 5 |
|---|---|---|---|---|---|---|
| $P(Y = y)$ | 0.67 | 0.05 | $a + 2b$ | $a - b$ | $2a + b$ | 0.04 |

   **b** **i** Show that $4a + 2b = 0.24$.
   **ii** Given that E($Y$) = 1, find $a$ and $b$.
   **c** Find which student is more likely to pass the test.

Mathematics SL November 2010 Paper 2 Q9

**24** The distances thrown by Josie in an athletics competition is modelled by a normal distribution with mean 40 m and standard deviation 5 m. Any distance less than 40 m gets 0 points. Any distance between 40 m and 46 m gets 1 point. Any distance above 46 m gets 4 points.
   **a** Find the expected number of points Josie gets if she throws
      **i** once
      **ii** twice.
   **b** What assumptions have you made in **a ii**? Comment on how realistic these assumptions are.

**25** When a fair six-sided dice is rolled $n$ times, the probability of getting no 6s is 0.194, correct to three significant figures. Find the value of $n$.

**26** Find the smallest number of times that a fair coin must be tossed so that the probability of getting no heads is smaller than 0.001.

**27** The probability of obtaining 'tails' when a biased coin is tossed is 0.57. The coin is tossed 10 times. Find the probability of obtaining
   **a** **at least** four tails
   **b** the fourth tail on the 10th toss.

Mathematics SL May 2012 Paper 2 TZ1 Q7

**28** A group of 100 people are asked about their birthdays. Find the expected number of dates on which no people have a birthday. You may assume that there are 365 days in a year and that people's birthdays are equally likely to be on any given date.

**29** A private dining chef sends out invitations to an exclusive dinner club. From experience he knows that only 50% of those invited turn up. He can only accommodate four guests. On the first four guests he makes $50 profit per guest; however, if more than four guests turn up he has to turn the additional guests away, giving them a voucher allowing them to have their next dinner for free, costing him $100 per voucher.
   **a** Assuming that responses are independent, show that his expected profit if he invites five people is $120 to 3 significant figures.
   **b** How many invitations should he send out?

# 9 Core: Differentiation

## ESSENTIAL UNDERSTANDINGS

- Differentiation describes the rate of change between two variables. Understanding these rates of change allows us to model, interpret and analyse real-world problems and situations.
- Differentiation helps us understand the behaviour of functions and allows us to interpret features of their graphs.

## In this chapter you will learn...

- how to estimate the value of a limit from a table or graph
- informally about the gradient of a curve as a limit
- about different notation for the derivative
- how to identify intervals on which functions are increasing or decreasing
- how to interpret graphically $f'(x) > 0$, $f'(x) = 0$, $f'(x) < 0$
- how to differentiate functions of the form $f(x) = ax^n + bx^m$, where $n, m \in \mathbb{Z}$
- how to find the equations of the tangent and normal to a function at a given point.

## CONCEPTS

The following concepts will be addressed in this chapter:

- The derivative may be represented physically as a rate of **change** and geometrically as the gradient or slope function.
- Differentiation allows you to find a **relationship** between a function and its gradient.
- This relationship can have a graphical or algebraic **representation**.

## LEARNER PROFILE – Knowledgeable

What types of mathematical knowledge are valued by society? There are few jobs where you will be using differentiation daily, so knowing lots of facts is not really the most important outcome of a mathematics education. However, there are lots of skills fundamental to mathematics which are desirable in everyday life: the ability to think logically, communicate precisely and pay attention to small details.

■ **Figure 9.1** How can we model the changes represented in these photos?

**Differentiation** is the process of establishing the rate of change of the $y$-coordinate of a graph when the $x$-coordinate is changed. You are used to doing this for a straight line, by finding the gradient. In that case the rate of change is the same no matter where you are on the graph, but for non-linear graphs this is not the case – the rate of change is different at different values of $x$.

Differentiation has wide-ranging applications, from calculating velocity and acceleration in physics, to the rate of a reaction in chemistry, to determining the optimal price of a quantity in economics.

## Starter Activity

Look at the pictures in Figure 9.1. In small groups, describe what examples of change you can see. Can you think of any more examples?

**Now look at this problem:**

In a drag race, the displacement, $d$ metres, of a car along a straight road is modelled by $d = 4t^2$, where $t$ is the time in seconds.

Find the average speed of the car:

a   in the first 5 seconds

b   between 1 and 5 seconds

c   between 4 and 5 seconds.

What do you think the reading on the car's speedometer is at 5 seconds?

# 9A Limits and derivatives

## ■ Introduction to the concept of a limit

Suppose we are not sure what the value of f(a) is. If f(x) gets closer and closer to a particular value when x gets closer and closer to a, then it is said to have a limit. For example, in the Starter Activity problem you might have found that the speed of the car gets closer and closer to $40\,\mathrm{m\,s^{-1}}$.

You might reasonably ask why we do not just calculate f(a) directly. The main situation when we cannot do this is if, when we try to find f(a), we have to do a division by zero. If only the denominator is zero then the answer is undefined; however, if both the numerator and denominator are zero then it is possible that it has a limit.

### TOK Links

Some mathematicians have tried to create a new number system where division by zero is defined – often as infinity. These systems have usually been rejected within mathematical communities. Who should decide whether these new number systems are acceptable, and what criteria should they use?

**WORKED EXAMPLE 9.1**

Use technology to suggest the limit of $\dfrac{\sin x}{\left(\dfrac{\pi x}{180}\right)}$ as x tends to zero, where x is in degrees.

We can use a spreadsheet to look at what happens when x gets very small. We could also use a graphical calculator to sketch the function

| $x$ | $\dfrac{\sin x}{\left(\dfrac{\pi x}{180}\right)}$ |
|---|---|
| 10 | 0.994 931 |
| 5 | 0.998 731 |
| 1 | 0.999 949 |
| 0.1 | 0.999 999 |
| 0.01 | 1 |

We can then interpret the spreadsheet. Notice that even though the spreadsheet has an output of 1, it does not mean that the value of the function here is exactly 1. It just means that it is 1 within the degree of accuracy the spreadsheet is using

As x gets closer to 0, the function gets closer to 1, so this is the limit.

## Links to: Physics

The function $\dfrac{\sin x}{x}$ turns out to be very important in optics, where it is used to give the amplitude of the light hitting a screen in a double slit experiment. This investigation of limits can be used to prove that the central line is brightest.

## ■ The derivative interpreted as a gradient function and as a rate of change

One of the main uses of limits in calculus is in finding the **gradient of a curve.**

**KEY POINT 9.1**

The gradient of a curve at a point is the gradient of the **tangent to the curve** at that point.

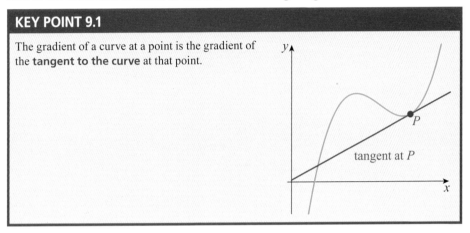

We could try to draw tangents on a graph and measure their gradients, but if you try this you will see that it is quite hard to do accurately. There is one important insight that helps us to calculate the gradient of the tangent.

**KEY POINT 9.2**

The gradient of a tangent at a point is the limit of the gradient of smaller and smaller chords from that point.

In Worked Example 9.2 the gradient of the chord was written as $\frac{\Delta y}{\Delta x}$ where '$\Delta$' is the Greek equivalent of 'd'. The German philosopher and mathematician Gottfried Leibniz came up with the notation of replacing '$\Delta$' for small changes with 'd' for infinitesimally small changes, that is, the limit as the change tends towards zero.

■ **Figure 9.2** Gottfried Leibniz

## WORKED EXAMPLE 9.2

On the graph $y = x^2$ a chord is drawn from $P(4, 16)$ to the point $Q(x, y)$.

Copy and complete the table below to find the gradient of various chords from $(4, 16)$ to point $Q$.

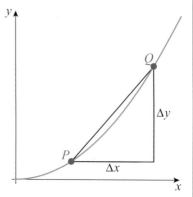

| $x$ | $y$ | $\Delta x$ | $\Delta y$ | Gradient of $PQ$ |
|---|---|---|---|---|
| 5 | 25 | 1 | 9 | 9 |
| 4.5 | | | | |
| 4.1 | | | | |
| 4.01 | | | | |

Hence, estimate the limit of the gradient of the chord from $(4, 16)$ as the chord gets very small. What does this limit tell you?

Fill in the table using:

$y = x^2$

$\Delta x = x - 4$

$\Delta y = y - 16$

Gradient $= \dfrac{\Delta y}{\Delta x}$

| $x$ | $y$ | $\Delta x$ | $\Delta y$ | Gradient of $PQ$ |
|---|---|---|---|---|
| 5 | 25 | 1 | 9 | 9 |
| 4.5 | 20.25 | 0.5 | 4.25 | 8.5 |
| 4.1 | 16.81 | 0.1 | 0.81 | 8.1 |
| 4.01 | 16.0801 | 0.01 | 0.0801 | 8.01 |

Look at what the gradient is tending to

The gradient of the chord is tending to 8.
This suggests that the gradient of the curve at $(4, 16)$ is 8.

## TOK Links

This approach of gathering evidence and suggesting a result is really a method of scientific induction rather than mathematical proof. In the analysis and approaches HL book, there is a formalization of this method called 'differentiation from first principles' which does prove these results; however, some people find scientific induction gives them a better understanding of what is going on. Is proof always the best way to explain something? Does explanation mean different things in maths and science?

**Links to: Physics**

The notation $f'(x)$ is called Lagrange's notation. There is another notation popular in physics called Newton's notation, where $\dot{x}$ is used to represent $\dfrac{dx}{dt}$.

The process of finding the gradient of a curve is called differentiation. Each point of the curve can have a different value for the gradient. The function which gives the gradient at any point on a curve is called the **derivative**, **slope function** or **gradient function**. There are several different notations for the derivative. If the curve is expressed as $y$ in terms of $x$, we would write the derivative as $\dfrac{dy}{dx}$. However, there is nothing special about $y$ and $x$ here. If there was a graph of variable G against variable $t$ then the derivative would be $\dfrac{dG}{dt}$.

If the function being differentiated is $f(x)$ then the derivative is $f'(x)$. Much of the rest of this chapter will look at how we can calculate the derivative of a function, but it is also important to think about why we want to do this. As well as representing gradients of graphs, derivatives have another interpretation. If the graph is $y$ against $x$ then the gradient tells you the instantaneous amount $y$ is increasing for each unit that $x$ increases. This is called the rate of change of $y$ with respect to $x$. In many real-life situations, we have ideas about the rates of change of variables, which allows us to create models called differential equations.

**TOOLKIT:** Modelling

There are many textbooks about solving differential equations, but increasingly technology can be used to solve them. One of the main skills modern mathematicians need is creating differential equations which capture all the important features of a real-world scenario.

### WORKED EXAMPLE 9.3

The rate of growth of a bacterial population of size $N$ in a Petri dish over time $t$ is proportional to the population size. Write an equation that describes this information.

Interpret the rate of growth as a derivative. We are looking at $N$ changing with respect to $t$

$$\frac{dN}{dt} \propto N$$

With direct proportion we can turn it into an equation with an unknown constant factor

$$\frac{dN}{dt} = kN$$

### CONCEPTS – CHANGE

It is easy to think that 'rate of change' means **change** over time, but it could be the rate at which the height of a ball changes as the distance travelled changes, or the rate at which side-effects increase as the dose of a medicine increases.

Ancient Greek mathematicians came very close to 'discovering' calculus. Not realizing that the issue of zero divided by zero could be resolved by considering limits was one of the main hurdles they failed to overcome. They were aware of many difficulties concerned with dealing correctly with limits, the most famous of which are called Zeno's paradoxes.

## Exercise 9A

In this exercise all angles are in degrees.

For questions 1 to 5, use the method demonstrated in Worked Example 9.1 to suggest the limit of the following functions as $x$ tends to zero.

1  a  $\dfrac{3x}{5x}$        2  a  $\dfrac{3x + 5x^2}{2x}$        3  a  $\dfrac{\sin 2x}{\left(\dfrac{\pi x}{180}\right)}$        4  a  $\dfrac{\tan x}{\left(\dfrac{\pi x}{180}\right)}$        5  a  $\dfrac{2^x - 1}{x}$

  b  $\dfrac{7x^2}{2x}$        b  $\dfrac{x^2 + 3x}{2x^2 + x}$        b  $\dfrac{\sin 3x}{\left(\dfrac{\pi x}{180}\right)}$        b  $\dfrac{\tan 2x}{\left(\dfrac{\pi x}{90}\right)}$        b  $\dfrac{3^x - 1}{x}$

For questions 6 to 10, use a graphical calculator to sketch the graph of the function and hence suggest the limit as $x$ tends to zero.

6  a  $\dfrac{x}{5x}$        7  a  $\dfrac{x + x^3}{2x}$        8  a  $\dfrac{\sin 5x}{\left(\dfrac{\pi x}{36}\right)}$        9  a  $\dfrac{\cos x - 1}{\left(\dfrac{\pi x}{180}\right)^2}$        10  a  $\dfrac{5^x - 1}{2x}$

  b  $\dfrac{7x^2}{10x}$        b  $\dfrac{x^2 + x}{x^2 + 2x}$        b  $\dfrac{\sin 10x}{\left(\dfrac{\pi x}{180}\right)}$        b  $\dfrac{\cos x - 1}{\left(\dfrac{\pi x}{180}\right)}$        b  $\dfrac{4^x - 1}{4x}$

For questions 11 to 15, use the method demonstrated in Worked Example 9.2 to suggest a value for the gradient of the tangent to the given curve at point $P$.

11  a  The curve is $y = x^2$ and $P$ is $(0, 0)$.

| $x$ | $y$ | $\Delta x$ | $\Delta y$ | Gradient of $PQ$ |
|---|---|---|---|---|
| 1 | 1 | 1 | 1 | 1 |
| 0.5 | | | | |
| 0.1 | | | | |
| 0.01 | | | | |

  b  The curve is $y = x^2$ and $P$ is $(2, 4)$.

| $x$ | $y$ | $\Delta x$ | $\Delta y$ | Gradient of $PQ$ |
|---|---|---|---|---|
| 3 | 9 | 1 | 5 | 5 |
| 2.5 | | | | |
| 2.1 | | | | |
| 2.01 | | | | |

12  a  The curve is $y = 2x^3$ and $P$ is $(1, 2)$.

| $x$ | $y$ | $\Delta x$ | $\Delta y$ | Gradient of $PQ$ |
|---|---|---|---|---|
| 2 | 16 | 1 | 14 | 14 |
| 1.5 | | | | |
| 1.1 | | | | |
| 1.01 | | | | |

  b  The curve is $y = 3x^4$ and $P$ is $(0, 0)$.

| $x$ | $y$ | $\Delta x$ | $\Delta y$ | Gradient of $PQ$ |
|---|---|---|---|---|
| 1 | 3 | 1 | 3 | 3 |
| 0.5 | | | | |
| 0.1 | | | | |
| 0.01 | | | | |

13  a  The curve is $y = \sqrt{x}$ and $P$ is $(1, 1)$.

| $x$ | $y$ | $\Delta x$ | $\Delta y$ | Gradient of $PQ$ |
|---|---|---|---|---|
| 2 | 1.414 | 1 | 0.414 | 0.414 |
| 1.5 | | | | |
| 1.1 | | | | |
| 1.01 | | | | |

  b  The curve is $y = \sqrt[3]{x}$ and $P$ is $(8, 2)$.

| $x$ | $y$ | $\Delta x$ | $\Delta y$ | Gradient of $PQ$ |
|---|---|---|---|---|
| 9 | 2.0801 | 1 | 0.0801 | 0.0801 |
| 8.5 | | | | |
| 8.1 | | | | |
| 8.01 | | | | |

**14** a The curve is $y = \dfrac{1}{x}$ and $P$ is $(1, 1)$.

| $x$ | $y$ | $\Delta x$ | $\Delta y$ | Gradient of $PQ$ |
|---|---|---|---|---|
| 2 | 0.5 | 1 | $-0.5$ | $-0.5$ |
| 1.5 | | | | |
| 1.1 | | | | |
| 1.01 | | | | |

b The curve is $y = \dfrac{1}{x^2}$ and $P$ is $(1, 1)$.

| $x$ | $y$ | $\Delta x$ | $\Delta y$ | Gradient of $PQ$ |
|---|---|---|---|---|
| 2 | 0.25 | 1 | $-0.75$ | $-0.75$ |
| 1.5 | | | | |
| 1.1 | | | | |
| 1.01 | | | | |

**15** a The curve is $y = \ln x$ and $P$ is $(1, 0)$.

| $x$ | $y$ | $\Delta x$ | $\Delta y$ | Gradient of $PQ$ |
|---|---|---|---|---|
| 2 | 0.693 | 1 | 0.693 | 0.693 |
| 1.5 | | | | |
| 1.1 | | | | |
| 1.01 | | | | |

b The curve is $y = 2^x$ and $P$ is $(0, 1)$.

| $x$ | $y$ | $\Delta x$ | $\Delta y$ | Gradient of $PQ$ |
|---|---|---|---|---|
| 1 | 2 | 1 | 1 | 1 |
| 0.5 | | | | |
| 0.1 | | | | |
| 0.01 | | | | |

For questions 16 to 20, write the given expressions as a derivative.

**16** a The rate of change of $z$ as $v$ changes.

b The rate of change of $a$ as $b$ changes.

**17** a The rate of change of $p$ with respect to $t$.

b The rate of change of $b$ with respect to $x$.

**18** a How fast $y$ changes when $n$ is changed.

b How fast $t$ changes when $f$ is changed.

**19** a How quickly the height of an aeroplane ($h$) changes over time ($t$).

b How quickly the weight of water ($w$) changes as the volume ($v$) changes.

**20** a The rate of increase in the average wage ($w$) as the unemployment rate ($u$) increases.

b The rate of increase in the rate of reaction ($R$) as the temperature ($T$) increases.

For questions 21 to 26, use the method demonstrated in Worked Example 9.3 to write the given information as an equation.

**21** a The gradient of the graph of $y$ against $x$ at any point is equal to the $y$-coordinate.

b The gradient of the graph of $y$ against $x$ at any point is equal to half of the $x$-coordinate.

**22** a The gradient of the graph of $s$ against $t$ is proportional to the square of the $t$-coordinate.

b The gradient of the graph of $q$ against $p$ is proportional to the square root of the $q$-coordinate.

**23** a The growth rate of a plant of size $P$ over time $t$ is proportional to the amount of light it receives, $L(P)$.

b The rate at which companies' profits ($P$) increase over time ($t$) is proportional to the amount spent on advertising, $A(P)$.

**24** a The rate of increase of the height ($h$) of a road with respect to the distance along the road ($x$) equals the reciprocal of the height.

b The rate at which distance away from the origin ($r$) of a curve ($r$) increases with respect to the angle ($\theta$) with the positive $x$-axis is a constant.

**25** a The rate at which $s(t)$ is increasing as $t$ increases is 7.

b The rate at which $q(x)$ increases with respect to $x$ equals $7x$.

**26** a The rate of change of acidity (pH) of a solution as the temperature changes ($T$) is a constant.

b In economics, the marginal cost is the amount the cost of producing each item ($C$) increases by as the number of items ($n$) increases. In an economic model the marginal cost is proportional to the number of items produced below 1000.

27  The rate of increase of voltage ($V$) with respect to time ($t$) through an electrical component is equal to $V + 1$.

   a  Express this information as an equation involving a derivative.

   b  Find the gradient of the graph of $V$ against $t$ when $V = 4$.

 28  Use technology to suggest the value of the limit of $\dfrac{\sin x^2}{\left(\dfrac{\pi x}{180}\right)^2}$ as $x$ tends to zero.

 29  Suggest the value of the limit of $\dfrac{\ln x}{x - 1}$ as $x$ tends to 1.

30  a  Find the gradient of the chord connecting $(1, 1)$ to $(x, x^2)$.

   b  Find the limit of the gradient of this chord as $x$ tends towards 1. What is the significance of this value?

 31  Use technology to suggest a value for the limit of $\sqrt{x^2 + 6x} - x$ as $x$ gets very large.

32  The rate of change of $x$ with respect to $y$ is given by $3x$. Find the gradient on the graph of $y$ against $x$ when $x = 2$.

**You are the Researcher**

There are many models in natural sciences and social sciences which are described in terms of derivatives. Examples you could research include models such as the SIR model for epidemics, the von Bertalanffy model for animal growth or the Black–Scholes model for financial instruments. However, none of these models are perfect and current research involves looking at their assumptions and how they can be improved.

## 9B Graphical interpretation of derivatives

A good way to get an understanding of derivatives before getting too involved in algebra, is to focus on how they relate to the graph of the original function.

### Increasing and decreasing functions

**TOOLKIT:** Modelling

You might see models in the news reporting things like: 'The government predicts that the national debt will rise to a value of 1.3 trillion then start falling'. Five years later, it turns out that the debt actually rose to 1.5 trillion and then started falling. Does this mean the model was wrong? Mathematical modellers know that there are quantitative results from a model – that is, numerical predictions – which are very uncertain (but they are often the things which make the headlines!). Often the more useful thing to take away from a model is a more qualitative result, such as 'the debt will rise then fall'. Try to find a model in the news and look for qualitative and quantitative results associated with it – which do you think is more useful?

When we are sketching a curve we often do not care whether the gradient at a point is 5 or 6 – all that matters is that it we get the direction roughly right. There is some terminology which we use to describe this.

---

**KEY POINT 9.3**

If when $x$ increases f($x$) increases (so that $f'(x) > 0$), the function is *increasing*.

If when $x$ increases f($x$) decreases (so that $f'(x) < 0$), the function is *decreasing*.

---

### TOK Links

Do you think that this terminology is obvious? Perhaps you are from a country where text is read from left to right. If you are used to the reading the other way, would that change the natural definition? If you look at a cross section of a mountain, is it obvious which way is uphill and which way is downhill? Some of the hardest assumptions to check are the ones you do not know you are making.

A function is not necessarily always increasing or decreasing. We might have to identify intervals in which it is increasing or decreasing.

---

**WORKED EXAMPLE 9.4**

For the graph on the right, use inequalities in $x$ to describe the regions in which the function is increasing.

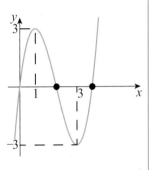

The function starts off increasing until it gets to the point where the gradient is zero (at $x - 1$). The gradient is then negative until it hits zero again at $x - 3$ .............. $x < 1$ or $x > 3$

---

We can use ideas of increasing and decreasing functions to sketch the derivative function of a given graph.

---

**WORKED EXAMPLE 9.5**

Sketch the derivative of this function.

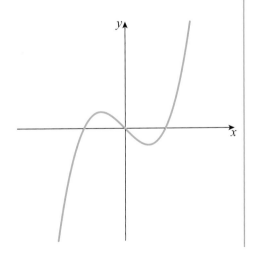

Consider the points $A$, $B$, $C$ and $D$
on the graph of $y = f(x)$

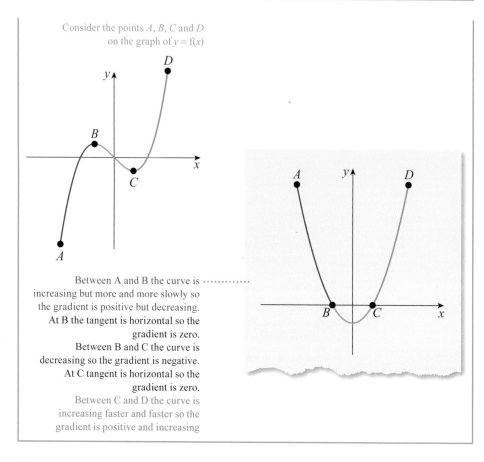

Between A and B the curve is
increasing but more and more slowly so
the gradient is positive but decreasing.
At B the tangent is horizontal so the
gradient is zero.
Between B and C the curve is
decreasing so the gradient is negative.
At C tangent is horizontal so the
gradient is zero.
Between C and D the curve is
increasing faster and faster so the
gradient is positive and increasing

## ■ Graphical interpretation of $f'(x) > 0$, $f'(x) = 0$ and $f'(x) < 0$

One very important skill in mathematics is the ability to think backwards. As well as looking at graphs of functions and finding out about their derivatives, you might be given a graph of the derivative and have to make inferences about the original function.

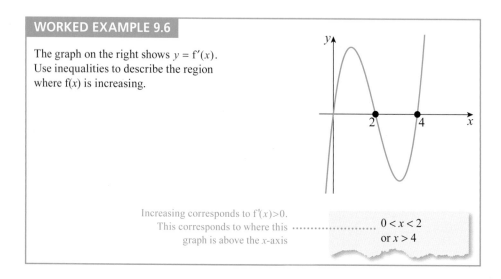

**WORKED EXAMPLE 9.6**

The graph on the right shows $y = f'(x)$.
Use inequalities to describe the region
where $f(x)$ is increasing.

Increasing corresponds to $f'(x) > 0$.
This corresponds to where this
graph is above the $x$-axis

$0 < x < 2$
or $x > 4$

If we have information about the derivative, that can also help us to sketch the original function.

---

**WORKED EXAMPLE 9.7**

The graph shows the derivative of a function.
Sketch a possible graph of the original function.

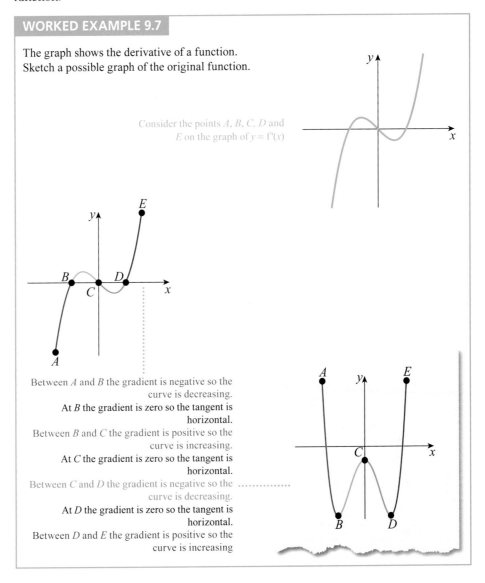

Consider the points *A*, *B*, *C*, *D* and
*E* on the graph of $y = f'(x)$

Between *A* and *B* the gradient is negative so the curve is decreasing.
At *B* the gradient is zero so the tangent is horizontal.
Between *B* and *C* the gradient is positive so the curve is increasing.
At *C* the gradient is zero so the tangent is horizontal.
Between *C* and *D* the gradient is negative so the curve is decreasing.
At *D* the gradient is zero so the tangent is horizontal.
Between *D* and *E* the gradient is positive so the curve is increasing

---

Note that in Worked Example 9.7 there was more than one possible graph that could have been drawn, depending on where the sketch started.

In Chapter 10 you will learn more about this ambiguity when you learn how to 'undo' differentiation.

## Exercise 9B

For questions 1 to 3, use the method demonstrated in Worked Example 9.4 to write an inequality to describe where the function f(x) is increasing. The graphs all represent $y = f(x)$.

**1 a**

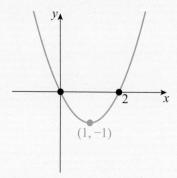

**b**

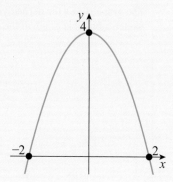

**2 a**

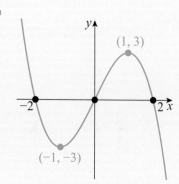

**b**

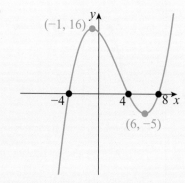

**3 a**

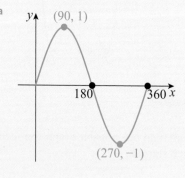

**b**

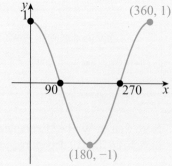

For questions 4 to 9, use the method demonstrated in Worked Example 9.5 to sketch the graph of the derivative of the function shown, labelling any intercepts with the $x$-axis.

4   a
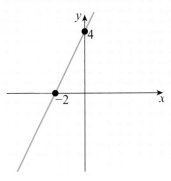

    b
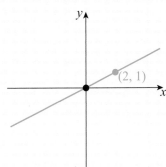

5   a
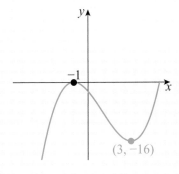

    b
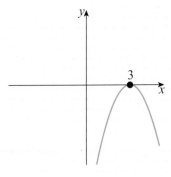

6   a
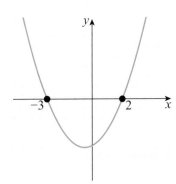

    b
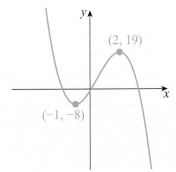

7  a

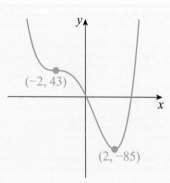

   b

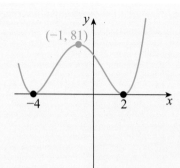

8  a

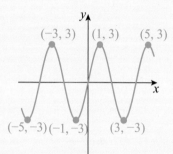

   b

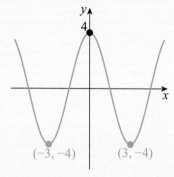

9  a

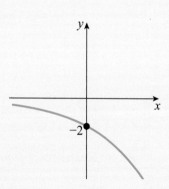

   b

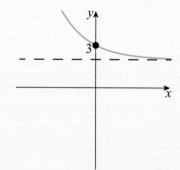

For questions 10 to 12, use the method demonstrated in Worked Example 9.6 to write inequalities in $x$ to describe where the function f($x$) is increasing. The graphs all represent $y = $ f$'(x)$.

10  a

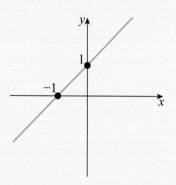

    b

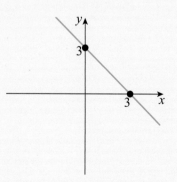

11  a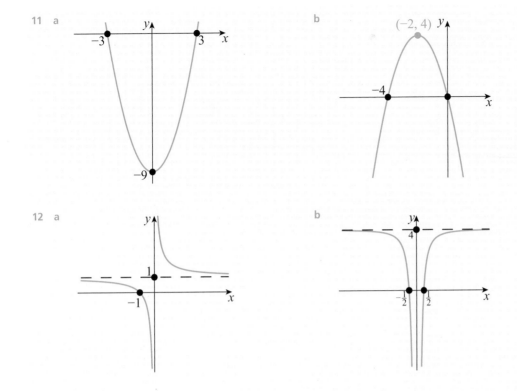

For questions 13 to 18, use the method demonstrated in Worked Example 9.7 to sketch a possible graph for $y = f(x)$, given that the graph shows $y = f'(x)$. Mark on any points on $y = f(x)$ where the gradient of the tangent is zero.

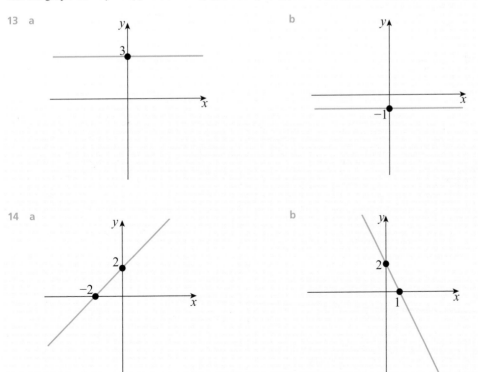

15  a                                    b
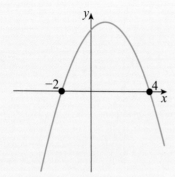
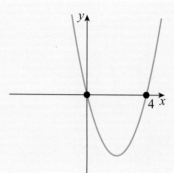

16  a                                    b
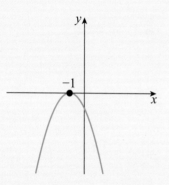
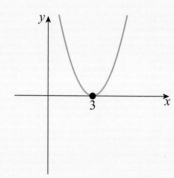

17  a                                    b
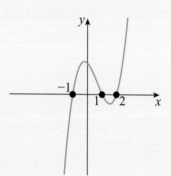
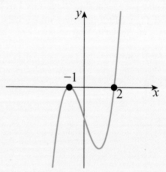

18  a                                    b

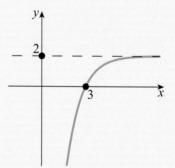

19  Sketch a function where $f(x)$ is always positive and $f'(x)$ is always positive.

20  Sketch a function where $f(x)$ is always increasing but $f'(x)$ is always decreasing.

**21** For the following graph:

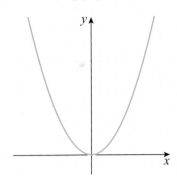

   a  If the graph represents $y = f(x)$, sketch a graph of $y = f'(x)$.

   b  If the graph represents $y = f'(x)$, sketch two possible graphs of $y = f(x)$.

**22** For the following graph:

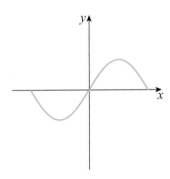

   a  If the graph represents $y = f(x)$, sketch a graph of $y = f'(x)$.

   b  If the graph represents $y = f'(x)$, sketch a possible graph of $y = f(x)$.

   c  Explain why the answer to part (b) is not unique.

**23**  a  Use technology to sketch the graph of $y = f(x)$ where $x^2(x^2 - 1)$.

   b  Find the interval in which $f(x) > 0$.

   c  Find the interval in which $f'(x) < 0$.

**24** For the graphs below put them into pairs of a function and its derivative.

   a                      b                          c

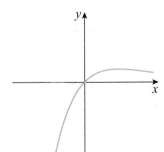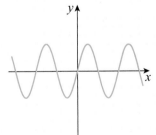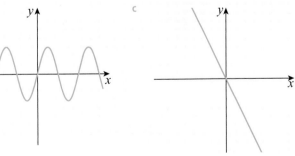

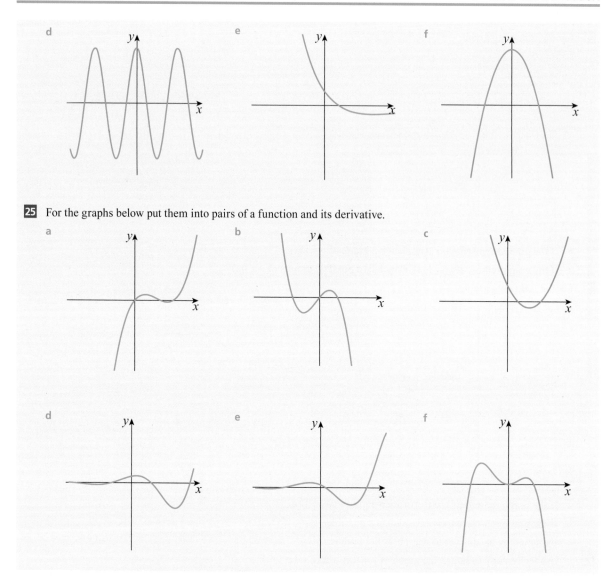

25  For the graphs below put them into pairs of a function and its derivative.

**TOOLKIT:** Problem Solving

The points where the gradient of a function is zero are called stationary points. Investigate, for different types of function, the relationship between the number of stationary points on $y = f(x)$ and the number of stationary points on $y = f'(x)$.

# 9C Finding an expression for the derivative

You saw in the previous two sections that the derivative of a function is also a function, and that we could sketch it from the original function. In this section we shall look at how to find an expression for the derivative of some types of function.

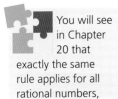
You will see in Chapter 20 that exactly the same rule applies for all rational numbers, not just integers.

## The derivative of $f(x) = ax^n$

If a function is of the form $f(x) = x^n$, there is a rule for finding the derivative. This can be proved using the idea of limits, but for now we shall just quote it:

If $f(x) = x^n$, where $n \in \mathbb{Z}$, then $f'(x) = nx^{n-1}$.

---

**WORKED EXAMPLE 9.8**

Given $f(x) = x^3$, find $f'(x)$.

Use $f'(x) = nx^{n-1}$ .......... $f'(x) = 3x^{3-1} = 3x^2$

---

**WORKED EXAMPLE 9.9**

A curve has equation $y = x^{-4}$. Find $\dfrac{dy}{dx}$.

Use $\dfrac{dy}{dx} = nx^{n-1}$ .......... $\dfrac{dy}{dx} = -4x^{-4-1} = -4x^{-5}$

---

This basic rule is unaffected by multiplication by a constant.

---

**KEY POINT 9.4**

If $f(x) = ax^n$, where $a$ is a real constant and $n \in \mathbb{Z}$, then $f'(x) = anx^{n-1}$.

---

**WORKED EXAMPLE 9.10**

Find the derivative of $f(t) = 2t^5$.

Differentiate $t^5$ and then multiply by 2 .......... $f'(t) = 2 \times 5t^4 = 10t^4$

---

You will often need to use the laws of indices before you can differentiate.

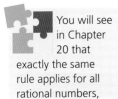
See Chapter 1 if you need to review the rules of indices.

**WORKED EXAMPLE 9.11**

Differentiate $y = \dfrac{6}{x^2}$.

Use the laws of indices to rewrite in the form $ax^n$ .......... $y = \dfrac{6}{x^2} = 6x^{-2}$

Then differentiate $x^{-2}$ and multiply by 6 .......... $y' = 6(-2)x^{-3} = -12x^{-3}$

## ■ The derivative of $f(x) = ax^n + bx^m + \ldots$

Key Point 9.4 can be extended to deal with a sum of functions of the form $ax^n$. In this case you just differentiate each term in turn.

---

**KEY POINT 9.5**

If $f(x) = ax^n + bx^m + \ldots$, where $a$ and $b$ are real constants and $n, m \in \mathbb{Z}$, then

$f'(x) = anx^{n-1} + bmx^{m-1} + \ldots$

---

**Tip**

Worked Example 9.12 shows that the derivative of $ax$ will always be $a$ and the derivative of $a$ will always be zero. You can use these results without having to show the working each time.

---

**WORKED EXAMPLE 9.12**

Find $\dfrac{dy}{dx}$ for $y = 7x - 2$.

Write $7x$ as $7x^1$ and $-2$ as $-2x^0$ .......... $f(x) = 7x^1 - 2x^0$

Differentiate .......... $f'(x) = 7 \times 1x^{1-1} - 2 \times 0x^{0-1}$
each term separately
$= 7x^0 - 0 = 7$

---

None of the Key Points in this section provide a rule for differentiating products or quotients of functions. So, before you can differentiate these, you will need to convert them into terms of the form $ax^n$.

---

**WORKED EXAMPLE 9.13**

If $a$ is a constant, find the rate of change of $y = (x + 3)(x - a)$ with respect to $x$.

Expand the brackets .......... $y = (x + 3)(x - a)$
$= x^2 + 3x - ax - 3a$

Finding the rate of
change is another way
of saying differentiate. .......... $\dfrac{dy}{dx} = 2x + 3 - a$
The constant $a$ can be
treated just like any other
numerical constant

---

In fact, there is a rule for differentiating products and quotients of functions, which you will meet in Chapter 20.

---

**WORKED EXAMPLE 9.14**

Differentiate $f(x) = \dfrac{x^2 - 6}{2x}$.

$$f(x) = \frac{x^2 - 6}{2x}$$

Use the laws of indices
to rewrite as a sum of .......... $= \dfrac{x^2}{2x} - \dfrac{6}{2x}$
functions of the form $ax^n$

$$= \frac{1}{2}x - 3x^{-1}$$

Then differentiate .......... $f'(x) = \dfrac{1}{2} - 3(-1)x^{-2}$

$$= \frac{1}{2} + 3x^{-2}$$

## Be the Examiner 9.1

Given $y = \dfrac{6x^2 - 5x}{x^3}$, find $\dfrac{dy}{dx}$.

Which is the correct solution? Identify the errors made in the incorrect solutions.

| Solution 1 | Solution 2 | Solution 3 |
|---|---|---|
| $y = \dfrac{6x^2 - 5x}{x^3}$ | $y = \dfrac{6x^2 - 5x}{x^3}$ | $y = \dfrac{6x^2 - 5x}{x^3}$ |
| $y' = \dfrac{12x - 5}{3x^2}$ | $= 6x^{-1} - 5x^{-2}$ | $= x^{-3}(6x^2 - 5x)$ |
|  | $y' = -6x^{-2} + 10x^{-3}$ | $y' = -3x^{-4}(12x - 5)$ |

You can use the rules of differentiation from this section together with the idea of increasing and decreasing functions from Section 9B.

### WORKED EXAMPLE 9.15

Find the interval of values of $x$ for which the function $f(x) = 5x^2 - 12x + 3$ is decreasing.

For any question about the increasing or decreasing of functions, start by finding f'(x) · · · · · · · · · · · · · · · $f'(x) = 10x - 12$

$$f'(x) < 0$$

A decreasing function has negative gradient · · · · · · · · · · · · · · · $10x - 12 < 0$

$$x < 1.2$$

## Be the Examiner 9.2

Is the function $f'(x) = 2x^3 - 3x$ increasing or decreasing at $x = 1$?

Which is the correct solution? Identify the errors made in the incorrect solutions.

| Solution 1 | Solution 2 | Solution 3 |
|---|---|---|
| $f(1) = 2 \times 1^3 - 3 \times 1$ | $f'(x) = 6x^2 - 3$ | $f(0) = 0$ and $f(2) = 10$ |
| $= 2 - 3$ | $f'(1) = 6 \times 1^2 - 3$ | $\therefore$ f is increasing |
| $= -1 < 0$ | $= 6 - 3$ |  |
| $\therefore$ f is decreasing. | $= 3 > 0$ |  |
|  | $\therefore$ f is increasing. |  |

# Exercise 9C

For questions 1 to 4, use the method demonstrated in Worked Examples 9.8 and 9.9 to find the derivative of the given function or graph.

1  a $f(x) = x^4$

   b $g(x) = x^6$

2  a $h(u) = u^{-1}$

   b $z(t) = t^{-4}$

3  a $y = x^8$

   b $p = q$

4  a $z = t^{-5}$

   b $s = r^{-10}$

For questions 5 to 7, use the method demonstrated in Worked Example 9.10 to find the derivative of the given function or graph.

5  a $f(x) = -4x$

   b $g(x) = 7x^2$

6  a $y = 6x$

   b $y = 3x^5$

7  a $y = 7$

   b $y = -6$

For questions 8 to 11, use the method demonstrated in Worked Example 9.11 to find the derivative of the given function or graph.

8  a  $g(x) = \dfrac{1}{x}$  9  a  $z = \dfrac{3}{x^2}$  10  a  $y = \dfrac{1}{4x^4}$  11  a  $f(x) = -\dfrac{3}{2x}$

   b  $h(x) = \dfrac{1}{x^3}$     b  $y = -\dfrac{10}{t^5}$     b  $y = \dfrac{1}{5x^5}$     b  $f(x) = -\dfrac{5}{3x^3}$

For questions 12 to 15, use the method demonstrated in Worked Example 9.12 to find the derivative of the given function or graph.

12  a  $f(x) = x^2 - 4x$  13  a  $y = 3x^3 - 5x^2 + 7x - 2$  14  a  $y = \dfrac{1}{2}x^4$  15  a  $y = 3x - \dfrac{1}{4}x^3$

   b  $g(x) = 2x^2 - 5x + 3$     b  $y = -x^4 + 6x^2 - 2x$     b  $y = -\dfrac{3}{4}x^6$     b  $y = 3 - 2x^3 + \dfrac{1}{4}x^4$

For questions 16 to 18, use the method demonstrated in Worked Example 9.13 to expand the brackets and hence find the derivative of the given function.

16  a  $f(x) = x^3(2x - 5)$  17  a  $g(x) = (x + 3)(x - 1)$  18  a  $h(x) = \left(1 + \dfrac{1}{x}\right)^2$

   b  $f(x) = x(x^2 + 3x - 9)$     b  $f(x) = (x - 2)(x + 1)$     b  $g(x) = \left(2x - \dfrac{3}{x}\right)^2$

For questions 19 to 25, use the method demonstrated in Worked Example 9.13 to find the rate of change of $y$ with respect to $x$ when $a, b$ and $c$ are constants.

19  a  $y = ax + b$  20  a  $y = ax^2 + (3 - a)x$  21  a  $y = x^2 + a^2$  22  a  $y = x^{2a} + a^{2a}$

   b  $y = ax^2 + bx + c$     b  $y = x^3 + b^2x$     b  $y = a^2x^3 + a^2b^3$     b  $y = x^{-a} - x^{-b} + ab$

23  a  $y = 7ax - \dfrac{3b}{x^2}$  24  a  $y = (ax)^2$  25  a  $y = (x + a)(x + b)$

   b  $y = 5b^2x^2 + \dfrac{3a}{cx}$     b  $y = (3ax)^2$     b  $y = (ax + b)(bx + a)$

For questions 26 to 28, use the method demonstrated in Worked Example 9.14 to differentiate the expression with respect to $x$.

26  a  $\dfrac{x + 10}{x}$  27  a  $\dfrac{2 + x}{x^2}$  28  a  $\dfrac{x^2 + 3x^3}{2x}$

   b  $\dfrac{2x - 5}{x}$     b  $\dfrac{1 + 2x}{x^3}$     b  $\dfrac{7x^5 + 2x^9}{4x}$

For questions 29 to 30, use the method demonstrated in Worked Example 9.15 to find the interval of values of $x$ for which the given function is decreasing.

29  a  $f(x) = 3x^2 - 6x + 2$                30  a  $f(x) = -x^2 + 12x - 9$

   b  $f(x) = x^2 + 8x + 10$                   b  $f(x) = 10 - x^2$

**31**  Differentiate $d = 6t - \dfrac{4}{t}$ with respect to $t$.

**32**  Find the rate of change of $q = m + 2m^{-1}$ as $m$ varies.

**33**  A model for the energy of a gas particle ($E$) at a temperature $T$ suggests that $E = \dfrac{3}{2}kT$ where $k$ is a constant.

   Find the rate at which $E$ increases with increasing temperature.

**34**  Find the interval in which $x^2 - x$ is an increasing function.

**35**  Find the interval in which $x^2 + bx + c$ is an increasing function, given that $b$ and $c$ are constants.

**36**  If $x + y = 8$ find $\dfrac{dy}{dx}$.

**37**  If $x^3 + y = x$ find $\dfrac{dy}{dx}$.

**38** In an astronomical model of gravity it is believed that potential energy ($V$) depends on distance from a star ($r$) by the rule:

$$V = \frac{k}{r}$$

where $k$ is a constant.

The force is defined as the rate of change of the potential energy with respect to distance.

  a  Find an expression for the force in terms of $k$ and $r$.

  b  Find an expression for the force in terms of $V$ and $k$.

  c  The star has two orbiting planets – Alpha and Omega. At a particular time Alpha lies exactly midway between the star and Omega. At this time, find the value of the ratio:

  $$\frac{\text{Force on Alpha}}{\text{Force on Omega}}$$

**39** A model for the reading age of a book ($A$) when the average sentence length is $L$ suggests that $A = 2 + qL + qL^2$.

  a  Find the rate of change of $A$ with respect to $L$.

  b  Explain why it might be expected that $A$ is an increasing function. What constraint does this place on $q$?

**40** Show that the function $f(x) = 4x^3 + 7x - 2$ is increasing for all $x$.

# 9D Tangents and normals at a given point and their equations

## ▌ Calculating the gradient at a given point

To calculate the gradient of a function at any particular point, you simply substitute the value of $x$ into the expression for the derivative.

---

**WORKED EXAMPLE 9.16**

Find the gradient of the graph of $y = 5x^3$ at the point where $x = 2$.

The gradient is given by the derivative ............................................... $\dfrac{dy}{dx} = 15x^2$

Substitute the given value for $x$ ............................................ When $x = 2$,

$$\frac{dy}{dx} = 15 \times 2^2 = 60$$

So, the gradient is 60

---

If you know the gradient of a graph at a particular point, you can find the value of $x$ at that point.

**WORKED EXAMPLE 9.17**

Find the values of $x$ for which the tangent to the graph of $y = x^3 + 6x^2 + 2$ has the gradient 15.

The gradient is given by the derivative ......... $\dfrac{dy}{dx} = 3x^2 + 12x$

You are told that $\dfrac{dy}{dx} = 15$ .........

$$3x^2 + 12x = 15$$
$$x^2 + 4x - 5 = 0$$
$$(x + 5)(x - 1) = 0$$
$$x = -5 \text{ or } 1$$

## ■ Analytic approach to finding the equation of a tangent

You already know from Section 9A that a tangent to a function at a given point is just a straight line passing through that point with the same gradient as the function at that point. Using the rules of differentiation in Section 9C, you can now find equations of tangents.

See Chapter 3 for a reminder of how to find the equation of a straight line from its gradient and a point on the line.

**KEY POINT 9.6**

The equation of the tangent to the curve $y = f(x)$ at the point where $x = a$ is given by $y - y_1 = m(x - x_1)$, where:

- $m = f'(a)$
- $x_1 = a$
- $y_1 = f(a)$

**WORKED EXAMPLE 9.18**

Find the equation of the tangent to the curve $y = x^2 + 4x^{-1} - 5$ at the point where $x = 2$.

Give your answer in the form $ax + by + c = 0$.

Find the gradient of the tangent by finding the value of $y'$ when $x = 2$ .........

$$y' = 2x - 4x^{-2}$$
When $x = 2$,
$$y' = 2 \times 2 - 4 \times 2^{-2}$$
$$= 4 - 1$$

$m$ is often used to denote the value of the gradient at a particular point .........

$$\therefore m = 3$$

Find the value of $y$ when $x = 2$ .........

When $x = 2$,
$$y = 2^2 + 4 \times 2^{-1} - 5$$
$$= 1$$

Substitute into $y - y_1 = m(x - x_1)$ .........

So, equation of tangent is:
$$y - 1 = 3(x - 2)$$
$$3x - y - 5 = 0$$

---

**CONCEPTS – RELATIONSHIPS AND REPRESENTATION**

There are lots of **relationships** going on when we are doing calculus:
- the $x$-coordinate of the curve and the $y$-coordinate of a point on the curve.
- the $x$-coordinate of the tangent and the $y$-coordinate of a point on the tangent.
- the $x$-coordinate of the curve and the value of the derivative at that point.
- the relationship between the graph of the curve the graph of the derivative.

It is very easy to use the same letters to **represent** different things and get very confused. Make sure when you write down an expression you know exactly what all your $x$s and $y$s represent.

You might have to interpret given information about the tangent and use it to find an unknown.

---

**WORKED EXAMPLE 9.19**

The tangent at a point on the curve $y = x^2 + 4$ passes through $(0, 0)$.

Find the possible coordinates of the point.

| | |
|---|---|
| You need to find the equation of the tangent at the unknown point | Let the point have coordinates $(p, q)$. |
| As the point lies on the curve, $(p, q)$ must satisfy $y = x^2 + 4$ | Then $q = p^2 + 4$ |
| | $\dfrac{dy}{dx} = 2x$ |
| The gradient of the tangent is given by $\dfrac{dy}{dx}$ with $x = p$ | When $x = p$, $\dfrac{dy}{dx} = 2p$ $\therefore m = 2x$ |
| Write down the equation of the tangent | Equation of the tangent is: $y - q = 2p(x - p)$ $y - (p^2 + 4) = 2p(x - p)$ |
| The tangent passes through the origin, so set $x = 0$ and $y = 0$ in the equation | Since the tangent passes through $(0, 0)$: $0 - (p^2 + 4) = 2p(0 - p)$ $-p^2 - 4 = -2p^2$ $p^2 = 4$ Hence, $p = 2$ or $-2$ |
| Now use $q = p^2 + 1$ to find the corresponding values of $q$ | When $p = 2$, $q = 8$ When $p = -2$, $q = 8$ So, the coordinates are $(2, 8)$ or $(-2, 8)$ |

## Analytic approach to finding the equation of a normal

The **normal to a curve** at a given point is a straight line which crosses the curve at that point and is perpendicular to the tangent at that point.

You can use the fact that for perpendicular lines $m_1 m_2 = -1$ to find the gradient of the normal.

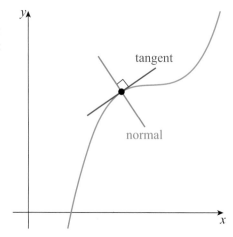

### KEY POINT 9.7

The equation of the normal to the curve $y = f(x)$ at the point where $x = a$ is given by

$$y - y_1 = m(x - x_1),$$

where:

- $m = -\dfrac{1}{f'(a)}$
- $x_1 = a$
- $y_1 = f(a)$

We will often denote the gradient of the tangent by $m_t$ and the gradient of the normal by $m_n$, so that $m_n = -\dfrac{1}{m_t}$.

### WORKED EXAMPLE 9.20

Find the equation of the normal to the curve $y = x^2 + 4x^{-1} - 5$ at the point where $x = 2$ (from Worked Example 9.18).

Give your answer in the form $ax + by + c = 0$, where $a$, $b$ and $c$ are integers.

| From Worked Example 9.18, you know that the gradient of the tangent at $x = 2$ is 3 | At $x = 2$, gradient of tangent is $m_t = 3$ |
|---|---|
| Find the gradient of the normal using $m_t m_n = -1$ | So, gradient of normal is $m_n = -\dfrac{1}{3}$. |
| Substitute into $y - y_1 = m(x - x_1)$. You know from Worked Example 9.18 that $y_1 = 1$ | Hence, equation of normal is: $y - 1 = -\dfrac{1}{3}(x - 2)$ $3y - 3 = -x + 2$ $x + 3y - 5 = 0$ |

■ Using technology to find the equation of tangents and normals

In some cases you will just be expected to use your calculator to find the gradient at a specific point, rather than needing to know how to differentiate the function.

---

**WORKED EXAMPLE 9.21**

Using technology, find the equation of the tangent to the curve $y = 4\ln x$ at the point where $x = 1$.

```
d/dx(4ln X,1)
                                    4



Solve  d/dx  d²/dx²  ∫·dx        ▷
```

Use your GDC to find $\cdots\cdots\cdots\cdots\cdots\cdots$ From GDC, $m = 4$
the gradient at $x = 1$

Find the value of $y$ when $x = 1$ $\cdots\cdots\cdots\cdots\cdots\cdots$ When $x = 1$,
$$y = 4\ln 1 = 0$$

So, equation of tangent is:

Substitute into $y - y_1 = m(x - x_1)$ $\cdots\cdots\cdots\cdots\cdots\cdots$ $y - 0 = 4(x - 1)$
$$y = 4x - 4$$

---

Exactly the same approach as in Worked Example 9.21 can be used to find the equation of a normal. Start as before by using your GDC to find the gradient at the given point, then use $m_1 m_2 = -1$ to find the gradient of the normal.

A slightly more advanced calculator trick is to get the calculator to plot the derivative of the function. This is particularly useful when being asked to find the $x$-value of a point with a given gradient.

---

**WORKED EXAMPLE 9.22**

Find the $x$-coordinate of the point on the graph of $y = xe^x$ with gradient 2.

This is not a function you know how to differentiate, so you need to get the calculator to sketch $y = f'(x)$. You also need to add the line $y = 2$ and find ............ where these two graphs intersect.

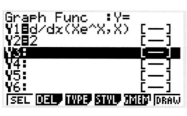

For working, you should make ............ a copy of the sketch

We can read off the graph the ............ So, the gradient is 2 when $x \approx 0.375$.
$x$-coordinate which has $f'(x) = 2$

---

# Exercise 9D

For questions 1 and 2, use the method demonstrated in Worked Example 9.16 to find the derivative at the given point.

1  a  If $y = 4x^2$, find $\dfrac{dy}{dx}$ when $x = 2$.
   b  If $y = x^3 - 7x$, find $\dfrac{dy}{dx}$ when $x = 3$.

2  a  If $A = 5 - 2h^{-1}$ find $\dfrac{dA}{dh}$ when $h = 4$.
   b  If $\phi = \theta^3 - \theta^{-2}$, find $\dfrac{d\phi}{d\theta}$ when $\theta = 0.5$.

For questions 3 and 4, use the method demonstrated in Worked Example 9.16 to find the gradient of the tangent to the curve at the given point.

3  a  The curve $y = x^4$ at the point where $x = 2$.
   b  The curve $y = 3x + x^2$ at the point where $x = -5$.

4  a  The curve $y = x^2 + \dfrac{1}{x}$ at the point where $x = \dfrac{3}{2}$.
   b  The curve $y = x^2 - \dfrac{4}{x}$ at the point where $x = 0.2$.

For questions 5 and 6, find and evaluate an appropriate derivative.

5  a  How quickly does $f = 4T^3$ change as $T$ changes when $T = 3$?
   b  How quickly does $g = 2y^3$ change as $y$ changes when $y = 1$?

6  a  What is the rate of increase of $W$ with respect to $q$ when $q$ is $-3$ if $W = -4q^2$?
   b  What is the rate of change of $M$ with respect to $c$ when $c = 4$ if $M = \dfrac{3}{c} + 5$?

For questions 7 to 9, use the method demonstrated in Worked Example 9.17 to find the value(s) of $x$ at which the curve has the given gradient.

7  a  gradient 6 on the graph of $y = x^4 + 2x$
   b  gradient 36 on the graph of $y = 3x^3$

8  a  gradient 2.25 on the graph of $y = x - 5x^{-1}$
   b  gradient $-27$ on the graph of $y = 5x + \dfrac{8}{x}$

9  a  gradient $-4$ on the graph of $y = x^3 + 6x^2 + 5x$
   b  gradient $-1.5$ on the graph of $y = x^3 - 3x + 2$.

For questions 10 to 12, use the method demonstrated in Worked Example 9.18 to find the equation of the tangent of the given curve at the given point in the form $ax + by = c$.

10  a  $y = x^2 + 3$ at $x = 2$                                b  $y = x^2 + x$ at $x = 0$

11  a  $y = 2x^3 - 4x^2 + 7$ at $x = 1$                        b  $y = x^3 + x^2 - 8x + 1$ at $x = -1$

12  a  $y = \dfrac{3}{x}$ at $x = 3$                           b  $y = 1 - \dfrac{2}{x}$ at $x = 1$

For questions 13 to 16, use the method demonstrated in Worked Example 9.20 to find the equation of the normal of the given curve at the given point in the form $y = mx + c$.

13  a  $y = x + 7$ at $x = 4$                                  b  $y = 3 - 0.5x$ at $x = 6$

14  a  $y = x^3 + x$ at $x = 0$                                b  $y = 2x^3 - 5x + 1$ at $x = -1$

15  a  $y = x^3 - 3x^2 + 2$ at $x = 1$                         b  $y = 2x^4 - x^3 + 3x^2 + x$ at $x = -1$

16  a  $y = \dfrac{2}{x}$ at $x = 2$                           b  $y = 1 + \dfrac{4}{x}$ at $x = -2$

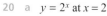

For questions 17 to 20, use the method demonstrated in Worked Example 9.21 (that is, use technology) to find the equation of the tangent of the given curve at the given point in the form $y = mx + c$.

17  a  $y = \sqrt{x}$ at $x = 4$                               b  $y = \dfrac{1}{\sqrt{x}}$ at $x = 4$

18  a  $y = \dfrac{1}{x + 1}$ at $x = 1$                       b  $y = \dfrac{1}{(x - 1)^2}$ at $x = 5$

19  a  $y = \ln x$ at $x = 1$                                  b  $y = x \ln x$ at $x = 1$

20  a  $y = 2^x$ at $x = 2$                                    b  $y = 3^x$ at $x = 0$

For questions 21 to 24, use technology to find the equation of the normal of the given curve at the given point.

21  a  $y = \sqrt{x + 1}$ at $x = 3$                           b  $y = \sqrt{2x + 1}$ at $x = 0$

22  a  $y = 2\sqrt{x} - \dfrac{2}{\sqrt{x}}$ at $x = 4$        b  $y = \dfrac{x^2 + 4}{\sqrt{x}}$ at $x = 4$

23  a  $y = \ln 3x$ at $x = 1$                                 b  $y = 4 + 2\ln 3x$ at $x = 1$

24  a  $y = xe^x$ at $x = 0$                                   b  $y = (x - 2)e^x$ at $x = 0$

For questions 25 to 27, use the method demonstrated in Worked Example 9.22 (that is, use technology) to find the $x$-coordinate of the point(s) on the curve with the given gradient.

25  a  gradient 4 on $y = \sqrt{x + 1}$                        b  gradient 2 on $\sqrt{1 - x^2}$

26  a  gradient 0.5 on $y = \ln x$                             b  gradient 2 on $y = e^x$

27  a  gradient $-1$ on $y = \dfrac{1}{x + 2}$                 b  gradient 0 on $y = x^2 e^x$

28  A graph is drawn of $y = x^4 - x$.

   a  Find an expression for $\dfrac{dy}{dx}$.

   b  What is the gradient of the tangent of the graph at $x = 0$?

   c  Hence, find the equation of the normal to the curve at $x = 0$.

29  A graph is drawn of $y = f(x)$ with $f(x) = x^3 + \dfrac{1}{x}$.

   a  Find $f'(x)$.

   b  What is the gradient of the tangent of the graph at $x = 1$?

   c  Hence, show that the tangent to the curve at $x = 1$ passes through the origin.

**30** Use technology to find the equation of the tangent to the curve $y = x\sqrt{x+1}$ at $x = 3$.

**31** Use technology to find the equation of the normal to the curve $y = \dfrac{1}{x+4}$ at $x = -2$.

**32** Find the equation of the normal to the curve $y = x^2 e^x$ at $x = 0$.

**33** Find the coordinates of the point at which the normal to the curve $y = x^2$ at $x = 1$ meets the curve again.

**34** Find the coordinates of the point at which the normal to the curve $y = \dfrac{1}{x}$ at $x = 2$ meets the curve again.

**35** Find the equation of the tangent to the curve $y = x^2$ which is parallel to $y + 2x = 10$.

**36** Find the equation of the normal to the curve $y = x^2 + 2x$ which is parallel to $y = \dfrac{x}{4}$.

**37** Find the equations of the normals to the curve $y = x^3$ which are perpendicular to the line $y = 3x + 5$.

**38** Find the coordinates of the points at which the tangent to the curve $y = x^3 - x$ is parallel to $y = 11x$.

**39** Find all points on the curve $y = x^2 + 4x + 1$ where the gradient of the tangent equals the $y$-coordinate.

**40** The tangent at point $P$ on the curve $y = \dfrac{1}{x}$ passes through $(4, 0)$. Find the coordinates of $P$.

**41** The tangent at point $P$ on the curve $y = \dfrac{4}{x}$ passes through $(1, 3)$. Find the possible $x$-coordinates of $P$.

**42** Find the coordinates of the points on the curve $y = x^2$ for which the tangent passes through the point $(2, 3)$.

**43** A tangent is drawn on the graph $y = \dfrac{1}{x}$ at the point where $x = a$ $(a > 0)$. The tangent intersects the $y$-axis at $P$ and the $x$-axis at $Q$. If $O$ is the origin, show that the area of the triangle $OPQ$ is independent of $a$.

**44** The point $P$ on the curve $y = \dfrac{a}{x}$ has $x$-coordinate 2. The normal to the curve at $P$ is parallel to $x - 5y + 10 = 0$. Find the constant $a$.

**45** Show that the tangent to the curve $y = x^3 + x + 1$ at the point with $x$-coordinate $k$ meets the curve again at a point with $x$-coordinate $-2k$

---

**TOOLKIT:** Modelling

Draw a circle centred on the origin. Measure the gradient of the tangent at several points, including an estimate of the error. How easy is it to manually find the gradient?

See if you can find a relationship between the $x$-coordinate of the point, the $y$-coordinate and the gradient of the tangent at that point. Does the radius of the circle matter? See if you can prove your conjecture.

## Checklist

- You should be able to estimate the value of a limit from a table or graph.

- You should have an informal understanding of the gradient of a curve as a limit:
  - The gradient of a curve at a point is the gradient of the tangent to the curve at that point.
  - The gradient of a tangent at a point is the limit of the gradient of smaller and smaller chords from that point.

- You should know about different notation for the derivative.

- You should be able to identify intervals on which functions are increasing or decreasing:
  - If when $x$ increases $f(x)$ increases (so that $f'(x) > 0$), the function is increasing.
  - If when $x$ increases $f(x)$ decreases (so that $f'(x) < 0$), the function is decreasing.

- You should be able to interpret graphically $f'(x) > 0$, $f'(x) = 0$, $f'(x) < 0$.

- You should be able to differentiate functions of the form $f(x) = ax^n + bx^m + \ldots$
  - If $f(x) = ax^n + bx^m + \ldots$, where $a$ and $b$ are real constants and $n$, $m \in \mathbb{Z}$, then $f'(x) = anx^{n-1} + bmx^{m-1} + \ldots$

- You should be able to find the equations of the tangent and normal to a function at a given point:
  - The equation of the tangent to the curve $y = f(x)$ at the point where $y = a$ is given by $y - y_1 = m(x - x_1)$, where
    - $m = f'(a)$
    - $x_1 = a$
    - $y_1 = f(a)$

- The equation of the normal to the curve $y = f(x)$ at the point where $x = a$ is given by
  $y - y_1 = m(x - x_1)$, where:
  - $m = -\dfrac{1}{f'(a)}$
  - $x_1 = a$
  - $y_1 = f(a)$

## ■ Mixed Practice

**1** A curve has equation $y = 4x^2 - x$.

  **a** Find $\dfrac{dy}{dx}$.

  **b** Find the coordinates of the point on the curve where the gradient equals 15.

**2 a** Use technology to sketch the curve $y = f(x)$ where $f(x) = x^2 - x$.

  **b** State the interval in which the curve is increasing.

  **c** Sketch the graph of $y = f'(x)$.

**3** The function f is given by $f(x) = 2x^3 + 5x^2 + 4x + 3$.

  **a** Find $f'(x)$.

  **b** Calculate the value of $f'(-1)$.

  **c** Find the equation of the tangent to the curve $y = f(x)$ at the point $(-1, 2)$.

**4** Use technology to suggest the limit of $\dfrac{\ln(1 + x) - x}{x^2}$ as $x$ gets very small.

**5** Find the equation of the tangent to the curve $y = x^3 - 4$ at the point where $y = 23$.

**6** Water is being poured into a water tank. The volume of the water in the tank, measured in m³, at time $t$ minutes is given by $V = 50 + 12t + 5t^2$.

  **a** Find $\dfrac{dV}{dt}$.

  **b** Find the values of $V$ and $\dfrac{dV}{dt}$ after 6 minutes. What do these values represent?

  **c** Is the volume of water in the tank increasing faster after 6 minutes or after 10 minutes?

**7** The accuracy of an x-ray ($A$) depends on the exposure time ($t$) according to
$A = t(2 - t)$ for $0 < t < 2$.
   **a** Find an expression for the rate of change of accuracy with respect to time.
   **b** At $t = 0.5$. find:
      **i** the accuracy of the x-ray
      **ii** the rate at which the accuracy is increasing with respect to time.
   **c** Find the interval in which $A$ is an increasing function.

**8** Consider the function $f(x) = 0.5x^2 - \dfrac{8}{x}$ , $x \neq 0$.
   **a** Find $f(-2)$.
   **b** Find $f'(x)$.
   **c** Find the gradient of the graph of f at $x = -2$.

Let $T$ be the tangent to the graph of f at $x = -2$.
   **d** Write down the equation of $T$.
   **e** Sketch the graph of f for $-5 \leqslant x \leqslant 5$ and $-20 \leqslant y \leqslant 20$.
   **f** Draw $T$ on your sketch.

The tangent, $T$, intersects the graph of f at a second point, $P$.
   **g** Use your graphic display calculator to find the coordinates of $P$.

Mathematical Studies SL May 2015 P2 TZ2

**9** A curve has equation $y = 2x^3 - 8x + 3$. Find the coordinates of two points on the curve where the gradient equals $-2$.

**10** A curve has equation $y = x^3 - 6x^2$.
   **a** Find the coordinates of the two points on the curve where the gradient is zero.
   **b** Find the equation of the straight line which passes through these two points.

**11** A curve has equation $y = 3x^2 + 6x$.
   **a** Find the equations of the tangents at the two points where $y = 0$.
   **b** Find the coordinates of the point where those two tangents intersect.

**12** The gradient of the curve $y = 2x^2 + c$ at the point $(p, 5)$ equals $-8$. Find the values of $p$ and $c$.

**13** A company's monthly profit, $\$P$, varies according to the equation $P(t) = 15t^2 - t^3$, where $t$ is the number of months since the foundation of the company.
   **a** Find $\dfrac{dP}{dt}$.
   **b** Find the value of $\dfrac{dP}{dt}$ when $t = 6$ and when $t = 12$.
   **c** Interpret the values found in part **b**.

**14** The diagram shows the graphs of the functions $f(x) = 2x + 1$ and $g(x) = x^2 + 3x - 4$.

   **a** **i** Find $f'(x)$.
      **ii** Find $g'(x)$.
   **b** Calculate the value of $x$ for which the gradient of the two graphs is the same.
   **c** On a copy of the diagram above, sketch the tangent to $y = g(x)$ for this value of $x$, clearly showing the property in part **b**.

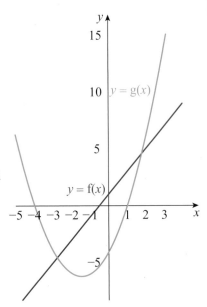

**15** Consider the graph on the right:

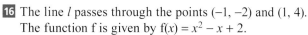

**a** If the graph represents $y = f(x)$:
  **i** use inequalities to describe the interval in which $f(x)$ is decreasing
  **ii** sketch $y = f'(x)$.
**b** If the graph represents $y = f'(x)$:
  **i** use inequalities to describe the interval in which $f(x)$ is decreasing
  **ii** sketch a possible graph of $y = f(x)$.

**16** The line $l$ passes through the points $(-1, -2)$ and $(1, 4)$.
The function f is given by $f(x) = x^2 - x + 2$.
**a** Find the gradient of $l$.
**b** Differentiate $f(x)$ with respect to $x$.
**c** Find the coordinates of the point where the tangent to $y = f(x)$ is parallel to the line $l$.
**d** Find the coordinates of the point where the tangent to $y = f(x)$ is perpendicular to the line $l$.
**e** Find the equation of the tangent to $y = f(x)$ at the point $(3, 8)$.
**f** Find the coordinates of the vertex of $y = f(x)$ and state the gradient of the curve at this point.

**17** Given that $f(x) = x^2 + x - 5$:
**a** write down $f'(x)$
**b** find the values of $x$ for which $f'(x) = f(x)$.

**18** The gradient of the curve $y = ax^2 + bx$ at the point $(2, -2)$ is 3. Find the values of $a$ and $b$.

**19** The gradient of the normal to the curve with equation $y = ax^2 + bx$ at the point $(1, 5)$ is $\frac{1}{3}$. Find the values of $a$ and $b$.

**20** For the curve with equation $y = 5x^2 - 4$, find the coordinates of the point where the tangent at $x = 1$ intersects the tangent at the point $x = 2$.

**21** Let $f(x) = \dfrac{\ln(4x)}{x}$, for $0 < x \leqslant 5$.

Points $P(0.25, 0)$ and $Q$ are on the curve of f. The tangent to the curve of f at $P$ is perpendicular to the tangent at $Q$. Find the coordinates of $Q$.

Mathematics SL May 2015 Paper 2 TZ1

## ESSENTIAL UNDERSTANDINGS

- Integration describes the accumulation of limiting areas. Understanding these accumulations allows us to model, interpret and analyse real-world problems and situations.

### In this chapter you will learn...

- how to integrate functions of the form $f(x) = ax^n + bx^m$, where $n, m \in \mathbb{Z}$, $n, m \neq -1$
- how to use your GDC to find the value of definite integrals
- how to find areas between a curve $y = f(x)$ and the $x$-axis, where $f(x) > 0$
- how to use integration with a boundary condition to determine the constant term.

### CONCEPTS

The following concept will be addressed in this chapter:
- Areas under curves can be **approximated** by the sum of areas of rectangles which may be calculated even more accurately using integration.

### PRIOR KNOWLEDGE

Before starting this chapter, you should already be able to complete the following:

1. Write $\dfrac{(x-2)^2}{x^4}$ in the form $ax^n + bx^m + cx^k$.
2. Differentiate the function $f(x) = 4x^{-3} + 2x - 5$.

■ **Figure 10.1** What is being accumulated in each of these photographs?

In many physical situations it is easier to model or measure the rate of change of a quantity than the quantity itself. For example, if you know the forces acting on an object that cause it to move, then you can find an expression for its acceleration, which is the rate of change of velocity. You cannot directly find an expression for the velocity itself, however, so to find that you now need to be able to undo the process of differentiation to go from rate of change of velocity to velocity.

## Starter Activity

Look at the images in Figure 10.1. Discuss how, in physical situations such as these, rate of change is related to accumulation of a quantity.

**Now look at this problem:**

Estimate the shaded area between the graph and the *x*-axis.

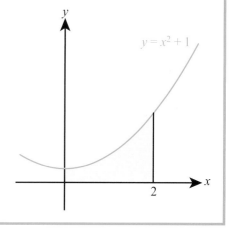

### LEARNER PROFILE – Thinkers
Have you ever tried making an 'essay plan' for a problem? Sometimes it is good to map out a route through a problem without getting bogged down in the algebraic and arithmetic details.

## 10A Anti-differentiation

 **Introduction to integration as anti-differentiation of functions of the form $f(x) = ax^n + bx^m + \ldots$**

If you are given the gradient function f(x), you can find the function f(x) by 'undoing' the differentiation. This process of reversing differentiation is known as **integration**.

However, it isn't possible to fully specify the original function f(x) without further information. For example, if you were given $f'(x) = 2x$ you would know that f(x) must contain $x^2$ since this differentiates to give $2x$, but you wouldn't know whether you had $f(x) = x^2 + 3$, $f(x) = x^2 - \dfrac{2}{5}$, or in fact $f(x) = x^2 + c$ for any constant c.

So, we say that if $f'(x) = 2x$ then $f(x) = x^2 + c$, or using the integration symbol,

$$\int 2x\,dx = x^2 + c$$

The $dx$ indicates that the variable you are integrating with respect to is $x$, just as in $\dfrac{dy}{dx}$ it indicates that the variable you are differentiating with respect to is $x$.

Reversing the basic rule for differentiation gives a basic rule for integration:

$$\int x^n\,dx = \frac{1}{n+1}x^{n+1} + c$$

where $n \in \mathbb{Z}$, $n \neq -1$ and $c \in \mathbb{R}$.

**Tip**

Note the condition $n \neq -1$, which ensures that you are not dividing by zero.

 You will see how to integrate $x^{-1}$ in Chapter 21.

---

**WORKED EXAMPLE 10.1**

If $f'(x) = x^{-4}$ find an expression for f(x).

Integrate to find f(x) $\cdots\cdots\cdots\cdots\cdots\cdots\cdots$ $f(x) = \displaystyle\int x^{-4}\,dx$

Use $\displaystyle\int x^n\,dx = \frac{1}{n+1}x^{n+1} + c$ $\cdots\cdots\cdots\cdots\cdots\cdots$ $= \dfrac{1}{-4+1}x^{-4+1} + c$

$= -\dfrac{1}{3}x^{-3} + c$

---

As with differentiation, the basic rule is unaffected by multiplication by a constant, and it can be extended to deal with sums.

**KEY POINT 10.1**

$$\int \left(ax^n + bx^m + \ldots\right)dx = \frac{a}{n+1}x^{n+1} + \frac{b}{m+1}x^{m+1} + \ldots + c$$

where $a$, $b$ and $c$ are real constants and $n$, $m \in \mathbb{Z}$, $n$, $m \neq -1$.

**WORKED EXAMPLE 10.2**

If $\dfrac{dy}{dx} = 4x^3 + 10x^4 + 3$ find an expression for $y$.

Integrate to find $y$ .................... $y = \displaystyle\int (4x^3 + 10x^4 + 3)\,dx$

Integrate term by term.
You can think of 3 as $3x^0$,
although after a while you ........... $= \dfrac{4}{3+1}x^{3+1} + \dfrac{10}{4+1}x^{4+1} + \dfrac{3}{0+1}x^{0+1} + c$
will probably just be able to
write the integral of a $k$ as
$kx$ without going through $= x^4 + 2x^5 + 3x + c$
this process every time

Again, as with differentiation, you don't have a rule for integrating products or quotients of functions. So, before you can integrate these you will need to convert them into sums of terms of the form $ax^n$.

 In fact, there is a rule for integrating some products and quotients, which you will meet in Chapter 20.

The integral sign was first use by the German mathematician Gottfried Leibniz in 1675. It is an old letter S, which originates from one of its interpretations as a sum of infinitely small rectangles.

**WORKED EXAMPLE 10.3**

Find $\displaystyle\int \dfrac{2x-5}{x^3}\,dx$.

Use the laws of exponents
to rewrite as a sum of ........... $\displaystyle\int \dfrac{2x-5}{x^3}\,dx = \int \left(\dfrac{2x}{x^3} - \dfrac{5}{x^3}\right)dx$
functions of the form $ax^n$
Then integrate ..................... $= \displaystyle\int \left(2x^{-2} - 5x^{-3}\right)$

$= \dfrac{2}{-1}x^{-1} - \dfrac{5}{-2}x^{-2} + c$

$= -2x^{-1} + \dfrac{5}{2}x^{-2} + c$

# Be the Examiner 10.1

Find $\displaystyle\int 2x\left(x^2 + 4\right)dx$.

Which is the correct solution? Identify the errors made in the incorrect solutions.

| Solution 1 | Solution 2 | Solution 3 |
|---|---|---|
| $\displaystyle\int 2x\left(x^2+4\right)dx = x^2\left(\dfrac{x^3}{3}+4x\right)+c$ $= \dfrac{x^5}{3}+4x^3+c$ | $\displaystyle\int 2x\left(x^2+4\right)dx = \int \left(2x^3+8x\right)dx$ $= \dfrac{1}{2}x^4+4x^2+c$ | $\displaystyle\int 2x\left(x^2+4\right)dx = 2x\int \left(x^2+4\right)dx$ $= 2x\left(\dfrac{x^3}{3}+4x\right)+c$ $= \dfrac{2}{3}x^4+8x^2+c$ |

## ■ Anti-differentiation with a boundary condition

You can determine the constant when you integrate if, alongside the gradient function f'($x$), you are also given the value of a function at a particular value of $x$. This extra information is known as a boundary condition.

**WORKED EXAMPLE 10.4**

If $\dfrac{dy}{dx} = x^2 + 1$ and $y = 5$ when $x = 3$, find an expression for $y$ in terms of $x$.

Integrate to find $y$ $\cdots\cdots\cdots\cdots\cdots\cdots\cdots\cdots\cdots$ $y = \displaystyle\int (x^2 + 1)\,dx$

$$= \frac{1}{3}x^3 + x + c$$

Substitute in the boundary condition $x = 3$, $y = 5$ in order to find $c$ $\cdots\cdots\cdots\cdots\cdots$ When $x = 3$, $y = 5$:

$$5 = \frac{1}{3}(3)^3 + 3 + c$$

$$5 = 12 + c$$

$$c = -7$$

Make sure you state $\cdots\cdots\cdots\cdots\cdots\cdots$ So, $y = \dfrac{1}{3}x^3 + x - 7$
the final answer

## Exercise 10A

For questions 1 to 3, use the method demonstrated in Worked Example 10.1 to find an expression for f($x$).

1 a $f'(x) = x^3$      2 a $f'(x) = x^{-2}$      3 a $f'(x) = 0$

   b $f'(x) = x^5$        b $f'(x) = x^{-3}$        b $f'(x) = 1$

For questions 4 to 14, use the method demonstrated in Worked Example 10.2 to find an expression for $y$.

4 a $\dfrac{dy}{dx} = 3x^2$    5 a $\dfrac{dy}{dx} = -7x^3$    6 a $\dfrac{dy}{dx} = \dfrac{3}{2}x^5$    7 a $\dfrac{dy}{dx} = -3x^{-4}$

  b $\dfrac{dy}{dx} = -5x^4$     b $\dfrac{dy}{dx} = 3x^7$     b $\dfrac{dy}{dx} = -\dfrac{5}{3}x^9$     b $\dfrac{dy}{dx} = 5x^{-6}$

8 a $\dfrac{dy}{dx} = 6x^{-5}$    9 a $\dfrac{dy}{dx} = -\dfrac{2}{5}x^{-2}$    10 a $\dfrac{dy}{dx} = 3x^2 - 4x + 5$    11 a $\dfrac{dy}{dx} = x - 4x^5$

  b $\dfrac{dy}{dx} = -4x^{-3}$     b $\dfrac{dy}{dx} = \dfrac{7}{4}x^{-8}$     b $\dfrac{dy}{dx} = 7x^4 + 6x^2 - 2$     b $\dfrac{dy}{dx} = 6x^3 - 5x^7$

12 a $\dfrac{dy}{dx} = \dfrac{3}{4}x^2 + \dfrac{7}{3}x$    13 a $\dfrac{dy}{dx} = 2x^{-5} + 3x$    14 a $\dfrac{dy}{dx} = \dfrac{5}{2}x^{-2} - \dfrac{10}{7}x^{-6}$

   b $\dfrac{dy}{dx} = \dfrac{4}{5} - \dfrac{2}{3}x^4$     b $\dfrac{dy}{dx} = 5 - 9x^{-4}$     b $\dfrac{dy}{dx} = \dfrac{4}{3}x^{-3} - \dfrac{2}{5}x^{-7}$

For questions 15 to 17, find the given integrals by first expanding the brackets.

15  a  $\int x^2 (x + 5)\,dx$     16  a  $\int (x + 2)(x - 3)\,dx$     17  a  $\int (3x - x^{-1})^2\,dx$

   b  $\int 3x(2 - x)\,dx$     b  $\int (4 - x)(x - 1)\,dx$     b  $\int (2x + x^{-3})^2\,dx$

For questions 18 and 19, use the method demonstrated in Worked Example 10.3 to find the given integrals by first writing the expression as a sum of terms of the form $ax^n$.

18  a  $\int \dfrac{3x^2 - 2x}{x}\,dx$     b  $\int \dfrac{5x^3 - 3x^2}{x^2}\,dx$

19  a  $\int \dfrac{4x - 7}{2x^4}\,dx$     b  $\int \dfrac{2x^3 - 3x^2}{4x^5}\,dx$

For questions 20 to 22, use the method demonstrated in Worked Example 10.4 to find an expression for $y$ in terms of $x$.

20  a  $\dfrac{dy}{dx} = 3x^2$ and $y = 6$ when $x = 0$     b  $\dfrac{dy}{dx} = 5x^4$ and $y = 5$ when $x = 0$

21  a  $y' = x^3 - 12x$ and $y = -10$ when $x = 2$     b  $y' = 3 - 10x^4$ and $y = 5$ when $x = 1$

22  a  $y' = \dfrac{6}{x^3} + 4x$ and $y = 16$ when $x = 3$     b  $y' = 9x^2 - \dfrac{2}{x^2}$ and $y = -9$ when $x = -1$

23  Find $\int \dfrac{4}{3t^2} - \dfrac{2}{t^5}\,dt$.

24  A curve has the gradient given by $\dfrac{dy}{dx} = 3x^2 - 4$ and the curve passes through the point (1, 4). Find the expression for $y$ in terms of $x$.

25  Find the equation of the curve with gradient $\dfrac{dy}{dx} = \dfrac{4}{x^2} - 3x^2$ which passes through the point (2, 0).

26  Find $\int (3x - 2)(x^2 + 1)\,dx$.

27  Find $\int z^2 \left(z + \dfrac{1}{z}\right)dz$.

28  Find $\int \dfrac{x^5 - 2x}{3x^3}\,dx$.

29  The rate of growth of the mass ($m$ kg) of a weed is modelled by
$\dfrac{dm}{dt} = kt + 0.1$ where $t$ is measured in weeks.
When $t = 2$ the rate of growth is 0.5 kg per week.

  a  Find the value of $k$.

  b  If the weed initially has negligible mass, find the mass after 5 weeks.

30  A bath is filling from a hot water cylinder. At time $t$ minutes the bath is filling at a rate given by $\dfrac{80}{t^2}$ litres per minute. When $t = 1$ the bath is empty.

  a  How much water is in the bath when $t = 2$?

  b  At what time will the bath hold 60 litres?

  c  The capacity of the bath is 100 litres. Using technology, determine whether the bath will ever overflow if this process is left indefinitely.

# 10B Definite integration and the area under a curve

 ### ■ Definite integrals using technology

So far you have been carrying out a process called **indefinite integration** – it is indefinite because you have the unknown constant each time.

A similar process, called **definite integration**, gives a numerical answer – no *x*s and no unknown constant.

You only need to be able to evaluate definite integrals on your GDC, but to do so you need to know the upper and lower limit of the integral: these are written at the top and bottom respectively of the integral sign.

You will see in Chapter 20 that definite integration is related to indefinite integration. Once you have done the indefinite integral you evaluate the expression at the upper and lower limit then find their difference.

**Tip**

In the expression $\int_{1}^{4} x^3 dx$, there is nothing special about *x*. It could be replaced with *y* or *t* or any other letter and the answer would be the same. However, your calculator might need you to turn any such expressions into one involving *x*.

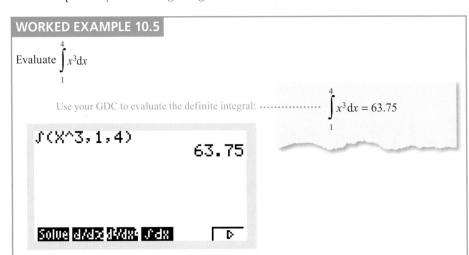

**WORKED EXAMPLE 10.5**

Evaluate $\int_{1}^{4} x^3 dx$

Use your GDC to evaluate the definite integral: ............ $\int_{1}^{4} x^3 dx = 63.75$

∫(X^3,1,4)
                        63.75

Solve  d/dx  ∫d/dx↑  ∫dx        ▷

### ■ Areas between a curve $y = f(x)$ and the *x*-axis where $f(x) > 0$

In the same way that the value of the derivative at a point has a geometrical meaning (it is the gradient of the tangent to the curve at that point) so does the value of a definite integral.

Perhaps surprisingly, the value of a definite integral is the area between a curve and the *x*-axis between the **limits of integration**.

## Tip

In a question where you are asked to find the area under a curve, it is important that you write down the correct definite integral before finding the value with your GDC.

**KEY POINT 10.2**

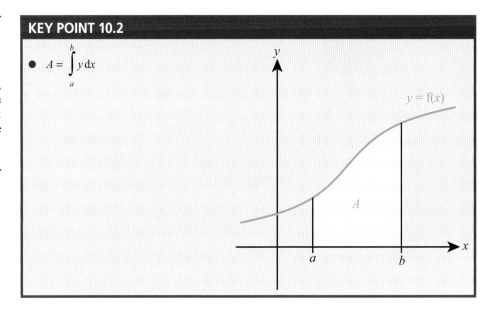

- $A = \int_a^b y \, \mathrm{d}x$

**WORKED EXAMPLE 10.6**

The diagram on the right shows the curve $y = x^2 - x^3$.

Find the shaded area.

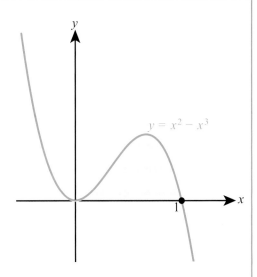

The area is the definite integral between the limits of 0 and 1. Write down the definite integral to be found ............... $\text{Area} = \int_0^1 (x^2 - x^3) \, \mathrm{d}x$

Use your GDC to evaluate the definite integral ............... $= 0.0833$

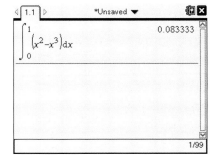

**TOOLKIT:** Problem Solving

Find an expression for the sum of the areas of the three rectangles bounded by the curve $y = x^2$ between $x = 1$ and $x = 2$.

Find an expression in terms of $n$ for $n$ rectangles of equal width between $x = 1$ and $x = 2$.

Use your GDC to find the limit of your expression as $n$ gets very large.

Compare this to the value of

$$\int_1^2 x^2 dx.$$

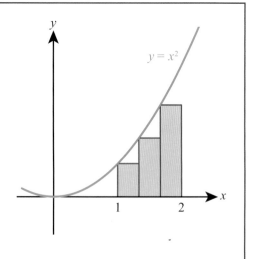

## CONCEPTS – APPROXIMATION

In the problem-solving task above, the area under the curve is not the area of the rectangle – it is only an **approximation**. However, this approximation gets better and better the more rectangles you have. How many would you need to have a 'good enough' approximation for a real-world situation?

It gets so good, that we use this approximation to define the area under the curve. The idea of 'perfect approximations' might make your head spin, but it was put on a firm foundation by the work of eminent mathematicians such as Augustin Cauchy and it forms the basis for modern calculus.

The problem-solving activity above suggests an important further interpretation of integration as the amount of material accumulated if you know the rate of accumulation. We can write this as shown below.

## KEY POINT 10.3

● $f(a) = f(b) + \int_b^a f'(t)dt$

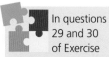
## WORKED EXAMPLE 10.7

Water is dripping into a bucket at a rate of $t^2$ cm$^3$ per minute, after a time of $t$ minutes. The volume in the bucket after 1 minute is 5cm$^3$. Find the volume in the bucket after 4 minutes.

The volume accumulated is the integral of the rate between 1 and 4 minutes. We use these as the limits on the integral

Volume accumulated $= \int_1^4 t^2 dt$

Use your GDC to evaluate this

$= 21$ cm$^3$

We need to add the volume accumulated to the volume already in the bucket

So total volume in the bucket is $21 + 5 = 26$ cm$^3$

## Exercise 10B

For questions 1 and 2, use the method demonstrated in Worked Example 10.5 to evaluate the given definite integrals with your GDC.

1  a  $\displaystyle\int_{-1}^{3} 2x^4\,dx$                    b  $\displaystyle\int_{1}^{4} 3x^{-2}\,dx$

2  a  $\displaystyle\int_{1}^{2} \left(2 - 7t^{-3}\right)dt$          b  $\displaystyle\int_{2}^{5} \left(t - 4t^2\right)dt$

For questions 3 and 4, use the method demonstrated in Worked Example 10.6 to find the shaded area.

3  a

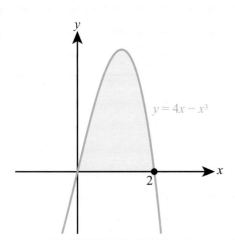

b

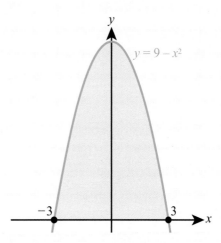

4  a

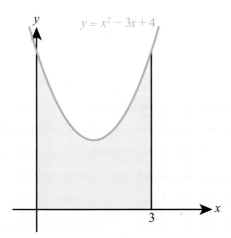

b

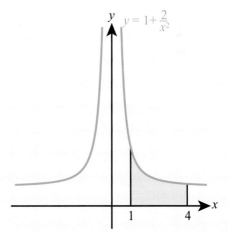

5  Evaluate $\displaystyle\int_{2}^{5} 2x + \frac{1}{x^2}\,dx$.

6  Evaluate $\displaystyle\int_{1}^{2} (x - 1)^3\,dx$.

**7** Find the shaded area.

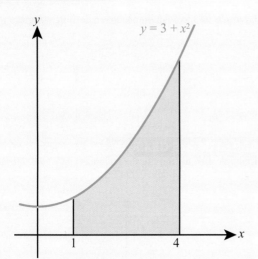

$y = 3 + x^2$

**8** Find the shaded area.

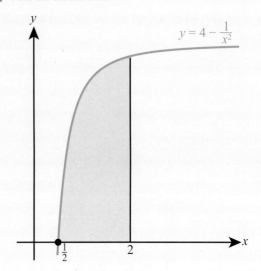

$y = 4 - \dfrac{1}{x^2}$

**9** Find the area enclosed between the $x$-axis and the graph of $y = -x^2 + 8x - 12$.

**10** Find the area enclosed by the part of the graph of $y = 9 - x^2$ for $x > 0$, the $x$-axis and the $y$-axis.

**11** The diagram shows the graphs with equations $y = 5x - x^2$ and $y = \dfrac{5x}{2}$.

   **a** Find the coordinates of the non-zero point of intersection of the two graphs.

   **b** Find the shaded area.

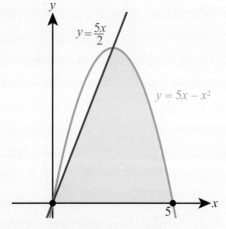

$y = \dfrac{5x}{2}$

$y = 5x - x^2$

**12** As a chemical filter warms up it can filter water more quickly. At a time $t$ after it starts, it filters water at a rate of $5 + kt^2$ litres per minute. After 2 minutes it filters at a rate of 9 litres per minute.

How much water is filtered in the first 2 minutes?

**13** For the first minute of activity, the rate at which a paint sprayer emits paint is directly proportional to the time it has been on. After 10 seconds it sprays paint at a rate of 20 grams per second. What is the total amount of paint it sprays in the first minute of activity?

**14** A model suggests that sand falls through a timer at a rate of $\dfrac{100}{t^3}$ g where $t$ is the time in seconds after it was started. When $t = 1$ there are $10\,$g of sand in the base of the timer.

   **a** Find the amount of sand in the base 5 seconds after it was started.

   **b** Use technology to estimate the amount of sand that will eventually fall into the base.

   **c** Provide one criticism of this model.

**15** The function f satisfies $f'(x) = x^2$ and $f(0) = 4$ Find $\displaystyle\int_0^3 f(x)\,dx$.

**16** Given that $\displaystyle\int f(x)\,dx = 4\left(x^3 + \dfrac{1}{x^2} + c\right)$, find an expression for $f(x)$.

**17** Find the area enclosed by the curve $y = x^2$, the $y$-axis and the line $y = 4$.

**18** A one-to-one function $f(x)$ satisfies $f(0) = 0$ and $f(x) \geqslant 0$

If $\displaystyle\int_0^a f(x)dx = A$ for $a > 0$, find an expression for $\displaystyle\int_0^{f(a)} f^{-1}(x)dx$.

## Checklist

▪ You should be able to integrate functions of the form $f(x) = ax^n + bx^m$, where $n, m \in \mathbb{Z}$, $n, m \neq -1$:

☐ $\displaystyle\int (ax^n + bx^m + \ldots)dx = \frac{a}{n+1}x^{n+1} + \frac{b}{m+1}x^{m+1} + \ldots + c$

where $a$, $b$ and $c$ are constants and $n, m \in \mathbb{Z}$, $n, m \neq -1$.

▪ You should be able to use integration with a boundary condition to determine the constant term.

▪ You should be able to use your GDC to find the value of definite integrals.

▪ You should be able to find areas between a curve $y = f(x)$ and the $x$-axis, where $f(x) > 0$:

☐ $\displaystyle A = \int_a^b y\, dx$

where the area is bounded by the lines $x = a$ and $x = b$

▪ You should be able to use integration to find the amount of material accumulated:

☐ $\displaystyle f(b) = f(a) + \int_a^b f'(t)dt$

## ■ Mixed Practice

**1** Find $\int x^3 - \dfrac{3}{x^2}\,dx$.

**2** Find $\int 4x^2 - 3x + 5\,dx$.

**3** Evaluate $\int\limits_{1}^{5} \dfrac{2}{x^4}\,dx$.

**4** A curve has gradient given by $\dfrac{dy}{dx} = 3x^2 - 8x$ and passes through the point $(1, 3)$. Find the equation of the curve.

**5** Find the value of $a$ such that $\int\limits_{2}^{a} 2 - \dfrac{8}{x^2}\,dx = 9$.

**6** The graph of $y = 9x - x^2 - 8$ is shown in the diagram.

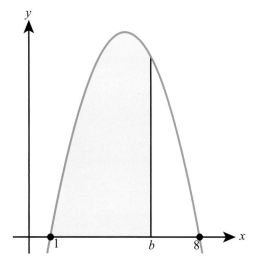

Given that the shaded area equals 42.7, find the value of $b$ correct to one decimal place.

**7** The diagram shows the graph of $y = -x^3 + 9x^2 - 24x + 20$.

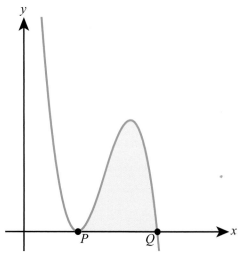

**a** Find the coordinates of $P$ and $Q$.

**b** Find the shaded area.

**8** The diagram shows the curve with equation $y = 0.2x^2$ and the tangent to the curve at the point where $x = 4$.

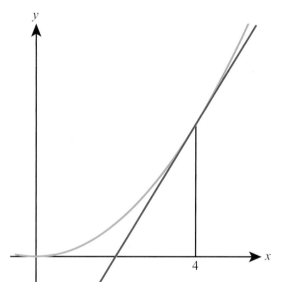

  **a** Find the equation of the tangent.
  **b** Show that the tangent crosses the $x$-axis at the point $(2, 0)$.
  **c** Find the shaded area.

**9** The diagram shows the graph of $y = \dfrac{1}{x^2}$ and the normal at the point where $x = 1$.

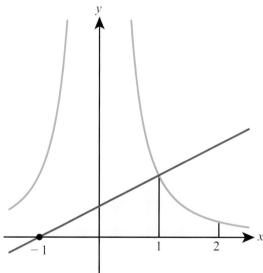

  **a** Show that the normal crosses the $x$-axis at $(-1, 0)$.
  **b** Find the shaded area.

**10** The gradient of the normal to the curve $y = f(x)$ at any point equals $x^2$. If $f(1) = 2$, find $f(2)$.

**11** The rate at which water is accumulated in a rainwater measuring device in a storm is given by $20t \, \text{cm}^3/\text{minute}$.

If the container is initially empty, find the volume of water in the container after 10 minutes.

**12** A puppy at age 6 months has a mass of 2.3 kg. Its growth after this point is modelled by

$\dfrac{A}{20} + c$ kg/month where $A$ is the puppy's age in months.

When the puppy is 10 months old it is growing at a rate of 1.5 kg per month. Use this model to estimate the mass of the puppy when it is 18 months old.

**13** The function $f(x)$ is such that $f'(x) = 3x^2 + k$ where $k$ is a constant. If $f(1) = 13$ and $f(2) = 24$, find $f(3)$.

**14** A nutritionist designs an experiment to find the energy contained in a nut. They burn the nut and

model the energy emitted as $\dfrac{k}{t^2}$ calories for $t > 1$ where $t$ is in seconds.

They place a beaker of water above the burning nut and measure the energy absorbed by the water using a thermometer.

**a** State one assumption that is being made in using this experimental setup to measure the energy content of the nut.

**b** When $t = 1$ the water has absorbed 10 calories of energy. When $t = 2$ the water has absorbed 85 calories of energy. Use technology to estimate the total energy emitted by the nut if it is allowed to burn indefinitely.

**15** The gradient at every point on a curve is proportional to the square of the $x$-coordinate at that point.

Find the equation of the curve, given that it passes through the points $(0, 3)$ and $\left(1, \dfrac{14}{3}\right)$.

**16** The function f satisfies $f'(x) = 3x - x^2$ and $f(4) = 0$. Find $\displaystyle\int_0^4 f(x)\,dx$.

**17** Given that $\displaystyle\int f(x)\,dx = 3x^2 - \dfrac{2}{x} + c$, find an expression for $f(x)$.

**18** The diagram shows a part of the curve with equation $y = x^2$. Point $A$ has coordinates $(0, 9)$.

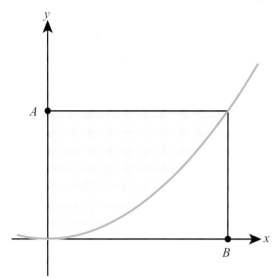

   **a** Find the coordinates of point $B$.
   **b** Find the shaded area.

# Core SL content: Review Exercise

**1**  An arithmetic sequence has first term 7 and third term 15.

    **a**  Find the common difference of the series.

    **b**  Find the 20th term.

    **c**  Find the sum of the first 20 terms.

**2**  Line $L$ has equation $2x - 3y = 12$.

    **a**  Find the gradient of $L$.

    **b**  The point $P(9, k)$ lies on $L$. Find the value of $k$.

    **c**  Line $M$ is perpendicular to $L$ and passes through the point $P$. Find the equation of $M$ in the form $ax + by = c$.

**3**  **a**  Sketch the graph of $y = (x + 2)e^{-x}$, labelling all the axis intercepts and the coordinates of the maximum point.

    **b**  State the equation of the horizontal asymptote.

    **c**  Write down the range of the function $f(x) = (x + 2)e^{-x}$, $x \in \mathbb{R}$.

**4**  A function is defined by $f(x) = \sqrt{x - a}$. The graph of $y = f(x)$ is shown in the diagram.

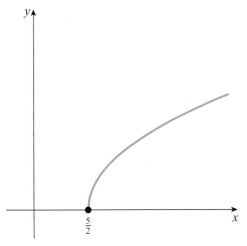

    **a**  Write down the value of $a$ and state the domain of f.

    **b**  Write down the range of f.

    **c**  Find $f^{-1}(4)$.

**5**  In triangle $ABC$, $AB = 12\,\text{cm}$, $BC = 16\,\text{cm}$ and $AC = 19\,\text{cm}$.

    **a**  Find the size of the angle $ACB$.

    **b**  Calculate the area of the triangle.

**6** A random variable $X$ follows a normal distribution with mean 26 and variance 20.25.

   **a** Find the standard deviation of $X$.

   **b** Find $P(21.0 < x < 25.3)$.

   **c** Find the value of $a$ such that $P(X > a) = 0.315$.

**7** The curve in the diagram has equation $y = (4 - x^2)e^{-x}$.

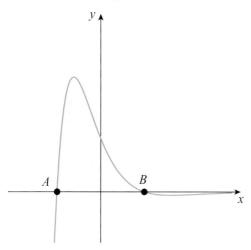

   **a** Find the coordinates of the points $A$ and $B$.

   **b** Find the shaded area.

**8** The IB grades attained by a group of students are listed as follows.

   6  4  5  3  7  3  5  4  2  5

   **a** Find the median grade.

   **b** Calculate the interquartile range.

   **c** Find the probability that a student chosen at random from the group scored at least a grade 4.

                                                  Mathematical Studies SL May 2015 Paper 1 TZ1 Q2

**9** *ABCDV* is a solid glass pyramid. The base of the pyramid is a square of side 3.2 cm. The vertical height is 2.8 cm. The vertex *V* is directly above the centre *O* of the base.

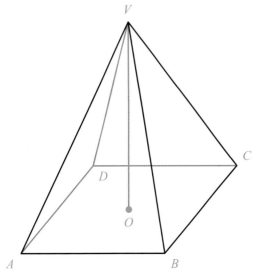

   **a** Calculate the volume of the pyramid.

   **b** The glass weighs 9.3 grams per cm³. Calculate the weight of the pyramid.

   **c** Show that the length of the sloping edge *VC* of the pyramid is 3.6 cm.

   **d** Calculate the angle at the vertex $B\hat{V}C$.

   **e** Calculate the total surface area of the pyramid.

Mathematical Studies SL November 2007 Paper 2 Q2

**10** **a** Sketch the graph of the function $f(x) = \dfrac{2x + 3}{x + 4}$, for $-10 \leqslant x \leqslant 10$, indicating clearly the axis intercepts and any asymptotes.

   **b** Write down the equation of the vertical asymptote.

   **c** On the same diagram sketch the graph of $g(x) = x + 0.5$.

   **d** Using your graphical display calculator write down the coordinates of one of the points of intersection on the graphs of f and g, giving your answer correct to five decimal places.

   **e** Write down the gradient of the line $g(x) = x + 0.5$.

   **f** The line *L* passes through the point with coordinates $(-2, -3)$ and is perpendicular to the line $g(x)$. Find the equation of *L*.

Mathematical Studies May 2008 Paper 2 TZ1 Q1

**11** Find the equation of the normal to the curve $y = x + 3x^2$ at the point where $x = -2$. Give your answer in the form $ax + by + c = 0$.

**12** **a** Find the value of $x$ given that $\log_{10} x = 3$.

   **b** Find $\log_{10} 0.01$.

**13** **a** Expand and simplify $(2x^2)^3(x - 3x^{-5})$.

   **b** Differentiate $y = (2x^2)^3(x - 3x^{-5})$.

**14** A geometric series has first term 18 and common ratio $r$.

  a Find an expression for the 4th term of the series.

  b Write down the formula for the sum of the first 15 terms of this series.

  c Given that the sum of the first 15 terms of the series is 26.28 find the value of $r$.

**15** Theo is looking to invest £200 for 18 months. He needs his investment to be worth £211 by the end of the 18 months. What interest rate does he need if the interest is compounded monthly? Give your answer as a percentage correct to one decimal place.

**16** The distribution of a discrete random variable $X$ is given by $P(X = x) = \dfrac{k}{x^2}$ for $X = 1, 2, 3, 4$. Find $E(X)$.

**17** A box holds 240 eggs. The probability that an egg is brown is 0.05.

  a Find the expected number of brown eggs in the box.

  b Find the probability that there are 15 brown eggs in the box.

  c Find the probability that there are at least 10 brown eggs in the box.

Mathematics SL May 2011 Paper 2 TZ1 Q5

**18** The vertices of quadrilateral $ABCD$ as shown in the diagram are $A(3, 1)$, $B(0, 2)$, $C(-2, 1)$ and $D(-1, -1)$.

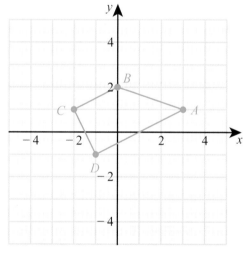

  a Calculate the gradient of line $CD$.

  b Show that line $AD$ is perpendicular to line $CD$.

  c Find the equation of line $CD$. Give your answer in the form $ax + by = c$ where $a, b, c \in \mathbb{Z}$.

Lines $AB$ and $CD$ intersect at point $E$. The equation of line $AB$ is $x + 3y = 6$.

  d Find the coordinates of $E$.

  e Find the distance between $A$ and $D$.

The distance between $D$ and $E$ is $\sqrt{20}$.

  f Find the area of triangle $ADE$.

Mathematical Studies May 2009 Paper 2 TZ1 Q3

**19** **a** Jenny has a cylinder with a lid. The cylinder has height 39 **cm** and diameter 65 **mm**.

    **i**   Calculate the volume of the cylinder **in cm³**. Give your answer correct to **two** decimal places.

The cylinder is used for storing tennis balls.

Each ball has a **radius** of 3.25 cm.

    **ii**  Calculate how many balls Jenny can fit in the cylinder if it is filled to the top.

    **iii** **I**  Jenny fills the cylinder with the number of balls found in part **ii** and puts the lid on. Calculate the volume of air inside the cylinder in the spaces between the tennis balls.

       **II**  Convert your answer to **iii I** into cubic metres.

**b** An old tower (*BT*) leans at 10° away from the vertical (represented by line *TG*).

The base of the tower is at *B* so that $M\hat{B}T = 100°$.

Leonardo stands at *L* on flat ground 120 m away from *B* in the direction of the lean.

He measures the angle between the ground and the top of the tower *T* to be $B\hat{L}T = 26.5°$.

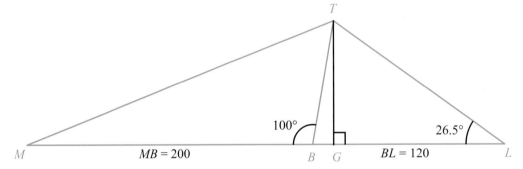

    **i**  **I**  Find the value of angle $B\hat{T}L$.

       **II**  Use triangle *BTL* to calculate the sloping distance *BT* from the base *B* to the top, *T* of the tower.

    **ii**  Calculate the vertical height *TG* of the top of the tower.

    **iii**  Leonardo now walks to point *M*, a distance 200 m from *B* on the opposite side of the tower. Calculate the distance from *M* to the top of the tower at *T*.

Mathematical Studies SL May 2007 Paper 2 Q2

**20** The following table shows the number of bicycles, $x$, produced daily by a factory and their total production cost, $y$, in US dollars (USD). The table shows data recorded over seven days.

|  | Day 1 | Day 2 | Day 3 | Day 4 | Day 5 | Day 6 | Day 7 |
|---|---|---|---|---|---|---|---|
| **Number of bicycles, $x$** | 12 | 15 | 14 | 17 | 20 | 18 | 21 |
| **Production cost, $y$** | 3900 | 4600 | 4100 | 5300 | 6000 | 5400 | 6000 |

a i Write down the Pearson's product–moment correlation coefficient, $r$, for these data.

   ii Hence comment on the result.

b Write down the equation of the regression line $y$ on $x$ for these data, in the form $y = ax + b$.

c Estimate the total cost, to the nearest USD, of producing 13 bicycles on a particular day.

All the bicycles that are produced are sold. The bicycles are sold for 304 USD each.

d Explain why the factory does not make a profit when producing 13 bicycles on a particular day.

e i Write down an expression for the total selling price of $x$ bicycles.

   ii Write down an expression for the profit the factory makes when producing $x$ bicycles on a particular day.

   iii Find the least number of bicycles that the factory should produce, on a particular day, in order to make a profit.

*Mathematical Studies May 2015 Paper 2 TZ2 Q6*

**21** It is known that, among all college students, the time taken to complete a test paper is normally distributed with mean 52 minutes and standard deviation 7 minutes.

a Find the probability that a randomly chosen student completes the test in less than 45 minutes.

b In a group of 20 randomly chosen college students, find the probability that:

   i exactly one completes the test in less than 45 minutes

   ii more than three complete the test in less than 45 minutes.

**22** In a class of 26 students, 15 study French, 14 study biology and 8 study history. All students study at least one of these subjects.

Of those students, 7 study both French and biology, 4 study French and history, and 3 study biology and history.

a Using a Venn diagram, or otherwise, find how many students study all three subjects.

b Find the probability that a randomly selected student studies French only.

c Given that a student studies French, what is the probability that they do not study biology?

d Two students are selected at random. What is the probability that at least one of them studies history?

## ESSENTIAL UNDERSTANDINGS

- Number and algebra allow us to represent patterns, show equivalences and make generalizations.
- Algebra is an abstraction of numerical concepts and employs variables to solve mathematical problems.

### In this chapter you will learn...

- about equations and identities
- how to construct simple deductive proofs.

### CONCEPTS

The following concepts will be addressed in this chapter:
- Numbers and formulae can appear in different but equivalent forms, or **representations**, which can help us to establish identities.
- Formulae are a **generalization** made on the basis of specific examples, which can then be extended to new examples.

### PRIOR KNOWLEDGE

Before starting this chapter, you should already be able to complete the following:

1 Expand and simplify $(2x - 1)(x + 4)$.

2 Evaluate $\dfrac{3}{4} + \dfrac{2}{5}$.

■ **Figure 11.1** Does the word 'proof' mean different things in different contexts?

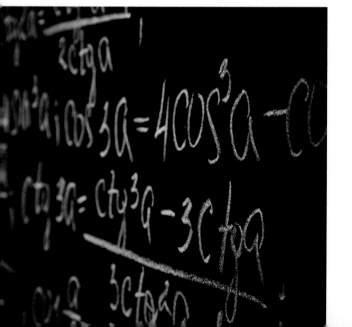

Proof is one of the most important elements of mathematics as it allows us to establish whether certain results that we might suspect are true, really are true.

In science, reasoning is often inductive; scientists base their theories on observation but have to be open to revising their ideas if new evidence comes to light.

In mathematics, deductive proof starts with a statement you know to be true and proceeds with a sequence of valid steps to arrive at a conclusion. As long as the starting point is true and you do not make any subsequent mistakes in working from one step to the next, you know for certain that the conclusion is true.

### TOK Links

Look at the pictures in Figure 11.1. Does 'proof' mean the same in different disciplines? What is the difference between proof in maths, science and law? To what extent can we talk about proof in relation to philosophy, or social sciences?

### Starter Activity

Necklaces are made by threading $n$ gold or silver beads onto a closed loop. For example, if $n = 2$, there are three possible necklaces that could be made: SS, GG, GS.

a  Find how many different necklaces can be made for

   i   $n = 0$

   ii  $n = 1$

   iii $n = 3$

b  On the basis of these results (including $n = 2$), suggest a formula for the number of necklaces that can be made with $n$ beads.

c  Check your formula with the result for $n = 4$.

### LEARNER PROFILE – Communicators
Is a proof the same as an explanation?

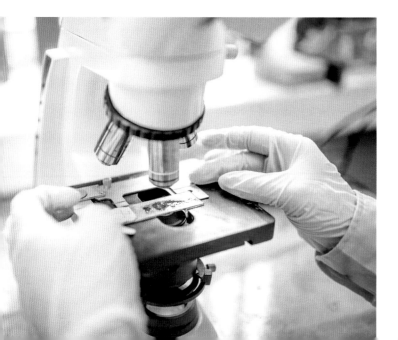

# 11A The structure of mathematical proof

## ■ The symbols and notation for equality and identity

**Equations** are one of the most frequently used mathematical structures. An equation is only true for some particular value(s) of a variable. For example, $x^2 = 4$ is true for $x = \pm 2$ but not for any other values of $x$.

An **identity** is true for all values of a variable. To emphasize that we are dealing with an identity and not just an equation, the symbol $\equiv$ is used instead of the $=$ symbol. For example, $x^2 - 4 \equiv (x - 2)(x + 2)$.

**Tip**

The identity symbol, $\equiv$, is often only used for emphasis. You will find lots of identities are written with an equals sign instead.

### WORKED EXAMPLE 11.1

$$x^2 + 6x + 2 \equiv (x + a)^2 + b$$

Find the values of the constants $a$ and $b$.

| | |
|---|---|
| The expression on the left-hand side (LHS) must be identical to that on the right-hand side (RHS) for all values of $x$. Multiply out the brackets so you can compare the two sides | $x^2 + 6x + 2 \equiv (x + a)^2 + b$ $\equiv x^2 + 2ax + a^2 + b$ |
| The coefficient of $x$ (the number in front of $x$) must be the same on both sides | Coefficient of $x$: $6 = 2a$, so $a = 3$ |
| The constant term must be the same on both sides | Constant term: $2 = a^2 + b$ |
| You already know $a = 3$ | So, $2 = 3^2 + b$ $b = -7$ |

### CONCEPTS – REPRESENTATIONS

Are the expressions $x^2 + 6x + 2$ and $(x + 3)^2 - 7$ identical? Why might you use one rather than the other? You will find out more about this in Chapter 16.

In Worked Example 11.1 we compared coefficients, but it is important to note that just because two expressions **represent** the same value, they do not have to have comparable parts. For example, if $\frac{a}{b} = \frac{1}{2}$ it does not mean that $a$ must equal 1 and $b$ must equal 2.

## ■ Simple deductive proof

To prove that a statement is true, you need to start with the given information and proceed by a sequence of steps to the conclusion.

Often proofs involve algebra as this is a tool that enables us to express ideas in general terms.

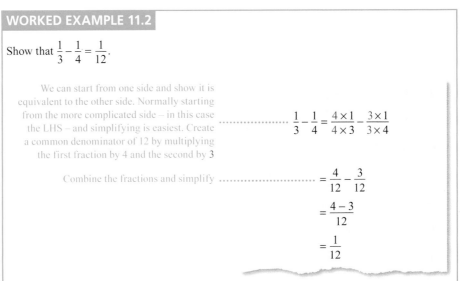

**WORKED EXAMPLE 11.2**

Show that $\dfrac{1}{3} - \dfrac{1}{4} = \dfrac{1}{12}$.

We can start from one side and show it is equivalent to the other side. Normally starting from the more complicated side – in this case the LHS – and simplifying is easiest. Create a common denominator of 12 by multiplying the first fraction by 4 and the second by 3

$$\frac{1}{3} - \frac{1}{4} = \frac{4 \times 1}{4 \times 3} - \frac{3 \times 1}{3 \times 4}$$

Combine the fractions and simplify

$$= \frac{4}{12} - \frac{3}{12}$$

$$= \frac{4 - 3}{12}$$

$$= \frac{1}{12}$$

### TOK Links

Would you use this proof to explain this fact to someone just meeting fractions? Would you need to include every step if you were proving it to a university professor? Whenever you are communicating knowledge, you must consider the audience. However, sometimes in maths, 'obvious' facts turn out to be very difficult to prove. For example, the 'fact' that parallel lines never meet has been challenged by some geometers. The philosopher Bertrand Russell spent many pages proving that $1 + 1 = 2$. Whenever we are producing a proof we need to know what the accepted facts are in our mathematical community.

**WORKED EXAMPLE 11.3**

Prove that $\dfrac{6a+5}{3} - \dfrac{4a+3}{2} \equiv \dfrac{1}{6}$.

In this case the LHS is more complicated so we start there and try to show it is equivalent to the RHS. Create a common denominator of 6 by multiplying the first fraction by 2 and the second by 3

$$\frac{6a+5}{3} - \frac{4a+3}{2} \equiv \frac{2(6a+5)}{2\times 3} - \frac{3(4a+3)}{3\times 2}$$

$$\equiv \frac{12a+10}{6} - \frac{12a+9}{6}$$

Combine the fractions and simplify the algebra

$$\equiv \frac{12a+10-12a-9}{6}$$

$$\equiv \frac{1}{6}$$

**WORKED EXAMPLE 11.4**

Prove that the sum of an even number and an odd number is always odd.

Define a general even number...

Let the even number be $2n$, for some $n \in \mathbb{Z}$.

...and a general odd number (you could also use $2m-1$) Don't use $2n+1$ (or $2n-1$) as this would be the next integer up (or down) from $2n$, which would be too specific

Let the odd number be $2m+1$, for some $m \in \mathbb{Z}$.

$$2n + (2m+1) = 2n + 2m + 1$$

Make it clear that the result is an odd number by writing the result in the form $2k+1$

$$= 2(n+m) + 1$$

$$= 2k+1, \text{ for some } k \in \mathbb{Z}$$

Make a conclusion

So, the sum is odd.

## CONCEPTS – GENERALIZATION

When solving a problem, lots of people start by trying some cases in order to convince themselves about what is happening. While this is a good initial tactic, no matter how many cases you try you will never create a mathematical proof (unless you can also argue that you have covered all possible cases). For example, there is a famous conjecture, due to Goldbach, that states:

'Every odd number bigger than 1 is either prime or can be written as a prime plus twice a square number.'

This does seem to work for a lot of the odd numbers – mathematicians have used computers to search for counterexamples up to $10^{13}$ and only found two: 5777 and 5993.

Another famous conjecture by Goldbach states that every even integer greater than 2 can be written as the sum of two primes. This has been tested up to $10^{18}$ and no counterexamples have been found, but this does not mean mathematicians believe it has been proven.

The big advantage of algebra is that it is not considering specific examples, but is used to represent all possible examples. This is what makes it so useful in proof.

# Be the Examiner 11.1

Prove that the product of an even and an odd number is always even.

Which is the correct solution? Identify the errors made in the incorrect solutions.

| Solution 1 | Solution 2 | Solution 3 |
|---|---|---|
| $2n(2n+1) = 2(2n^2 + n)$ $= 2k$ So, the product is even. | $2m(2n+1) = 2(2mn + m)$ $= 2k$ So, the product is even. | If $n$ is even, then $n+1$ is odd or vice versa. So, $n(n+1) = 2k$ So, the product is even. |

## Exercise 11A

**1** Given that $x^2 + 8x + 23 \equiv (x+a)^2 + b$, find the values of $a$ and $b$.

**2** Given that $x^2 - 12x - 1 \equiv (x-a)^2 - b$, find the values of $a$ and $b$.

**3** Find the values of $p$ and $q$ such that $(2x-5)(2x+5) \equiv px^2 - q^2$.

**4** Prove that $(n-1)^2 + n^2 + (n+1)^2 \equiv 3n^2 + 2$.

**5** Prove that $\dfrac{3x-2}{3} - \dfrac{2x-3}{2} \equiv \dfrac{5}{6}$.

**6** Prove that $\dfrac{7x+6}{12} - \dfrac{3x+5}{10} \equiv \dfrac{17x}{60}$.

**7** a Show that $\dfrac{1}{2} + \dfrac{1}{4} = \dfrac{3}{4}$.

   b Show that the algebraic generalization of this is $\dfrac{1}{n} + \dfrac{1}{2n} \equiv \dfrac{3}{2n}$.

**8** Which of the following equations are, in fact, identities?

   a $(x-5)(x+5) = x^2 - 25$

   b $(x+5)^2 = x^2 + 25$

   c $(x-5)^2 = x^2 - 10x + 25$

**9** Find the values of $a$ and $b$ so that $x^3 - 5x^2 + 3x + 9 \equiv (x+a)(x-b)^2$.

**10** Find the values of $p$ and $q$ such that $x^3 - 2x^2 + 5x - 10 \equiv (x^2 + p)(x-q)$.

**11** Prove that $\dfrac{1}{x-2} - \dfrac{1}{x+3} \equiv \dfrac{5}{x^2 + x - 6}$.

**12** a Show that $\dfrac{1}{3} - \dfrac{1}{5} = \dfrac{2}{15}$.

   b Show that the algebraic generalization of this is $\dfrac{1}{n-1} - \dfrac{1}{n+1} \equiv \dfrac{2}{n^2 - 1}$.

**13** Dina says that $x^3 - y^3 = (x-y)(x^2 + y^2)$ is an identity.

   a Find one example to show that Dina is wrong.

   b Find the value of $a$ such that $x^3 - y^3 \equiv (x-y)(x^2 + axy + y^2)$.

**14** Prove that $(2n+1)^2 - (2n+1)$ is an even number for all positive integer values of $n$.

**15** Prove that $(2n+1)^2 - (2n-1)^2$ is a multiple of 8 for all positive integer values of $n$.

**16** Prove algebraically that the difference between the squares of any two consecutive integers is always an odd number.

**17** Prove that the sum of the squares of two consecutive odd numbers is never a multiple of 4.

## Checklist

- You should be able to work with and understand the notation for identities.
  - ☐ An identity occurs when two expressions always take the same value for all allowed values of the variable.
  - ☐ When this is emphasized the symbol $\equiv$ is used in place of the equals sign.
  - ☐ If two polynomials are identical you can compare coefficients.
- You should be able to construct simple deductive proofs.
  - ☐ The usual method is to start from one side and show through a series of identities that it is equivalent to the other side.

## ■ Mixed Practice

**1** Which of the following are identities?

   **a** $(x+2)^3 = x^3 + 8$

   **b** $x^3 + 64 = (x+4)(x^2 - 4x + 16)$

   **c** $x^4 - 4 = (x^2 - 2)(x^2 + 2)$

**2** **a** Use an example to show that $x^2 + 5 = (x+1)(x+5)$ is not an identity.

   **b** Find the value of $k$ so that $x^2 + kx + 5 \equiv (x+1)(x+5)$.

**3** **a** Use an example to show that $x^2 + 4x + 9 = (x+3)^2$ is not an identity.

   **b** Given that $x^2 + 4x + 9 \equiv (x+p)^2 + q$ , find the values of $p$ and $q$.

**4** Find the values of $a$ and $b$ such that $x^2 - 8x \equiv (x-a)^2 - b$.

**5** Prove that

$$\frac{x-2}{2} - \frac{x-3}{3} \equiv \frac{x}{6}.$$

**6** Find the values of $A$ and $B$ such that

$$A(x+2) + B(2x-1) \equiv 5x.$$

**7** Prove that $(2n+3)^2 - (2n-3)^2$ is a multiple of 12 for all positive integer values of $n$.

**8** Prove algebraically that the difference between any two different odd numbers is an even number.

**9** The product of two consecutive positive integers is added to the larger of the two integers. Prove that the result is always a square number.

**10** Prove that $\dfrac{x}{(x-2)^2} - \dfrac{1}{x-2} \equiv \dfrac{2}{(x-2)^2}$.

**11** Prove that the sum of the squares of any three consecutive odd numbers is always 11 more than a multiple of 12.

**12** **a** Find the values of $a$ and $b$ such that $x^3 - 8 = (x-2)(x^2 + ax + b)$.

   **b** $n$ is a positive integer greater than 3. Prove that $n^3 - 8$ is never a prime number.

# 12 Analysis and approaches: Exponents and logarithms

## ESSENTIAL UNDERSTANDINGS

- Number and algebra allow us to represent patterns, show equivalences and make generalisations.
- Algebra is an abstraction of numerical concepts and employs variables to solve mathematical problems.

## In this chapter you will learn...

- how to extend the laws of exponents to general rational exponents
- how to work with logarithms to any (positive) base
- how to use the laws of logarithms
- how to change the base of a logarithm
- how to solve exponential equations.

### CONCEPTS

The following concepts will be addressed in this chapter:
- Numbers and formulae can appear in different but **equivalent** forms, or **representations**, which can help us to establish identities.
- **Patterns** in numbers inform the development of algebraic tools that can be applied to find unknowns.

### PRIOR KNOWLEDGE

Before starting this chapter, you should already be able to complete the following:

1 Simplify the following:

 a $\quad 3x^2y^4 \times 5x^7y$ 
 b $\quad \dfrac{8c^2d^3}{2c^3d}$ 
 c $\quad (3a^4b^{-2})^2$

2 Find the value of $y$ for which $\log_{10}(y+9) = 2$.

 3 Solve the equation $3^{x-5} = \left(\dfrac{1}{9}\right)^x$.

4 Solve the equation $e^{2x-1} = 11$.

■ **Figure 12.1** Why are logarithms used to describe the severity of an earthquake, the volume of a sound, the growth of bacteria and the acidity of a solution?

You have already seen how the laws of exponents allow you to manipulate exponential expressions and how this can be useful for solving some types of exponential equation.

In the same way, it is useful to have some laws of logarithms that will enable you to solve some more complicated exponential equations and equations involving different logarithm terms. It should be no surprise, given the relationship between exponents and logarithms, that these laws of logarithms follow from the laws of indices.

## Starter Activity

Look at the pictures in Figure 12.1. Investigate how logarithmic functions are used to measure or model these different phenomena.

**Now try this problem:**

By trying different positive values of $x$ and $y$, suggest expressions for the following in terms of $\ln x$ and $\ln y$.

a  $\ln(xy)$

b  $\ln\left(\dfrac{x}{y}\right)$

c  $\ln(x^y)$

Do your suggested relationships work for $\log_{10}$ as well?

---

**LEARNER PROFILE** – Reflective
Is being really good at arithmetic the same as being really good at mathematics?

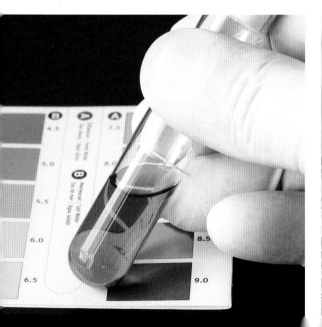

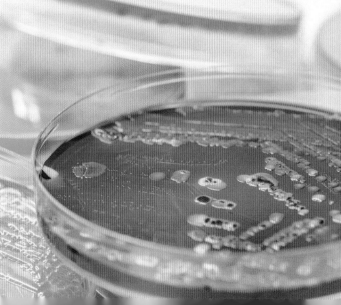

For a reminder of the laws of exponents for integer exponents, see Section 1A.

..........................

**Tip**

If $n$ is even then $\sqrt[n]{a}$ is defined to be positive, and it is only real if $a$ is also positive. If $n$ is odd, then $a$ can be negative, and $\sqrt[n]{a}$ has the same sign as $a$.

..........................

# 12A Laws of exponents with rational exponents

In order to extend the laws you know for integer exponents to non-integer rational exponents, you need a new law.

---

**KEY POINT 12.1**

$$a^{\frac{1}{n}} = \sqrt[n]{a}$$

---

**Proof 12.1**

Explain why $a^{\frac{1}{n}} = \sqrt[n]{a}$.

> We need a defining feature ................. $\sqrt[n]{a}$ is the number which equals $a$ when
> of $\sqrt[n]{a}$                                                raised to the exponent $n$.
>
> Use the fact that $(x^a)^b \equiv x^{ab}$ ................. We know that $\left(a^{\frac{1}{n}}\right)^n \equiv a^{\left(\frac{1}{n} \times n\right)} \equiv a^1$
>
>                                                            Therefore $a^{\frac{1}{n}}$ has the defining property of $\sqrt[n]{a}$

---

**CONCEPTS – REPRESENTATION**

You might ask why writing the same thing in a different notation has any benefit. Using an exponent **representation** of $\sqrt[n]{a}$ has distinct advantage as it allows us to use the laws of exponents on these expressions.

---

**WORKED EXAMPLE 12.1**

Evaluate $16^{\frac{1}{4}}$.

> Use $a^{\frac{1}{n}} = \sqrt[n]{a}$ ..................................... $16^{\frac{1}{4}} = \sqrt[4]{16}$
>
>                                                                        $= 2$

---

Combining the law in Key Point 12.1 with the law that states that $(a^m)^n \equiv a^{mn}$ allows us to cope with any rational exponent.

**WORKED EXAMPLE 12.2**

Evaluate $27^{-\frac{2}{3}}$.

| | |
|---|---|
| Use $(a^m)^n \equiv a^{mn}$ to split the exponent | $27^{-\frac{2}{3}} = \left(27^{\frac{1}{3}}\right)^{-2}$ |
| Remember that a negative exponent turns into 1 divided by the same expression with a positive exponent. $27^{\frac{1}{3}} = \sqrt[3]{27}$ | $= \dfrac{1}{\left(\sqrt[3]{27}\right)^2}$ |
| You should know small perfect squares and cubes. You can recognise that $\sqrt[3]{27} = 3$ | $= \dfrac{1}{3^2} = \dfrac{1}{9}$ |

You will also need to be able use this new law in an algebraic context.

**WORKED EXAMPLE 12.3**

Write $\dfrac{4x}{\sqrt[3]{x}}$ in the form $kx^n$.

| | |
|---|---|
| Use $a^{\frac{1}{n}} = \sqrt[n]{a}$ on the denominator | $\dfrac{4x}{\sqrt[3]{x}} = \dfrac{4x}{x^{\frac{1}{3}}}$ |
| Then use $\dfrac{a^m}{a^n} = a^{m-n}a$ | $= 4x^{1-\frac{1}{3}}$ |
| | $= 4x^{\frac{2}{3}}$ |

#  Exercise 12A

For questions 1 to 3, use the method demonstrated in Worked Example 12.1 to evaluate the expression without using a calculator.

1   a   $49^{\frac{1}{2}}$    2   a   $8^{\frac{1}{3}}$    3   a   $256^{\frac{1}{4}}$

   b   $25^{\frac{1}{2}}$       b   $27^{\frac{1}{3}}$       b   $625^{\frac{1}{4}}$

For questions 4 to 8, use the method demonstrated in Worked Example 12.2 to evaluate the expression without using a calculator.

4   a   $8^{\frac{2}{3}}$    5   a   $625^{\frac{3}{4}}$    6   a   $100^{-\frac{1}{2}}$

   b   $16^{\frac{3}{4}}$       b   $125^{\frac{2}{3}}$       b   $1000^{-\frac{1}{3}}$

7   a   $8^{-\frac{2}{3}}$    8   a   $32^{-\frac{2}{5}}$

   b   $27^{-\frac{2}{3}}$       b   $100000^{-\frac{3}{5}}$

For questions 9 to 13, use the method demonstrated in Worked Example 12.3 to simplify each expression, giving your answers in the form $ax^p$.

9   a   $x^2\sqrt{x}$    10   a   $\dfrac{x^2}{\sqrt[3]{x}}$    11   a   $\dfrac{4\sqrt{x}}{x^2}$

   b   $x\sqrt[3]{x}$       b   $\dfrac{x^2}{\sqrt{x}}$       b   $\dfrac{5\sqrt{x}}{x^3}$

12  a  $\dfrac{x^2}{5\sqrt{x}}$

    b  $\dfrac{x}{3\sqrt[3]{x}}$

13  a  $\sqrt[3]{x}\sqrt{x}$

    b  $x\sqrt[3]{x^2}$

**14**  Find the exact value of $8^{-\frac{4}{3}}$.

**15**  Find the exact value of $\left(\dfrac{1}{4}\right)^{-\frac{1}{2}}$.

**16**  Find the exact value of $\left(\dfrac{4}{9}\right)^{\frac{3}{2}}$.

**17**  Write $\sqrt[3]{x^2}\times\sqrt[4]{x}$ in the form $x^p$.

**18**  Write $\dfrac{3}{\sqrt[3]{x}}+2\sqrt{x}$ in the form $ax^p+bx^q$.

**19**  Write $\dfrac{1}{3\sqrt{x^3}}$ in the form $ax^p$.

**20**  Solve the equation $x^{\frac{3}{2}}=\dfrac{1}{8}$.

**21**  Solve the equation $x^{-\frac{1}{2}}=\dfrac{2}{5}$.

**22**  Write in the form $x^p$:
$\dfrac{x\sqrt{x}}{\sqrt[3]{x}}$

**23**  Write in the form $x^p$:
$\dfrac{x}{x^2\sqrt{x}}$

**24**  Write $\left(x\times\sqrt[3]{x}\right)^2$ in the form $x^k$.

**25**  Write $\left(\dfrac{1}{2\sqrt{x}}\right)^3$ in the form $ax^p$.

**26**  Write in the form $x^a+x^b$:
$\dfrac{x^2+\sqrt{x}}{x\sqrt{x}}$

**27**  Write in the form $x^a-x^b$:
$\dfrac{\left(x+\sqrt{x}\right)\left(x-\sqrt{x}\right)}{\sqrt{x}}$

**28**  Write in the form $ax^k$:
$\dfrac{1}{3x\sqrt{x}}$

**29**  Write in the form $ax^p+bx^q$:
$\dfrac{x^2+3\sqrt{x}}{2x}$

**30**  Given that $y=2\sqrt[3]{x^2}$, write $y^4$ in the form $ax^k$.

**31**  Given that $y=27\sqrt{x}$, write $\sqrt[3]{y}$ in the form $ax^k$.

**32**  Given that $\sqrt{x}=\sqrt[3]{y}$, write $y$ in the form $x^k$.

**33**  Given that $y=\dfrac{2}{3\sqrt{x}}$, write $y^3$ in the form $ax^k$.

**34**  Given that $y=x\sqrt{x}$, express $x$ in terms of $y$.

**35**  Solve the equation $x^{\frac{2}{3}}=9$.

**36**  Solve the equation $\sqrt{x}=2\sqrt[3]{x}$.

# 12B Logarithms

In Section 1C, you worked with logarithms to base 10 and logarithms to base e. Now you need to be able to work with any (positive) base.

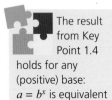 The result from Key Point 1.4 holds for any (positive) base: $a=b^x$ is equivalent to $\log_b a=x$.

---

**WORKED EXAMPLE 12.4**

Calculate the value of $\log_3\dfrac{1}{9}$.

Let the value of $\log_3\dfrac{1}{9}$ be $x$ ............................................ $x=\log_3\dfrac{1}{9}$

Use $a=b^x$ is equivalent ............................... Therefore, $3^x=\dfrac{1}{9}$
to $\log_b a=x$

You can see (or experiment ........................... $\log_3\dfrac{1}{9}=-2$
to find) that $x=-2$

**WORKED EXAMPLE 12.5**

Find the exact value of $y$ if $\log_5(y-3) = -1$.

Use $a = b^x$ is equivalent to $\log_b a = x$

$$\log_5(y-3) = -1$$

$$y - 3 = 5^{-1} = \frac{1}{5}$$

$$y = \frac{16}{5}$$

## ■ Laws of logarithms

The laws of exponents lead to a set of laws of logarithms.

**KEY POINT 12.2**

- $\log_b xy = \log_b x + \log_b y$
- $\log_b \dfrac{x}{y} = \log_b x - \log_b y$
- $\log_b x^m = m\log_b x$

where $b, x, y > 0$

**Tip**

Be careful not to apply invented, similar-looking rules of logarithms. For example, many students claim that $\log(x+y) = \log x + \log y$ or that $\log(x+y) = \log x \times \log y$. Neither of these are true, in general.

**CONCEPTS – PATTERNS**

Many rules like the ones in Key Point 12.2 are explored and discovered by systematically looking at **patterns** in numbers.

For example:

$\log_{10} 2 = 0.30103 \quad \log_{10} 20 = 1.30103 \quad \log_{10} 200 = 2.30103 \quad \log_{10} 2000 = 3.30103$

The first law of logarithms is proved below in Proof 12.2. The others can be proved similarly from the corresponding law of exponents.

**Proof 12.2**

Prove that $\log_b xy = \log_b x + \log_b y$.

Start by using the related law of exponents $b^m \times b^n = b^{m+n}$ with $m = \log_b x$ and $n = \log_b y$

$$b^{\log_b x} \times b^{\log_b y} = b^{\log_b x + \log_b y}$$

Use $b^{\log_b x} = x$ on both terms of the product on the LHS

$$xy = b^{\log_b x + \log_b y}$$

Take $\log_b$ of both sides.

$$\log_b xy = \log_b b^{\log_b x + \log_b y}$$

Use $\log_b b^x = x$ on the RHS

$$\log_b xy = \log_b x + \log_b y$$

**CONCEPTS – EQUIVALENCE**

The ability to go easily between representing equations using logs and using exponents allows us to turn our old rules into **equivalent** new rules. This is a very common and powerful technique in many areas of mathematics, for example, looking at how rules in differentiation apply to integration.

---

**WORKED EXAMPLE 12.6**

If $p = \log a$ and $q = \log b$, express $\log\left(\dfrac{a^3}{b}\right)$ in terms of $p$ and $q$.

Use $\log_b \dfrac{x}{y} = \log_b x - \log_b y$ .......... $\log\left(\dfrac{a^3}{b}\right) = \log a^3 - \log b$

Then use $\log_b x^m = m\log_b x$ .......... $= 3\log a - \log b$
on the first term

Now replace $\log a$ with .......... $= 3p - q$
$p$ and $\log a$ with $q$

---

One common application of the laws of logarithms is in solving log equations. The usual method is to combine all log terms into one.

---

**WORKED EXAMPLE 12.7**

Solve $\log_2(3x - 2) - \log_2(x - 4) = 3$.

Combine the log terms using .......... $\log_2(3x - 2) - \log_2(x - 4) = 3$
$\log_b x - \log_b y = \log_b \dfrac{x}{y}$

$\log_2\left(\dfrac{3x-2}{x-4}\right) = 3$

Remove the log .......... $\dfrac{3x-2}{x-4} = 2^3$
using $\log_b a = x$ is
equivalent to $a = b^x$

Solve for $x$ .......... $3x - 2 = 8(x - 4)$

$3x - 2 = 8x - 32$

$5x = 30$

$x = 6$

---

# Be the Examiner 12.1

Solve $\log_{10}(x + 10) + \log_{10} 2 = 2$.

Which is the correct solution? Identify the errors made in the incorrect solutions.

| Solution 1 | Solution 2 | Solution 3 |
|---|---|---|
| $\log_{10}(x + 12) = 2$ | $\log_{10}(2x + 10) = 2$ | $\log_{10}(2x + 20) = 2$ |
| $x + 12 = 100$ | $2x + 10 = 100$ | $2x + 20 = 100$ |
| $x = 88$ | $x = 45$ | $x = 40$ |

## Change of base of a logarithm

The laws of logarithms only apply when the bases of the logarithms involved are the same. So, it is useful to have a way of changing the base of a logarithm.

---

**KEY POINT 12.3**

- $\log_a x = \dfrac{\log_b x}{\log_b a}$

where $a, b, x > 0$

---

**Proof 12.3**

Prove that $\log_a x = \dfrac{\log_b x}{\log_b a}$.

| | |
|---|---|
| | Let $\log_a x = y$ |
| Write the first statement in the equivalent form without logs | Then $x = a^y$ |
| Then take $\log_b$ of both sides to introduce the new base | So, $\log_b x = \log_b a^y$ |
| Use $\log_b x^m = m\log_b x$ | $\log_b x = y\log_b a$ |
| Make $y$ the subject | $y = \dfrac{\log_b x}{\log_b a}$ |
| $y = \log_a x$ | That is, $\log_a x = \dfrac{\log_b x}{\log_b a}$ |

---

**WORKED EXAMPLE 12.8**

Write $\log_3 5$ in terms of natural logarithms.

| | |
|---|---|
| Use $\log_a x = \dfrac{\log_b x}{\log_b a}$ with $\ln$ as the new logarithm | $\log_3 5 = \dfrac{\ln 5}{\ln 3}$ |

**WORKED EXAMPLE 12.9**

Solve $\log_3 x = \log_x 81$.

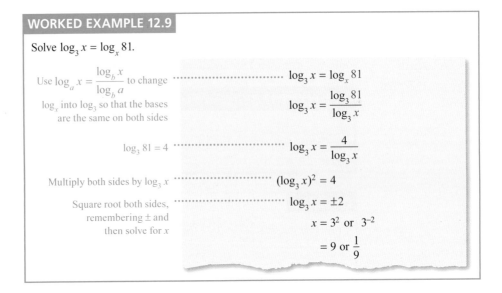

Use $\log_a x = \dfrac{\log_b x}{\log_b a}$ to change
$\log_x$ into $\log_3$ so that the bases
are the same on both sides

$$\log_3 x = \log_x 81$$

$$\log_3 x = \frac{\log_3 81}{\log_3 x}$$

$\log_3 81 = 4$

$$\log_3 x = \frac{4}{\log_3 x}$$

Multiply both sides by $\log_3 x$

$$(\log_3 x)^2 = 4$$

Square root both sides,
remembering $\pm$ and
then solve for $x$

$$\log_3 x = \pm 2$$

$$x = 3^2 \text{ or } 3^{-2}$$

$$= 9 \text{ or } \frac{1}{9}$$

## ■ Solving exponential equations

You have already met the technique of solving certain exponential equations by making the bases the same on both sides of the equation. The only difference now is that you might have to work with the new law of exponents introduced in Key Point 12.1.

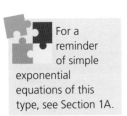

For a
reminder
of simple
exponential
equations of this
type, see Section 1A.

**WORKED EXAMPLE 12.10**

Solve the equation $\left(\sqrt[3]{2}\right)^x = 4^{x-1}$.

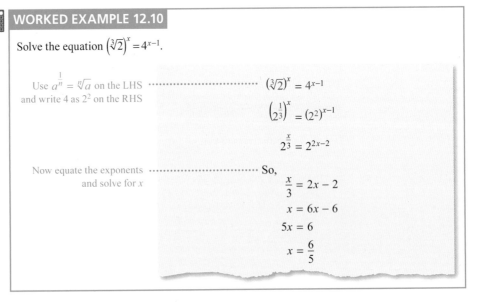

Use $a^{\frac{1}{n}} = \sqrt[n]{a}$ on the LHS
and write 4 as $2^2$ on the RHS

$$\left(\sqrt[3]{2}\right)^x = 4^{x-1}$$

$$\left(2^{\frac{1}{3}}\right)^x = (2^2)^{x-1}$$

$$2^{\frac{x}{3}} = 2^{2x-2}$$

Now equate the exponents
and solve for $x$

So,

$$\frac{x}{3} = 2x - 2$$

$$x = 6x - 6$$

$$5x = 6$$

$$x = \frac{6}{5}$$

Similarly, you have also met exponential equations that needed logs to solve them. Rather than being restricted to $\log_{10}$ or ln, you could now be required to use $\log_b$ for any $b > 0$.

For a reminder of exponential equations of this type to base 10 or base e, see Section 1C.

**WORKED EXAMPLE 12.11**

Solve the equation $5^{x+3} = 7$.

| | |
|---|---|
| Use $a = b^x$ is equivalent to $\log_b a = x$ | $5^{x+3} = 7$ |
| | $x + 3 = \log_5 7$ |
| Rearrange | $x = \log_5 7 - 3$ |
| Evaluate using your GDC | $= -1.79$ |

When $x$ appears in the exponent on both sides of the equation, but with different bases, you may need to start by taking a logarithm of both sides. You can choose any base; 10 and e are the two most common choices.

**WORKED EXAMPLE 12.12**

Solve the equation $7^{x-2} = 5^{x+3}$, giving your answer in terms of natural logarithms.

| | |
|---|---|
| Take natural logs of both sides | $\ln(7^{x-2}) = \ln(5^{x+3})$ |
| Use $\ln a^m = m \ln a$ | $(x-2)\ln 7 = (x+3)\ln 5$ |
| Expand the brackets | $x \ln 7 - 2 \ln 7 = x \ln 5 + 3 \ln 5$ |
| Group the $x$ terms and the number terms (remember that $\ln 5$ and $\ln 7$ are just numbers) | $x \ln 7 - x \ln 5 = 3 \ln 5 + 2 \ln 7$ |
| Factorize the LHS and divide | $x(\ln 7 - \ln 5) = 3 \ln 5 + 2 \ln 7$ |
| | $x = \dfrac{3 \ln 5 + 2 \ln 7}{\ln 7 - \ln 5}$ |

**Tip**

The final answer to Worked Example 12.12 can be written in a different form, by using laws of logarithms on the top and the bottom:

$$x = \frac{\ln(5^3 \times 7^2)}{\ln(7 \div 5)}$$

$$= \frac{\ln(6125)}{\ln(1.4)}.$$

## Exercise 12B

For questions 1 to 5, use the method demonstrated in Worked Example 12.4 to find the exact value of the logarithm.

1 a $\log_3 81$

  b $\log_4 16$

2 a $\log_2\left(\dfrac{1}{8}\right)$

  b $\log_5\left(\dfrac{1}{25}\right)$

3 a $\log_{64} 8$

  b $\log_{16} 2$

4 a $\log_{0.5} 2$

  b $\log_{0.2} 5$

5 a $\log_2 1$

  b $\log_7 1$

For questions 6 to 10, use the method demonstrated in Worked Example 12.5 to solve the equations.

6   a   $\log_3 x = -2$

     b   $\log_4 x = -2$

7   a   $\log_2 (y - 1) = -3$

     b   $\log_3 (y - 2) = -2$

8   a   $\log_5 (x + 2) = 2$

     b   $\log_4 (x + 3) = 3$

9   a   $\log_9 z = \dfrac{1}{2}$

     b   $\log_8 z = \dfrac{1}{3}$

10   a   $\log_4 (x - 1) = 1$

      b   $\log_5 (x + 3) = 0$

For questions 11 to 14, use the method demonstrated in Worked Example 12.6. Write each given expression in terms of $p$ and $q$, where $p = \log_5 a$ and $q = \log_5 b$.

11   a   $\log_5 \left( \dfrac{a^2}{b} \right)$

      b   $\log_5 \left( \dfrac{b^3}{a} \right)$

12   a   $\log_5 \left( \dfrac{a^2}{b^3} \right)$

      b   $\log_5 \left( \dfrac{b^4}{a^2} \right)$

13   a   $\log_5 \sqrt{a^3 b}$

      b   $\log_5 \sqrt{a^4 b^3}$

14   a   $\log_5 \left( \dfrac{25a}{b^2} \right)$

      b   $\log_5 \left( \dfrac{5a^2}{b^5} \right)$

For questions 15 to 18, use the method demonstrated in Worked Example 12.7 to solve the equations, giving your answers in an exact form.

15   a   $\log_{10} x + \log_{10} 2 = 3$

      b   $\log_6 x + \log_6 3 = 1$

16   a   $\log_5 (x + 3) + \log_5 2 = 2$

      b   $\log_{12} (x - 1) + \log_{12} 4 = 1$

17   a   $\log_2 (x + 1) - \log_2 (x - 2) = 3$

      b   $\log_3 (x + 1) - \log_3 (x - 1) = 2$

18   a   $\ln (x - 3) - \ln (x + 5) = 4$

      b   $\ln (x + 2) - \ln (x - 1) = 3$

For questions 19 to 23, use the change of base formula from Key Point 12.3, with the method demonstrated in Worked Example 12.8, to write each logarithm in the given base. Simplify your answer where possible.

19   a   $\log_2 7$ in terms of natural logarithms

      b   $\log_5 8$ in terms of natural logarithms

20   a   $\log_5 3$ in base 2

      b   $\log_7 5$ in base 3

21   a   $\ln 4$ in base 5

      b   $\ln 7$ in base 4

22   a   $\log_{16} 5$ in base 2

      b   $\log_{27} 2$ in base 3

23   a   $\log_2 25$ in base 5

      b   $\log_3 8$ in base 2

For questions 24 to 26, use the method demonstrated in Worked Example 12.9 to solve the equations.

24   a   $\log_2 x = \log_x 16$

      b   $\log_5 x = \log_x 625$

25   a   $2\log_5 x = \log_x 25$

      b   $5\log_2 x = \log_x 32$

26   a   $\log_2 x = 9\log_x 2$

      b   $\log_4 x = 16\log_x 4$

For questions 27 and 28, use the method demonstrated in Worked Example 12.10 to solve the equations.

27   a   $(\sqrt{3})^x = 9^{x-1}$

      b   $(\sqrt{2})^x = 4^{x+2}$

28   a   $(\sqrt[3]{2})^{2x} = 8^{x+1}$

      b   $(\sqrt[3]{3})^{4x} = 9^{x-3}$

For questions 29 to 31, use the method demonstrated in Worked Example 12.11 to solve the equations. Use your calculator and give your answers to three significant figures.

29   a   $3^x = 17$

      b   $8^x = 3$

30   a   $5^{x+1} = 11$

      b   $7^{x-1} = 12$

31   a   $2^{3x-1} = 11$

      b   $3^{4x+3} = 8$

For questions 32 and 33, use the method demonstrated in Worked Example 12.12 to solve the equations, giving your answer in terms of natural logarithms.

32   a   $3^{x-2} = 2^{x+1}$

      b   $3^{x-1} = 2^{x+2}$

33   a   $7^{2x-5} = 2^{x+3}$

      b   $7^{3x+1} = 2^{x+8}$

**34** Given that $x = \log_3 a$, $y = \log_3 b$ and $z = \log_3 c$, write the following in terms of $x$, $y$ and $z$:

   a   $\log_3\left(ab^4\right)$        b   $\log_3\left(\dfrac{a^2b}{c^5}\right)$        c   $\log_3\left(27a^2b^3\right)$

**35** Given that $x = \log_5 a$, $y = \log_5 b$ and $z = \log_5 c$, write the following in terms of $x$, $y$ and $z$:

   a   $\log_5\left(25\sqrt{a}\right)$        b   $\log_5\left(\dfrac{b}{5c^5}\right)$

**36** Write $2\ln a + 6\ln b$ as a single logarithm.

**37** Write $\dfrac{1}{3}\ln x - \dfrac{1}{2}\ln y$ as a single logarithm.

**38** a   Find the exact value of $\log_2\left(\dfrac{1}{\sqrt{2}}\right)$.

   b   Solve the equation $\log_x 27 = -3$.

**39** Solve $\log_2(x + 3) = 3$.

**40** Solve the equation $\log_3(2x - 3) = 4$.

**41** Use logarithms to solve these equations:

   a   $5^x = 10$        b   $2 \times 3^x + 6 = 20$

**42** Solve the equation $3 \times 1.1^x = 20$.

**43** Solve the equation $\log_x 32 = 5$.

**44** Solve the equation $\log_x 64 = 3$.

**45** a   Given that $5 \times 6^x = 12 \times 3^x$, write down the exact value of $2^x$.

   b   Hence find the exact value of $x$.

**46** a   Write $\log_x e$ in terms of natural logarithms.

   b   Hence solve the equation $16\ln x = \log_x e$.

**47** Solve the equation $8^{3x+1} = 4^{x-3}$.

**48** Solve the equation $5^{2x+3} = 9^{x-5}$, giving your answer in terms of natural logarithms.

**49** The radioactivity $(R)$ of a substance after a time $t$ days is modelled by $R = 10 \times 0.9^t$.

   a   Find the initial (i.e. $t = 0$) radioactivity.

   b   Find the time taken for the radioactivity to fall to half of its original value.

**50** The population of bacteria $(B)$ at time $t$ hours after being added to an agar dish is modelled by

$B = 1000 \times 1.1^t$.

   a   Find the number of bacteria

     i   initially

     ii   after 2 hours.

   b   Find an expression for the time it takes to reach 2000. Use technology to evaluate this expression.

**51** The population of penguins $(P)$ after $t$ years on two islands is modelled by:

first island: $P = 200 \times 1.1^t$

second island: $P = 100 \times 1.2^t$.

How many years are required before the population of penguins on both islands is equal?

**52** Find the exact solution of the equation $\log_4 x = 6\log_x 8$.

**53** Solve the equation $2^{5-3x} = 3^{2x-1}$, giving your answer in the form $\dfrac{\ln p}{\ln q}$, where $p$ and $q$ are integers.

**54** Moore's law states that the density of transistors on an integrated circuit doubles every 2 years. Write an expression for the time taken for the density to multiply by 10 and use technology to evaluate this expression.

**55** a   If $\log_a\left(x^2\right) = b$ find the product of all possible values of $x$.

   b   If $\left(\log_a x\right)^2 = b$ find the product of all possible values of $x$.

## Checklist

■ You should be able to extend the laws of exponents to general rational exponents:
$$a^{\frac{1}{n}} = \sqrt[n]{a}$$

■ You should be able to work with logarithms to any (positive) base.

■ You should know the laws of logarithms:
□ $\log_b xy = \log_b x + \log_b y$
□ $\log_b \dfrac{x}{y} = \log_b x - \log_b y$
□ $\log_b x^m = m\log_b x$
where $a$, $x$, $y > 0$

■ You should be able to change the base of a logarithm:
$$\log_a x = \frac{\log_b x}{\log_b a}$$
where $a$, $b$, $x > 0$

■ You should be able to solve exponential equations.

## ■ Mixed Practice

 **1 a** Find the exact value of $\left(\dfrac{4}{9}\right)^{-\frac{3}{2}}$.

  **b** Find $\log_2\left(\dfrac{1}{8}\right)$.

**2** Write $\ln 4 + 2\ln 3$ in the form $\ln k$.

 **3** Solve the equation $2 \times 3^{x-2} = 54$.

**4** Use technology to solve $1.05^x = 2$.

 **5** Solve the equation $100^{x+1} = 10^{3x}$.

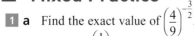 **6** Find the value of $x$ such that $\log_3(5x + 1) = 2$.

**7** Find the exact solution of the equation $3\ln x + 2 = 2(\ln x - 1)$.

**8** Given that $a = \log_3 x$, $b = \log_3 y$ and $c = \log_3 z$, write the following in terms of $a$, $b$ and $c$:

  **a** $\log_3\left(x^2 y\right)$       **b** $\log_3\left(\dfrac{x}{yz^3}\right)$       **c** $\log_3\left(\sqrt{zx^3}\right)$

**9** Find the value of each of the following, giving your answer as an integer.
  **a** $\log_6 36$       **b** $\log_6 4 + \log_6 9$       **c** $\log_6 2 - \log_6 12$

Mathematics SL May 2015 Paper 1 TZ1 Q2

**10** Given that $x = \log_2 a$ and $y = \log_2 b$, write in terms of $x$ and $y$:

  **a** $\log_2\left(\dfrac{a}{\sqrt{b}}\right)$       **b** $\log_2\left(\dfrac{a^2}{8b^3}\right)$

**11** Given that $x = \ln 2$ and $y = \ln 5$, write the following in terms of $x$ and $y$:
  **a** $\ln 10$       **b** $\ln 50$       **c** $\ln 0.08$

**12** Write $3 + 2\log 5 - 2\log 2$ as a single logarithm.

**13** Find the exact solution of the equation $3\ln x + \ln 8 = 5$, giving your answer in the form $Ae^k$ where $A$ and $k$ are fractions.

**14** Solve the equation $\log_3 x = 4\log_x 3$.

**15** Solve the equation $4^{3x+5} = 8^{x-1}$.

**16** Solve the simultaneous equations:

$\log_2 x + \log_3 y = 5$

$\log_2 x - 2\log_3 y = -1$

**17** Find the exact value of $x$ such that $\log_x 8 = 6$.

**18** Solve the equation $3^{2x} = 2e^x$, giving your answer in terms of natural logarithms.

**19** Solve the equation $5^{2x+1} = 7^{x-3}$.

**20** Find the solution of the equation $12^{2x} = 4 \times 3^{x+1}$ in the form $\dfrac{\log p}{\log q}$, where $p$ and $q$ are positive integers.

**21** The number of cells in a laboratory experiment satisfies the equation $N = 150e^{1.04t}$, where $t$ hours is the time since the start of the experiment.
   **a** What was the initial number of cells?
   **b** How many cells will there be after 3 hours?
   **c** How long will it take for the number of cells to reach 1000?

**22** A geometric sequence has first term 15 and common ratio 1.2. The $n$th term of the sequence equals 231 to the nearest integer. Find the value of $n$.

**23** Find the value of
   **a** $\log_2 40 - \log_2 5$;
   **b** $8^{\log_2 5}$

   *Mathematics SL May 2013 Paper 1 TZ1 Q7*

**24** **a** Find $\log_2 32$.
   **b** Given that $\log_2\left(\dfrac{32^x}{8^y}\right)$ can be written as $px + qy$, find the value of $p$ and of $q$.

   *Mathematics SL May 2009 Paper 1 TZ2 Q4*

**25** Let $\log_3 p = 6$ and $\log_3 q = 7$.
   **a** Find $\log_3 p^2$.
   **b** Find $\log_3\left(\dfrac{p}{q}\right)$.
   **c** Find $\log_3(9p)$.

   *Mathematics SL May 2013 Paper 1 TZ2 Q3*

**26** Solve the equation $8^{x-1} = 6^{3x}$. Express your answer in terms of $\ln 2$ and $\ln 3$.

   *Mathematics HL May 2014 Paper 1 TZ2 Q2*

**27** Solve the equation $3^{x+1} = 3^x + 18$.

**28** Given that $x = \ln a$ and $y = \ln b$, express $\log_b \sqrt{a}$ in terms of $x$ and $y$.

**29** Given that $x = \log_b a$, express the following in terms of $x$:
   **a** $\log_b\left(\dfrac{a^2}{b^3}\right)$
   **b** $\log_a b^2$.

**30** Write in the form $k \ln x$, where $k$ is an integer:

$\ln x + \ln x^2 + \ln x^3 + \cdots + \ln x^{20}$

**31** Find the exact value of

$\log_3\left(\dfrac{1}{3}\right) + \log_3\left(\dfrac{3}{5}\right) + \log_3\left(\dfrac{5}{7}\right) + \cdots + \log_3\left(\dfrac{79}{81}\right)$

# 13 Analysis and approaches: Sequences and series

## ESSENTIAL UNDERSTANDINGS

- Algebra is an abstraction of numerical concepts and employs variables to solve mathematical problems.

### In this chapter you will learn...

- how to find the sum of infinite convergent geometric sequences
- how to use the binomial theorem to expand expressions of the form $(a + b)^n$, where $n \in \mathbb{Z}^+$
- how to use Pascal's triangle to find the coefficients in a binomial expansion
- the formula for ${}^nC_r$, which can be used to find the coefficients in a binomial expansion.

### CONCEPTS

The following concept will be addressed in this chapter:
- The binomial theorem is a **generalization** which provides an efficient method for expanding binomial expressions.

### PRIOR KNOWLEDGE

Before starting this chapter, you should already be able to complete the following:

1 Find the sum of the first 20 terms of the geometric sequence 80, 20, 5, 1.25, …

2 Use technology to find the limit of the function $f(x) = 0.5^x$ as $x$ gets larger.

3 Find the range of values of $x$ for which $|x| < 1$.

4 Expand and simplify $(3 - 4x)^2$.

■ **Figure 13.1** How do mathematicians reconcile the fact that some conclusions conflict with intuition?

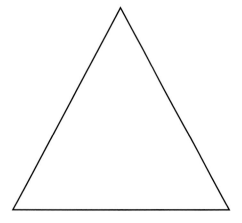

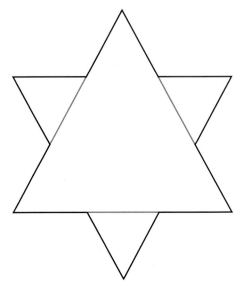

It might seem natural that if you continue adding terms that are all positive in a sequence the sum will just get larger and larger, and that if you continue adding terms that are all negative the sum will just get more and more negative. This is certainly the case with many sequences. It is also possible in some sequences, however, that no matter how many positive (or negative) terms you add, the sum can never exceed a certain value. In particular, this will happen with certain geometric sequences.

## Starter Activity

Figure 13.1 shows the first four stages of building a Koch snowflake. At each stage the external sides are split into three and an equilateral triangle is added to the middle. What happens to the perimeter of this shape as the stages progress? What happens to the area?

**Now try this problem:**

Using technology, investigate the value, $S$, of the infinite geometric series

$$S = 1 + r + r^2 + r^3 + \dots$$

when

a    $r = 2$

b    $r = 1$

c    $r = 0.25$

d    $r = -0.25$

e    $r = -1$

f    $r = -2$.

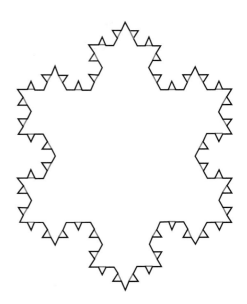

# 13A The sum of infinite convergent geometric sequences

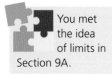

You met the sum of a geometric sequence in Section 2B.

You met the idea of limits in Section 9A.

You know that you can find the sum of the first $n$ terms of a geometric sequence with common ratio $r$ and first term $u_1$ using the formula $S_n = \dfrac{u_1(1 - r^n)}{1 - r}$. Sometimes this sum will just increase (or decrease, if negative) the more terms you add.

However, if $r$ is between 1 and $-1$ then as $n$ gets very large, $r^n$ tends towards 0.

As a result, $\dfrac{u_1(1 - r^n)}{1 - r}$ tends towards $\dfrac{u_1(1 - 0)}{1 - r} = \dfrac{u_1}{1 - r}$.

So, in this situation the sum of infinitely many terms converges to a finite limit – this is called the **sum to infinity**, $S_\infty$, of the geometric sequence.

> **KEY POINT 13.1**
>
> For a geometric sequence with common ratio $r$,
>
> $S_\infty = \dfrac{u_1}{1 - r}$  if $|r| < 1$

## Tip

Remember that $|r| < 1$ is just another way of writing $r$ is between $-1$ and 1.

---

**WORKED EXAMPLE 13.1**

Find the value of the infinite geometric series $2 + \dfrac{4}{3} + \dfrac{8}{9} + \dfrac{16}{27} + \ldots$

This is a geometric series with first term 2 and common ratio $\dfrac{2}{3}$ ............................ $u_1 = 2, \; r = \dfrac{2}{3}$

Use $S_\infty = \dfrac{u_1}{1 - r}$ ............................ $S_\infty = \dfrac{2}{1 - \dfrac{2}{3}}$

$= 6$

---

**TOOLKIT:** Proof

Does the following diagram prove that

$\dfrac{1}{4} + \dfrac{1}{16} + \dfrac{1}{64} \ldots = \dfrac{1}{3}$?

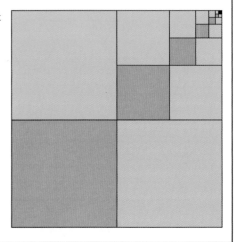

**You are the Researcher**

The ancient Greeks had some real difficulties working with limits of series. Some of them thought, incorrectly, that if each term in the sum got smaller and smaller, then the series would reach a limit. However, this is not always true. You might like to explore the harmonic series, $\frac{1}{2} + \frac{1}{3} + \frac{1}{4}...$ and see why that does not converge.

There are many tests that mathematicians use to decide if a series will converge, for example, the ratio test. You might like to learn more about these and how they work.

**WORKED EXAMPLE 13.2**

The geometric series $(x+4) + (x+4)^2 + (x+4)^3 + ...$ converges.

Find the range of possible values of $x$.

State the common ratio ............................ $r = x + 4$

You know that $|r| < 1$ ............................ Since the series converges,

$$|x+4| < 1$$
$$-1 < x + 4 < 1$$
$$-5 < x < -3$$

## Be the Examiner 13.1

Find the sum to infinity of the geometric series,

$$\frac{1}{2} - \frac{3}{4} + \frac{9}{8} - \frac{27}{16} + ...$$

Which is the correct solution? Identify the errors made in the incorrect solutions.

| Solution 1 | Solution 2 | Solution 3 |
|---|---|---|
| $S_\infty = \dfrac{1}{1 - \dfrac{1}{2}}$ $= 2$ | $S_\infty = \dfrac{\dfrac{1}{2}}{1 - \left(-\dfrac{3}{2}\right)}$ $= \dfrac{1}{5}$ | $r = -\dfrac{3}{4} \div \dfrac{1}{2} = -\dfrac{3}{2}$ so it doesn't exist. |

**TOOLKIT: Problem Solving**

You may not have realized it, but you have already met infinite geometric series in your previous work. Explain how you can find the following sum using methods from your prior learning. Confirm the answer by using the formula for the geometric series.

$$\sum_{r=1}^{\infty} \frac{3}{10^r}$$

What happens to your argument when applied to the sum below?

$$\sum_{r=1}^{\infty} \frac{9}{10^r}$$

**TOOLKIT:** Proof

Intuition for sums tends to break down when dealing with infinite sums. For example, consider

$$S = 1 - 1 + 1 - 1 + 1 - 1 \ldots$$

Consider the following attempts to evaluate $S$:

## Attempt A:

$$S = (1 - 1) + (1 - 1) + (1 - 1) \ldots$$

$$= 0 + 0 + 0 \ldots$$

$$= 0$$

## Attempt B:

$$S = 1 + (-1 + 1) + (-1 + 1) + (-1 + 1) \ldots$$

$$= 1 + 0 + 0 + 0 \ldots$$

$$= 1$$

## Attempt C:

$$S = 1 - 1 + 1 - 1 + 1 - 1 \ldots$$

$$S = \quad +1 - 1 + 1 - 1 + 1 - 1 \ldots$$

Adding these together gives:

$$2S = 1 + 0 + 0 + 0 \ldots$$

$$S = \frac{1}{2}$$

All of these look plausible, but none are considered correct because you cannot generally group or rearrange terms in an infinite sum.

# Exercise 13A

For questions 1 to 5, use the method demonstrated in Worked Example 13.1 to find the sum of the infinite geometric series.

1   a   $\dfrac{1}{4} + \dfrac{1}{16} + \dfrac{1}{64} + \ldots$

    b   $\dfrac{1}{2} + \dfrac{1}{4} + \dfrac{1}{8} + \ldots$

2   a   $\dfrac{1}{3} + \dfrac{2}{9} + \dfrac{4}{27} + \ldots$

    b   $\dfrac{1}{3} + \dfrac{2}{15} + \dfrac{4}{75} + \ldots$

3   a   $4 + 1 + \dfrac{1}{4} + \ldots$

    b   $6 + 2 + \dfrac{2}{3} + \ldots$

4   a   $\dfrac{1}{2} - \dfrac{1}{4} + \dfrac{1}{8} - \dfrac{1}{16} + \ldots$

    b   $\dfrac{1}{3} - \dfrac{1}{9} + \dfrac{1}{27} - \dfrac{1}{81} + \ldots$

5   a   $15 - 9 + \dfrac{27}{5} - \dfrac{81}{25} + \ldots$

    b   $16 - 12 + 9 - \dfrac{27}{4} + \ldots$

For questions 6 to 10, use the method demonstrated in Worked Example 13.2 to find the range of values of $x$ for which the infinite geometric series converges.

6   a   $(x - 2) + (x - 2)^2 + (x - 2)^3 + \ldots$

    b   $(x + 3) + (x + 3)^2 + (x + 3)^3 + \ldots$

7   a   $1 + 2x + 4x^2 + \ldots$

    b   $1 + 3x + 9x^2 + \ldots$

8   a   $1 + \dfrac{x}{2} + \dfrac{x^2}{4} + \ldots$

    b   $1 + \dfrac{x}{5} + \dfrac{x^2}{25} + \ldots$

9   a   $(x + 4) - (x + 4)^2 + (x + 4)^3 - \ldots$

    b   $(x - 1) - (x - 1)^2 + (x - 1)^3 - \ldots$

10   a   $1 - \dfrac{3x}{2} + \dfrac{9x^2}{4} - \ldots$

     b   $1 - \dfrac{4x}{3} + \dfrac{16x^2}{9} - \ldots$

**11** A geometric series has first term 3 and common ratio $\frac{1}{4}$. Find the sum to infinity of the series.

**12** Find the sum to infinity of the geometric series with first term 5 and common ratio $-\frac{1}{4}$.

**13** Find the sum to infinity of the series $2 - \frac{2}{3} + \frac{2}{9} - \frac{2}{27} + \ldots$

**14** An infinite geometric series has the first term 8 and its sum to infinity is 6. Find the common ratio of the series.

**15** The first term of a geometric series is 3. Given that the sum to infinity of the series is 4, find the value of the common ratio.

**16** The sum to infinity of a geometric series is 3 and the common ratio is $\frac{1}{3}$. Find the first three terms of the series.

**17** The second term of a geometric series is 2 and the sum to infinity is 9. Find two possible values of the common ratio.

**18** A geometric series is given by $\sum_{r=0}^{\infty} \left(\frac{2}{5}\right)^r$.

   a  Write down the first three terms of the series.

   b  Find the sum of the series.

**19** Evaluate $\sum_{r=0}^{\infty} \frac{2}{3^r}$.

**20**  a  Find the range of values of $x$ for which the geometric series $3 - \frac{x}{3} + \frac{x^2}{27} - \frac{x^3}{243} + \ldots$ converges.

   b  Find the sum of the series when $x = -2$.

**21** A geometric series is given by $5 + 5(x-3) + 5(x-3)^2 + \ldots$

   a  Find the range of values of $x$ for which the series converges.

   b  Find the expression, in terms of $x$, for the sum of the series.

**22** For the geometric series $2 + 4x + 8x^2 + \ldots$

   a  Find the range of values of $x$ for which the series converges.

   b  Find the expression, in terms of $x$, for the sum to infinity of the series.

**23** The second term of a geometric series is $-\frac{6}{5}$ and the sum to infinity is 5. Find the first term of the series.

**24** Given that $x$ is a positive number,

   a  Find the range of values of $x$ for which the geometric series $x + 4x^3 + 16x^5 + \ldots$ converges.

   b  Find an expression, in terms of $x$, for the sum to infinity of the series.

**25**  a  Find the value of $x$ such that $\sum_{r=0}^{\infty} \frac{x^{r+1}}{2^r} = 3$.

   b  Find the range of values of $x$ for which the series converges.

**26** An infinite geometric series has sum to infinity of 27 and sum of the first three terms equal to 19. Find the first term.

# 13B The binomial expansion

## ■ The binomial theorem

If you expand $(a + b)^2$, $(a + b)^3$, $(a + b)^4$ and so on, by multiplying out brackets repeatedly, you will get:

$$(a + b)^2 = a^2 + 2ab + b^2$$
$$(a + b)^3 = a^3 + 3a^2b + 3ab^2 + b^3$$
$$(a + b)^4 = a^4 + 4a^3b + 6a^2b^2 + 4ab^3 + b^4$$

You can see that there are patterns in the powers of $a$ and $b$ in each expansion – the power of $a$ decreases by one as the power of $b$ increases by one between terms.

There are also symmetrical coefficients (in red) – these are often referred to as the **binomial coefficients.** The coefficient in the term containing $b^r$ of the expansion of $(a + b)^n$ is called $^nC_r$. They can be found using the $^nC_r$ button on your calculator. For example, the coefficients in the expansion of $(a + b)^4$ are found using:

$$^4C_0 = 1 \quad ^4C_1 = 4 \quad ^4C_2 = 6 \quad ^4C_3 = 4 \quad ^4C_4 = 1$$

### Tip

Note that since $^4C_0 = 1$ and $^4C_4 = 1$, these coefficients do not need to be written in the expansion (this is the same in the first and last terms of all of these expansions).

### Tip

Remember that $n \in \mathbb{Z}^+$ means that $n$ is a positive integer.

**KEY POINT 13.2**

$(a + b)^n = a^n + {}^nC_1 a^{n-1}b + \ldots + {}^nC_r a^{n-r}b^r + \ldots + b^n$

where $n \in \mathbb{Z}^+$.

**WORKED EXAMPLE 13.3**

Find the first three terms in increasing powers of $x$ in the expansion of $(2 + x)^5$.

Use
$(a + b)^n = a^n + {}^nC_1 a^{n-1}b + {}^nC_2 a^{n-2}b^2 + \ldots$ $\qquad$ $(2 + x)^5 = 2^5 + {}^5C_1 2^4 x^1 + {}^5C_2 2^3 x^2 + \ldots$

Evaluate the coefficients and simplify $\qquad$ $= 32 + 5(16)(x) + 10(8)(x^2) + \ldots$

$= 32 + 80x + 80x^2 + \ldots$

**CONCEPTS – GENERALIZATION**

If you had tried to multiply out $(2 + x)^5$ manually you would have seen that the binomial expansion is a much more efficient way to get to the same result. You might have thought that with technology such as algebraic computer systems we no longer really need to study these algebraic **generalizations**; however, there are many reasons why we still do so. In many technological situations millions of algebraic calculations are required – for example, when modelling the air flowing around an aeroplane or animating droplets of water in a cartoon. In these situations, using the most efficient algebraic generalizations can massively improve performance, as the computer does not need to go back to first principles in every calculation.

**Tip**

Make sure to read carefully whether the question wants the coefficient, $-15\,120$, or the term, $-15\,120y^3$.

**WORKED EXAMPLE 13.4**

Find the coefficient of $y^3$ in the expansion of $(2 - 3y)^7$.

Use the general term, ${}^nC_r a^{n-r}b^r$, $\qquad$ The relevant term is
with $n = 7$, $r = 3$, $a = 2$, $b = -3$ $\qquad$ ${}^7C_3 2^4(-3y)^3$

$= (35)(16)(-27y^3)$

$= -15\,120y^3$

You are asked for the coefficient $\qquad$ So, the coefficient of $y^3$ is $-15\,120$
(the number in front of $y^3$)

# Be the Examiner 13.2

Expand and simplify $(5 - 4x)^3$.

Which is the correct solution? Identify the errors made in the incorrect solutions.

| Solution 1 | Solution 2 | Solution 3 |
|---|---|---|
| $(5 - 4x)^3 = 5^3 + {}^3C_1 5^2(-4x)$ $+ {}^3C_2 5^1(-4x)^2 + (-4x)^3$ $= 125 - 100x + 80x^2 - 64x^3$ | $(5 - 4x)^3 = 5^3 + {}^3C_1 5^2(-4x)$ $+ {}^3C_2 5^1(-4x)^2 + (-4x)^3$ $= 125 - 300x - 60x^2 - 4x^3$ | $(5 - 4x)^3 = 5^3 + {}^3C_1 5^2(-4x)$ $+ {}^3C_2 5^1(-4x)^2 + (-4x)^3$ $= 125 - 300x + 240x^2 - 64x^3$ |

## ■ Use of Pascal's triangle and the factorial formula to find $^nC_r$

As well as getting the binomial coefficients from the $^nC_r$ button on your calculator, you can also get them from Pascal's triangle, which is formed by adding pairs of adjacent numbers to give the number between them in the row below:

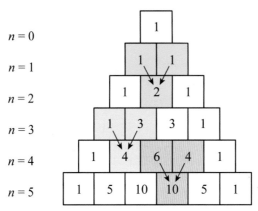

$n = 0$

$n = 1$

$n = 2$

$n = 3$

$n = 4$

$n = 5$

This enables you to do binomial expansions for small values of $n$ without your GDC.

### WORKED EXAMPLE 13.5

Expand and simplify $(1 - 2x)^4$.

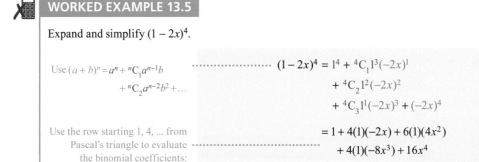

Use $(a + b)^n = a^n + {}^nC_1 a^{n-1}b$
$\qquad + {}^nC_2 a^{n-2}b^2 + \dots$

$$(1 - 2x)^4 = 1^4 + {}^4C_1 1^3(-2x)^1$$
$$+ {}^4C_2 1^2(-2x)^2$$
$$+ {}^4C_3 1^1(-2x)^3 + (-2x)^4$$

Use the row starting 1, 4, ... from
Pascal's triangle to evaluate
the binomial coefficients:

$$= 1 + 4(1)(-2x) + 6(1)(4x^2)$$
$$+ 4(1)(-8x^3) + 16x^4$$
$$= 1 - 8x + 24x^2 - 32x^3 + 16x^4$$

1
1  1
1  2  1
1  3  3  1
1  4  6  4  1

Although in the Western world this famous pattern is attributed to the French mathematician Blaise Pascal (1623–1662), there is evidence that it was known much earlier in Chinese work, such as that of Yang Hui (1238–1298).

There is also a formula for the binomial coefficients. This involves the **factorial** function:

$$n! = n \times (n - 1) \times \dots \times 2 \times 1$$

Note that 0! is defined to be 1.

> **KEY POINT 13.3**
>
> - $^{n}C_{r} = \dfrac{n!}{r!(n-r)!}$

## Tip

In some other resources, and some past paper questions, you might see $^{n}C_{r}$ referred to using the notation $\binom{n}{r}$.

 This formula can be proved in various ways. If you study Mathematics: analysis and approaches HL you will learn about ways to prove it using induction (Chapter 5) and consider it as a way of choosing options (Chapter 1).

 **WORKED EXAMPLE 13.6**

Evaluate $^{12}C_{10}$.

Use $^{n}C_{r} = \dfrac{n!}{r!(n-r)!}$ .................... $^{12}C_{10} = \dfrac{12!}{10!(12-10)!}$

$$= \dfrac{12!}{10!2!}$$

Note that
$12! = 12 \times 11 \times 10 \times 9 \ldots \times 2 \times 1$ .................... $= \dfrac{12 \times 11 \times 10!}{10!2!}$

Cancel $10!$ and evaluate .................... $= \dfrac{12 \times 11}{2}$

$$= 6 \times 11$$

$$= 66$$

 **WORKED EXAMPLE 13.7**

Using technology, find the possible values of $x$ when $^{10}C_{x} = 210$.

Use the table function to
generate values of $^{10}C_{x}$ for .................... From GDC,
different values of $x$ .................... $x = 4$ or $6$

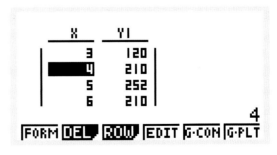

**TOOLKIT:** Problem solving

There are many identities associated with binomial coefficients. Investigate and see if you can justify expressions for:

$$^{n-1}C_r + {}^{n-1}C_{r-1}$$

$$\sum_{r=0}^{n} {}^{n}C_r$$

$$\sum_{r=0}^{n} \left({}^{n}C_r\right)^2$$

**You are the Researcher**

You might want to investigate how the binomial expansion generalizes to more complicated expansions, as described by the multinomial theorem.

## Exercise 13B

For questions 1 to 8, use the method demonstrated in Worked Example 13.3 to find the first four terms of the binomial expansion in increasing powers of $x$ or $b$.

1 a $(2+x)^6$
  b $(2+x)^7$

2 a $(3-x)^5$
  b $(3-x)^4$

3 a $(1+2x)^6$
  b $(1+2x)^7$

4 a $(1-5x)^4$
  b $(1-5x)^5$

5 a $(2+3x)^{10}$
  b $(2+3x)^9$

6 a $(2-3x)^5$
  b $(2-3x)^6$

7 a $(a+2b)^7$
  b $(a+2b)^8$

8 a $(3a-2b)^5$
  b $(3a-2b)^6$

For questions 9 to 14, use the method demonstrated in Worked Example 13.4 to find the required coefficient.

9 a $y^4$ in $(2+y)^8$
  b $y^4$ in $(2+y)^9$

10 a $y^5$ in $(3-y)^9$
   b $y^5$ in $(3-y)^{10}$

11 a $x^6$ in $(1-2x)^{10}$
   b $x^6$ in $(1-2x)^9$

12 a $x^5$ in $(2+3x)^{12}$
   b $x^5$ in $(2+3x)^{11}$

13 a $z^2$ in $(2z-1)^{20}$
   b $z^2$ in $(2z-1)^{19}$

14 a $z^7$ in $(2z-3)^{12}$
   b $z^7$ in $(2z-3)^{11}$

For questions 15 to 17, use Pascal's triangle from Worked Example 13.5 to expand the brackets.

15 a $(1+2x)^4$
   b $(2+x)^4$

16 a $(3+x)^3$
   b $(1+3x)^3$

17 a $(x+2)^5$
   b $(x+1)^5$

For questions 18 to 22, use the method demonstrated in Worked Example 13.6 to evaluate these binomial coefficients:

18 a $^6C_1$
   b $^8C_1$

19 a $^7C_2$
   b $^6C_2$

20 a $^9C_3$
   b $^8C_3$

21 a $^{10}C_9$
   b $^{13}C_{12}$

22 a $^{15}C_{13}$
   b $^{11}C_9$

For questions 23 to 25, use technology, as in Worked Example 13.7, to find the unknown value (or values).

23 a $^{14}C_r = 2002$
   b $^{15}C_r = 6435$

24 a $^{20}C_r = 1140$
   b $^{17}C_r = 6188$

25 a $^nC_5 = 792$
   b $^nC_8 = 43\,758$

26 Expand and simplify $(10-3x)^4$.

27 Expand and simplify $(2x-1)^5$.

28 Find the coefficient of $x^5$ in the expansion of $(5+3x)^7$.

29 Find the coefficient of $x^8$ in the expansion of $(3x-2)^{12}$.

30 Find the value of $n$ such that $^nC_4 = 495$.

**31** a  Find the first three terms in the expansion of $(2 + 3x)^{10}$ in ascending powers of $x$.

  b  Hence find an approximate value of $2.003^{10}$, correct to the nearest integer.

**32** a  Find the first four terms, in ascending powers of $x$, in the expansion of $(3 - 2x)^9$.

  b  Hence find the value of $2.98^9$ correct to one decimal place.

**33** a  Find the first three terms, in ascending power of $x$, in the expansion of $(2 + x)^7$.

  b  Hence expand $(2 + x)^7 (3 - x)$ up to and including the term in $x^2$.

**34** a  Find the first three terms in the expansion of $(2 - 3x)^5$.

  b  Hence find the coefficient of $x^2$ in the expansion of $(2 - 3x)^5(1 + 2x)$.

**35**  Find the coefficient of $x^2$ in the expansion of $(2 - x)^5(1 + 2x)^6$.

**36** a  Expand and simplify $(2 + x)^4 - (2 - x)^4$.

  b  Hence find the exact value of $2.01^4 - 1.99^4$.

**37**  The coefficient of $x^{n-3}$ in the expansion of $(x + 2)^n$ is 1760. Find the value of $n$.

**38**  The coefficient of $x^4$ in the expansion of $(x + 3)^n$ is 153 090. Find the value of $n$.

**39**  Expand and simplify $\left(x + \dfrac{2}{x}\right)^4$.

**40**  Expand and simplify $(x^2 + 3x)^5$.

**41**  Find the coefficient of $x^{27}$ in the expansion of $(x^2 + 3x)^{15}$.

**42**  Find the coefficient of $x^5$ in the expansion of $\left(2x^2 - \dfrac{3}{x}\right)^{10}$.

**43**  Find the coefficient of $x^9$ in $(x - 1)^7(x + 1)^7$.

## Checklist

- You should be able to find the sum of infinite geometric sequences:
  - For a geometric sequence with common ratio $r$,

$$S_\infty = \frac{u_1}{1-r} \quad \text{if } |r| < 1$$

- You should be able to use the binomial theorem to expand expressions of the form $(a + b)^n$, where $n \in \mathbb{Z}^+$:

$$(a + b)^n = a^n + {}^nC_1 a^{n-1}b + \ldots + {}^nC_r a^{n-r}b^r + \ldots + b^n$$

- You should be able to use Pascal's triangle to find the coefficients in a binomial expansion.

- You should be able to use the formula for ${}^nC_r$ to find the coefficients in a binomial expansion:

$${}^nC_r = \frac{n!}{r!(n-r)!}$$

---

## ■ Mixed Practice

**1**  Expand and simplify $(2 + x)^4$.

**2**  Find the sum of the infinite geometric series $\dfrac{1}{3}, \dfrac{1}{9}, \dfrac{1}{27}, \ldots$

**3**  A geometric series has first term 7 and common ratio $-\dfrac{5}{9}$. Find the sum to infinity of the series.

**4**  Find the first term of the geometric series with common ratio $\dfrac{3}{4}$ and sum to infinity 12.

**5**  The coefficient of $x^6$ in the expansion of $(x + 1)^n$ is 3003. Find the value of $n$.

**6** In the expansion of $(3x - 2)^{12}$, the term in $x^5$ can be expressed as $^{12}\text{C}_r \times (3x)^p \times (-2)^q$.
  **a** Write down the value of $p$, of $q$ and of $r$.
  **b** Find the coefficient of the term in $x^5$.

Mathematics SL May 2013 Paper 2 TZ1 Q3

**7** Consider the expansion of $(x + 3)^{10}$.
  **a** Write down the number of terms in this expansion.
  **b** Find the term containing $x^3$.

Mathematics SL May 2014 Paper 2 TZ1 Q2

**8** **a** Find the first three terms in the expansion of $(1 + 2x)^{10}$.
  **b** Use your expansion to find an approximate value of $1.002^{10}$.

**9** An infinite geometric series is given by
$$(2 - 3x) + (2 - 3x)^2 + (2 - 3x)^3 + \ldots$$
  **a** Find the range of values of $x$ for which the series converges.
  **b** Given that the sum of the series is $\dfrac{1}{2}$, find the value of $x$.
  **c** Show that the sum of the series cannot equal $-\dfrac{2}{3}$.

**10** The sum to infinity of a geometric series is three times larger than the first term. Find the common ratio of the series.

**11** Find the coefficient of $x^4$ in the expansion of $(2 - x)^{10}$.

**12** Given that $(3 + x)^n = 81 + kx + \ldots$
  **a** Find the value of $n$.
  **b** Find the value of $k$.

**13** **a** Write down an expression, in terms of $n$, for $^n\text{C}_1$.
  **b** Given that $(x + 2)^n = x^n + 18x^{n-1} + bx^{n-2} + \ldots$
    **i** Find the value of $n$.
    **ii** Find the value of $b$.

**14** The third term in the expansion of $(2x + p)^6$ is $60x^4$. Find the possible values of $p$.

Mathematics SL November 2012 Paper 2 Q4

**15** The sum of the first three terms of a geometric sequence is 62.755, and the sum of the infinite sequence is 440. Find the common ratio.

Mathematics SL May 2013 Paper 2 TZ2 Q5

**16** Evaluate
$$\sum_{r=0}^{\infty} \frac{3^r + 4^r}{5^r}$$

**17** **a** Explain why the geometric series $e^{-x} + e^{-2x} + e^{-3x} + \ldots$ converges for all positive values of $x$.
  **b** Find an expression for the sum to infinity of the series.
  **c** Given that the sum to infinity of the series is 2, find the exact value of $x$.

**18** A geometric series has sum to infinity 27, and the sum from (and including) the fourth term to infinity is 1. Find the common ratio of the series.

**19** Given that $\left(1 + \frac{2}{3}x\right)^n (3 + nx)^2 = 9 + 84x + \ldots$, find the value of $n$.

Mathematics SL May 2012 Paper 1 TZ2 Q7

**20** Find the coefficient of $x^{-2}$ in the expansion of $(x - 1)^3 \left(\frac{1}{x} + 2x\right)^6$.

Mathematics HL May 2014 Paper 2 TZ2 Q5

**21** Let $\{u_n\}$, $n \in \mathbb{Z}^+$, be an arithmetic sequence with first term equal to $a$ and common difference of $d$, where $d \neq 0$. Let another sequence $\{v_n\}$, $n \in \mathbb{Z}^+$, be defined by $v_n = 2^{u_n}$.

  **a**  **i**  Show that $\dfrac{v_{n+1}}{v_n}$ is a constant.

      **ii**  Write down the first term of the sequence $\{v_n\}$.

      **iii**  Write down a formula for $v_n$ in terms of $a$, $d$ and $n$.

Let $S_n$ be the sum of the first $n$ terms of the sequence $\{v_n\}$.

  **b**  **i**  Find $S_n$, in terms of $a$, $d$ and $n$.

      **ii**  Find the values of $d$ for which $\displaystyle\sum_{i=1}^{\infty} v_i$ exists.

    You are now told that $\displaystyle\sum_{i=1}^{\infty} v_i$ does exist and is denoted by $S_\infty$.

      **iii**  Write down $S_\infty$ in terms of $a$ and $d$.

      **iv**  Given that $S_\infty = 2^{a+1}$ find the value of $d$.

Let $\{w_n\}$, $n \in \mathbb{Z}^+$, be a geometric sequence with first term equal to $p$ and common ratio $q$, where $p$ and $q$ are both greater than zero. Let another sequence $\{z_n\}$ be defined by $z_n = \ln(w_n)$.

  **c**  Find $\displaystyle\sum_{i=1}^{n} z_i$ giving your answer in the form $\ln k$ with $k$ in terms of $n$, $p$ and $q$.

Mathematics HL May 2015 Paper 1 TZ1 Q12

# Analysis and approaches: Functions

## ESSENTIAL UNDERSTANDINGS

- Creating different representations of functions to model relationships between variables, visually and symbolically, as graphs, equations and tables represents different ways to communicate mathematical ideas.

### In this chapter you will learn...

- how to form composite functions
- when composite functions exist
- what the identity function is
- how to find inverse functions
- when inverse functions exist.

### CONCEPTS

The following concept will be addressed in this chapter:
- Functions **represent** mappings that assign to each value of the independent variable (input) one and only one dependent variable (output).

### PRIOR KNOWLEDGE

Before starting this chapter, you should already be able to complete the following:

1  If $f(x) = 3 - 4x$, calculate $f(-2)$.

2  Find the largest possible domain of $f(x) = \dfrac{1}{x+3}$.

3  a  Find the range of $f(x) = x^2 + 4x + 1$, $x \in \mathbb{R}$.

   b  State whether $f(x)$ is a one-to-one or many-to-one function.

■ **Figure 14.1** Does the order in which we do things always matter?

4 Sketch the inverse function of the following graph:

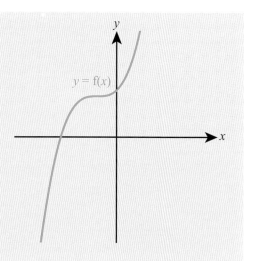

$y = f(x)$

5 Make $x$ the subject of the following:

a $y = \dfrac{3x+1}{x-1}$

b $y = e^x - 2$

Functions can be combined in many different ways: they can be added, multiplied, divided or we can take one function and apply it to another function. Sometimes the order in which functions are combined does not matter, but at other times we get different results depending on the order in which the functions have been combined.

## Starter Activity

Look at the pictures in Figure 14.1. In small groups label each part of the process of baking a cake shown. Does the order in which they occur matter? Can each process be reversed?

**Now look at this problem:**

Ann thinks of a number and squares it. She then adds five to the result.

Bill thinks of a number and adds five to it. He then squares the result.

Both Ann and Bill end up with the same answer.

a What was the answer?

b Will Ann and Bill always end up with the same answer?

# 14A Composite functions

If, after applying one function, g, to a number you then apply a second function, f, to the result, you have a composite function.

This is written as $f(g(x))$ or $(f \circ g)(x)$ or simply $fg(x)$.

**WORKED EXAMPLE 14.1**

$f(x) = 2x - 3$ and $g(x) = x^2$

Find

a $f(g(x))$              b $g(f(x))$.

Replace $x$ in $f(x)$ with $g(x)$ ·········· a  $f(g(x)) = f(x^2)$

$= 2x^2 - 3$

Replace $x$ in $g(x)$ with $f(x)$ ·········· b  $g(f(x)) = g(2x - 3)$

$= (2x - 3)^2$

## Be the Examiner 14.1

$f(x) = x + 4$ and $g(x) = 5x$

Find $(f \circ g)(3)$.

Which is the correct solution? Identify the errors made in the incorrect solutions.

| Solution 1 | Solution 2 | Solution 3 |
|---|---|---|
| $f(3) = 3 + 4 = 7$ | $(f \circ g)(3) = 5(3 + 4)$ | $(f \circ g)(3) = 5 \times 3 + 4$ |
| $g(3) = 5 \times 3 = 15$ | $= 5 \times 7$ | $= 15 + 4$ |
| So, $(f \circ g)(3) = 7 \times 15 = 105$ | $= 35$ | $= 19$ |

**TOOLKIT: Problem Solving**

If $f(g(x)) \equiv g(f(x))$ then the two functions are said to be commutative.

Find examples of two functions $f(x)$ and $g(x)$ which commute. Can you make and prove any general conjectures about some types of functions which always commute with each other?

Can you find a function $f(x)$ which commutes with any other function?

For a composite function $f(g(x))$ to exist, all possible outputs from g must be allowed as inputs to f, i.e. the range of g must lie entirely within the domain of f.

You met domain and range in Section 3A.

**WORKED EXAMPLE 14.2**

$f(x) = \ln(x+2)$, $x > -2$

$g(x) = x+1$

Find the largest possible domain of $(f \circ g)(x)$.

The only thing restricting the domain is that the range of g($x$) must fit into the domain of f($x$) ·················· Largest possible domain of $(f \circ g)$ occurs when g($x$) > −2

The domain of f is $x > -2$. So, the largest the range of g can be is g($x$) > −2

Solve the inequality for $x$ ································ $x + 1 > -2$
$$x > -3$$
This is the largest possible domain.

## Exercise 14A

For questions 1 to 8, use the method demonstrated in Worked Example 14.1 to find an expression for the composite function $(f \circ g)(x)$.

1   a   $f(x) = 3x - 1$, $g(x) = x^2$

    b   $f(x) = 4x + 2$, $g(x) = x^2$

2   a   $f(x) = x^2$, $g(x) = 2x + 1$

    b   $f(x) = x^2$, $g(x) = 3x - 2$

3   a   $f(x) = 3x^2 + 2x$, $g(x) = 2x$

    b   $f(x) = 5x^2 - 3x$, $g(x) = 2x$

4   a   $f(x) = 3e^x$, $g(x) = 2x + 5$

    b   $f(x) = 4e^x$, $g(x) = 3x + 1$

5   a   $f(x) = 3x + 1$, $g(x) = 4e^x$

    b   $f(x) = 2x + 5$, $g(x) = 3e^x$

6   a   $f(x) = x^3 - 2x$, $g(x) = e^x$

    b   $f(x) = x^3 + 4x$, $g(x) = e^x$

7   a   $f(x) = \dfrac{1}{x+1}$, $g(x) = 3x + 2$

    b   $f(x) = \dfrac{1}{x-2}$, $g(x) = 2x + 5$

8   a   $f(x) = 3x - 2x^2$, $g(x) = \dfrac{1}{x}$

    b   $f(x) = 4x + 3x^2$, $g(x) = \dfrac{1}{x}$

For questions 9 to 14, use the method demonstrated in Worked Example 14.2 to find the largest possible domain for the function $(f \circ g)(x)$. Where the domain of f is not given, assume the largest possible real domain.

9   a   $f(x) = \ln(x + 3)$, $x > -3$ and $g(x) = x + 5$

    b   $f(x) = \ln(x + 8)$, $x > -8$ and $g(x) = x + 1$

10   a   $f(x) = \ln(4 - x)$, $x < 4$ and $g(x) = x + 3$

     b   $f(x) = \ln(1 - x)$, $x < 1$ and $g(x) = x + 1$

11   a   $f(x) = \sqrt{2x + 1}$, $x \geqslant -\dfrac{1}{2}$ and $g(x) = 3x - 2$

     b   $f(x) = \sqrt{3x + 1}$, $x \geqslant \dfrac{1}{3}$ and $g(x) = 2x - 3$

12   a   $f(x) = \sqrt{2x - 1}$ and $g(x) = 4 - x$

     b   $f(x) = \sqrt{3x - 1}$ and $g(x) = 2 - x$

13   a   $f(x) = \dfrac{1}{x-2}$, $x \neq 2$ and $g(x) = 4x + 1$

     b   $f(x) = \dfrac{1}{x+5}$, $x \neq -5$ and $g(x) = 2x - 7$

14   a   $f(x) = \dfrac{1}{x-3}$ and $g(x) = e^x$

     b   $f(x) = \dfrac{1}{x-7}$ and $g(x) = e^x$

15   Let $f(x) = 3x - 1$ and $g(x) = 4 - 3x$.

    a   Find and simplify and expression for $(f \circ g)(x)$.

    b   Solve the equation $(g \circ f)(x) = 4$.

16   Let $f(x) = x^2 + 1$ and $g(x) = x - 1$.

    a   Find and simplify an expression for $f(f(x))$.

    b   Solve the equation $f(g(x)) = g(f(x))$.

17   Given that $f(x) = 2x^3$, find a simplified expression for $(f \circ f)(x)$.

18  Given that $f(x) = \sqrt{x-4}$, $x \geqslant 4$ and $g(x) = 3x + 10$,

a   Find the largest possible domain for the function $(f \circ g)$.

b   Solve the equation $(f \circ g)(x) = 5$.

19  Let $f(x) = \ln x (x > 0)$ and $g(x) = x - 5$.

a   Write down the exact values of $(f \circ g)(8)$ and $(g \circ f)(8)$.

b   Solve the equation $fg(x) = 8$.

20  Functions f and g both have domain {1, 2, 3, 4, 5} and their values are given in the following table.

| $x$ | 1 | 2 | 3 | 4 | 5 |
|-----|---|---|---|---|---|
| $f(x)$ | 5 | 4 | 3 | 2 | 1 |
| $g(x)$ | 3 | 1 | 4 | 5 | 2 |

a   Find

i   $(f \circ g)(3)$                      ii   $(g \circ f)(4)$

b   Solve the equation $(f \circ g)(x) = 1$

21  Some of the values of the functions f and g are given in the following table.

| $x$ | 1 | 3 | 5 | 7 | 9 |
|-----|---|---|---|---|---|
| $f(x)$ | 3 | 7 | 5 | 1 | 9 |
| $g(x)$ | 5 | 7 | 9 | 3 | 1 |

a   Find

i   $(f \circ g)(3)$                      ii   $(g \circ f)(9)$

b   Solve the equation $(f \circ g)(x) = 5$

22  Let $f(x) = \ln x$, $x > 0$ and $g(x) = x - 3$.

a   Find the largest possible domain for the function $(f \circ g)$.

b   Solve the equation $(f \circ g)(x) = 1$.

c   Solve the equation $(g \circ f)(x) = 1$.

d   Solve the equation $(f \circ g)(x) = (g \circ f)(x)$.

23  Given that $f(x) = \dfrac{1}{x+2}$ and $g(x) = \dfrac{1}{x-3}$,

a   Find and simplify an expression for $(f \circ g)(x)$.

b   Find the largest possible domain for the function $(f \circ g)(x)$.

c   Solve the equation $(f \circ g)(x) = 2$.

24  Let $f(x) = \dfrac{4}{x}$, $x \neq 0$ and $g(x) = 2x^2$, $x \in \mathbb{R}$. Prove that $gf(x) \equiv k\,fg(x)$ for some constant $k$, which you should find.

25  Let $f(x) = \dfrac{1}{3x+2}$ for $x \neq -\dfrac{2}{3}$.

a   Find the largest possible domain for $(f \circ f)(x)$.

b   Use technology to find the range of $(f \circ f)(x)$ for the domain from part a.

c   Solve the equation $(f \circ f)(x) = 1$.

# 14B Inverse functions

## ◼ Identity function

You have already met the idea that the inverse, $f^{-1}$, of a function f reverses the effect of f. This can be expressed more formally in terms of composite functions.

---
**KEY POINT 14.1**

$(f \circ f^{-1})(x) = (f^{-1} \circ f)(x) = x$

---

A function that has no effect on any value in its domain is called the **identity function**. So, the functions $(f \circ f)^{-1}$ and $(f^{-1} \circ f)$ are both the identity function.

## ◼ Finding the inverse function

In order to find an expression for $f^{-1}$ from f you need to rearrange f to find the input $(x)$ in terms of the output $(y)$.

---
**KEY POINT 14.2**

To find an expression for $f^{-1}(x)$:
1  Let $y = f(x)$.
2  Rearrange to make $x$ the subject.
3  State $f^{-1}(x)$ by replacing any $y$s with $x$s.

---

---
**WORKED EXAMPLE 14.3**

$f(x) = \dfrac{3x - 5}{2}$

Find the inverse function, $f^{-1}$.

Let $y = f(x)$ $\cdots\cdots\cdots\cdots\cdots\cdots\cdots\cdots$ $y = \dfrac{3x - 5}{2}$

Rearrange to make $x$ the subject $\cdots\cdots\cdots\cdots$ $2y = 3x - 5$

$$2y + 5 = 3x$$

$$x = \frac{2y + 5}{3}$$

Write the resulting $\cdots\cdots\cdots\cdots$ So, $f^{-1}(x) = \dfrac{2x + 5}{3}$
function in terms of $x$

---

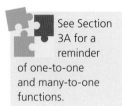

See Section 3A for a reminder of one-to-one and many-to-one functions.

# ■ The existence of an inverse for one-to-one functions

For $f^{-1}(x)$ to be a function, it must map each input value to a single output value. But since the graph $y = f^{-1}(x)$ is a reflection in $y = x$ of the graph $y = f(x)$, it follows that for $f(x)$ each output must come from a single input, i.e. $f(x)$ must be one-to-one.

If $f(x)$ is many-to-one, $f^{-1}(x)$ would not be a function as there would be input values that map to more than one output value.

### KEY POINT 14.3

A function has to be one-to-one to have an inverse.

Any function can be made one-to-one (and therefore to have an inverse) by restricting its domain.

### WORKED EXAMPLE 14.4

Use technology to find the largest possible domain of the function $f(x) = x\ln x$ of the form $x \geqslant k$ for which the inverse $f^{-1}$ exists.

$f^{-1}$ will only exist if f is one-to-one. Use the GDC to sketch the graph and find the minimum point. Eliminating values of $x$ to the left of the minimum will leave f being one-to-one

Largest possible domain of f for which $f^{-1}$ exists:

$x \geqslant 0.368$

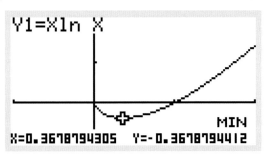

Y1=Xln X

X=0.3678794305     Y=-0.3678794412     MIN

---

## Be the Examiner 14.2

Find the inverse function of $f(x) = x^2 - 4$, $x < 0$.

Which is the correct solution? Identify the errors made in the incorrect solutions.

| Solution 1 | Solution 2 | Solution 3 |
|---|---|---|
| $y = x^2 - 4$ <br> $x^2 = y + 4$ <br><br> $x = \sqrt{y+4}$ <br><br> So, $f^{-1}(x) = \sqrt{x+4}$ | $y = x^2 - 4$ <br> $x^2 = y + 4$ <br><br> $x = -\sqrt{y+4}$ <br><br> So, $f^{-1}(x) = -\sqrt{x+4}$ | f isn't one-to-one, so the inverse function doesn't exist. |

## Exercise 14B

For questions 1 to 10, use the method demonstrated in Worked Example 14.3 to find an expression for the inverse function, $f^{-1}(x)$.

1  a  $f(x) = \dfrac{4x+1}{2}$

   b  $f(x) = \dfrac{3x+4}{5}$

2  a  $f(x) = 4x - 3$

   b  $f(x) = 5x + 1$

3  a  $f(x) = e^{4x}$

   b  $f(x) = e^{3x}$

4  a  $f(x) = 3e^{x-2}$

   b  $f(x) = 2e^{x+3}$

5  a  $f(x) = \log_2(3x + 1)$

   b  $f(x) = \log_3(4x - 1)$

6  a  $f(x) = \sqrt{x - 2}$

   b  $f(x) = \sqrt{x + 3}$

7  a  $f(x) = (x + 2)^3$

   b  $f(x) = (x - 3)^3$

8  a  $f(x) = x^3 - 2$

   b  $f(x) = x^3 + 5$

9  a  $f(x) = \dfrac{x+3}{x-2}$

   b  $f(x) = \dfrac{x+1}{x-3}$

10  a  $f(x) = \dfrac{2x+1}{3x-2}$

    b  $f(x) = \dfrac{3x-1}{2x-3}$

For questions 11 to 17, use the method demonstrated in Worked Example 14.4. For each function f find the largest possible domain of the given form such that f has an inverse function.

11  a  $f(x) = (x - 2)^2,\ x \geqslant a$

    b  $f(x) = (x + 5)^2,\ x \geqslant a$

12  a  $f(x) = (x + 1)^2,\ x \leqslant b$

    b  $f(x) = (x - 3)^2,\ x \leqslant b$

13  a  $f(x) = x^3 - 3x^2,\ x \geqslant c$

    b  $f(x) = x^3 + 2x^2,\ x \geqslant c$

14  a  $f(x) = x^3 - 3x,\ c \leqslant x \leqslant d$

    b  $f(x) = x^3 - 12x,\ c \leqslant x \leqslant d$

15  a  $f(x) = xe^x,\ x \leqslant a$

    b  $f(x) = xe^{-x},\ x \leqslant a$

16  a  $f(x) = xe^{\frac{x}{2}},\ x \geqslant a$

    b  $f(x) = xe^{-\frac{x}{3}},\ x \geqslant a$

17  a  $f(x) = \dfrac{x^2+1}{x^2-4},\ x < b$

    b  $f(x) = \dfrac{x^2-2}{x^2-1},\ x < b$

18  For the function $f(x) = \dfrac{4}{3-x}\ (x \neq 3)$:

    a  find $f^{-1}(3)$

    b  find an expression for $f^{-1}(x)$.

19  Given that $f(x) = 3e^{5x}$, find an expression for $f^{-1}(x)$.

20  The diagram shows the graph of $y = f(x)$, where $f(x) = \dfrac{1}{5}x^3 - 3$.

    a  Copy the graph and, on the same axes, sketch $y = f^{-1}(x)$.

    b  Find an expression for $y = f^{-1}(x)$.

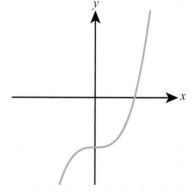

21  Let $f(x) = e^{\frac{x}{2}} - 1$ for all $x \in \mathbb{R}$. The diagram shows the graph of $y = f(x)$.

    a  Sketch the graph of $y = f^{-1}(x)$.

    b  Find an expression for $y = f^{-1}(x)$ and state its domain.

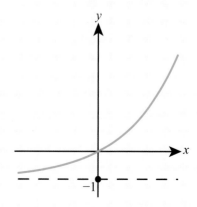

**22** Given that $f(x) = \dfrac{x^2+1}{x^2-4}$ for $x > 2$,

   a  Find an expression for $f^{-1}(x)$.

   b  State the range of $f^{-1}$.

**23** If $f(x) = \dfrac{1}{4+\sqrt{x}}$, $x > 0$ and $g(x) = 2x + 1$, solve $(g^{-1} \circ f^{-1})(x) = 4$.

**24** Let $f(x) = e^{\frac{x}{2}} + x - 5$ for $x \in \mathbb{R}$. The diagram shows a part of the graph of $y = f(x)$.

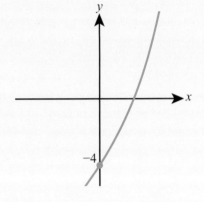

   a  Copy the graph and sketch $y = f^{-1}(x)$ on the same axes.

   b  Find the exact solution of the equation $f(x) = f^{-1}(x)$.

**25** Let $f(x) = x^2 + 3$ for $x \leqslant a$.

   a  Find the largest possible value of $a$ such that f has an inverse function.

   b  For this domain, find an expression for $f^{-1}(x)$.

**26** a  For the function $g(x) = 9(x-5)^2$, find the largest possible domain of the form $x \leqslant k$ such that g has an inverse function.

   b  For this domain, find an expression for $g^{-1}(x)$.

**27** A function f is called *self-inverse* if $f(x) \equiv f^{-1}(x)$. Find the value of $a$ such that $f(x) = \dfrac{ax+3}{x-4}$ is a self-inverse function.

## Checklist

■ You should be able to form composite functions.

■ You should know when composite functions exist.

■ You should know what the identity function is and how it is related to inverse functions
   □ $(f \circ f^{-1})(x) = (f^{-1} \circ f)(x) = x$

■ You should be able to find inverse functions
   □ To find an expression for $f^{-1}(x)$:
     1  Let $y = f(x)$
     2  Rearrange to make $x$ the subject
     3  State $f^{-1}(x)$ by replacing any $y$s with $x$s

■ You should know when inverse functions exist
   □ A function has to be one-to-one to have an inverse.

# ▓ Mixed Practice

**1** **a** For the function $f(x) = 3x - 1$, find the inverse function, $f^{-1}(x)$.
  **b** Verify that $(f \circ f^{-1})(x) = x$ for all $x$.

**2** A function is defined by $h(x) = \sqrt{5 - x}$ for $x \leqslant a$.
  **a** State the largest possible value of $a$.
  **b** Find $h^{-1}(3)$.

**3** Given that $f(x) = e^{3x}$ evaluate $f^{-1}(4)$.

**4** The table shows some values of the function $f(x)$.

| $x$ | 0 | 1 | 2 | 3 | 4 |
|---|---|---|---|---|---|
| $f(x)$ | 3 | 4 | 0 | 1 | 2 |

  **a** Find $(f \circ f)(2)$.
  **b** Find $f^{-1}(4)$.

**5** Let $f(x) = x + 3$ and $g(x) = e^{2x}$. Solve the equation $(g \circ f)(x) = 1$.

**6** Given that $g(x) = 3\ln(x - 2)$,
  **a** Find the largest possible domain for g.
  **b** Using the domain for part **a**, find an expression for $g^{-1}(x)$ and state its range.

**7** Given the functions $f(x) = 3x + 1$ and $g(x) = x^3$, find $(f \circ g)^{-1}(x)$.

**8** Let $f(x) = 2x + 3$ and $g(x) = x^3$.
  **a** Find $(f \circ g)(x)$.
  **b** Solve the equation $(f \circ g)(x) = 0$.

Mathematics SL November 2014 Paper 2 Q1

**9** The following diagram shows the graph of $y = f(x)$, for $-4 \leqslant x \leqslant 5$.

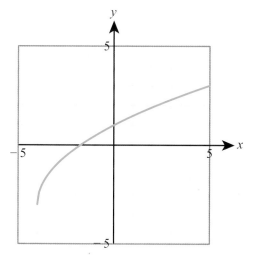

  **a** Write down the value of
    **i** $f(-3)$;
    **ii** $f^{-1}(1)$.
  **b** Find the domain of $f^{-1}$.
  **c** Sketch the graph of $f^{-1}$.

Mathematics SL May 2014 Paper 1 TZ2 Q3

**10** For the function $f(x) = x - 2$ $(x \in \mathbb{R})$ and $g(x) = \dfrac{1}{x-1}$ $(x \neq 1)$,

    **a** Find $(f \circ g)(x)$ and state its domain.

    **b** Verify that $(f \circ g)^{-1}(x) = (g^{-1} \circ f^{-1})(x)$ for all $x$.

**11** Given that $f(x) = \dfrac{3x-1}{x+4}$ for $x \neq -4$, find an expression for $f^{-1}(x)$.

**12** Let the function $g(x) = x + e^x$ be defined for all real numbers $x$.

    **a** By sketching a graph, or otherwise, show that g has an inverse function.

    **b** Solve the equation $g^{-1}(x) = 2$.

**13** Let $f(x) = \sqrt{x}$ $(x \geqslant 0)$ and $g(x) = 9^x (x \in \mathbb{R})$.

    **a** Evaluate $(g \circ f)\left(\dfrac{1}{4}\right)$.

    **b** Solve the equation $(f^{-1} \circ g)(x) = \dfrac{1}{3}$.

**14** Let $h(x) = \ln(x-2)$ for $x > 2$, and $g(x) = e^x$ for $x \in \mathbb{R}$.

    **a** Find $h^{-1}(x)$ and state its range.

    **b** Find $(g \circ h)(x)$, giving your answer in the form $ax + b$, where $a, b \in \mathbb{Z}$.

**15** Function f is defined by $f(x) = (2x + 1)^2$ for $x \leqslant a$.

    **a** By using a graph, or otherwise, find the largest value of $a$ for which f has an inverse function.

    **b** Find an expression for $f^{-1}(x)$.

**16** Let $f(x) = x^2 + 3$, $x \geqslant 1$ and $g(x) = 12 - x$, $x \in \mathbb{R}$.

    **a** Evaluate $f(5)$.

    **b** Find and simplify an expression for $gf(x)$.

    **c** **i** State the geometric relationship between the graphs of $y = f(x)$ and $y = f^{-1}(x)$.

       **ii** Find an expression for $f^{-1}(x)$.

       **iii** Find the range of $f^{-1}(x)$.

**17** Let $f(x) = 3x + 1$, $x \in \mathbb{R}$ and $g(x) = \dfrac{x+4}{x-1}$, $x \neq 1$.

    **a** Find and simplify

       **i** $f(7)$

       **ii** $fg(x)$

       **iii** $ff(x)$

    **b** State the range of f. Hence, explain why $gf(x)$ does not exist.

    **c** **i** Show that $g^{-1}(x) = g(x)$ for all $x \neq 1$.

       **ii** State the range of $g^{-1}(x)$.

**18** Let $f(x) = \sqrt{x - 5}$, for $x \geqslant 5$.

    **a** Find $f^{-1}(2)$.

    **b** Let g be a function such that $g^{-1}$ exists for all real numbers. Given that $g(30) = 3$, find $(f \circ g^{-1})(3)$.

Mathematics SL May 2013 Paper 1 TZ1 Q5

**19** Let $f(x) = 3x - 2$ and $g(x) = \dfrac{5}{3x}$, for $x \neq 0$.

    **a** Find $f^{-1}(x)$.

    **b** Show that $(g \circ f^{-1})(x) = \dfrac{5}{x+2}$.

Let $h(x) = \dfrac{5}{x+2}$, for $x \geqslant 0$. The graph of h has a horizontal asymptote at $y = 0$.

**c i** Find the $y$-intercept of the graph of h.

    **ii** Hence, sketch the graph of h.

**d** For the graph of $h^{-1}$,

    **i** write down the $x$-intercept;

    **ii** write down the equation of the vertical asymptote.

**e** Given that $h^{-1}(a) = 3$, find the value of $a$.

Mathematics SL November 2013 Paper 1 Q8

**20** For the functions $f(x) = e^{2x}$ and $g(x) = \ln(x - 2)$, verify that $(f \circ g)^{-1}(x) = (g^{-1} \circ f^{-1})(x)$.

**21** Let $f(x) = \dfrac{1}{1 + \sqrt{x}}$ and $g(x) = x + 7$. Solve $f^{-1}(g^{-1}(x)) = 9$.

**22** Let $f(x) = 2 + x - x^3$ for $x \geqslant a$.

**a** Find the smallest value of $a$ so that f has an inverse function.

**b** State the geometric relationship between the graphs of $y = f(x)$ and $y = f^{-1}(x)$.

**c** Find the exact solution of the equation $f^{-1}(x) = f(x)$.

**23** Consider the functions given below.

$f(x) = 2x + 3, \ x \in \mathbb{R}$

$g(x) = \dfrac{1}{x}, \ x \neq 0$

**a i** Find $(g \circ f)(x)$ and write down the domain of the function.

    **ii** Find $(f \circ g)(x)$ and write down the domain of the function.

**b** Find the coordinates of the point where the graph of $y = f(x)$ and the graph of $y = (g^{-1} \circ f \circ g)(x)$ intersect.

Mathematics HL May 2011 Paper 1 TZ1 Q8

- Creating different representations of functions to model relationships between variables, visually and symbolically, as graphs, equations and tables represents different ways to communicate mathematical ideas.

## In this chapter you will learn...

- how to recognize the shape and main features of quadratic graphs
- how to complete the square
- how to solve quadratic equations by factorizing, completing the square and using the quadratic formula
- how to solve quadratic inequalities
- how to use the discriminant to determine the number of roots of a quadratic equation.

### CONCEPTS

The following concepts will be addressed in this chapter:

- The parameters in a function or equation correspond to geometrical features of a graph and can represent physical **quantities** in spatial dimensions.
- Equivalent **representations** of quadratic functions can reveal different characteristics of the same relationship.

### PRIOR KNOWLEDGE

Before starting this chapter, you should already be able to complete the following:

1 Expand and simplify $(3x - 2)(x + 4)$.

2 Factorize $x^2 - 9x - 10$.

3 Express $\sqrt{12}$ in the form $k\sqrt{3}$.

4 Solve the inequality $4x - 5 > x + 2$.

■ **Figure 15.1** How can quadratics help us to understand the real world?

A frequently used type of function is a quadratic. Quadratic graphs are relatively simple graphs that have vertices (maximum or minimum points), and since so many problems in applied mathematics centre on finding the maximum or minimum of a function, quadratics are often used to model these situations. Such problems arise frequently in economics and business, science and engineering.

The characteristic 'U-shaped' parabola of a quadratic function can be observed in many natural processes – such as the trajectory of a projectile – and has been used by scientists and engineers to design and create solutions to real-world problems. Examples include the shape of the cables in a suspension bridge and parabolic reflectors, such as satellite dishes or car headlights.

## Starter Activity

Look at the pictures in Figure 15.1. In small groups discuss the characteristic features of the shapes shown.

**Now look at this problem:**

Consider the following sales data for a particular product.

| Selling price (£) | Quantity sold per week |
|---|---|
| 10 | 50 |
| 15 | 40 |
| 20 | 30 |
| 25 | 20 |

Each item costs £10 to produce.

a  Suggest a formula for the quantity sold ($q$) in terms of the selling price ($s$).

b  Write down an expression for the total profit per week ($P$) in terms of $q$ and $s$.

c  Hence find an expression for $P$ in terms of $s$.

d  Using your GDC, find the selling price that gives the maximum profit.

# 15A Graphs of quadratic functions

A quadratic function is a function of the form $f(x) = ax^2 + bx + c$, where $a, b, c \in \mathbb{R}$.

The graph of a quadratic function is a curve called a **parabola**, which can be one of two possible shapes.

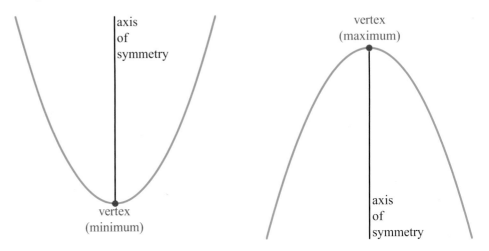

Different features of the parabola can be identified by expressing the quadratic function in different forms.

## ■ The quadratic function $f(x) = ax^2 + bx + c$

In this form you can identify the shape of the graph and the $y$-intercept.

**KEY POINT 15.1**

For the quadratic function $f(x) = ax^2 + bx + c$,
● the shape of the graph $y = f(x)$ depends on the coefficient $a$:
  □ If $a > 0$

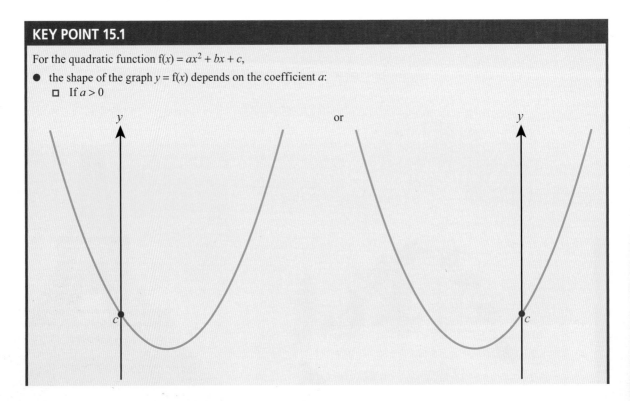

□ If $a < 0$

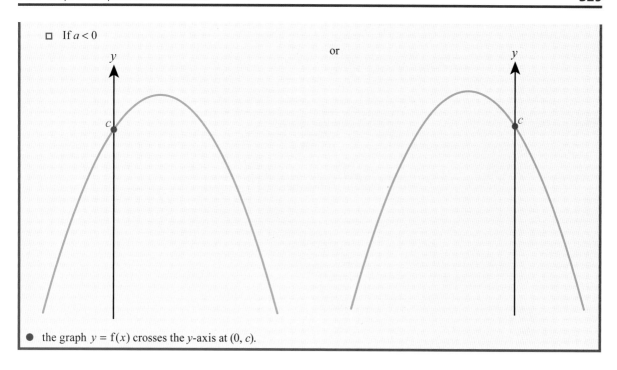

or

● the graph $y = f(x)$ crosses the $y$-axis at $(0, c)$.

**CONCEPTS – QUANTITY**

Use technology to investigate what happens to the graph $y = ax^2 + bx + c$ when you change one of the parameters – $a$, $b$ or $c$ – and keep the others fixed.

**WORKED EXAMPLE 15.1**

Match each equation to the corresponding graph.

a $y = -3x^2 + 4x + 2$

b $y = 2x^2 + 3x - 4$

c $y = -x^2 + 5x - 3$

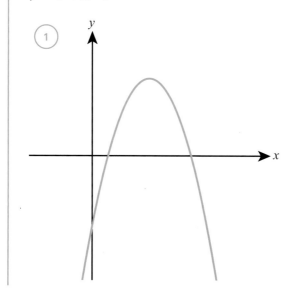

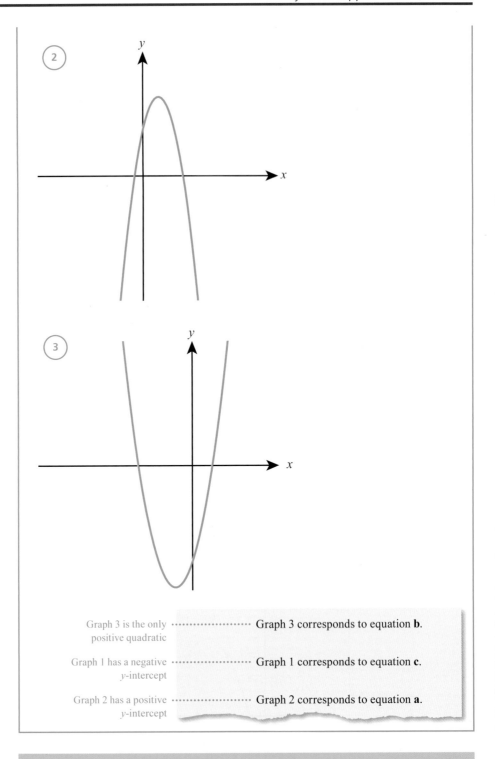

Graph 3 is the only ···················· Graph 3 corresponds to equation **b**.
positive quadratic

Graph 1 has a negative ···················· Graph 1 corresponds to equation **c**.
*y*-intercept

Graph 2 has a positive ···················· Graph 2 corresponds to equation **a**.
*y*-intercept

## CONCEPTS – QUANTITY

How many points do you need to uniquely define a straight line?

How many different parabolae go through all three of the points (0, 0), (1, 2) and (2, 6)? Find their equation(s). Can you explain your answer in terms of the number of parameters in a quadratic function?

## ■ The form f(x) = a(x − p)(x − q)

In this form you can identify the *x*-intercepts, since it shows directly two *x* values which make the expression equal to zero.

### KEY POINT 15.2

The quadratic graph $y = a(x - p)(x - q)$ crosses the *x*-axis at $(p, 0)$ and $(q, 0)$.

To express the function in this form you need to factorize.

### Tip

One very common type of factorization you should watch out for is 'the difference of two squares' which states that
$a^2 - b^2 = (a - b)(a + b)$.

### WORKED EXAMPLE 15.2

Sketch the graph $y = 2x^2 - 16x + 30$, labelling any axis intercepts.

The *y*-intercept of ⋯⋯⋯⋯⋯⋯ *y*-intercept: (0, 30)
$ax^2 + bx + c$ is (0, *c*)

First take out a factor of 2 ⋯⋯⋯⋯⋯⋯ $2x^2 - 16x + 30 = 2(x^2 - 8x + 15)$

Then factorize into two ⋯⋯⋯⋯⋯⋯⋯⋯⋯ $= 2(x - 3)(x - 5)$
brackets. Look for two
numbers which multiply to
give +15 and add to give −8

Comparing this to $a(x - p)$ ⋯⋯⋯⋯⋯⋯ *x*-intercepts: (3, 0), (5, 0)
$(x - q)$, *p* = 3 and *q* = 5

Since $a > 0$, this is a ⋯⋯⋯⋯⋯
positive quadratic shape
Do not forget the *y*-intercept

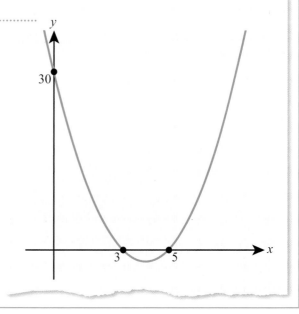

**WORKED EXAMPLE 15.3**

Find the equation of the graph below in the form $y = ax^2 + bx + c$.

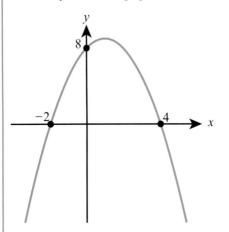

| | |
|---|---|
| *x*-intercepts at $p$ and $q$ mean the equation is $y = a(x - p)(x - q)$ | $\cdots\cdots\cdots\cdots\cdots\cdots$ $y = a(x - 4)(x + 2)$ |
| Substitute in the point $(0, -8)$ to find $a$ | $\cdots\cdots\cdots\cdots\cdots$ When $x = 0$, $y = 8$: |
| | $8 = a(0 - 4)(0 + 2)$ |
| | $8 = -8a$ |
| | $a = -1$ |
| Multiply out to give the equation in the required form | $\cdots\cdots\cdots\cdots$ So, $y = -(x - 4)(x + 2)$ |
| | $y = -(x^2 - 2x - 8)$ |
| | $y = -x^2 + 2x + 8$ |

# ■ The form $\mathrm{f}(x) = a(x - h)^2 + k$

In this form you can find the coordinates of the vertex of the graph (and therefore the equation of the line of symmetry too).

**KEY POINT 15.3**

The quadratic graph $y = a(x - h)^2 + k$ has
- vertex $(h, k)$
- line of symmetry $x = h$.

To express the function in this form you need a process called **completing the square**.

**KEY POINT 15.4**

Completing the square:

$$x^2 + bx + c \equiv \left(x + \frac{b}{2}\right)^2 - \left(\frac{b}{2}\right)^2 + c$$

**Tip**

Key Point 15.4 just says that you halve the coefficient of $x$ to find out what goes in the bracket and then subtract that term squared outside the bracket.

**WORKED EXAMPLE 15.4**

a  Express $f(x) = x^2 - 10x + 7$ in the form $(x - h)^2 + k$.

b  Hence, find the coordinates of the vertex of $y = f(x)$.

Halve the coefficient of $x$ ⋯⋯ **a**  $x^2 - 10x + 7 = (x - 5)^2 - (-5)^2 + 7$

Subtract $(-5)^2$ $= (x - 5)^2 - 25 + 7$

$= (x - 5)^2 - 18$

Use the fact that the quadratic
graph $y = (x - h)^2 + k$ ⋯⋯ **b**  Coordinates of vertex: $(5, -18)$
has vertex $(h, k)$

If the coefficient of $x^2$ is not 1, you need to first factorize it before completing the square.

**WORKED EXAMPLE 15.5**

a  Express $f(x) = 3x^2 + 9x + 4$ in the form $a(x - h)^2 + k$

b  Hence, find the equation of the line of symmetry of $y = f(x)$.

Factorize 3 from the ⋯⋯ **a**  $3x^2 + 9x + 4 = 3\,[x^2 + 3x] + 4$
terms involving $x$

Then complete the square ⋯⋯ $= 3\left[\left(x + \dfrac{3}{2}\right)^2 - \left(\dfrac{3}{2}\right)^2\right] + 4$
inside the brackets:
halve the coefficient of $x$

Subtract $\left(\dfrac{3}{2}\right)^2$ ⋯⋯ $= 3\left[\left(x + \dfrac{3}{2}\right)^2 - \dfrac{9}{4}\right] + 4$

Multiply the factor of 3 back in ⋯⋯ $= 3\left(x + \dfrac{3}{2}\right)^2 - \dfrac{27}{4} + 4$

$= 3\left(x + \dfrac{3}{2}\right)^2 - \dfrac{11}{4}$

Use the fact that the equation ⋯⋯ **b**  Line of symmetry: $x = -\dfrac{3}{2}$
of the line of symmetry of
$y = a(x - h)^2 + k$ is $x = h$

## Tip

There is no point in factorizing the coefficient of $x^2$ from the constant term, as that term isn't involved in the process of completing the square.

## TOK Links

Just because something looks like a parabola, does that mean it must be so? A rainbow is actually part of a circle. A hanging chain is actually a shape called a catenary with an equation something like $y = 2^x + 2^{-x}$. However, these things are still often modelled using quadratic functions. Is it more important to be right or to be useful?

Which of the **representations** of a quadratic function would you choose to use if you knew each of the following characteristics?
- the vertex
- the axis intercepts
- three random points on the graph.

# Be the Examiner 15.1

Express $-x^2 + 8x - 3$ in the form $a(x - h)^2 + k$.

Which is the correct solution? Identify the errors made in the incorrect solutions.

| Solution 1 | Solution 2 | Solution 3 |
|---|---|---|
| $-x^2 + 8x - 3 = -(x-4)^2 - 16 - 3$ $= -(x-4)^2 - 19$ | $-x^2 + 8x - 3 = -[x^2 - 8x] - 3$ $= -[(x-4)^2 - 16] - 3$ $= -(x-4)^2 + 16 - 3$ $= -(x-4)^2 + 13$ | Multiplying by $-1$: $x^2 - 8x + 3$ Completing the square: $x^2 - 8x + 3 = (x-4)^2 - 16 + 3$ $= (x-4)^2 - 13$ |

 **WORKED EXAMPLE 15.6**

Find the equation of the parabola shown below.

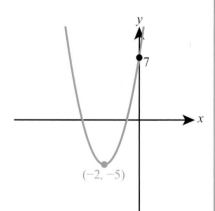

If the vertex is at $(h, k)$ then the most useful form is $y = a(x - h)^2 + k$ ......... $y = a(x + 2)^2 - 5$

Substitute in the point $(0, 7)$ to find $a$ ......... When $x = 0$, $y = 7$:
$$7 = a(0 + 2)^2 - 5$$
$$7 = 4a - 5$$
$$a = 3$$

You are not asked to give the equation in the form $y = ax^2 + bx + c$ so it can be left as it is ......... So, $y = 3(x + 2)^2 - 5$

## Exercise 15A

For questions 1 to 3, use the method demonstrated in Worked Example 15.1 to match each equation with the corresponding graph.

1   a   i   $y = 3x^2 + x - 2$   ii   $y = x^2 + 3x + 1$   iii   $y = -2x^2 + x + 1$

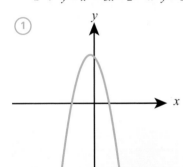 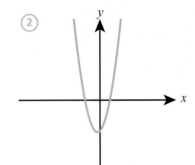 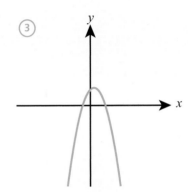

   b   i   $y = x^2 - 3x - 2$   ii   $y = 2x^2 + x + 2$   iii   $y = -2x^2 - x + 3$

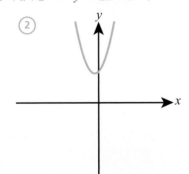 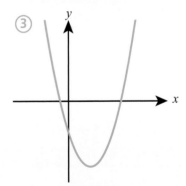

2   a   i   $y = 3x^2 + x + 2$   ii   $y = -x^2 + 3x + 1$   iii   $y = -2x^2 + x - 1$

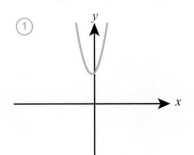 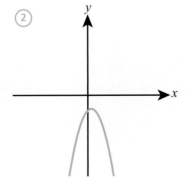 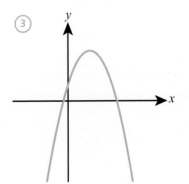

b  i  $y = x^2 - 3x + 2$   ii  $y = -2x^2 + x + 2$   iii  $y = -2x^2 - x - 3$

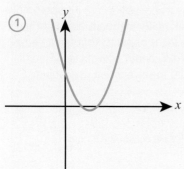 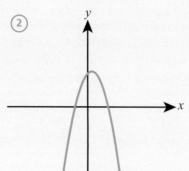 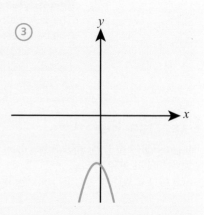

3  a  i  $y = x^2 - 3x + 2$   ii  $y = 2x^2 + x - 2$   iii  $y = 2x^2 - x$

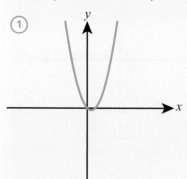 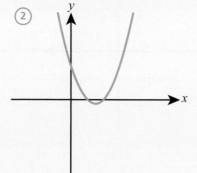 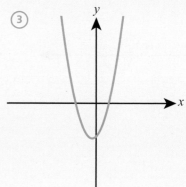

b  i  $y = x^2 - 3x$   ii  $y = 2x^2 + x + 2$   iii  $y = x^2 - 3x - 1$

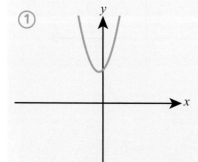 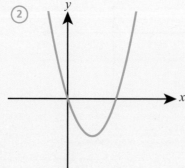 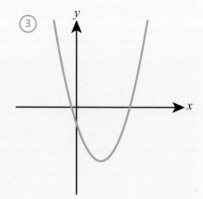

For questions 4 to 10, use the method demonstrated in Worked Example 15.2 to sketch the graph, labelling all the axis intercepts.

4  a  $y = x^2 - 5x + 6$

   b  $y = x^2 - 6x + 8$

5  a  $y = x^2 + x - 6$

   b  $y = x^2 - x - 12$

6  a  $y = 2x^2 + 10x + 8$

   b  $y = 2x^2 + 10x + 12$

7  a  $y = 3x^2 - 6x - 45$

   b  $y = 3x^2 + 3x - 60$

8  a  $y = -2x^2 - 4x - 3$

   b  $y = -3x^2 - 15x - 18$

9  a  $y = -3x^2 + 3x + 18$

   b  $y = -3x^2 - 3x - 6$

10  a  $y = -x^2 - x + 6$

    b  $y = -x^2 + x + 12$

For questions 11 to 16, use the method demonstrated in Worked Example 15.3 to find the equation of each quadratic graph in the form $y = a(x - p)(x - q)$.

11  a

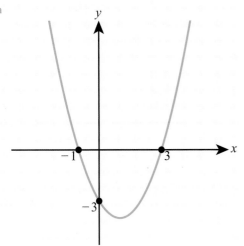

   b

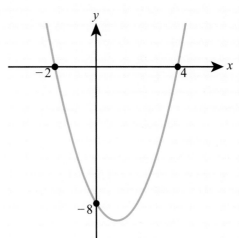

12  a

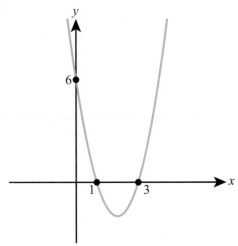

   b

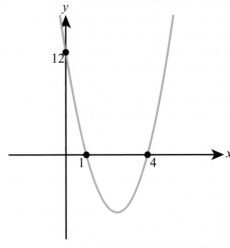

13  a

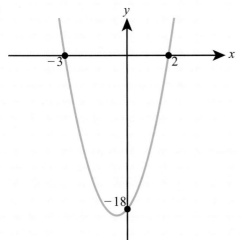

   b

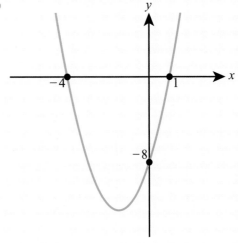

14   a

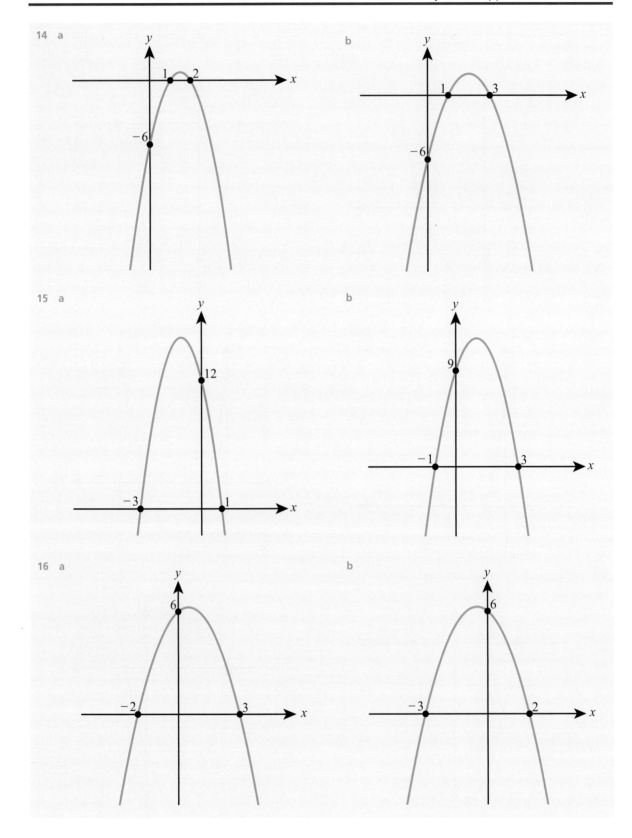

b

15   a

b

16   a

b

For questions 17 to 24, use the method demonstrated in Worked Example 15.4 to complete the square and then find the coordinates of the vertex of the graph.

17   a   $y = x^2 - 6x + 15$

      b   $y = x^2 - 8x + 18$

18   a   $y = x^2 + 4x + 7$

      b   $y = x^2 + 6x + 15$

19   a   $y = x^2 + 4x + 1$

      b   $y = x^2 + 10x + 5$

20   a   $y = x^2 - 4x + 1$

      b   $y = x^2 - 8x + 3$

21   a   $y = x^2 + 4x - 5$

      b   $y = x^2 + 2x - 8$

22   a   $y = x^2 + 3x + 5$

      b   $y = x^2 + 5x + 8$

23   a   $y = x^2 - 3x + 1$

      b   $y = x^2 - 7x + 3$

24   a   $y = x^2 - \dfrac{1}{2}x - \dfrac{3}{2}$

      b   $y = x^2 - \dfrac{1}{3}x - \dfrac{1}{3}$

For questions 25 to 31, use the method demonstrated in Worked Example 15.5 to complete the square and then find the equation of the line of symmetry of the graph.

25   a   $y = 2x^2 + 8x + 15$

      b   $y = 2x^2 + 12x + 23$

26   a   $y = 3x^2 - 6x + 10$

      b   $y = 3x^2 - 12x + 5$

27   a   $y = 2x^2 + 4x - 1$

      b   $y = 2x^2 + 8x - 3$

28   a   $y = 2x^2 - 2x + 1$

      b   $y = 2x^2 - 6x + 5$

29   a   $y = 3x^2 + x + 2$

      b   $y = 3x^2 - x + 1$

30   a   $y = -2x^2 + 4x + 1$

      b   $y = -2x^2 + 8x - 1$

31   a   $y = -x^2 - 4x - 2$

      b   $y = -x^2 - 6x + 3$

For questions 32 to 37, use the method demonstrated in Worked Example 15.6 to find the equation of each graph in the form $y = a(x - h)^2 + k$.

32   a

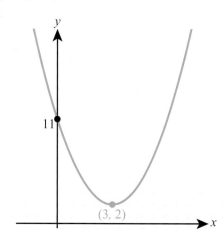

     b

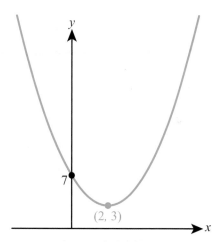

33   a

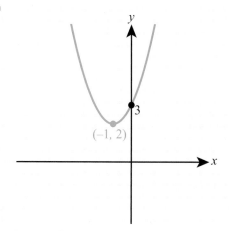

     b

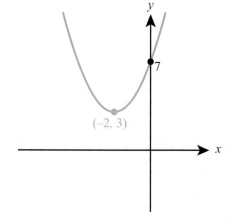

34 a

b

35 a

b

36 a

b

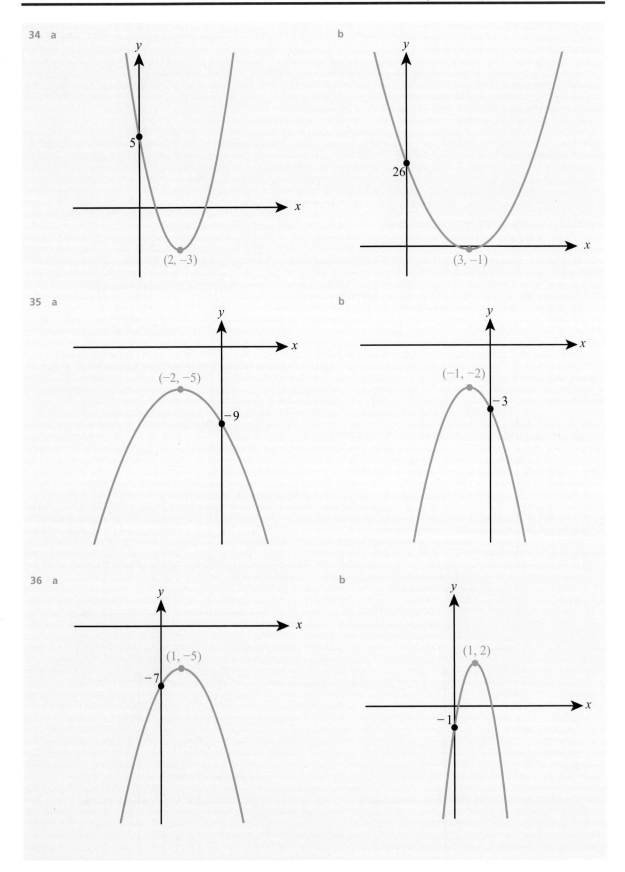

37 a

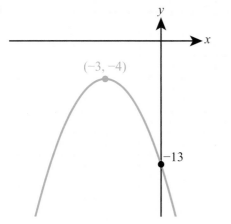

b

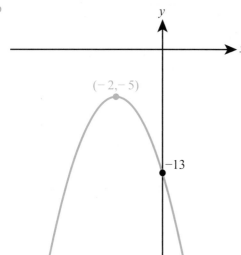

38 The diagram shows the graph with equation $y = x^2 - 5x - 24$.

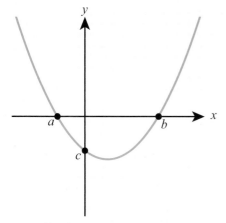

a Write down the value of $c$.

b Find the values of $a$ and $b$.

39

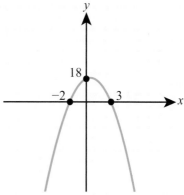

The equation of the graph shown in the diagram can be written as $y = a(x + p)(x - q)$.

a Write down the values of $p$ and $q$.

b Find the value of $a$.

c Hence find the equation of the graph in the form $y = ax^2 + bx + c$.

40 a Write $x^2 - 5x + 1$ in the form $(x - h)^2 - k$.

b Hence write down the equation of the line of symmetry of the graph with equation $y = x^2 - 5x + 1$.

41 A graph has equation $y = 2x^2 + 12x + 23$.

a Write down the coordinates of the point where the graph crosses the $y$-axis.

b Write $2x^2 + 12x + 23$ in the form $a(x + h)^2 + k$.

c Hence find the coordinates of the vertex of the graph.

42

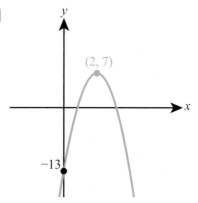

The graph shown in the diagram has equation $y = a(x - h)^2 + k$.

a Write down the values of $h$ and $k$.

b Find the value of $a$.

c Write the equation of the graph in the form $y = ax^2 + bx + c$.

43 a Write $3x^2 + 6x - 2$ in the form $a(x + h)^2 - k$.

b Hence find the minimum value of $3x^2 + 6x - 2$.

44 a Write $5x^2 - 10x + 3$ in the form $a(x - h)^2 + k$.

b Hence find the range of the function $f(x) = 5x^2 - 10x + 3, \ x \in \mathbb{R}$.

# 15B Solving quadratic equations and inequalities

A quadratic equation is an equation of the form $ax^2 + bx + c = 0$. There are three main techniques for finding the solutions to such an equation without using technology.

## ■ Factorizing

If the quadratic can be factorized, this is often the easiest way of solving the equation.

> **WORKED EXAMPLE 15.7**
>
> By first factorizing, find the roots of the equation $x^2 + 5x - 6 = 0$.
>
> Factorize the expression: ............ $(x - 1)(x + 6) = 0$
> look for two numbers which
> multiply to −6 and add up to 5
>
> If the product of the two brackets ............ $x - 1 = 0$ or $x + 6 = 0$
> is zero, then one or other of
> the brackets must be zero
>
> Solve each linear equation for $x$ ............ $x = 1$ or $x = -6$

## ■ Completing the square

If the quadratic does not factorize, one option is to complete the square and then rearrange to solve the equation in the new form.

> **WORKED EXAMPLE 15.8**
>
> a  Express $4x^2 + 24x + 31$ in the form $a(x - h)^2 + k$.
> b  Hence solve the equation $4x^2 + 24x + 31 = 0$, giving your answer in the form $\dfrac{p \pm \sqrt{q}}{r}$ where $p, q, r \in \mathbb{Z}$.
>
> Complete the square ......... **a**  $4x^2 + 24x + 31 = 4(x^2 + 6x) + 31$
> using the method from
> Worked Example 15.5                    $= 4\left[(x + 3)^2 - 9\right] + 31$
>
>                                        $= 4(x + 3)^2 - 5$
>
> Re-write the equation in the ......... **b**  $4x^2 + 24x + 31 = 0$
> form $a(x - h)^2 + k = 0$
>                                        $4(x + 3)^2 - 5 = 0$
>
> Solve by rearranging to ............ $4(x + 3)^2 = 5$
> make $x$ the subject
>                                        $(x + 3)^2 = \dfrac{5}{4}$
>
> Done forget ± when taking ............ $x + 3 = \pm\sqrt{\dfrac{5}{4}}$
> the square root of both sides
>
> Use $\sqrt{4} = 2$ ............ $x + 3 = \pm\dfrac{\sqrt{5}}{2}$
>
>                                        $x = -3 \pm \dfrac{\sqrt{5}}{2}$
>
> Write the answer as ............ $x = \dfrac{-6 \pm \sqrt{5}}{2}$
> a single fraction

## ■ The quadratic formula

If the quadratic doesn't factorize, an alternative to completing the square is to use the quadratic formula.

---

**KEY POINT 15.5**

If $ax^2 + bx + c = 0$, then $x = \dfrac{-b \pm \sqrt{b^2 - 4ac}}{2a}$

---

**Proof 15.1**

Prove that if $ax^2 + bx + c = 0$, then

$$x = \frac{-b \pm \sqrt{b^2 - 4ac}}{2a}.$$

| | |
|---|---|
| Factorize $a$ from the first two terms | $ax^2 + bx + c = 0$ |
| | $a\left[x^2 + \dfrac{b}{a}x\right] + c = 0$ |
| Complete the square inside the brackets | $a\left[\left(x + \dfrac{b}{2a}\right)^2 - \dfrac{b^2}{4a^2}\right] + c = 0$ |
| Multiply $a$ back in | $a\left(x + \dfrac{b}{2a}\right)^2 - \dfrac{b^2}{4a} + c = 0$ |
| Start rearranging to leave $x$ as the subject | $a\left(x + \dfrac{b}{2a}\right)^2 = \dfrac{b^2}{4a} - c$ |
| | $a\left(x + \dfrac{b}{2a}\right)^2 = \dfrac{b^2 - 4ac}{4a}$ |
| Divide by $a$ | $\left(x + \dfrac{b}{2a}\right)^2 = \dfrac{b^2 - 4ac}{4a^2}$ |
| Square root both sides | $x + \dfrac{b}{2a} = \pm\sqrt{\dfrac{b^2 - 4ac}{4a^2}}$ |
| | $x + \dfrac{b}{2a} = \dfrac{\pm\sqrt{b^2 - 4ac}}{2a}$ |
| | $x = \dfrac{-b \pm \sqrt{b^2 - 4ac}}{2a}$ |

---

 Although quadratic equations (in a geometric context at least) were known and solved by Ancient Greek mathematicians, they had different methods for solving equations of the form $x^2 + ax = b$ to $x^2 + b = ax$. It was really the Indian mathematician Brahmagupta, who formalized the study of negative numbers and zero, who allowed all quadratic equations to be treated in the same way.

Your calculator might have a polynomial solver. A quadratic is a special type of polynomial of degree 2, so you should also be able to solve quadratics using technology.

**WORKED EXAMPLE 15.9**

Find the solutions of the equation $2x^2 - 4x - 3 = 0$.

Identify the coefficients $a$, $b$ and $c$ ·········· $a = 2$, $b = -4$, $c = -3$

Substitute into ·········· $x = \dfrac{-(-4) \pm \sqrt{(-4)^2 - 4(2)(-3)}}{2(2)}$
$x = \dfrac{-b \pm \sqrt{b^2 - 4ac}}{2a}$

$= \dfrac{4 \pm \sqrt{16 + 24}}{4}$

$= \dfrac{4 \pm \sqrt{40}}{4}$

Simplify the surd: ·········· $= \dfrac{4 \pm 2\sqrt{10}}{4}$
$\sqrt{40} = \sqrt{4}\sqrt{10} = 2\sqrt{10}$

Use $\sqrt{4} = 2$ ·········· $= \dfrac{2 \pm \sqrt{10}}{2}$

**TOK Links**

Do you have a preference for $\dfrac{4 \pm \sqrt{40}}{4}$, $\dfrac{2 \pm \sqrt{10}}{2}$, $1 \pm \sqrt{\dfrac{5}{2}}$ or 2.58, −0.581? What criteria do you use for deciding which form to leave an answer in? What is the role of aesthetics in mathematics?

## ■ Inequalities

To solve quadratic inequalities, sketch the graph in order to identify the required region(s).

**WORKED EXAMPLE 15.10**

Solve the inequality $x^2 - 3x - 10 > 0$.

Start by finding the roots ·········· $x^2 - 3x - 10 = 0$
of the equation – in this
case it will factorize          $(x + 2)(x - 5) = 0$

                                $x = -2$ or 5

Sketch the graph and highlight
the part where the graph
is positive (where $y > 0$)

There are two parts to ·········· So, $x < -2$ or $x > 5$
the region, so we need
to describe both

Remember what you learned in Chapter 3: the roots of an equation give the $x$-intercepts of the graph.

**Tip**

Normally the inequalities in the solution will be strict (i.e. > or <) if the question has a strict inequality and non-strict (⩽ or ⩾) if the question has a non-strict inequality.

## Exercise 15B

For questions 1 to 6, use the method demonstrated in Worked Example 15.7 to solve the equation by factorizing.

1   a   $x^2 - 5x + 6 = 0$         2   a   $x^2 + x - 6 = 0$         3   a   $x^2 - 6x + 9 = 0$
   b   $x^2 - 8x + 15 = 0$         b   $x^2 - x - 12 = 0$         b   $x^2 + 10x + 25 = 0$

4   a   $2x^2 + 8x + 6 = 0$       5   a   $2x^2 - 2x - 12 = 0$     6   a   $4x^2 + 24x + 36 = 0$
   b   $3x^2 + 9x + 6 = 0$         b   $3x^2 + 3x - 60 = 0$        b   $5x^2 - 20x + 20 = 0$

7   a   $x^2 - 5x = 0$           8   a   $3x^2 + 12x = 0$        9   a   $x^2 - 16 = 0$
   b   $x^2 + 3x = 0$            b   $3x^2 - 9x = 0$          b   $x^2 - 49 = 0$

10   a   $5x^2 - 20 = 0$
    b   $5x^2 - 80 = 0$

For questions 11 to 16, use the method demonstrated in Worked Example 15.8 to solve the equation by completing the square.

11   a   $x^2 - 6x + 3 = 0$       12   a   $x^2 + 10x - 3 = 0$     13   a   $x^2 - 3x + 1 = 0$
    b   $x^2 - 4x - 2 = 0$        b   $x^2 + 8x + 4 = 0$        b   $x^2 - 5x - 2 = 0$

14   a   $x^2 + x - 5 = 0$       15   a   $4x^2 - 24x + 3 = 0$    16   a   $4x^2 + 2x - 3 = 0$
    b   $x^2 + 3x - 1 = 0$        b   $4x^2 - 8x - 1 = 0$       b   $4x^2 + 6x + 1 = 0$

For questions 17 to 24, use the method demonstrated in Worked Example 15.9 to solve the equation by using the quadratic formula. Give your answer in surd form, simplified as far as possible.

17   a   $x^2 + 7x + 3 = 0$       18   a   $x^2 + 5x - 2 = 0$       19   a   $x^2 - 5x + 2 = 0$
    b   $x^2 + 5x + 3 = 0$        b   $x^2 + 3x - 5 = 0$        b   $x^2 - 9x + 4 = 0$

20   a   $x^2 + 8x - 3 = 0$       21   a   $2x^2 + 5x + 1 = 0$     22   a   $5x^2 - 7x - 2 = 0$
    b   $x^2 + 6x - 4 = 0$        b   $2x^2 + 7x + 2 = 0$       b   $5x^2 - 3x - 4 = 0$

23   a   $3x^2 - 10x + 1 = 0$     24   a   $4x^2 + 8x - 3 = 0$
    b   $3x^2 - 8x + 2 = 0$       b   $4x^2 + 4x - 1 = 0$

For questions 25 to 28, use the method demonstrated in Worked Example 15.10 to solve the quadratic inequality.

25   a   $x^2 - 10x + 21 < 0$     26   a   $x^2 - 2x - 15 \leqslant 0$    27   a   $x^2 + 5x + 6 > 0$
    b   $x^2 - 8x + 12 < 0$      b   $x^2 - 3x - 10 \leqslant 0$     b   $x^2 + 7x + 12 > 0$

28   a   $x^2 - 6x \geqslant 0$
    b   $x^2 - 7x \geqslant 0$

29   a   Factorize $x^2 - x - 12$.
    b   Hence solve the equation $x^2 - x - 12 = 0$.

30   a   Write $x^2 - 6x - 2$ in the form $(x - h)^2 - k$.
    b   Find the exact solutions of the equation $x^2 - 6x - 2 = 0$.

31   a   Write $x^2 + 10x$ in the form $(x + h)^2 - k$.
    b   Find the exact roots of the equation $x^2 + 10x = 7$.

32   Solve $x^2 + 5 < 6x$.

33   Solve $x^2 \geqslant b^2$ where $b$ is a positive number.

34   Solve the equation $x^2 - 4x = 21$.

35   A square has side length $x$ cm. The area is $A$ cm$^2$ and the perimeter is $P$ cm. For what values of $x$ is it true that $A > P$?

36   Find the exact solution of the equation $3x^2 = 4x + 1$, simplifying your answer as far as possible.

37   Solve $x^2 - 2xp + p^2 > q^2$ where $q$ is a positive number.

38   Two numbers differ by 3 and their product is 40. Let the smaller of the two numbers be $x$.
    a   Show that $x^2 + 3x - 40 = 0$.
    b   Hence find the possible values of the two numbers.

**39** The length of a rectangle is 5 cm more than its width and the area of the rectangle is 24 cm². Let the width of the rectangle be $x$ cm.

   a  Show that $x^2 + 5x - 24 = 0$.

   b  Hence find the perimeter of the rectangle.

**40** Find the coordinates of the points of intersection of the line $y = x + 2$ with the parabola $y = x^2 - 4$.

**41** Solve the inequality $x^2 + 9 \leqslant 6x$.

**42** While playing golf, Daniel chips a ball onto the green. The ball and the green are both at ground level. The height, $h$ metres, of the ball $t$ seconds after being hit is modelled by $h = 10t - 5t^2$ until it lands.

   a  When does the ball land?

   b  Find the maximum height reached by the ball.

   c  For how long is the ball more than 1 m above the ground?

**43** Find the coordinates of the intersection of the line $y = x - 3$ with the circle $x^2 + y^2 = 16$.

**44** Find the exact solutions of the equation $\dfrac{6}{x-1} - 3x = 4$.

**45** Find all values of $x$ for which $4 < x^2 < 9$.

**46** Find all values of $x$ for which $x^2 \leqslant 3x$ and $5x > x^2 + 4$.

**47** If $2x^2 + y^2 = 3xy$ find $y$ in terms of $x$.

**48** If $x^2 + k^2 - 2kx - x + k = 0$ find $x$ in terms of $k$.

**49** Solve the inequality $x^2 + y^2 + 1 \leqslant 2y$.

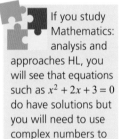

If you study Mathematics: analysis and approaches HL, you will see that equations such as $x^2 + 2x + 3 = 0$ do have solutions but you will need to use complex numbers to find them.

# 15C The discriminant

If you try to solve the equation $x^2 + 2x + 3 = 0$ using the quadratic formula you get

$$x = \frac{-2 \pm \sqrt{2^2 - 4(1)(3)}}{2(1)} = \frac{-2 \pm \sqrt{-8}}{2}$$

Since the square root of a negative number is not a real number, there are no real solutions of this equation. This will happen whenever the expression under the square root, $b^2 - 4ac$, is negative.

This corresponds to a graph that has no $x$-intercepts:

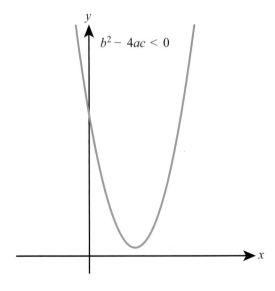

In any equation where $b^2 - 4ac = 0$, the quadratic formula gives

$$x = \frac{-b \pm \sqrt{0}}{2a} = -\frac{b}{2a}$$

which means there is only the one solution.

This corresponds to a graph that touches the $x$-axis at one point:

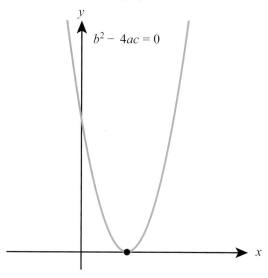

In the remaining case, where $b^2 - 4ac > 0$, there will be two real solutions, corresponding to a graph with two $x$-intercepts:

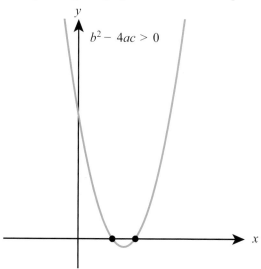

The expression $b^2 - 4ac$ is called the **discriminant,** and is often denoted by Greek capital delta, $\Delta$.

---

**KEY POINT 15.6**

For the quadratic equation $ax^2 + bx + c = 0$,
- if $\Delta > 0$ the equation has two distinct real roots
- if $\Delta = 0$ the equation has two equal real roots
- if $\Delta < 0$ the equation has no real roots

where $\Delta = b^2 - 4ac$.

**WORKED EXAMPLE 15.11**

Find the set of values of $k$ for which the equation $kx^2 + (k-3)x + 1 = 0$ has no real roots.

No real roots means $\Delta < 0$ ················· $b^2 - 4ac < 0$

$a = k$, $b = k - 3$, $c = 1$ ················· $(k-3)^2 - 4k(1) < 0$

$$k^2 - 6k + 9 - 4k < 0$$
$$k^2 - 10k + 9 < 0$$

Solve the quadratic ················· $k^2 - 10k + 9 = 0$
inequality. First find the
roots of the equation    $(k-1)(k-9) = 0$

$$k = 1 \text{ or } 9$$

Then sketch the graph and
highlight the part where the ·········
graph is negative (where $y < 0$)

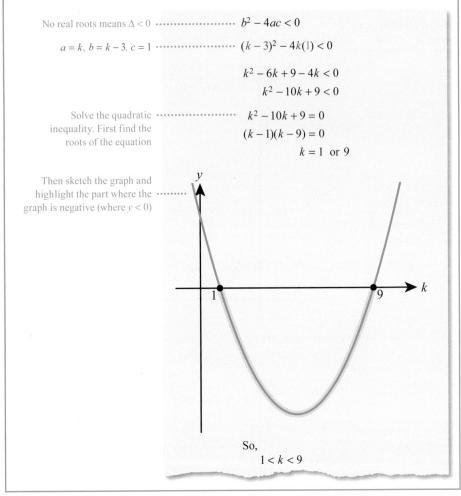

So,
$$1 < k < 9$$

## TOK Links

The discriminant is probably the first time in maths you have met something called an existence proof – establishing when a solution exists, without needing to find exactly what that solution is. This idea is hugely important in many areas of mathematics, physics and economics. Can you think of some situations when you care about a solution existing, even if you do not need to know the details? Does this challenge your preconceptions about what it means when a mathematician says that they have 'found a solution to a problem'?

# Exercise 15C

For questions 1 to 4, evaluate the discriminant and hence determine the number of distinct real solutions of each quadratic equation.

1    a   $x^2 + 4x + 10 = 0$          2    a   $x^2 - 4x + 4 = 0$          3    a   $4x^2 + 3x = 0$

     b   $x^2 + 5x + 2 = 0$               b   $x^2 - 10x + 25 = 0$             b   $5x^2 - 7x = 0$

4    a   $2x^2 - 7 = 0$

     b   $3x^2 + 5 = 0$

For questions 5 to 9, use the method demonstrated in Worked Example 15.11 to find the set of values of $k$ for which the quadratic equation has the given number of distinct real roots.

5    a   $x^2 + 8x + k = 0$, two roots          6    a   $3x^2 - 5x + k = 0$, no roots          7    a   $kx^2 + 12x + 3 = 0$, one root

     b   $x^2 + 6x + k = 0$, two roots               b   $4x^2 - 2x + k = 0$, no roots               b   $kx^2 + 20x + 5 = 0$, one root

8    a   $12x^2 + kx + 3 = 0$, one root          9    a   $8x^2 + kx + 2 = 0$, no roots

     b   $20x^2 + kx + 5 = 0$, one root               b   $3x^2 + kx + 3 = 0$, no roots

10   a   Find the value of $k$ for which the quadratic equation $3x^2 + 8x + k = 0$ has a repeated root.

     b   For this value of $k$, find the root of the equation.

11   The equation $5x^2 + kx + 20 = 0$ has two equal roots. Find the possible values of $k$.

12   Find the set of values of $k$ for which the equation $kx^2 - 3x + 2 = 0$ has two distinct real roots.

13   Find the set of values of $k$ for which the equation $3x^2 - 5x + 2k = 0$ has no real roots.

14   The quadratic equation $2x^2 + kx + (k - 2) = 0$ has equal roots.

     a   Show that $k^2 - 8k + 16 = 0$.

     b   Hence find the values of $k$.

15   The equation $kx^2 + (k + 3)x - 1 = 0$ has one real root.

     a   Show that $k^2 + 10k + 9 = 0$

     b   Hence find the possible values of $k$.

16   Find the set of values of $k$ for which the equation $2x^2 + kx + 2 = 0$ has two distinct real roots.

17   Find the set of values of $a$ for which the equation $ax^2 + ax + 3 = 0$ has at least one real root.

18   The graph of $y = x^2 + ax + 9$ has its vertex on the $x$-axis. Find the possible values of $a$.

19   Find the set of values of $c$ for which the graph of $y = 2x^2 - 3x + c$ does not cross the $x$-axis.

20   Given that the graph of $x^2 + bx + 2b = 0$ lies entirely above the $x$-axis, find the set of possible values of $b$.

21   Find the set of values of $a$ for which the equation $3x^2 + (a + 1)x + 4 = 0$ has two distinct real roots.

22   Find the set of values of $k$ for which the equation $kx^2 - (k - 3)x + k = 0$ has no real roots.

23   Find the range of values of $a$ for which the equation $x + \dfrac{a}{x} = 3$ has real roots.

24   For what values of $a$ is the expression $x^2 + ax + 4$ always positive?

25   The graph of $y = ax^2 + 4x + (a - 3)$ lies entirely below the $x$-axis. Find the set of possible values of $a$.

26   Find the value of $k$ such that the line $y = x + k$ is tangent to the parabola $y = x^2 + 7$.

27   Find the set of values of $k$ for which the line $y = 2x + k$ intersects the circle with equation $x^2 + y^2 = 5$.

28   Show that the equation $3x^2 + (a + 2)x - 2 = 0$ has two distinct roots for all values of $a$.

29   If the equation $x^2 + bx + 9 = 0$ has no solutions for $x$, prove that $x^2 + 9x + b = 0$ has two solutions for $x$.

**TOOLKIT:** Modelling

A section of a stream is modelled by the quadratic function $y = x^2$ for $0 \leq x \leq 1$ where $x$ and $y$ are measured in kilometres. A stick on the surface of the water is observed travelling along the stream at $1\,\mathrm{m\,s^{-1}}$. By approximating the model using a series of straight lines, estimate the time taken for the stick to travel through this section. Is your answer likely to be an overestimate or an underestimate? Why? How can you improve the accuracy of your estimate? What assumptions are you making in your model? Is the accuracy limited more by the mathematical approximation or the modelling assumptions?

## Checklist

■ You should be able to recognize the shape and main features of quadratic graphs.

■ For the quadratic function $f(x) = ax^2 + bx + c$,
  □ the shape of the graph $y = f(x)$ depends on the coefficient $a$:
  □ $a > 0$                                                    □ $a < 0$

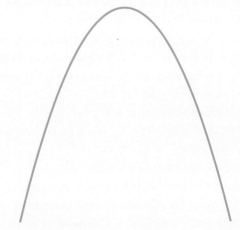

  □ the graph $y = f(x)$ crosses the $y$-axis at $(0, c)$.

■ The quadratic graph $y = a(x - p)(x - q)$ crosses the $x$-axis at $(p, 0)$ and $(q, 0)$.

■ The quadratic graph $y = a(x - h)^2 + k$ has
  □ vertex $(h, k)$
  □ line of symmetry $x = h$

■ You should be able to complete the square:
$$x^2 + bx + c \equiv \left(x + \frac{b}{2}\right)^2 - \left(\frac{b}{2}\right)^2 + c$$

■ You should be able to solve quadratic equations by factorizing, completing the square and using the quadratic formula:
  □ If $ax^2 + bx + c = 0$, then
$$x = \frac{-b \pm \sqrt{b^2 - 4ac}}{2a}$$

■ You should be able to solve quadratic inequalities.

■ You should be able to use the discriminant, $\Delta$, to determine the number of roots of a quadratic equation:
  □ For the quadratic equation $ax^2 + bx + c = 0$,
  □ if $\Delta > 0$ the equation has two distinct real roots
  □ if $\Delta = 0$ the equation has two equal real roots
  □ if $\Delta < 0$ the equation has no real roots
  where $\Delta = b^2 - 4ac$.

## ■ Mixed Practice

**1** **a** Factorize $x^2 + 7x - 18$.
   **b** Hence, write down the coordinates of the point where the graph of $y = x^2 + 7x - 18$ crosses the $x$-axis.

**2** Find the exact coordinates of the points where the graph of $y = 4x^2 - 9$ crosses the $x$-axis.

**3** **a** Write $-2x^2 + 8x - 3$ in the form $a(x - h)^2 + k$.
   **b** Hence, find the coordinates of the vertex of the parabola with equation $y = -2x^2 + 8x - 3$.

**4** Match each equation to the corresponding graph.
   **a** $y = 2x^2 - 4x - 3$
   **b** $y = x^2 + x + 5$
   **c** $y = -2x^2 - x + 3$

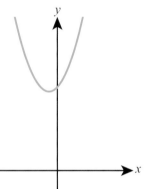
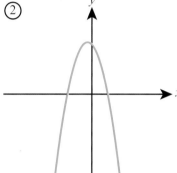
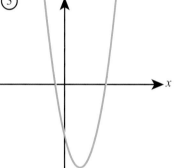

**5** The graph shown in the diagram has equation $y = (x - a)^2 - b$.

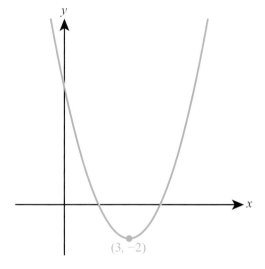

   **a** Write down the value of $a$ and the value of $b$.
   **b** Find the coordinates of the points where the graph crosses the $x$-axis.

**6** A parabola has equation $y = x^2 - 2x - 15$.
   **a** Find the coordinates of the points where the graph intersects the $x$-axis.
   **b** Find the equation of the line of symmetry of the graph.
   **c** Find the minimum value of $x^2 - 2x - 15$.

**7** For the quadratic equation $2x^2 - x - 5 = 0$,
  **a** Calculate the discriminant.
  **b** State the number of distinct real roots of the equation.

**8** Find the set of values of $k$ for which the equation $kx^2 + 4x + 5 = 0$ has no real roots.

**9** $y = f(x)$ is a quadratic function. The graph of $f(x)$ intersects the $y$-axis at the point $A(0, 6)$ and the $x$-axis at the point $B(1, 0)$. The vertex of the graph is at the point $C(2, -2)$.
  **a** Write down the equation of the axis of symmetry.
  **b** Sketch the graph of $y = f(x)$ for $0 \leqslant x \leqslant 4$. Mark clearly on the sketch the points $A$, $B$ and $C$.
  The graph of $y = f(x)$ intersects the $x$-axis for a second time at point $D$.
  **c** Write down the $x$-coordinate of point $D$.

Mathematical Studies SL May 2012 Paper 1 TZ1 Q4

**10 a** Write $3x^2 - 6x + 10$ in the form $a(x - h)^2 + k$.
  **b** Hence find the range of the function $f(x) = 3x^2 - 6x + 10$, $x \in \mathbb{R}$.

**11 a** Write $-x^2 - 4x + 3$ in the form $a(x - h)^2 + k$.
  **b** Hence find the range of the function $f(x) = -x^2 - 4x + 3$, $x \in \mathbb{R}$.

**12** The equation $2x^2 - (k + 2)x + 3 = 0$ has equal roots. Find the possible values of $k$, giving your answer in simplified surd form.

**13** A rectangle has perimeter $12\,cm$. Let the width of the rectangle be $x\,cm$.
  **a** Show that the area of the rectangle is given by $6x - x^2$.
  **b** By completing the square, find the maximum possible value of the area.

**14** Find the set of values of $k$ for which the equation $3x^2 - kx + 6 = 0$ has at least one real root.

**15** The $x$-coordinate of the minimum point of the quadratic function $f(x) = 2x^2 + kx + 4$ is $x = 1.25$.
  **a i** Find the value of $k$.
     **ii** Calculate the $y$-coordinate of this minimum point.
  **b** Sketch the graph of $y = f(x)$ for the domain $-1 \leqslant x \leqslant 3$.

Mathematical Studies SL November 2011 Paper 1 Q13

**16** The following diagram shows part of the graph of a quadratic function f.

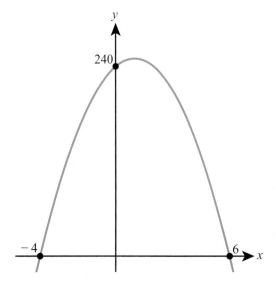

The $x$-intercepts are at $(-4, 0)$ and $(6, 0)$, and the $y$-intercept is at $(0, 240)$.
  **a**  Write down f$(x)$ in the form f$(x) = -10(x - p)(x - q)$.
  **b**  Find another expression for f$(x)$ in the form f$(x) = -10(x - h)^2 + k$.
  **c**  Show that f$(x)$ can also be written in the form f$(x) = 240 + 20x - 10x^2$.

Mathematics SL May 2011 Paper 1 TZ2 Q9

**17**  Let f$(x) = px^2 + (10 - p)x + \dfrac{5}{4}p - 5$.

  **a**  Show that the discriminant of f$(x)$ is $100 - 4p^2$.
  **b**  Find the values of $p$ so that f$(x) = 0$ has two **equal** roots.

Mathematics SL May 2015 Paper 1 TZ1 Q6

**18 a**  Write $x^2 - 6x + 15$ in the form $(x - a)^2 + b$.
   **b**  Hence find the maximum value of $\dfrac{3}{x^2 - 6x + 15}$.

**19**  Find the values of $c$ for which the line with equation $y = 2x + c$ is tangent to the circle $x^2 + y^2 = 3$.

**20**  Find the set of values of $a$ for which the equation $ax + \dfrac{1}{x} = 2$ has no real roots.

**21**  Show that the parabola with equation $y = 2x^2 - 4kx + 3k^2$ lies entirely above the $x$-axis for all values of $k$.

# **16** Analysis and approaches: Graphs

## ESSENTIAL UNDERSTANDINGS

■ Creating different representations of functions to model relationships between variables, visually and symbolically, as graphs, equations and tables represents different ways to communicate mathematical ideas.

### In this chapter you will learn...

■ how to apply single transformations to graphs: translations, stretches or reflections
■ how to apply composite transformations to graphs
■ about the reciprocal function $f(x) = \dfrac{1}{x}$ and its graph
■ about rational functions of the form $f(x) = \dfrac{ax + b}{cx + d}$ and their graphs
■ about exponential functions and their graphs
■ about logarithmic functions and their graphs.

### CONCEPTS

The following concept will be addressed in this chapter:

■ The parameters in a function or equation correspond to geometrical features of a graph and can represent physical quantities in **spatial** dimensions.

## PRIOR KNOWLEDGE

Before starting this chapter, you should already be able to complete the following:

1 For the function $f(x) = \sqrt{x + 1}$, state

   a the largest possible domain

   b the range for the domain in part **a**.

2 State the equation of the vertical and horizontal asymptotes of the function shown on the right.

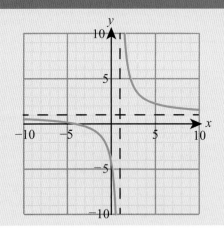

■ **Figure 16.1** How many types of symmetry are there?

3 Simplify

 a $3^{2\log_3 x}$

 b $\ln(e^x)^4$

4 The graph $y = f(x)$ is shown on the right.
 Sketch the graph of $y = f^{-1}(x)$.

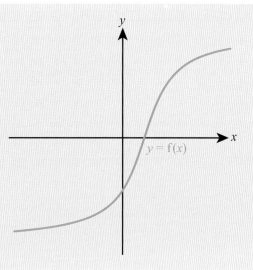

$y = f(x)$

Graphs are an alternative way of representing functions. Being able to switch between an algebraic representation of a function and a graphical representation can open up a much wider variety of tools for approaching some mathematical problems.

One of the most useful skills is being able to transform the graphs of simple, familiar functions to graphs of more complicated functions by understanding the effect of changing the function in different ways.

## Starter Activity

Look at the pictures in Figure 16.1. In small groups discuss what type of symmetry you see.

**Now look at this problem:**

Use your GDC or a dynamic graphing package to investigate transformations of $f(x) = x^2 - 4x$.

a Draw the graph of $y = f(x)$

b Draw each of the following, and describe their relationship to $y = f(x)$:

  i $y = f(x) + 3$    ii $y = f(x + 3)$    iii $y = 2f(x)$    iv $y = f(2x)$

c What happens if you apply i and then ii? Does the order matter?

d What happens if you apply i and then iii? Does the order matter?

# 16A Transformations of graphs

It is useful to be able to take a familiar graph such as $y = \dfrac{1}{x}$ and transform it into a more complicated graph. You need to be familiar with three types of transformation and how they affect the equation of a graph: **reflections** in the $x$-axis and $y$-axis, **translations** and **stretches**.

🌐 The standard Cartesian coordinate system of $x$ and $y$ axes meeting at an origin is usually attributed to the French mathematician René Descartes. Modern philosophers have suggested that this way of describing graphs is actually a reflection of Western European language and philosophy, where position tends to be described relative to the observer (for example, 'that stool is 5 metres in front of me'). It has been suggested that in other cultures this is not the convention and positions tend to be described relative to both the observer and the audience. For example, in Maori culture it would be usual to say 'that stool is 5 metres away from me and 2 metres away from you'. For this reason, it has been suggested that the natural coordinate system used in these cultures would require two origins.

## ▮ Translations

Adding a constant to a function translates its graph vertically:

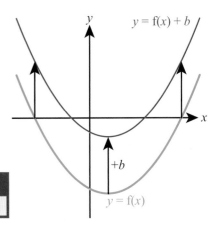

**Tip**

Remember that $b$ can be negative. This results in a downward translation.

---

**KEY POINT 16.1**

$y = \mathrm{f}(x) + b$ is a vertical translation by $b$ of $y = \mathrm{f}(x)$.

---

Replacing $x$ with $x + c$ in a function translates its graph horizontally:

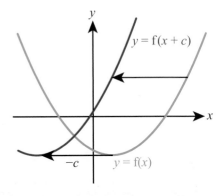

**Tip**

Notice that $y = \mathrm{f}(x + c)$ moves **to the left** by $c$, and not to the right by $c$.

---

**KEY POINT 16.2**

$y = \mathrm{f}(x + c)$ is a horizontal translation by $-c$ of $y = \mathrm{f}(x)$.

**Proof 16.1**

Prove that $y = f(x + c)$ is a horizontal translation by $-c$ of $y = f(x)$.

Define a point $(x_1, y_1)$ on the original curve and a point $(x_2 + c, y_2)$ on the transformed curve ............ Let,

$$y_1 = f(x_1)$$
$$y_2 = f(x_2 + c)$$

One way to make the two points have the same height $(y)$ is to set $x_1 = x_2 + c$ ............ If $x_1 = x_2 + c$, then $y_1 = y_2$.

Rearrange to make $x_2$ the subject ............ Equivalently, if $x_2 = x_1 - c$, then $y_1 = y_2$.

Interpret this graphically ............ So, the equivalent point to $(x_1, y)$ on $y = f(x)$ is $(x_1 - c, y)$ on $y = f(x + c)$, which means the points on $y = f(x)$ have been translated horizontally by $-c$ from $y = f(x + c)$.

**WORKED EXAMPLE 16.1**

The graph of $y = f(x)$ is shown below.

Sketch the graph $y = f(x) - 4$.

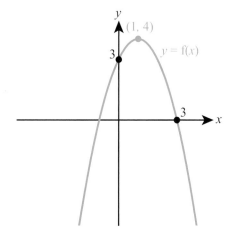

$y = f(x) - 4$ is a vertical translation by $-4$

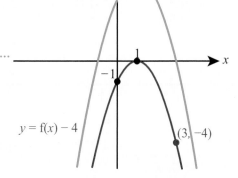

**WORKED EXAMPLE 16.2**

The graph of $y = x^2 - 3x + 5$ is translated to the left by 2 units.

Find the equation of the resulting graph in the form $y = ax^2 + bx + c$.

Let $f(x) = x^2 - 3x + 5$

A translation to the left ............................ Then the new graph is
(or horizontally by $-2$)                                    $y = f(x + 2)$
means $y = f(x + 2)$                                            $= (x + 2)^2 - 3(x + 2) + 5$

Expand and simplify ...............................            $= x^2 + 4x + 4 - 3x - 6 + 5$

$= x^2 + x + 3$

■ Stretches

Multiplying a function by a constant stretches its graph vertically:

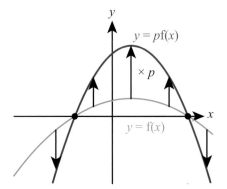

**Tip**

Remember that $p$ can be a fraction less than one.

**KEY POINT 16.3**

$y = pf(x)$ is a vertical stretch with scale factor $p$ of $y = f(x)$.

Replacing $x$ with $qx$ in a function stretches its graph horizontally:

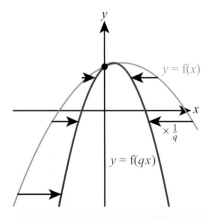

**KEY POINT 16.4**

$y = f(qx)$ is a horizontal stretch with scale factor $\frac{1}{q}$ of $y = f(x)$.

**WORKED EXAMPLE 16.3**

The graph of $y = f(x)$ is shown below.

Sketch the graph $y = f\left(\frac{1}{3}x\right)$.

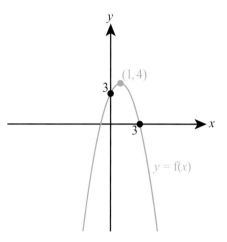

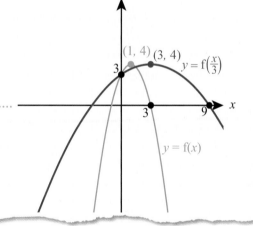

$y = f\left(\frac{1}{3}x\right)$ is a horizontal stretch with scale factor $\dfrac{1}{\left(\frac{1}{3}\right)} = 3$

**WORKED EXAMPLE 16.4**

Describe the transformation that maps the graph of $y = 2x^2 - 8x - 5$ to the graph of $y = x^2 - 4x - 2.5$.

Let $f(x) = 2x^2 - 8x - 5$

Relate the second equation to the first

Then,
$$y = x^2 - 4x - 2.5$$
$$= \frac{1}{2}(2x^2 - 8x - 5)$$

Express this in function notation

$$= \frac{1}{2}f(x)$$

State the transformation

Vertical stretch with scale factor $\dfrac{1}{2}$.

## ■ Reflections

Multiplying a function by −1 reflects its graph in the *x*-axis:

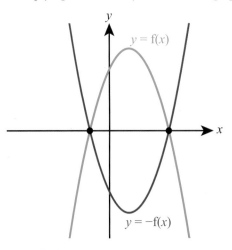

$y = -\text{f}(x)$ is a reflection in the *x*-axis of $y = \text{f}(x)$.

Replacing *x* with −*x* in a function reflects its graph in the *y*-axis:

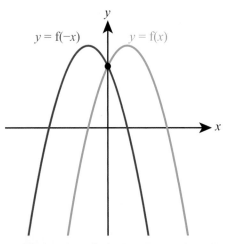

**Tip**

Notice that the negative sign inside the brackets affects the *x*-coordinates. This is consistent with translations and stretches where any alterations inside the brackets result in horizontal movements of the graph.

$y = \text{f}(-x)$ is a reflection in the *y*-axis of $y = \text{f}(x)$.

**WORKED EXAMPLE 16.5**

The graph of $y = \text{f}(x)$ has a single vertex at (5, −2).

Find the coordinates of the vertex on the graph of $y = \text{f}(-x)$.

$y = \text{f}(-x)$ is a reflection in the
*y*-axis. This will just multiply ·····················Vertex of $y = \text{f}(-x)$ is at (−5, −2)
the *x*-coordinate by −1

## Composite transformations

In Chapter 14, you saw how the order in which the functions in a composite function are applied can affect the result. The same is true for the order in which transformations are applied.

If you study Mathematics: analysis and approaches HL, you will see what happens when two horizontal transformations are applied.

### KEY POINT 16.7

- When two vertical transformations are applied, the order matters: $y = pf(x) + b$ is a vertical stretch with scale factor $p$ followed by a vertical translation by $b$.
- When one vertical and one horizontal transformation are applied, the order does not matter.

### WORKED EXAMPLE 16.6

The diagram shows the graph of $y = f(x)$.

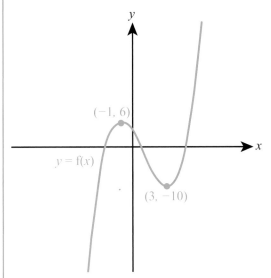

On separate sets of axes, sketch the graphs of

a   $y = 3f(x - 2)$                    b   $y = -f(x) + 4.$

There is one vertical and one horizontal transformation, so it does not matter which order you apply them in

Translate horizontally by +2…

**a**

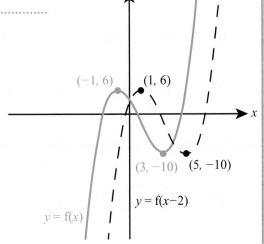

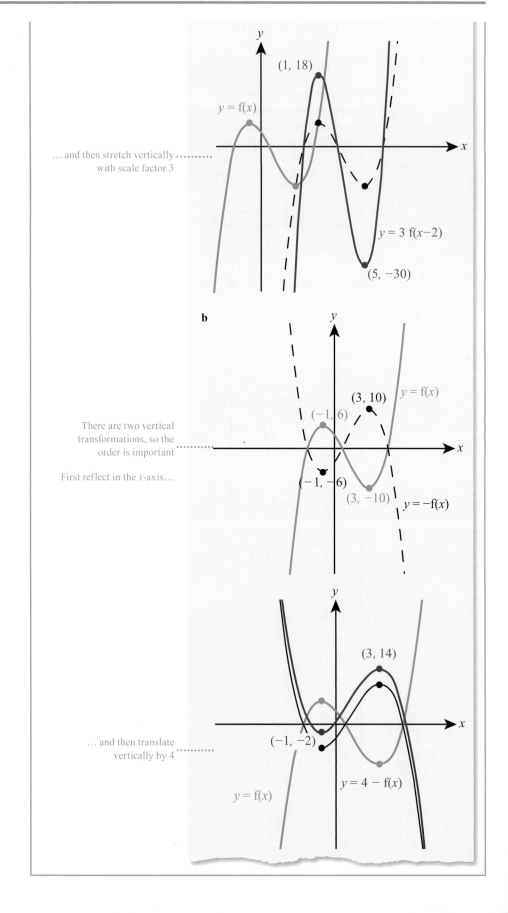

... and then stretch vertically with scale factor 3

$y = f(x)$

(1, 18)

$y = 3\,f(x-2)$

(5, −30)

**b**

There are two vertical transformations, so the order is important

First reflect in the *x*-axis...

(−1, 6)

(3, 10)

$y = f(x)$

(−1, −6)

(3, −10)

$y = -f(x)$

... and then translate vertically by 4

(3, 14)

(−1, −2)

$y = 4 - f(x)$

$y = f(x)$

**WORKED EXAMPLE 16.7**

Describe a sequence of transformations that maps the graph of $y = x^3 - x$ to the graph of $y = 2x^3 - 2x - 1$.

Let $f(x) = x^3 - x$

Express the second equation ·················· Then,
in function notation,
related to the first
$$y = 2x^3 - 2x - 1$$
$$= 2(x^3 - x) - 1$$
$$= 2f(x) - 1$$

State the transformation, ·················· Vertical stretch with scale factor 2,
making sure the stretch followed by vertical translation by $-1$
comes before the translation

**You are the Researcher**

Some graphs do not change under transformations. For example, the graph $y = x$ looks exactly the same after a horizontal stretch factor 2 and a vertical stretch factor 2. Graphs with these properties are studied in an area of maths that you might be interested in researching, called fractals. Fractals have found a wide variety of applications from data storage to art.

## Exercise 16A

For questions 1 to 4, use the method demonstrated in Worked Example 16.1 to sketch the required graph.
The graph of $y = f(x)$ is given below.

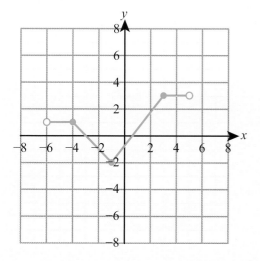

1  a  $y = f(x) + 3$
   b  $y = f(x) + 4$
4  a  $y = f(x - 3)$
   b  $y = f(x - 2)$

2  a  $y = f(x) - 2$
   b  $y = f(x) - 5$

3  a  $y = f(x + 4)$
   b  $y = f(x + 1)$

For questions 5 to 8, use the method demonstrated in Worked Example 16.2 to find the equation of the graph after the given transformation is applied.

5   a   $y = 3x^2$ after a translation of 3 units vertically up

    b   $y = 2x^3$ after a translation of 5 units vertically up

6   a   $y = 8x^2 - 7x + 1$ after a translation of 5 units vertically down

    b   $y = 8x^2 - 7x + 1$ after a translation of 2 units vertically down

7   a   $y = 4x^2$ after a translation of 3 units to the right

    b   $y = 3x^3$ after a translation of 6 units to the right

8   a   $y = x^2 + 6x + 2$ after a translation of 3 units to the left

    b   $y = x^2 + 5x + 4$ after a translation of 2 units to the left

The graph of $y = \mathrm{f}(x)$ is given below.

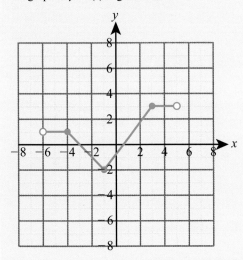

For questions 9 to 12, use the method demonstrated in Worked Example 16.3 to sketch the required graph.

9   a   $y = 2\mathrm{f}(x)$          10   a   $y = \dfrac{1}{2}\mathrm{f}(x)$          11   a   $y = \mathrm{f}(3x)$          12   a   $y = \mathrm{f}\left(\dfrac{x}{2}\right)$

    b   $y = 3\mathrm{f}(x)$             b   $y = \dfrac{1}{3}\mathrm{f}(x)$            b   $y = \mathrm{f}(2x)$             b   $y = \mathrm{f}\left(\dfrac{x}{3}\right)$

For questions 13 to 16, use the method demonstrated in Worked Example 16.4 to describe the transformation that maps the graph of $y = 12x^2 - 6x$ to the graph with the given equation.

13   a   $y = 36x^2 - 18x$      14   a   $y = 6x^2 - 3x$      15   a   $y = 48x^2 - 12x$      16   a   $y = 3x^2 - 3x$

     b   $y = 24x^2 - 12x$        b   $y = 4x^2 - 2x$         b   $y = 108x^2 - 18x$       b   $y = \dfrac{4}{3}x^2 - 2x$

For questions 17 to 20, use the method demonstrated in Worked Example 16.5.

17   a   The graph of $y = \mathrm{f}(x)$ has a vertex at (2, 3). Find the coordinates of the vertex of the graph of $y = -\mathrm{f}(x)$.

    b   The graph of $y = \mathrm{f}(x)$ has a vertex at (5, 1). Find the coordinates of the vertex of the graph of $y = -\mathrm{f}(x)$.

18   a   The graph of $y = \mathrm{f}(x)$ has a vertex at (−2, −4). Find the coordinates of the vertex of the graph of $y = -\mathrm{f}(x)$.

    b   The graph of $y = \mathrm{f}(x)$ has a vertex at (−3, −1). Find the coordinates of the vertex of the graph of $y = -\mathrm{f}(x)$.

19   a   The graph of $y = \mathrm{f}(x)$ has a vertex at (2, 3). Find the coordinates of the vertex of the graph of $y = \mathrm{f}(-x)$.

    b   The graph of $y = \mathrm{f}(x)$ has a vertex at (1, 5). Find the coordinates of the vertex of the graph of $y = \mathrm{f}(-x)$.

20 a The graph of $y = f(x)$ has a vertex at $(-2, -3)$. Find the coordinates of the vertex of the graph of $y = f(-x)$.

   b The graph of $y = f(x)$ has a vertex at $(-5, -1)$. Find the coordinates of the vertex of the graph of $y = f(-x)$.

For questions 21 to 24, use the method demonstrated in Worked Example 16.6 to sketch the required graph.

The graph of $y = f(x)$ is given below.

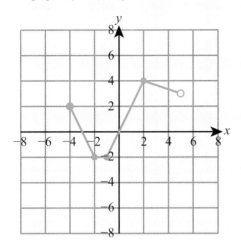

| | | | |
|---|---|---|---|
| 21 a $y = f(x - 2) + 3$ | 22 a $y = 3f(2x)$ | 23 a $y = 4f(-x)$ | 24 a $y = 5 - 3f(x)$ |
| b $y = f(x + 2) - 3$ | b $y = 2f(3x)$ | b $y = 2f(-x)$ | b $y = 3 - 2f(x)$ |

For questions 25 to 28, use the method demonstrated in Worked Example 16.7 to describe a sequence of transformation that maps the graph of $y = 4x^2 + x$ to the graph with the given equation.

25 a $y = 12x^2 + 3x - 2$      26 a $y = x^2 + \dfrac{1}{2}x - 1$      27 a $y = 4(x - 2)^2 + (x - 2) + 5$

   b $y = 20x^2 + 5x + 1$         b $y = x^2 + \dfrac{1}{2}x + 3$         b $y = 4(x + 3)^2 + (x + 3) - 4$

28 a $y = 36x^2 - 3x$

   b $y = 16x^2 - 2x$

29 The graph of $y = f(x)$ is shown below.

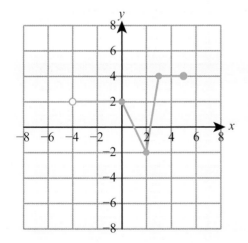

On separate diagrams, sketch the graph of

a $y = f(x - 2)$                    b $y = 2f(x)$                    c $y = f\left(\dfrac{1}{2}x\right) - 1$

**30** The graph of $y = 3x^2 - 4x$ is translated 3 units in the negative $x$-direction and then stretched vertically with scale factor 4. Find the equation of the resulting graph, giving your answer in the form $y = ax^2 + bx + c$.

**31** The graph of $y = e^x$ is translated horizontally 2 units in the positive direction and then stretched vertically with scale factor 3. Find the equation of the resulting graph.

**32** a Write $x^2 - 10x + 11$ in the form $(x - h)^2 + k$.

   b Hence describe a sequence of two transformations that map the graph of $y = x^2$ to the graph of $y = x^2 - 10x + 11$.

**33** a Write $5x^2 + 30x + 45$ in the form $a(x + h)^2$.

   b Hence describe a sequence of two transformations that map the graph of $y = x^2$ to the graph of $y = 5x^2 + 30x + 45$.

**34** The graph of $y = x^2$ is stretched vertically with scale factor 9.

   a Write down the equation of the resulting graph.

   b Find the scale factor of a horizontal stretch that maps the graph from part **a** back to the graph of $y = x^2$.

**35** a Find the equation of the graph obtained from the graph of $y = 2x^3$ by stretching it vertically with scale factor 8.

   b Find the scale factor of a horizontal stretch that has the same effect on the graph of $y = 2x^3$.

**36** The graph of $y = 2x^3 - 5x^2$ is reflected in the $x$-axis and then reflected in the $y$-axis. Find the equation of the resulting graph.

**37** The graph of $y = f(x)$ is stretched vertically with scale factor 5 and then translated 3 units up forming the graph of $y = g(x)$.

   a Write down the equation of the resulting graph.

   b The graph of $y = g(x)$ is instead formed from the graph of $y = f(x)$ by translating upward first and then stretching vertically. Describe fully the translation and the stretch.

**38** The graph of $y = f(x)$ is stretched vertically by a factor of 2 away from the line $y = 1$ (that is, $y = 1$ is kept invariant rather than the $x$-axis). Find the equation of the resulting graph.

# 16B Rational functions

A linear rational function is any function that can be written as a fraction where both the numerator and denominator are linear expressions.

## ■ The reciprocal function, $f(x) = \dfrac{1}{x}$

One particularly important function to be familiar with is the reciprocal function, $f(x) = \dfrac{1}{x}$. The graph $y = \dfrac{1}{x}$ is a curve called a **hyperbola**.

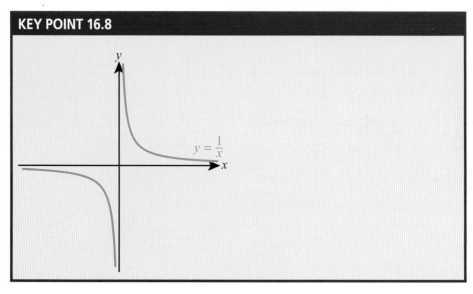

**KEY POINT 16.8**

$y = \dfrac{1}{x}$

You met asymptotes in Section 3B.

The domain of $f(x) = \dfrac{1}{x}$ is $x \in \mathbb{R}$, $x \neq 0$.

The range of $f(x) = \dfrac{1}{x}$ is $f(x) \in \mathbb{R}$, $f(x) \neq 0$.

Notice that the curve has asymptotes at $x = 0$ (the $y$-axis) and $y = 0$ (the $x$-axis).

Notice also that the graph is symmetrical about the line $y = x$, that is, if you reflect $y = \dfrac{1}{x}$ in the line $y = x$ you get $y = \dfrac{1}{x}$ again.

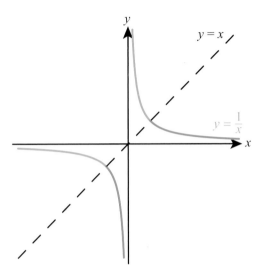

Since any inverse function is the reflection of the original function in the line $y = x$, this means that the inverse function of $f(x) = \dfrac{1}{x}$ is $f^{-1}(x) = \dfrac{1}{x}$. We say that $f(x) = \dfrac{1}{x}$ is a **self-inverse function**.

## ■ Rational functions of the form $f(x) = \dfrac{ax+b}{cx+d}$

The graph of any function of the form $f(x) = f(x) = \dfrac{ax+b}{cx+d}$ is the same shape as the graph of $f(x) = \dfrac{1}{x}$ (it is a hyperbola) but it will be in a different position on the axes as determined by the coefficients $a$ $b$, $c$ and $d$.

■ The $y$-intercept occurs when $x = 0$, which gives $y = \dfrac{b}{d}$.

■ The $x$-intercept occurs when $y = 0$, which gives $x = -\dfrac{b}{a}$.

■ The vertical asymptote occurs when the denominator is 0, which gives $x = -\dfrac{d}{c}$.

■ The horizontal asymptote occurs for large (positive or negative) values of $x$. The coefficients $b$ and $d$ become insignificant and so $\dfrac{ax+b}{cx+d}$ tends to $\dfrac{ax}{cx} = \dfrac{a}{c}$.

**KEY POINT 16.9**

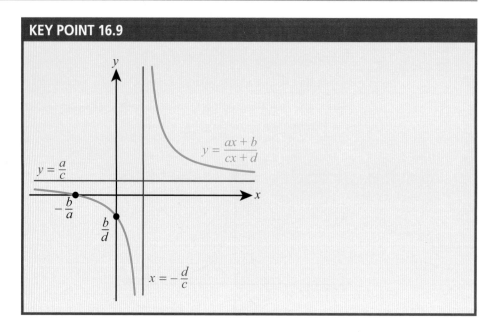

 **WORKED EXAMPLE 16.8**

Sketch the graph of $y = \dfrac{2x + 3}{x + 4}$, clearly labelling all asymptotes and axis intercepts.

Start by putting in the axis intercepts and asymptotes

The axis intercepts are at $\left(-\dfrac{b}{a}, 0\right)$ and $\left(0, \dfrac{b}{d}\right)$ and the asymptotes are $x = -\dfrac{d}{c}$ and $y = \dfrac{a}{c}$, where $a = 2, b = 3, c = 1, d = 4$

You know that the shape is a hyperbola, so you can now fit the curve in accordingly

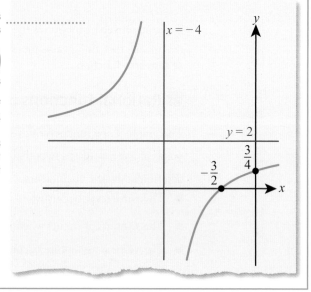

 **Tip**

When people use their calculators to sketch these type of functions, they often miss out the top left branch. You should remember that hyperbolas have two branches.

# Exercise 16B

For questions 1 to 4, use the method demonstrated in Worked Example 16.8 to sketch the graphs of these rational functions, indicating the asymptotes and the axis intercepts.

1   a   $y = \dfrac{2x + 3}{x + 1}$

   b   $y = \dfrac{3x + 1}{x + 2}$

2   a   $y = \dfrac{3x - 1}{2x + 1}$

   b   $y = \dfrac{4x - 3}{3x + 1}$

3   a   $y = \dfrac{3}{2x + 1}$

   b   $y = \dfrac{2}{3x - 1}$

4   a   $y = \dfrac{x}{x - 5}$

   b   $y = \dfrac{x}{x + 3}$

5   a   Sketch the graph of $y = \dfrac{1}{x}$, stating the equation of its asymptotes.

   b   Describe fully a transformation that transforms the graph of $y = \dfrac{1}{x}$ to the graph of $y = \dfrac{1}{x} + 2$.

   c   Hence sketch the graph of $y = \dfrac{1}{x} + 2$, stating the equations of its asymptotes.

6   a   State the equations of the asymptotes of the graph of $y = \dfrac{1}{x}$.

   b   Describe fully a transformation that transforms the graph of $y = \dfrac{1}{x}$ to the graph of $y = \dfrac{1}{x - 3}$.

   c   Hence state the equations of the asymptotes of the graph of $y = \dfrac{1}{x - 3}$.

7   a   Describe fully a transformation that transforms the graph of $y = \dfrac{1}{x}$ to the graph of $y = \dfrac{1}{x + 2}$.

   b   Hence sketch $y = \dfrac{1}{x}$ and $y = \dfrac{1}{x + 2}$ on the same diagram, indicating clearly the equations of any asymptotes.

8   For the graph of $y = \dfrac{x + 3}{x - 2}$,

   a   find the axis intercepts

   b   find the equations of the asymptotes

   c   sketch the graph, including all the information from parts **a** and **b**.

9   For the graph of $y = \dfrac{2x + 3}{x + 1}$

   a   find the equations of the asymptotes       b   find the $x$-intercept.

10   Sketch the graph of $y = \dfrac{3x - 1}{2x + 3}$, stating the equations of all asymptotes and coordinates of axis intercepts.

11   Sketch the graph of $y = \dfrac{4x + 3}{3x - 1}$, stating the equations of all asymptotes and coordinates of axis intercepts.

12   a   Write $\dfrac{2x + 5}{x}$ in the form $a + \dfrac{b}{x}$.

   b   Describe a sequence of two transformations that transform the graph of $y = \dfrac{1}{x}$ to the graph of $y = \dfrac{2x + 5}{x}$.

   c   Hence state the equations of the asymptotes of the graph of $y = \dfrac{2x + 5}{x}$.

13   a   Show that $\dfrac{2x - 5}{x - 3} = 2 + \dfrac{1}{x - 3}$.

   b   Describe a sequence of two transformations that transform the graph of $y = \dfrac{1}{x}$ to the graph of $y = \dfrac{2x - 5}{x - 3}$.

   c   Hence sketch the graph of $y = \dfrac{2x - 5}{x - 3}$, indicating the positions of any asymptotes.

14   a   Write $\dfrac{5x - 1}{x}$ in the form $a - \dfrac{b}{x}$.

   b   Describe a sequence of two transformations that transform the graph of $y = \dfrac{1}{x}$ to the graph of $y = \dfrac{5x - 1}{x}$.

   c   Hence sketch the graph of $y = \dfrac{5x - 1}{x}$, stating the axis intercepts and equations of the asymptotes.

15   a   Simplify $\dfrac{x^2 + 2x}{x^2 - 4}$.

   b   Hence sketch the graph of $y = \dfrac{x^2 + 2x}{x^2 - 4}$.

16   Sketch $y = \dfrac{x - a}{x - b}$ in the case where

   a   $0 < a < b$       b   $0 < b < a$

# 16C Exponential and logarithmic functions

The graph of $y = a^x$, $a > 0$, has one of two shapes depending on whether $a > 1$ or $0 < a < 1$.

**KEY POINT 16.10**

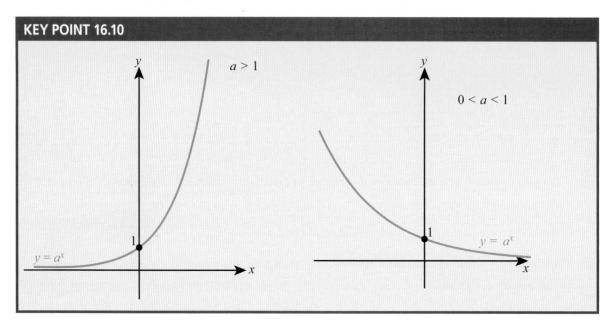

## Tip

Actually, there is another graph where a = 1 but this is a special case which is not considered an exponential graph.

Notice that in both cases the graph passes through the point (0, 1) since $a^0 = 1$ for all $a$. Also the $x$-axis is an asymptote.

Larger values of $a > 1$ result in a graph that grows more quickly:

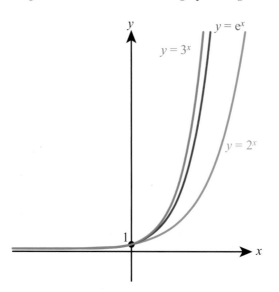

**WORKED EXAMPLE 16.9**

Sketch the graph of $y = 4^{-x}$, stating the equation of any axis intercepts and asymptotes.

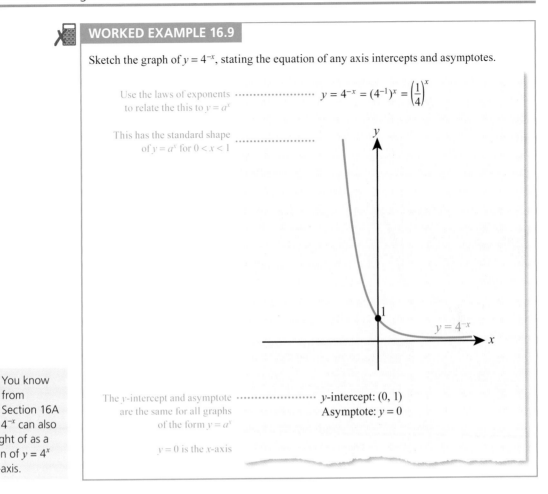

Use the laws of exponents to relate the this to $y = a^x$ ⋯⋯⋯⋯⋯⋯⋯⋯⋯⋯ $y = 4^{-x} = (4^{-1})^x = \left(\frac{1}{4}\right)^x$

This has the standard shape of $y = a^x$ for $0 < x < 1$ ⋯⋯⋯⋯⋯⋯⋯⋯⋯⋯

The *y*-intercept and asymptote are the same for all graphs of the form $y = a^x$ ⋯⋯⋯⋯⋯⋯⋯⋯⋯⋯ *y*-intercept: $(0, 1)$
Asymptote: $y = 0$

$y = 0$ is the *x*-axis

You know from Section 16A that $y = 4^{-x}$ can also be thought of as a reflection of $y = 4^x$ in the *y*-axis.

The graph of $y = \log_a x$, where $a, x > 0$ has many similar properties.

Again, it has one of two shapes depending on whether $a > 1$ or $0 < a < 1$.

**KEY POINT 16.11**

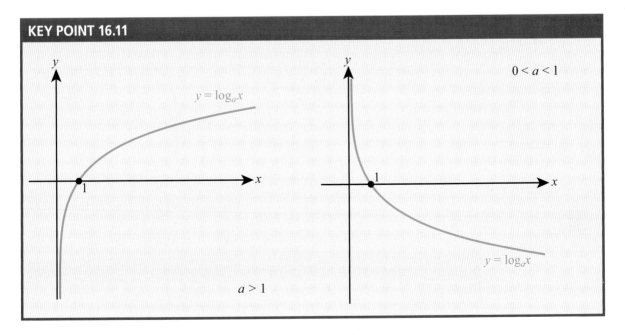

Notice that in both cases the graph passes through the point (1, 0) since $\log_a 1 = 0$ for all $a$, and that the $y$-axis is an asymptote.

Larger values of $a > 1$ result in a graph that grows less quickly:

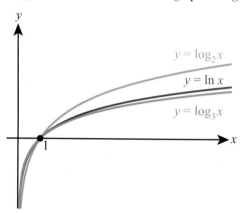

You met these relationships in Key Point 1.5, and you know from Key Point 14.1 that this means one is the inverse function of the other since $(f^{-1} \circ f)(x) = x$.

It should not be too surprising that there are many similarities between the graphs of $y = a^x$ and $y = \log_a x$ as you can see that they are inverse functions.

**KEY POINT 16.12**

$a^{\log_a x} = x$ and $\log_a a^x = x$

**WORKED EXAMPLE 16.10**

On the same set of axes, sketch the graphs $y = 3^x$ and $y = \log_3 x$, clearly showing the relationship between them.

$y = 3^x$ and $y = \log_3 x$ are inverse functions so one is a reflection of the other in the line $y = x$ .................

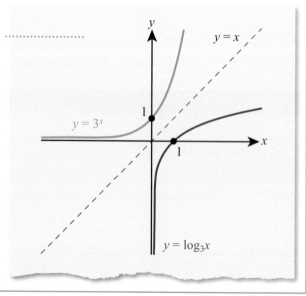

You saw this idea used when changing the base of a logarithm in Section 12B.

Sometimes it can be useful to change an exponential from base $a$ to base e.

**KEY POINT 16.13**

$a^x = e^{x \ln a}$

**WORKED EXAMPLE 16.11**

a  Show that $a^x = e^{x\ln a}$ for $a > 0$.

b  Hence write $5^x$ in the form $e^{kx}$, giving the value of $k$ to 3 significant figures.

Use $x\ln a = \ln a^x$. ........... **a**  $e^{x\ln a} = e^{\ln(a^x)}$

$e^x$ and $\ln x$ are inverse ...........     $= a^x$
functions: $e^{\ln y} = y$

Use $a^x = e^{x\ln a}$ with $a = 5$. ........... **b**  $5^x = e^{x\ln 5}$

$\qquad\qquad\qquad = e^{1.61x}$

$\qquad\qquad$ So, $k = 1.61$

# Be the Examiner 16.1

The graph of $y = \ln x$ has the following transformations applied:

**1**  translation up by 5

**2**  horizontal stretch with scale factor 2

**3**  reflection in the $x$-axis.

Find the equation of the resulting graph.

Which is the correct solution? Identify the errors made in the incorrect solutions.

| Solution 1 | Solution 2 | Solution 3 |
|---|---|---|
| Translation up by 5: $y = \ln x + 5$ Horizontal stretch with scale factor 2: $y = \ln(2x) + 5$ Reflection in the $x$-axis: $y = -\ln 2x + 5$ | Translation up by 5: $y = \ln x + 5$ Horizontal stretch with scale factor 2: $y = \ln\left(\frac{x}{2}\right) + 5$ Reflection in the $x$-axis: $y = -\ln\left(\frac{x}{2}\right) - 5$ | Translation up by 5: $y = \ln x + 5$ Horizontal stretch with scale factor 2: $y = \ln\left(\frac{x}{2}\right) + 5$ Reflection in the $x$-axis: $y = -\ln\left(\frac{x}{2}\right) + 5$ |

# Exercise 16C

In this exercise, do not use a calculator unless the question tells you.

For questions 1 to 4, use the method demonstrated in Worked Example 16.9 to sketch the graphs.

1  a  $y = 2^x$
   b  $y = 3^x$

2  a  $y = 0.4^x$
   b  $y = 0.3^x$

3  a  $y = 2^{-x}$
   b  $y = 3^{-x}$

4  a  $y = e^{-x}$
   b  $y = 10^{-x}$

For questions 5 to 8, use the method demonstrated in Worked Example 16.10 to sketch the pairs of graphs on the same axes.

5  a  $y = 2^x$ and $y = \log_2 x$
   b  $y = 4^x$ and $y = \log_4 x$

6  a  $y = 10^x$ and $y = \log x$
   b  $y = e^x$ and $y = \ln x$

7  a  $y = 0.5^x$ and $y = \log_{0.5} x$
   b  $y = 0.2^x$ and $y = \log_{0.2} x$

8  a  $y = 5^{-x}$ and $y = \log_{0.2} x$
   b  $y = 4^{-x}$ and $y = \log_{0.25} x$

For questions 9 to 12, use the method demonstrated in Worked Example 16.11 to write the given expression in the form $e^{kx}$, giving the value of $k$ to 3 significant figures. You may use a calculator.

9  a  $2.1^x$                 10  a  $4.2^x$                 11  a  $0.6^x$                 12  a  $\left(\dfrac{1}{3}\right)^x$

   b  $1.7^x$                     b  $5.1^x$                     b  $0.5^x$                     b  $\left(\dfrac{1}{4}\right)^x$

13  Match each graph with its equation:

  i  $y = 2^x$                         ii  $y = e^x$                         iii  $y = 0.5^x$

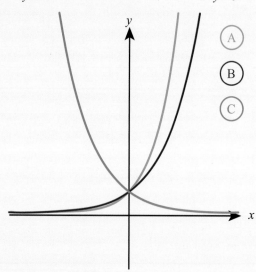

14  Sketch the graph of $y = 2^x$ and $y = \left(\dfrac{1}{2}\right)^x$ on the same set of axes, showing their asymptotes and axis intercepts.

15  Sketch the graphs of $y = e^x$ and $y = e^x + 2$ on the same diagram, showing their asymptotes and $y$-intercepts.

16  Sketch the graphs of $y = e^x$ and $y = 3e^x$ on the same diagram, showing their asymptotes and $y$-intercepts.

17  Sketch the graphs of $y = 6^x$ and $y = \log_6 x$ on the same set of axes, clearly showing the relationship between them.

18  Match each graph with its equation:

  i  $y = \ln x$                         ii  $y = \log_2 x$                         iii  $y = \log_5 x$

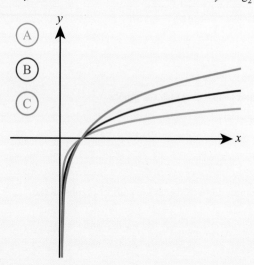

19 Match each graph with its equation:

i $y = \log_2 x$

ii $y = \log_{0.5} x$

iii $y = \log_{0.2} x$

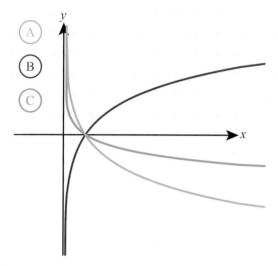

20 Sketch the graphs of $y = \ln x$ and $y = \ln(x - 2)$ on the same diagram, showing their asymptotes and $y$-intercepts.

21 The diagram shows the graph of $y = a + be^x$.

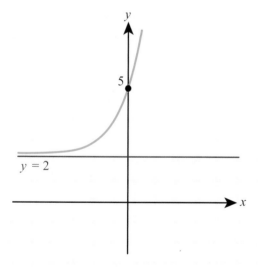

a Write down the value of $a$.

b By considering the $y$-intercept of the graph, find the value of $b$.

22 The graph of $y = e^x$ is translated vertically by 3 units and then stretched vertically with scale factor 2. Find

a the $y$-intercept

b the equation of the horizontal asymptote of the resulting graph.

23 The graph of $y = \ln x$ is translated 5 units in the negative $x$-direction.

a Write down the equation of the vertical asymptote of the resulting graph.

b Sketch the graph, showing the asymptote and the axis intercepts.

**24** The curve in the diagram has equation $y = Ca^x$.

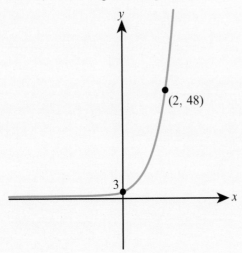

(2, 48)

3

Find the value of $C$ and the value of $a$.

**25** The diagram shows the curve with equation $y = e^{kx} - c$.

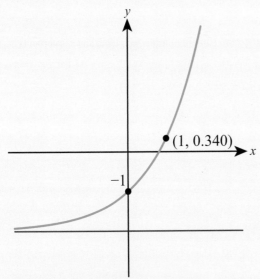

(1, 0.340)

−1

a Write down the value of $c$.

b Find the value of $k$, correct to two decimal places.

**26** a Write $5.2^x$ in the form $e^{kx}$, giving the value of $k$ correct to three significant figures.

b Hence describe the transformation that transforms the graph of $y = e^x$ to the graph of $y = 5.2^x$.

**27** Let $f(x) = 3e^x - 2$ for $x \in \mathbb{R}$.

a Find the range of f.

b Find $f^{-1}(x)$ and state its domain.

c Sketch $y = f(x)$ and $y = f^{-1}(x)$ on the same diagram, showing the relationship between them.

**28** a The graph of $y = \ln x$ is stretched horizontally with scale factor $\frac{1}{3}$. Write down the equation of the resulting graph.

b The graph of $y = \ln x$ can also be mapped to the graph from part **a** by a different single transformation. Describe this transformation fully.

**29** a Describe fully the transformation that maps the graph of $y = e^x$ to the graph of $y = 5e^x$.

b Describe fully a different transformation that maps the graph of $y = e^x$ to the graph of $y = 5e^x$.

## Checklist

- You should be able to apply single transformations to graphs:
  - ☐ $y = f(x) + b$ is a vertical translation by $b$
  - ☐ $y = f(x + c)$ is a horizontal translation by $-c$
  - ☐ $y = pf(x)$ is a vertical stretch with scale factor $p$
  - ☐ $y = f(qx)$ is a horizontal stretch with scale factor $\frac{1}{q}$
  - ☐ $y = -f(x)$ is a reflection in the $x$-axis
  - ☐ $y = f(-x)$ is a reflection in the $y$-axis

- You should be able to apply composite transformations to graphs.
  - ☐ When two vertical transformations are applied, the order matters: $y = pf(x) + b$ is a horizontal stretch with scale factor $p$ followed by a vertical translation by $b$.
  - ☐ When one vertical and one horizontal transformation are applied, the order does not matter.

- You should know about the reciprocal function $f(x) = \frac{1}{x}$ and its graph

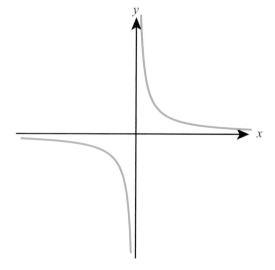

- You should know about rational functions of the form $f(x) = \frac{ax + b}{cx + d}$ and their graphs

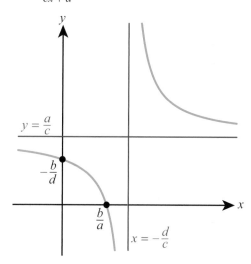

- You should know about exponential functions and their graphs

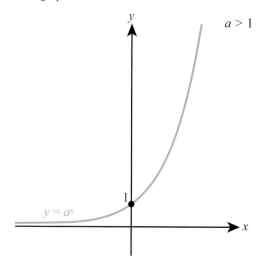

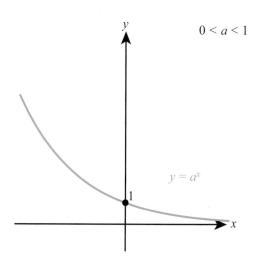

■ You should know about logarithmic functions and their graphs

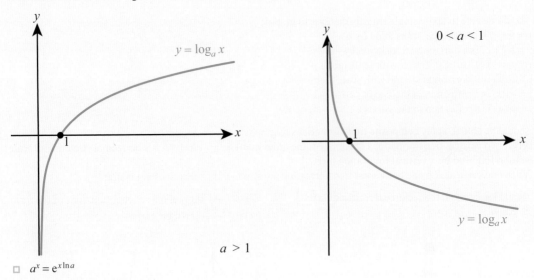

□  $a^x = e^{x \ln a}$

## ▓ **Mixed Practice**

**1** The graph of $y = f(x)$ is shown.

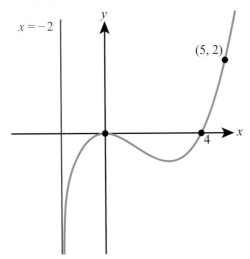

Sketch the following graphs, indicating the positions of asymptotes and $x$-intercepts.

**a**  $y = 2f(x - 3)$

**b**  $y = -f(2x)$

**c**  $y = f\left(\dfrac{x}{3}\right) - 2$

**2** The graph of $y = x^3 - 2x$ is translated 3 units to the right and then stretched vertically with scale factor 2. Find the equation of the resulting graph in the form $y = ax^3 + bx^2 + cx + d$.

**3 a**   Write $x^2 + 4x + 9$ in the form $(x + h)^2 + k$.

   **b**   Hence describe a sequence of two transformations that transform the graph of $y = x^2$ to the graph of $y = x^2 + 4x + 9$.

**4**   The graph of $y = \dfrac{1}{x}$ is translated 2 units in the negative $x$-direction and then stretched vertically with scale factor 3.

   **a**   Write down the equation of the resulting graph.

   **b**   Sketch the graph, indicating any asymptotes and intercepts.

**5 a**   Show that $2 + \dfrac{1}{x - 5} = \dfrac{2x - 9}{x - 5}$.

   **b**   Hence describe two transformations that map the graph of $y = \dfrac{1}{x}$ to the graph of $y = \dfrac{2x - 9}{x - 5}$.

   **c**   State the equations of the asymptotes of the graph of $y = \dfrac{2x - 9}{x - 5}$.

**6**   Sketch the following graphs. In each case, indicate clearly the positions of the vertical asymptote and the $x$-intercept.

   **a**   $y = \ln x$

   **b**   $y = 3\ln(x - 2)$

   **c**   $y = 5 - \ln(3x)$

**7**   The graph of $y = ax + b$ is transformed using the following sequence of transformations:

      Translation 3 units to the right.

      Vertical stretch with scale factor 7.

      Reflection in the $x$-axis.

    The resulting graph has equation $y = 35 - 21x$. Find the values of $a$ and $b$.

**8**   Find two transformations that transform the graph of $y = 9(x - 3)^2$ to the graph of $y = 3(x + 2)^2$.

**9**   The graph of $y = \ln x$ is translated 2 units to the right, then translated 3 units up and finally stretched vertically with scale factor 2. Find the equation of the resulting graph, giving your answer in the form $y = \ln(g(x))$.

**10**   For the graph of $y = \dfrac{4x - 3}{2x + 7}$,

   **a**   Write down the equations of the asymptotes.

   **b**   Find the axis intercepts.

   **c**   Hence sketch the graph.

**11**   Let $f(x) = \dfrac{3x - 1}{x + 5}$.

   **a**   Write down the equations of the asymptotes of the graph of $y = f(x)$.

   **b**   Hence state the domain and the range of $f(x)$.

**12 a**   Describe fully the transformation that maps the graph of $y = \ln x$ to the graph of $y = \ln(x + 3)$.

   **b**   On the same diagram, sketch the graphs of $y = \ln(x + 3)$ and $y = \ln(x^2 + 6x + 9)$ for $x > -3$.

**13**   The quadratic function $f(x) = p + qx - x^2$ has a maximum value of 5 when $x = 3$.

   **a**   Find the value of $p$ and the value of $q$.

   **b**   The graph of $f(x)$ is translated 3 units in the positive direction parallel to the $x$-axis. Determine the equation of the new graph.

Mathematics HL May 2011 Paper 1 TZ1 Q1

**14** Let $f(x) = p + \dfrac{9}{x-q}$, for $x \neq q$. The line $x = 3$ is a vertical asymptote to the graph of f.

**a** Write down the value of $q$.

The graph of f has a $y$-intercept at $(0, 4)$.

**b** Find the value of $p$.

**c** Write down the equation of the horizontal asymptote of the graph of f.

Mathematics SL November 2014 P1 Q5

**15** The number of fish, $N$, in a pond is decreasing according to the model

$$N(t) = ab^{-t} + 40, t \geqslant 0$$

where $a$ and $b$ are positive constants, and $t$ is the time in months since the number of fish in the pond was first counted.

At the beginning 840 fish were counted.

**a** Find the value of $a$.

After 4 months 90 fish were counted.

**b** Find the value of $b$.

The number of fish in the pond will **not** decrease below $p$.

**c** Write down the value of $p$.

Mathematical Studies SL May 2015 Paper 1 TZ2 Q14

**16** Let $f(x) = \ln x$. The graph of f is transformed into the graph of the function g by a translation of 3 to the right and 2 down, followed by a reflection in the $x$-axis. Find an expression for $g(x)$, giving your answer as a single logarithm.

Mathematics HL May 2012 Paper 2 TZ1 Q6

**17** Let $f(x) = \dfrac{2x - 9}{x - 5}$.

**a** State the equations of the vertical and horizontal asymptotes on the graph of $y = f(x)$.

**b** Find the values of $\alpha$ and $\beta$ if $f(x) \equiv \alpha + \dfrac{\beta}{x - 5}$.

**c** State two consecutive transformations that can be applied to the graph of $y = \dfrac{1}{x}$ to create the graph of $y = f(x)$.

**d** Find an expression for $f^{-1}(x)$ and state its domain.

**e** Describe the transformation that maps $y = f(x)$ to $y = f^{-1}(x)$.

**18** Prove that the graph of $y = 2^x$ can be created by stretching the graph of $y = 4^x$ and describe fully the stretch.

**19** Describe fully the stretch that maps the graph $y = \ln x$ to the graph $y = \log_{10} x$.

**20** Sketch the graph of $y = \log(x^2 - 8x + 16)$.

**21** The graph below shows $y = f(x)$. Sketch $y = xf(x)$.

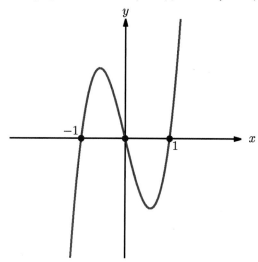

**22** The graph $y = f(x)$ is reflected in the line $y = 1$. Find the equation of the new graph.

**23** Prove that if $f(x^2) = x^2f(x)$ for all $x$ then $y = f(x)$ has the $y$-axis as a line of reflection symmetry.

## ESSENTIAL UNDERSTANDINGS

■ Creating different representations of functions to model relationships between variables, visually and symbolically, as graphs, equations and tables represents different ways to communicate mathematical ideas.

## In this chapter you will learn…

■ how to combine previous methods to solve more complicated equations analytically
■ how to use your GDC to solve equations graphically
■ how to apply analytical and graphical methods to solve equations that relate to real-life situations.

### CONCEPTS

The following concepts will be addressed in this chapter:

■ The parameters in a function or equation correspond to geometrical features of a graph and can **represent** physical quantities.
■ Moving between different forms to **represent** functions allows for a deeper understanding and provides different approaches to problem solving.
■ Our **spatial** frame of reference affects the visible part of a function and by changing this 'window' we can show more or less of the function to best suit our needs.
■ **Equivalent** representations of quadratic functions can reveal different characteristics of the same relationship.

### LEARNER PROFILE – Risk-takers

Do you ever start a maths problem without knowing whether you will be able to finish it? Are you happy to try new methods even if you are not sure you have fully mastered them? Good mathematical problem solvers often take a reasoned attitude towards risk and know that even if they fail in solving one problem, they can learn from it and do better with the next one!

■ **Figure 17.1** Optical illusions – can you believe your eyes? Who do you see? Which table is bigger? Which square is darker?

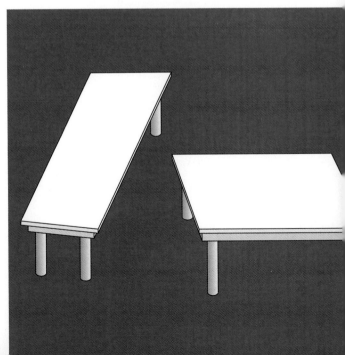

## PRIOR KNOWLEDGE

Before starting this chapter, you should already be able to complete the following:

1  Solve the equation $3x^2 = 2x$.

2  Solve the equation $x^2 - 3x - 28 = 0$.

3  Solve the equation $e^{x+2} = 7$.

4  Solve the equation $\log_3(x + 4) - \log_3 x = 2$.

5  £100 is invested in a bank account that pays 2.5% compound interest per annum. Find how much money is in the account after 5 years.

There are many equations you will meet that simply cannot be solved analytically, and so the only option is to use technology to find an approximate solution. Often this will be done graphically.

There are other equations that might look at first as though they cannot be solved analytically, or at least look as if they do not fall into any category of equation that you know how to solve analytically. However, with some algebraic manipulation, which will often involve a substitution or change of variable, the equation turns into one that you can solve analytically.

## Starter Activity

Look at the pictures in Figure 17.1. In small groups discuss how your perspective changes what you see. To what extent can you trust a visual analysis of a situation?

**Now look at this problem:**

How many possible intersections are there between the line $y = c$, $c \in \mathbb{R}$ and

a  a circle?

b  the graph $y = x^2$ ?

c  the graph $y = x^3 - x$ ?

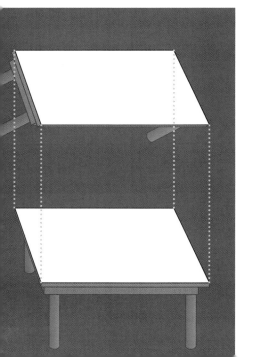
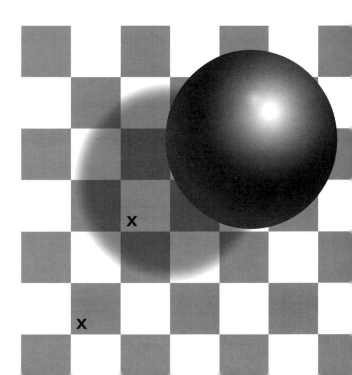

# 17A Solving equations analytically

You know how to solve various equations: quadratic, exponential, logarithmic. We are now going to combine those techniques to solve more-complicated equations.

## ▋ Factorizing

A key idea that can be used to solve some very tricky equations is to make one side zero and factorize. We know that if an expression equals zero then at least one of its factors must be zero.

 **WORKED EXAMPLE 17.1**

Solve the equation $xe^x = 3x$.

Move everything to one
side and factorize ................................................ $xe^x = 3x$
$$xe^x - 3x = 0$$
$$x(e^x - 3) = 0$$

Either one factor equals zero... ................................ $x = 0$

or

...or the other factor equals zero ................................ $e^x - 3 = 0$
$$e^x = 3$$
$$x = \ln 3$$

List the solutions ................................ So, $x = 0, \ln 3$

---

 **TOOLKIT: Proof**

What is wrong with the following proof that all numbers are equal?

Choose two numbers, $a$ and $b$, that are equal:

$$a = b$$

Multiply by $b$:

$$ab = b^2$$

Subtract $a^2$

$$ab - a^2 = b^2 - a^2$$

Factorize:

$$a(b - a) = (b - a)(a + b)$$

Cancel the factor of $b - a$:

$$a = a + b$$

But $a = b$

$$a = a + a = 2a$$

Cancel the factor of $a$:

$$1 = 2$$

Add on 1 to both sides

$$2 = 3$$

And so on…

**TOOLKIT:** Problem Solving

**a** Factorize $xy + x + y + 1$

**b** If $x$ and $y$ are integers, solve $xy + x + y = 12$

## ◼ Disguised quadratics

You will often come across equations that can be turned into quadratics by a substitution.

### WORKED EXAMPLE 17.2

Solve the equation $x^4 - 7x^2 - 18 = 0$.

$x^4 = (x^2)^2$ ............... $x^4 - 7x^2 - 18 = 0$

$(x^2)^2 - 7x^2 - 18 = 0$

You saw how to solve quadratic equations by factorization in Section 15B.

Making the substitution $y = x^2$ turns this into a quadratic equation ............... Let $y = x^2$

$y^2 - 7y - 18 = 0$

Solve the equation for $y$ ............... $(y - 9)(y + 2) = 0$

$y = 9$ or $-2$

Replace $y$ with $x^2$ and solve for $x$ ............... So, $x^2 = 9$

$x = \pm 3$

or

$x^2 = -2$ (no solutions)

List the solutions ............... So, $x = \pm 3$

### WORKED EXAMPLE 17.3

Solve the equation $e^{2x} - 6e^x + 5 = 0$, giving exact answers.

$e^{2x} = (e^x)^2$ ............... $e^{2x} - 6e^x + 5 = 0$

$(e^x)^2 - 6e^x + 5 = 0$

Making the substitution $y = e^x$ turns this into a quadratic equation ............... Let $y = e^x$

$y^2 - 6y + 5 = 0$

Solve the equation for $y$ ............... $(y - 5)(y - 1) = 0$

$y = 5$ or $1$

Replace $y$ with $e^x$ and solve for $x$ ............... So, $e^x = 5$

$x = \ln 5$

or $e^x = 1$

$x = 0$

List the solutions ............... So, $x = 0, \ln 5$

## Tip

Geometry, sequences and logs are all common topics which end up in quadratic equations.

You saw how to use the laws of logarithms to solve equations of this type in Section 12B.

## ■ Equations leading to quadratics

Other situations end up in a standard quadratic equation after manipulation.

### WORKED EXAMPLE 17.4

Solve the equation $\log_2 x + \log_2 (x+4) = 5$.

Use $\log_b x + \log_b y = \log_b xy$ ......... $\log_2 x + \log_2(x+4) = 5$

$$\log_2(x(x+4)) = 5$$

Remove the log using $\log_b a = x$ is equivalent to $a = b^x$ ......... $x(x+4) = 2^5$

This is now a quadratic equation. Solve by factorizing ......... $x^2 + 4x - 32 = 0$

$$(x+8)(x-4) = 0$$

$$x = -8 \ \text{or} \ 4$$

However, checking both possible solutions in the original equation, you can see that $x = -8$ is not valid as you cannot have a log of a negative number ......... So, $x = 4$

### CONCEPTS – EQUIVALENCE

When solving equations, you have to be very clear which lines are **equivalent** to previous lines. Just because a line follows on from the line directly preceding it, does not mean it is equivalent to all previous lines. For example:

| | |
|---|---|
| $x = 3$ | [A] |
| $x^2 = 9$ | [B] |
| $x = 3$ or $-3$ | [C] |

You should see that line C is equivalent to line B which follows from line A, but the argument does not work in reverse. Just because B is true does not guarantee that A is true.

This is why, when we solve equations, we should check our answer. It is not just in case we have made a mistake, but because our flow of logic does not always guarantee that the final answer is equivalent to the original equation, as seen in Worked Example 17.4.

## Exercise 17A

For questions 1 to 6, use the method demonstrated in Worked Example 17.1 to solve the equations.

1  a  $xe^x = 4x$
   b  $xe^x = 5x$

2  a  $x \ln x = 2x$
   b  $x \ln x = 3x$

3  a  $e^x \ln x = 2 \ln x$
   b  $e^x \ln x = 3 \ln x$

4  a  $x^3 = x$
   b  $x^3 = 9x$

5  a  $x\sqrt{x+1} = 3x$
   b  $x\sqrt{x+2} = 5x$

6  a  $xe^x \ln x = 5x \ln x$
   b  $3xe^x \ln(x+3) = 20x \ln(x+3)$

For questions 7 to 11, use the method demonstrated in Worked Example 17.2 to solve the equations.

7  a  $x^4 - 5x^2 + 4 = 0$
   b  $x^4 - 13x^2 + 36 = 0$

8  a  $x^6 - 5x^3 + 6 = 0$
   b  $x^6 - 8x^3 + 7 = 0$

9  a  $x^4 - x^2 - 12 = 0$
   b  $x^4 - 7x^2 - 18 = 0$

10  a  $x - 7\sqrt{x} + 10 = 0$
    b  $x - 7\sqrt{x} + 12 = 0$

11  a  $x - \sqrt{x} - 12 = 0$
    b  $x - 2\sqrt{x} - 15 = 0$

For questions 12 to 16, use the method demonstrated in Worked Example 17.3 to solve the equations.

12    a   $e^{2x} - 7e^x + 6 = 0$          13    a   $e^{2x} - 8e^x + 15 = 0$         14    a   $e^{2x} + 2e^x - 8 = 0$

      b   $e^{2x} - 4e^x + 3 = 0$              b   $e^{2x} - 7e^x + 12 = 0$           b   $e^{2x} + 3e^x - 10 = 0$

15    a   $e^{2x} - 6e^x + 9 = 0$          16    a   $e^{2x} + 5e^x + 6 = 0$

      b   $e^{2x} - 8e^x + 16 = 0$            b   $e^{2x} + 8e^x + 15 = 0$

For questions 17 and 18, use the method demonstrated in Worked Example 17.4 to solve the equations.

17    a   $\log_2 x + \log_2(x + 6) = 4$       18    a   $\log_3(x - 2) + \log_3(x + 6) = 2$

      b   $\log_2 x + \log_2(x + 7) = 3$           b   $\log_3(x - 2) + \log_3(x + 4) = 3$

**19**   Find the exact solutions of the equation $x^3 = 5x$.

**20**   Solve the equation $5x \ln x = 8 \ln x$.

**21**   Find the exact roots of the equation $x^4 - 10x^2 + 24 = 0$.

**22**   Find the exact solutions of the equation $(\ln x)^2 - 2 \ln x = 8$.

**23**   Solve the equation $e^{2x} - 10e^x + 21 = 0$.       **24**   Find the roots of the equation $e^{3x} - 9e^x = 0$.

**25**   Solve the equation $e^{2x} + e^x - 6 = 0$.          **26**   Solve the equation $9^x - 3^x = 72$.

**27**   Solve the equation $x^{\frac{2}{3}} - 2x^{\frac{1}{3}} = 3$.          **28**   Solve the equation $6^x - 8 \times 3^x = 0$.

**29**   Solve the equation $x\sqrt{x - 1} = 3x$.          **30**   Solve the equation $e^{(x-3)\ln(x-1)} = 1$.

**31**   Find the roots of the equation $\log_2 x = 3\log_x 2 + 2$.

**32**   Find the exact solution of the equation $3 \times 4^x - 2 \times 12^x = 0$.

**33**   Find all real solutions of the equation $(x - 7)^{x^2 - 16} = 1$.

 # 17B Solving equations graphically

You may be faced with an equation that you are not expected to be able to solve analytically, or even one for which no analytical method exists at all.

In such cases you can use a graphical approach with your GDC.

If the equation is f($x$) = 0, this is equivalent to looking at where the graph of $y$ = f($x$) crosses the $x$-axis.

---

**WORKED EXAMPLE 17.5**

Solve the equation $x^4 - 3x^3 + 2x + 1 = 0$.

Graph the function and use 'root' $\cdots\cdots\cdots\cdots\cdots\cdots\cdots\cdots$

Sketch the graph as a part of your working

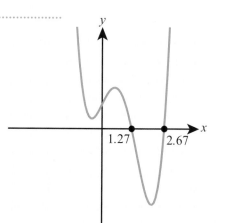

Move the cursor to the right to $\cdots\cdots\cdots\cdots\cdots\cdots\cdots$ $x = 1.27$ or $2.67$
identify the second solution

The equation in Worked Example 17.5 is called a quartic and a general method for solving it was found by the Italian renaissance mathematician Lodovico Ferrari (1522–1565), after another mathematician Niccolò Fontana Tartaglia revealed a similar method for solving cubic equations to him and his master, Gerolamo Cardano, in the form of a poem. Despite being sworn to secrecy, Cardano published both Tartaglia's method for solving cubic equations and his apprentice's related method for solving quartic equations, leading to a long feud between Tartaglia on the one hand and Cardano and Ferrari on the other. Ferrari went on to develop a greater understanding of cubic and quartic equations, and later defeated Tartaglia in a public mathematics competition (the like of which were common in sixteenth-century Italy and drew large crowds), winning instant fame and several lucrative job offers.

If the equation is of the form $f(x) = g(x)$ we could rearrange it into the form $f(x) - g(x) = 0$ and solve using the method above. Alternatively, we could find the intersections of $y = f(x)$ and $y = g(x)$.

## WORKED EXAMPLE 17.6

Solve the equation $\ln 3x = 2^x - x^2$.

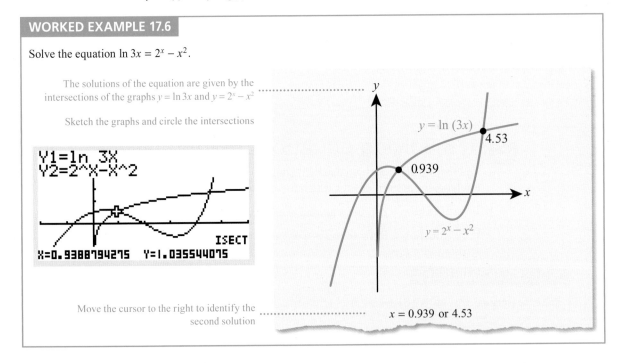

The solutions of the equation are given by the intersections of the graphs $y = \ln 3x$ and $y = 2^x - x^2$

Sketch the graphs and circle the intersections

Move the cursor to the right to identify the second solution

$x = 0.939$ or $4.53$

## TOK Links

How much do you trust graphical solutions to equations compared to analytic solutions? What criteria do you use to judge a mathematical solution? One controversial case you might like to investigate is the proof of the 'Four colour theorem' for colouring a map. It was proved using a computer program, so mathematicians are still debating whether or not this a valid proof.

CONCEPTS – SPACE

You might wonder why we ever need to use the method in Worked Example 17.6, rather than just making one side equal zero and using the method in Worked Example 17.5. The key issue is that finding all solutions can be tricky. You need to know which window to look at on your calculator. Sometimes two solutions are very close to each other and you need to zoom in. Sometimes the two solutions are very far apart and you need to zoom out. Your **spatial** frame of reference will change which solutions you see. Although some calculators have tools to help find an appropriate zoom, the more familiar you are with the functions you are sketching the more likely you are to find all the solutions. Most people are more comfortable with the graphs of $y = e^x$ and $y = x$ than $y = x - e^x$.

## Exercise 17B

For questions 1 to 4, use the method demonstrated in Worked Example 17.5 to find all solutions of the equation.

1. a $x^3 - x + 2 = 0$
   b $x^3 + 2x^2 - 5 = 0$

2. a $x^4 - 2x^3 - 3 = 0$
   b $x^4 + 3x^3 - 1 = 0$

3. a $x^5 - 8x^2 + 6 = 0$
   b $x^5 + 6x^2 - 3 = 0$

4. a $e^x - 8x = 0$
   b $9x - 2e^x = 0$

For questions 5 to 8, use the method demonstrated in Worked Example 17.6 to find all solutions of the equation.

5. a $4\ln x = 2 - x^2$
   b $\ln(x - 2) = 5 - x^2$

6. a $5x\,e^{x+1} = -2$
   b $(x - 2)\ln x = 3$

7. a $e^{x-1} = x^2 - 3$
   b $e^{-x} = x^2 - 1$

8. a $\dfrac{6}{x^2 + 1} = \dfrac{1}{x - 1}$
   b $\dfrac{15}{x^2 + 4} = \dfrac{1}{x - 3}$

9. Solve the equation $\dfrac{e^x}{x} = 4 - x^2$.

10. Find the roots of the equation $x - 5\ln(x - 2) = 0$.

11. a Sketch the graphs of $y = e^{-x}$ and $y = 2 - x^2$ on the same diagram.
    b Hence state the number of solutions of the equation $e^{-x} = 2 - x$.

12. Solve the equation $e^x = x + 2$.

13. a On the same diagram, sketch the graphs of $y = \ln(x - 2)$ and $y = \dfrac{1}{x - 3}$.
    b Hence state the number of solutions of the equation $\ln(x - 2) = \dfrac{1}{x - 3}$.

14. Find all roots of the equation $3\ln(x^2 + 2) = x + 2$.

15. Find the set of values of $k$ such that the equation $x^3 - 4x = k$ has exactly three real roots.

16. Solve the equation $\dfrac{2e^x + 3x}{e^x + 1} = 2$.

17. Find the set of values of $k$ for which the equation $ke^{-x} + 3x = 2$ has no real roots.

18. The value of $k$ is such that the equation $\ln x^3 = kx$ has exactly one real root. Find the number of real roots of the following equations:
    a $\ln x^4 = kx$
    b $\ln \sqrt{x} = kx$

# 17C Applications of equations

You may have to apply either a graphical or analytical approach to problems that arise in a range of mathematical contexts or a real-life setting. Equations are used in situations such as: determining pH, describing the motion of projectiles, predicting population growth, radioactive decay and calculating compound or production possibilities.

## Links to: Biology

The apparent value of using exponential equations to model population growth can be easily illustrated by imagining a type of bacteriaum that reproduces itself once an hour. If we start with one bacterium, after an hour we will have two bacteria, after two hours there will be four, after three hours there will be eight, and so on. But this model has its limits – or rather, it lacks limits. If the bacteria continue to reproduce at the same rate, after a day there will be more than 16 million, and within a few days there would be a thick layer of bacteria covering the entire surface of the earth!

Thankfully, real-life population growth is rarely this straightforward. If you are studying biology, you may have come across the notion of limiting factors – circumstances that restrict the unhindered growth of a population. Examples include the availability of food, water and space, as well as predation, disease and competition for resources.

The equation given in Worked Example 17.7 is an example of a logistic equation, an equation that models this kind of restricted growth by incorporating a *carrying capacity*, the maximum population size that can be supported by an environment or ecosystem.

---

### WORKED EXAMPLE 17.7

The population of a colony of ants after $t$ days is given by $P = \dfrac{1000}{1 + 99e^{-t}}$

Use technology to find how long it takes for the population to reach 850.

Graph $y = \dfrac{1000}{1 + 99e^{-t}}$ and $y = 850$, and find the $x$ value of their intersection point

$$\frac{1000}{1 + 99e^{-t}} = 850$$

$$t = 6.33 \text{ days}$$

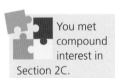

---

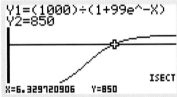 You met compound interest in Section 2C.

### WORKED EXAMPLE 17.8

£500 is invested for $n$ years in a bank account that pays 2% compound interest per annum.

Show that the balance of the account equals £600 when $n = \dfrac{\ln 1.2}{\ln 1.02}$.

A 2% interest corresponds to multiplication by 1.02 ............ After 1 year balance = $500 \times 1.02$
After 2 years balance = $500 \times 1.02^2$

Expressions for the first couple of years lead to an ............ After $n$ years balance = $500 \times 1.02^n$
expression for $n$ years

The balance after $n$ ............ $500 \times 1.02^n = 600$
years must equal £600                    $1.02^n = 1.2$

Take the natural log of both sides (this would ............ $\ln 1.02^n = \ln 1.2$
work with any base)

Use $\log_b x^n = n\log_b x$ ............ $n \ln 1.02 = \ln 1.2$

Rearrange for $n$ ............ $n = \dfrac{\ln 1.2}{\ln 1.02}$

## Exercise 17C

**1** A ball is thrown vertically up so that its height above ground $t$ seconds later is given by $h = 3.6 + 5.2t - 4.9t^2$. How long does it take for the ball to reach the ground?

**2** The size of a population of fish after $t$ months is given by $P = \dfrac{180e^t + 200}{e^t + 4}$. How long does it take for the population to reach 150 fish? Give your answer in months, correct to one decimal place.

**3** The speed of a falling object, $v\,\text{ms}^{-1}$, is given by the equation $v = \dfrac{15(1 - e^{-t})}{1 + 3e^{-t}}$, where time $t$ is measured in seconds. How long does it take for the speed to equal $10\,\text{ms}^{-1}$?

**4** The number of bacteria in a laboratory experiment is modelled by the equation $N = 1500 - 1000e^{-0.07t}$, where $t$ hours is the time since the start of the experiment. How long does it take for the number of bacteria to reach 1200?

**5** The numerical value of the volume of a cube in cm³ equals the sum of the length of all of its edges in cm. Find the surface area of the cube.

**6** The two shorter sides of a right-angled triangle are 5 cm and $x$ cm. The numerical value of the area in cm² equals the perimeter in cm. Find the value of $x$.

**7** A geometric series has the third term equal to 1 and the sum to infinity equal to 12. Find the possible values of the common ratio.

**8** £600 is invested in a bank account which pays compound interest of 3% per annum. Let $C(n)$ be the amount of money in the account at the end of $n$ years.

   a Write an equation for $C(n)$ in terms of $n$.

   b Show that the amount of money in the account equals £750 when $n = \dfrac{\ln 1.25}{\ln 1.03}$.

**9** Asher invests \$200 in a bank account with an interest rate of 2.6% per annum. Elsa invests \$300 in an account which pays \$10 interest every year. After how many whole years will Asher have at least as much money as Elsa?

**10** The approximate stopping distance (in metres) of a car travelling at $v$ km per hour on a dry road is modelled by $\dfrac{v}{5} + \dfrac{v^2}{150}$.

   a What is the stopping distance of a car travelling at 30 km per hour?

   b By using a substitution $u = \dfrac{v}{5}$ find the speed that results in a stopping distance of 36 m.

   c The approximate stopping distance of a car on an icy road is modelled by

     $\dfrac{v^2}{25} + \dfrac{v^4}{6\,250\,000}$

     Find the speed which results in a stopping distance of 101 m.

**11** The height, $y$ m, of a hanging chain is modelled by $y = e^x + e^{-x}$.

   a Find the exact $x$-coordinate of each of the points with height 4 m.

   b Prove that the average $x$-coordinate of the two points found in part a is zero.

**12** The fourth term of a geometric series is $-3$ and the sum to infinity is 20.

   a Show that $20r^4 - 20r^3 - 3 = 0$, where $r$ is the common ratio of the series.

   b Find the first term of the series.

**13** Given that $\displaystyle\sum_{k=0}^{\infty} e^{-(2k+1)x} = \dfrac{2}{3}$,

   a show that $2e^{2x} - 3e^x - 2 = 0$

   b hence, find the exact value of $x$.

## Checklist

- You should be able to combine previous methods to solve more complicated equations analytically:
  - ☐ factorizing
  - ☐ disguised quadratics
  - ☐ equations leading to quadratics.

- You should be able to use your GDC to solve equations graphically.

- You should be able to apply analytical and graphical methods to solve equations that relate to real-life situations.

---

### ■ Mixed Practice

**1** Find the exact solutions of the equation $x \ln(x + 1) = 5x$.

**2** Solve the equation $x \ln(x + 2) = 3$.

**3** The concentration (measured in suitable units) of a drug in the bloodstream, $t$ hours after an injection is given, varies according to the equation $c = \dfrac{100e^{0.2t}}{2 + e^{0.3t}}$.
  **a** Find the maximum concentration of the drug.
  **b** Find the two times when the concentration is 40 units.
  **c** How long does it take for the concentration to fall below 5 units?

**4** Find all real roots of the equation $x^4 - 2x^2 = 63$.

**5** The height of a projectile above the ground is modelled using the formula $h = t(1 - 5t)$ where $h$ is the height in metres and $t$ is the time in seconds.

  **a** Show that the projectile lands after 0.2 seconds.

  **b** The model is improved to take spin into account. The new formula is $t(1 - 5t) + \ln(t + 1)$. How much longer is the projectile in the air when spin is included? Give your answer to 2 significant figures.

**6 a** On the same diagram sketch the graphs of $y = 4 - x^2$ and $y = \ln x$.
  **b** Hence state the number of real roots of the equation $x^2 + \ln x = 4$.

**7** The population of big cats in Africa is increasing at a rate of 5% per year. At the beginning of 2004 the population was 10 000.
  **a** Write down the population of big cats at the beginning of 2005.
  **b** Find the population of big cats at the beginning of 2010.
  **c** Find the number of years, from the beginning of 2004, it will take the population of big cats to exceed 50 000.

Mathematical Studies SL November 2009 Paper 1 Q12

**8** Solve $e^x = 11 \ln x$.

**9** Solve the equation $x - 6\sqrt{x} + 5 = 0$.

**10** Find the exact roots of the equation $e^{2x} - 7e^x + 10 = 0$.

**11** Solve the equation $\log_3 x + \log_3(x + 6) = 3$.

**12** Let $f(x) = \ln(x - 1) + x^2 - 1$. By sketching suitable graphs, find the number of solutions of the equation $f(x) = 0$.

**13** Find the set of values of $k$ for which the equation $50e^{2x} = k(2 + 3e^{3x})$ has two real roots.

**14** The graph of $y = f(x)$ is shown in the diagram.

Find the number of real solutions of the following equations:

**a** $f(x) = 8$

**b** $f(x - 3) = 8$

**c** $3f(x) = 50$

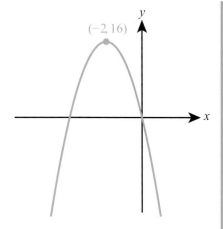

**15** Find the equation of the vertical asymptote of the graph of $y = \ln(2x^3 - 5x + 4)$.

**16** Find the equation of the vertical asymptote of the graph of $y = \dfrac{3}{e^{3x} - 9e^x}$, giving your answer in an exact form.

**17** The seventh term, $u_7$, of a geometric sequence is 108. The eighth term, $u_8$, of the sequence is 36.

**a** Write down the common ratio of the sequence.

**b** Find $u_1$.

The sum of the first $k$ terms in the sequence is 118 096.

**c** Find the value of $k$.

Mathematical Studies SL May 2011 Paper 1 TZ1 Q11

**18** Solve the equation $4^{x-1} = 2^x + 8$.

Mathematics HL May 2010 Paper 1 TZ1 Q4

**19** Solve the equation $4^x + 32 = 3 \times 2^{x+2}$.

**20** If $e^x + e^{-x} = 2k$ where $k > 1$, find and simplify an expression for $x$ in terms of $k$.

**21** A geometric sequence has first term $a$, common ratio $r$ and sum to infinity 76. A second geometric sequence has first term $a$, common ratio $r^3$ and sum to infinity 36.

Find $r$.

Mathematics HL May 2013 Paper 1 TZ2 Q6

**22** The real root of the equation $x^3 - x + 4 = 0$ is $-1.796$ to three decimal places. Determine the real root for each of the following.

**a** $(x - 1)^3 - (x - 1) + 4 = 0$          **b** $8x^3 - 2x + 4 = 0$

Mathematics HL November 2009 Paper 1 Q5

**23** The equation $e^x = kx$ with $k > 0$ has exactly one solution. How many solutions do the following equations have:

**a** $e^{2x} = kx$                                     **b** $e^x = -kx$

**24** Prove that there is only one right-angled triangle that has sides with lengths that are consecutive integers.

**25** What is the smallest possible fraction of the perimeter which can be taken up by the hypotenuse of a right-angled triangle?

## ESSENTIAL UNDERSTANDINGS

- Trigonometry allows us to quantify the physical world.
- This topic provides us with the tools for analysis, measurement and transformation of quantities, movements and relationships.

## In this chapter you will learn...

- about different units for measuring angles, called radians
- how to find the length of an arc of a circle
- how to find the area of a sector of a circle
- how to define the sine and cosine functions in terms of the unit circle
- how to define the tangent function
- how to work with exact values of trigonometric ratios of certain angles
- about the ambiguous case of the sine rule
- about the Pythagorean identity $\cos^2\theta + \sin^2\theta \equiv 1$
- about double angle identities for sine and cosine
- how to find the value of one trigonometric function given the value of another
- how to sketch the graphs of trigonometric functions
- about transformations of the graphs of trigonometric functions
- about trigonometric functions in real-life contexts
- how to solve trigonometric equations graphically
- how to solve trigonometric equations analytically
- how to solve trigonometric equations that lead to quadratic equations.

### CONCEPTS

The following concepts will be addressed in this chapter:
- **Equivalent** measurement systems, such as degrees and radians, can be used for angles to facilitate our ease of calculation.
- Different **representations** of the values of trigonometric relationships, such as exact or **approximate**, may not be **equivalent** to one another.
- The trigonometric functions of angles may be defined on the unit circle, which can visually and algebraically **represent** the periodic or symmetric nature of their values.

■ **Figure 18.1** Can trigonometric functions model the motion of a Ferris wheel? What about a double Ferris wheel?

**PRIOR KNOWLEDGE**

Before starting this chapter, you should already be able to complete the following:

1  Find the angle $\theta$.

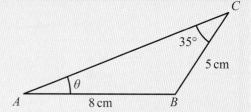

2  Solve the equation $x^2 + 5x + 6 = 0$.

3  What is the equation of the graph $y = x^2 + 2x$ after being translated 3 units to the right?

4  Write $\dfrac{1}{x-1} + \dfrac{1}{x+1}$ as a single, simplified fraction.

Up until now you have used trigonometric functions (sine, cosine and tangent) to find side lengths and angles in triangles. However, these functions can actually cope not only with angles larger than 180° but also with negative angles, which allows them to be used for many applications other than triangles.

Extending the domain of these functions beyond $0° < \theta < 180°$ also makes them many-to-one, so there is more than one value of $\theta$ for certain values of these functions. This means that careful methods are needed when solving equations that include trigonometric functions.

## Starter Activity

Look at the pictures in Figure 18.1. In small groups try to sketch a graph of the height of a passenger on each of the rides against time.

**Now look at this problem:**

1  How many triangles can you draw given the following information:
$AB = 15$cm, $BC = 10$cm, $A\hat{C}B = 40°$?

2  What if the other information is as given above except that $BC = 18$cm?

# 18A Radian measure of angles

Although degrees are the units you are familiar with for measuring angles, they are not always the best unit to use. Instead, **radians** are far more useful in many branches of mathematics.

Radians relate the size of an angle at the centre of the unit circle (a circle with radius 1) to the distance a point moves round the circumference of that circle:

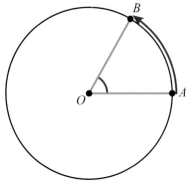

The length of the arc $AB$ is equal to the size of angle $AOB$ in radians.

Since the circumference of the unit circle is $2\pi \times 1 = 2\pi$, there are $2\pi$ radians in one full turn.

> **KEY POINT 18.1**
>
> $360° = 2\pi$ radians

> **WORKED EXAMPLE 18.1**
>
> a   Convert 75° to radians.
>
> b   Convert 1.5 radians to degrees.
>
> $360° = 2\pi$ radians.
> so $1° = \dfrac{2\pi}{360}$ radians
>
> **a**  $75° = \dfrac{2\pi}{360} \times 75$
>
> $= \dfrac{75\pi}{180}$
>
> $= \dfrac{5\pi}{12}$ radians
>
> $2\pi$ radians = 360°,
> so 1 radian $= \left(\dfrac{360}{2\pi}\right)°$
>
> **b**  1.5 radians $= \dfrac{360}{2\pi} \times 1.5$
>
> $= 85.9°$

Using the unit circle, we can define any positive angle as being between the positive x-axis and the radius from a point $P$ as $P$ moves anti-clockwise around the circle:

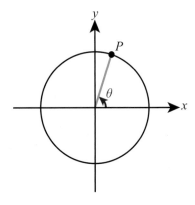

If the angle is greater than $2\pi$, $P$ just goes around the circle again:

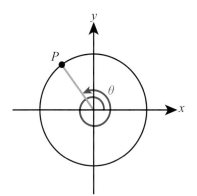

If the angle is negative, $P$ moves clockwise from the positive $x$-axis:

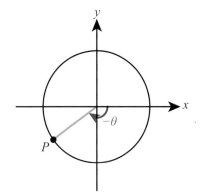

---

**WORKED EXAMPLE 18.2**

Mark on the unit circle the points corresponding to these angles:

**A** $\dfrac{4\pi}{3}$ **B** $-3\pi$ **C** $\dfrac{5\pi}{2}$ **D** $\dfrac{25\pi}{6}$

A: $\dfrac{4\pi}{3} = \dfrac{2}{3} \times 2\pi$ so it is $\dfrac{2}{3}$ of a whole turn

B: $-3\pi = -2\pi - \pi$ so it is a whole turn 'backwards' (clockwise) followed by half a turn in the same negative direction

C: $\dfrac{5\pi}{2} = 2\pi + \dfrac{\pi}{2}$ so it is a whole turn followed by $\dfrac{1}{4}$ of a turn

D: $\dfrac{25\pi}{6} = 4\pi + \dfrac{\pi}{6}$ so it is two whole turns followed by $\dfrac{1}{12}$ of a turn

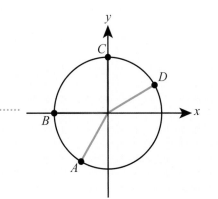

## ◼ Length of an arc

The arc *AB* **subtends** an angle $\theta$ at the centre of the circle.

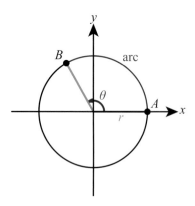

Since the ratio of the arc length, *s*, to the circumference will be the same as the ratio of $\theta$ to $2\pi$ radians, this gives us:

$$\frac{s}{2\pi r} = \frac{\theta}{2\pi}$$

Rearranging gives the formula for arc length when $\theta$ is measured in radians.

---

**KEY POINT 18.2**

The length of an arc is

$$s = r\theta$$

where *r* is the radius of the circle and $\theta$ is the angle subtended at the centre measured in radians.

---

**WORKED EXAMPLE 18.3**

Find the length of the arc *AB* in the circle shown.

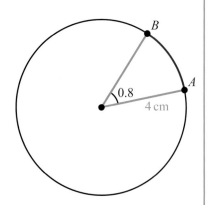

Use $s = r\theta$ with $\theta$ in radians  $s = r\theta$

$$= 4 \times 0.8$$

$$= 3.2\,\text{cm}$$

## ▧ Area of a sector

By a very similar argument to that above, we can obtain a formula for the area of a sector.

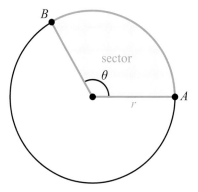

Since the ratio of the area of a sector, A, to the area of the circle will be the same as the ratio of $\theta$ to $2\pi$ radians, this gives us:

$$\frac{A}{\pi r^2} = \frac{\theta}{2\pi}$$

Rearranging gives the formula for the area of a sector when $\theta$ is measured in radians.

---

**KEY POINT 18.3**

The area of a sector is

$$A = \frac{1}{2}r^2\theta$$

where $r$ is the radius of the circle and $\theta$ is the angle subtended at the centre measured in radians.

---

**WORKED EXAMPLE 18.4**

Find the area of the sector $AOB$ in the circle shown.

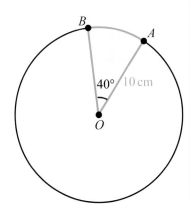

First convert 40° to radians  $\cdots\cdots\cdots\cdots\cdots\cdots\cdots\cdots$  $40° = \frac{2\pi}{360} \times 40$

$$= \frac{2\pi}{9} \text{ radians}$$

Then use $A = \frac{1}{2}r^2\theta$  $\cdots\cdots\cdots\cdots\cdots\cdots\cdots$  $A = \frac{1}{2} \times 10^2 \times \frac{2\pi}{9}$

$$= 34.9 \text{ cm}^2$$

## CONCEPTS – EQUIVALENCE

You have seen various quantities that can be measured in different units. For example, lengths can be measured in feet or metres and temperature in degrees Fahrenheit or Celsius. There is no mathematical reason to prefer one over the other for them because the units that we use to measure in do not change any formulae used; they are **equivalent**. You might have expected it to be the same for this new unit of angle measurement, however, as seen in Worked Example 18.4, in this case the units used actually do change the formulae. This has some very important consequences when it comes to differentiating trigonometric functions, as you will see in Chapter 20.

You met the sine rule, cosine rule, and area of a triangle formula in Section 5B.

## Tip

If you are working with $\sin\theta$, $\cos\theta$ or $\tan\theta$ with $\theta$ in radians, you must set your calculator to radian (not degree) mode.

You will often need to combine the results for arc length and area of sector with the results you know for triangles.

## WORKED EXAMPLE 18.5

The diagram shows a sector of a circle with radius 8 cm and angle 0.7 radians.

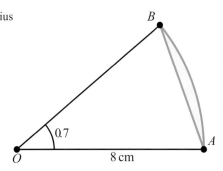

For the shaded region, find:

a   the perimeter
b   the area.

Use $s = r\theta$ to find the arc length

**a**   $s = 8 \times 0.7$
$\quad = 5.6$

Use the cosine rule to find the length of the chord $AB$. Remember to make sure your calculator is in radian mode

By cosine rule,

$AB^2 = 8^2 + 8^2 - 2 \times 8 \times 8\cos 0.7$
$\quad = 30.1002$

So,   $AB = 5.4864$

The perimeter of the shaded region is the length of the arc plus the length of the chord

Hence, $p = 5.6 + 5.49$
$\quad = 11.1\,\text{cm}$

Use $A = \frac{1}{2}r^2\theta$ to find the area of the sector

**b**   Area of sector $= \frac{1}{2} \times 8^2 \times 0.7$
$\quad = 22.4$

Use $A = \frac{1}{2}ab\sin C$ with $a = b = r$ and $C = \theta$ to find the area of the triangle

Area of triangle $= \frac{1}{2} \times 8^2 \sin 0.7$
$\quad = 20.615$

The area of the shaded region is the area of the sector minus the area of the triangle

$A = 22.4 - 20.6$
$\quad = 1.79\,\text{cm}^2$

## Exercise 18A

For questions 1 to 12, use the method demonstrated in Worked Example 18.1 to convert between degrees and radians.

Convert to radians, giving your answers in terms of π:

1   a   60°         2   a   150°         3   a   90°

    b   45°                b   120°               b   270°

Convert to radians, giving your answer to three significant figures:

4   a   28°         5   a   67°         6   a   196°

    b   36°                b   78°                b   236°

Convert to degrees, giving your answers to one decimal place where appropriate:

7   a   0.62 radians      8   a   1.26 radians      9   a   4.61 radians

    b   0.83 radians          b   1.35 radians          b   5.24 radians

10   a   $\dfrac{\pi}{5}$ radians      11   a   $\dfrac{7\pi}{12}$ radians      12   a   $\dfrac{7\pi}{3}$ radians

     b   $\dfrac{\pi}{8}$ radians           b   $\dfrac{4\pi}{15}$ radians           b   $\dfrac{11\pi}{6}$ radians

For questions 13 to 18, use the method demonstrated in Worked Example 18.2 to mark on the unit circle the points corresponding to these angles:

13   a   $\dfrac{2\pi}{3}$        14   a   $\dfrac{5\pi}{6}$        15   a   $-\dfrac{\pi}{2}$

     b   $\dfrac{3\pi}{4}$           b   $\dfrac{7\pi}{4}$           b   $-\dfrac{3\pi}{2}$

16   a   $-\dfrac{\pi}{3}$       17   a   $\dfrac{8\pi}{3}$        18   a   $\dfrac{11\pi}{2}$

     b   $-\dfrac{\pi}{4}$          b   $\dfrac{11\pi}{4}$         b   $\dfrac{19\pi}{2}$

For questions 19 to 21, use the methods demonstrated in Worked Examples 18.3 and 18.4 to find the length of the arc *AB* that subtends the given angle (in radians), and the area of the corresponding sector.

19   a

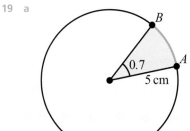

20   a

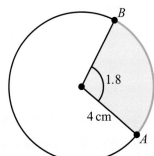

21   a

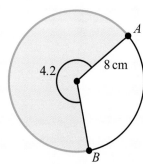

    b

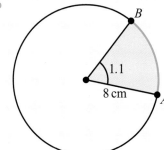

    b

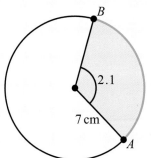

    b

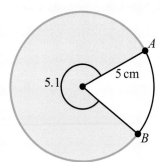

For questions 23 to 24, use the method demonstrated in Worked Example 18.5 to find the area and the perimeter of the shaded region.

22 a

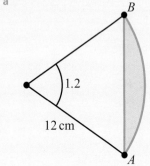

23 a

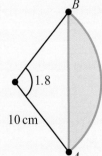

24 a

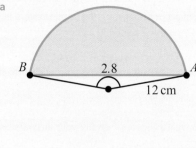

b

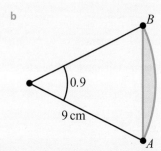

b

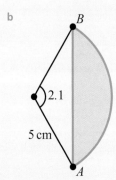

b

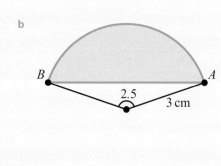

25 The diagram shows a sector *AOB* of a circle of radius 8 cm. The angle of the centre of the sector is 0.6 radians.

Find the perimeter and the area of the sector.

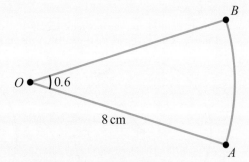

26 A circle has centre *O* and radius 6.2 cm. Points *A* and *B* lie on the circumference of the circle so that the arc *AB* subtends an angle of 2.5 radians at the centre of the circle. Find the perimeter and the area of the sector *AOB*.

27 An arc of a circle has length 12.3 cm and subtends an angle of 1.2 radians at the centre of the circle. Find the radius of the circle.

28 The diagram shows a sector of a circle of radius 5 cm. The length of the arc *AB* is 7 cm.

a Find the value of $\theta$.

b Find the area of the sector.

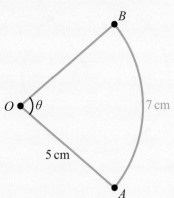

**29** A circle with centre $O$ has radius 23 cm. Arc $AB$ subtends angle $\theta$ radians at the centre of the circle. Given that the area of the sector $AOB$ is 185 cm$^2$, find the value of $\theta$.

**30** A sector of a circle has area 326 cm$^2$ and an angle at the centre of 2.7 radians. Find the radius of the circle.

**31** The diagram shows a sector of a circle. The area of the sector is 87.3 cm$^2$.

a Find the radius of the circle.

b Find the perimeter of the sector.

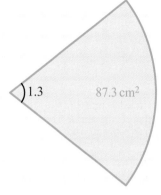

**32** The figure shown in the diagram consists of a rectangle and a sector of a circle.

Calculate the area and the perimeter of the figure.

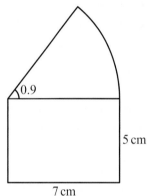

**33** The diagram shows a sector of a circle.

The perimeter of the sector is 26 cm. Find the radius of the circle.

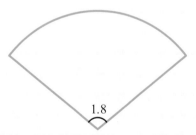

**34** A sector of a circle has area 18 cm$^2$ and perimeter 30 cm. Find the possible values of the radius of the circle.

 **35** The diagram shows an equilateral triangle and three arcs of circles with centres that are the vertices of the triangle.

The length of the sides of the triangle are 12 cm. Find the exact perimeter of the figure.

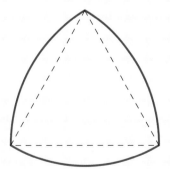

**36** A circle has centre $O$ and radius 8 cm. Chord $PQ$ subtends angle 0.9 radians at the centre. Find the difference between the length of the arc $PQ$ and the length of the chord $PQ$.

**37** The arc $AB$ of a circle of radius 12 cm subtends an angle of 0.6 radians at the centre. Find the perimeter and the area of the shaded region.

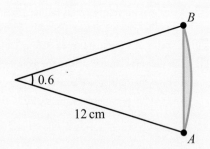

**38** A circle has centre $O$ and radius 4 cm. Chord $PQ$ subtends angle $\theta$ radians at the centre. The area of the shaded region is 6 cm$^2$.

a   Show that $\theta - \sin\theta = 0.75$.

b   Find the value of $\theta$.

c   Find the perimeter of the shaded region.

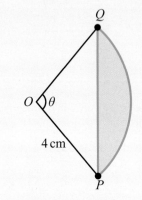

**39** The diagram shows two circular sectors with angle 1.2 radians at the centre. The radius of the smaller circle is 15 cm and the radius of the larger circle is $x$ cm larger.

Find the value of $x$ so that the area of the shaded region is 59.4 cm$^2$.

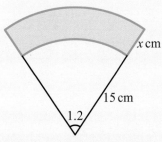

**40** A piece of paper has a shape of a circular sector with radius $r$ cm and angle $\theta$ radians. The paper is rolled into a cone with height 22 cm and base radius 8 cm.

Find the values of $r$ and $\theta$.

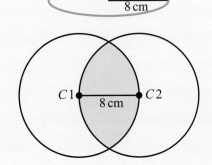

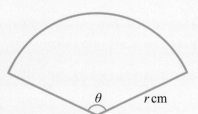

**41** Two identical circles each have radius 8 cm. They overlap in such a way that the centre of each circle lies on the circumference of the other, as shown in the diagram.

Find the perimeter and the area of the shaded region.

# 18B Trigonometric functions

## ▨ Definition of cos θ and sin θ in terms of the unit circle

Until now you have only used sin θ and cos θ in triangles, where θ had to be less than 180° (or π radians). However, using the unit circle we can define these functions so that θ can be any size, positive or negative.

**KEY POINT 18.4**

For a point, $P$, on the unit circle at an angle $\theta$ to the positive $x$-axis:
- $\sin\theta$ is the $y$-coordinate of the point $P$
- $\cos\theta$ is the $x$-coordinate of the point $P$.

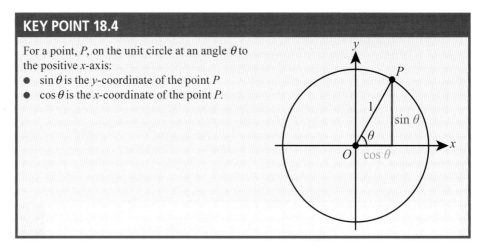

**WORKED EXAMPLE 18.6**

Given that $\sin\theta = 0.75$, find the value of:

a $\sin(\pi - \theta)$ 　　　　 b $\cos\left(\theta + \dfrac{\pi}{2}\right)$

Mark the points that relate to ⋯⋯⋯ **a**
θ and π − θ on the unit circle

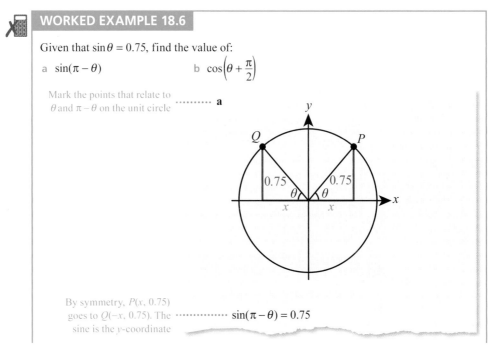

By symmetry, $P(x, 0.75)$
goes to $Q(-x, 0.75)$. The ⋯⋯⋯⋯ $\sin(\pi - \theta) = 0.75$
sine is the $y$-coordinate

Mark the points that relate to
$\theta$ and $\theta + \dfrac{\pi}{2}$ on the unit circle ·········· **b**

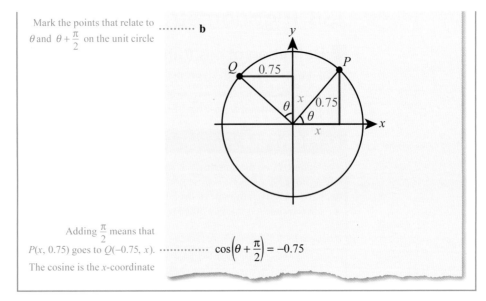

Adding $\dfrac{\pi}{2}$ means that
$P(x, 0.75)$ goes to $Q(-0.75, x)$. ·············· $\cos\!\left(\theta + \dfrac{\pi}{2}\right) = -0.75$
The cosine is the $x$-coordinate

Certain relationships for sine are very common, so rather than having to resort to the unit circle every time to work them out, it is worth learning them.

### KEY POINT 18.5

- $\sin(-\theta) = -\sin\theta$
- $\sin(\pi - \theta) = \sin\theta$
- $\sin(\theta + \pi) = -\sin\theta$
- $\sin(\theta + 2\pi) = \sin\theta$

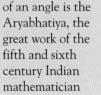 The first
work to
explicitly refer to
sine as a function
of an angle is the
Aryabhatiya, the
great work of the
fifth and sixth
century Indian
mathematician
Aryabhata.

Similar relationships hold for cosine.

### KEY POINT 18.6

- $\cos(-\theta) = \cos\theta$
- $\cos(\pi - \theta) = -\cos\theta$
- $\cos(\theta + \pi) = -\cos\theta$
- $\cos(\theta + 2\pi) = \cos\theta$

You need to be able to apply these relationships in both radians and degrees.

 **WORKED EXAMPLE 18.7**

Given that $\cos 140° = -0.766$, find

a  $\cos(-220°)$

b  $\cos 400°$

Use $\cos\theta = \cos(\theta + 360°)$ ·········· **a**  $\cos(-220°) = \cos(-220° + 360°)$
$= \cos 140°$
$= -0.766$

First use $\cos(\theta + 360°) = \cos\theta$ ·········· **b**  $\cos 400° = \cos 40°$
Then $\cos\theta = -\cos(180° - \theta)$ ·············· $= -\cos 140°$
$= 0.766$

## Definition of $\tan\theta$ as $\dfrac{\sin\theta}{\cos\theta}$

The tangent function is defined as the ratio of the sine function to the cosine function.

> **KEY POINT 18.7**
>
> - $\tan\theta = \dfrac{\sin\theta}{\cos\theta}$

Relationships for tangent can be worked out from the definition together with the relevant relationships for sine and cosine.

> **WORKED EXAMPLE 18.8**
>
> Using the definition of the tangent function, show that $\tan(\theta + \pi) = \tan\theta$.
>
> Use the definition of tan $\cdots\cdots\cdots\cdots\cdots\cdots$ $\tan(\theta + \pi) = \dfrac{\sin(\theta + \pi)}{\cos(\theta + \pi)}$
>
> Use $\sin(\theta + \pi) = -\sin\theta$ $\cdots\cdots\cdots\cdots\cdots\cdots$ $= \dfrac{-\sin\theta}{-\cos\theta}$
> and $\cos(\theta + \pi) = -\cos\theta$
>
> $= \dfrac{\sin\theta}{\cos\theta}$
>
> Use the definition $\cdots\cdots\cdots\cdots\cdots\cdots$ $= \tan\theta$
> of tan again

Other relationships for tangent can be worked out in a similar way, but again there are a couple of commonly occurring ones that are worth remembering.

> **KEY POINT 18.8**
>
> - $\tan(-\theta) = -\tan\theta$
> - $\tan(\theta + \pi) = \tan\theta$

> **WORKED EXAMPLE 18.9**
>
> Given that $\tan\dfrac{2\pi}{5} = 3.08$, find
>
> a $\quad\tan\dfrac{17\pi}{5}$ $\qquad\qquad$ b $\quad\tan\left(-\dfrac{7\pi}{5}\right)$.
>
> Note that $\dfrac{17\pi}{5} = \dfrac{2\pi}{5} + 3\pi$ $\cdots\cdots\cdots$ a $\quad\tan\dfrac{17\pi}{5} = \tan\dfrac{2\pi}{5}$
>
> Use $\tan(\theta + \pi) = \tan\theta$. This also means that $\qquad\qquad\qquad = 3.08$
> adding on any integer multiple of $\pi$ also does
> no change the value, so $\tan(\theta + 3\pi) = \tan\theta$
>
> Use $\tan(\theta + \pi) = \tan\theta$ $\cdots\cdots\cdots$ b $\quad\tan\left(-\dfrac{7\pi}{5}\right) = \tan\left(-\dfrac{2\pi}{5}\right)$
>
> Then use $\tan(-\theta) = -\tan\theta$ $\cdots\cdots\cdots\cdots\cdots$ $= -\tan\dfrac{2\pi}{5}$
>
> $= -3.08$

You saw in Chapter 5 that the equation of a straight line through the origin is $y = x\tan\theta$, where $\theta$ is the angle formed between the line and the positive $x$-axis.

# ■ Exact values of trigonometric ratios $0, \dfrac{\pi}{6}, \dfrac{\pi}{4}, \dfrac{\pi}{3}, \dfrac{\pi}{2}$ and their multiples

Values of trigonometric functions of most angles are difficult to find without a calculator, but the values of certain angles can be found.

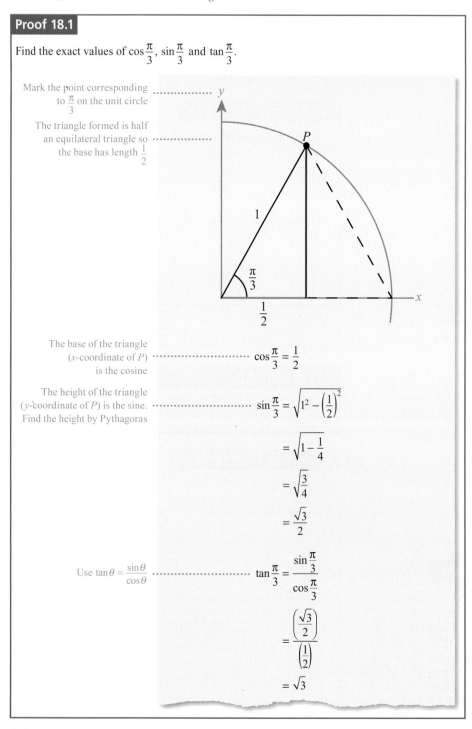

**Proof 18.1**

Find the exact values of $\cos\dfrac{\pi}{3}$, $\sin\dfrac{\pi}{3}$ and $\tan\dfrac{\pi}{3}$.

Mark the point corresponding to $\dfrac{\pi}{3}$ on the unit circle

The triangle formed is half an equilateral triangle so the base has length $\dfrac{1}{2}$

The base of the triangle ($x$-coordinate of $P$) is the cosine

$$\cos\dfrac{\pi}{3} = \dfrac{1}{2}$$

The height of the triangle ($y$-coordinate of $P$) is the sine. Find the height by Pythagoras

$$\sin\dfrac{\pi}{3} = \sqrt{1^2 - \left(\dfrac{1}{2}\right)^2}$$

$$= \sqrt{1 - \dfrac{1}{4}}$$

$$= \sqrt{\dfrac{3}{4}}$$

$$= \dfrac{\sqrt{3}}{2}$$

Use $\tan\theta = \dfrac{\sin\theta}{\cos\theta}$

$$\tan\dfrac{\pi}{3} = \dfrac{\sin\dfrac{\pi}{3}}{\cos\dfrac{\pi}{3}}$$

$$= \dfrac{\left(\dfrac{\sqrt{3}}{2}\right)}{\left(\dfrac{1}{2}\right)}$$

$$= \sqrt{3}$$

Other values can be worked out in the same way each time, but they occur so often that you should learn them.

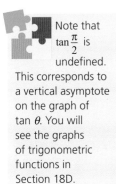

Note that $\tan\frac{\pi}{2}$ is undefined. This corresponds to a vertical asymptote on the graph of $\tan\theta$. You will see the graphs of trigonometric functions in Section 18D.

**KEY POINT 18.9**

| Radians | 0 | $\frac{\pi}{6}$ | $\frac{\pi}{4}$ | $\frac{\pi}{3}$ | $\frac{\pi}{2}$ | $\pi$ |
|---|---|---|---|---|---|---|
| Degrees | 0 | 30 | 45 | 60 | 90 | 180 |
| $\sin\theta$ | 0 | $\frac{1}{2}$ | $\frac{\sqrt{2}}{2}$ | $\frac{\sqrt{3}}{2}$ | 1 | 0 |
| $\cos\theta$ | 1 | $\frac{\sqrt{3}}{2}$ | $\frac{\sqrt{2}}{2}$ | $\frac{1}{2}$ | 0 | $-1$ |
| $\tan\theta$ | 0 | $\frac{1}{\sqrt{3}}$ | 1 | $\sqrt{3}$ | Not defined | 0 |

Values greater than $\frac{\pi}{2}$ can be worked out from these values using the relationships for sin, cos and tan in Key Points 18.5, 18.6 and 18.8 as shown in Proof 18.1.

 **WORKED EXAMPLE 18.10**

Find the exact value of $\sin\frac{4\pi}{3}$.

Relate $\frac{4\pi}{3}$ to one of the values in the table in Key Point 18.9 ......... $\sin\left(\frac{4\pi}{3}\right) = \sin\left(\frac{\pi}{3} + \pi\right)$

Use Key Point 18.5: $\sin(\theta + \pi) = -\sin\theta$ ......... $= -\sin\left(\frac{\pi}{3}\right)$

Use the value of $\sin\left(\frac{\pi}{3}\right)$ from the table ......... $= -\frac{\sqrt{3}}{2}$

**CONCEPTS – APPROXIMATION**

For an architect, is it more useful to know that $\sin 60° = \frac{\sqrt{3}}{2}$ or 0.866?

# Be the Examiner 18.1

Find the exact value of $\cos\frac{5\pi}{4}$.

Which is the correct solution? Identify the errors made in the incorrect solutions.

| Solution 1 | Solution 2 | Solution 3 |
|---|---|---|
| $\cos\frac{5\pi}{4} = \cos\left(\frac{\pi}{4} + \pi\right)$ | $\cos\frac{5\pi}{4} = \cos\left(\frac{\pi}{4} + \pi\right)$ | $\cos\frac{5\pi}{4} = \cos\left(\frac{\pi}{4} + \pi\right)$ |
| $= \cos\frac{\pi}{4} + \cos\pi$ | $= -\cos\frac{\pi}{4}$ | $= \cos\left(-\frac{\pi}{4}\right)$ |
| $= \frac{\sqrt{2}}{2} + (-1)$ | $= -\frac{\sqrt{2}}{2}$ | $= \cos\frac{\pi}{4}$ |
| $= \frac{\sqrt{2} - 2}{2}$ | | $= \frac{\sqrt{2}}{2}$ |

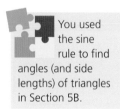

You used the sine rule to find angles (and side lengths) of triangles in Section 5B.

## Tip

Note that there is not a corresponding relationship for cos, so you do not get this second possibility when using the cosine rule.

## ■ Extension of the sine rule to the ambiguous case

You know from Key Point 18.5 that $\sin(\pi - \theta) = \sin\theta$. In other words, there are two angles between 0 and $\pi$ that both have the same sine: $\theta$ and $\pi - \theta$.

This has an immediate implication for finding angles in triangles using the sine rule (where angles will usually be measured in degrees).

### KEY POINT 18.10

When using the sine rule to find an angle, there may be two possible solutions: $\theta$ and $180 - \theta$.

Be aware that just because there is the possibility of a second value for an angle, this does not necessarily mean that a second triangle exists. You always need to check whether the angle sum is less than 180°.

### WORKED EXAMPLE 18.11

Find the size of the angle $\theta$ in the following triangle.

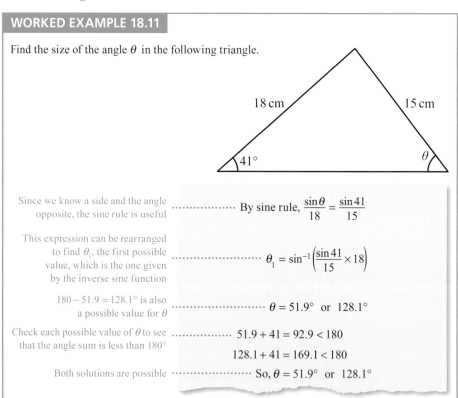

Since we know a side and the angle opposite, the sine rule is useful ·········· By sine rule, $\dfrac{\sin\theta}{18} = \dfrac{\sin 41}{15}$

This expression can be rearranged to find $\theta_1$, the first possible value, which is the one given by the inverse sine function ·········· $\theta_1 = \sin^{-1}\left(\dfrac{\sin 41}{15} \times 18\right)$

$180 - 51.9 = 128.1°$ is also a possible value for $\theta$ ·········· $\theta = 51.9°$ or $128.1°$

Check each possible value of $\theta$ to see ·········· $51.9 + 41 = 92.9 < 180$
that the angle sum is less than 180°

$128.1 + 41 = 169.1 < 180$

Both solutions are possible ·········· So, $\theta = 51.9°$ or $128.1°$

The diagram below shows both possible triangles. Note that the 41° angle must remain opposite the 15 cm side, as given.

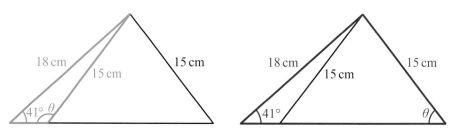

### Be the Examiner 18.2

In triangle $ABC$, $AB = 10\,\text{cm}$, $AC = 12\,\text{cm}$ and $A\hat{B}C = 70°$.

Find the size of $A\hat{C}B$.

Which is the correct solution? Identify the errors made in the incorrect solutions.

| Solution 1 | Solution 2 | Solution 3 |
|---|---|---|
| $\dfrac{\sin\theta}{10} = \dfrac{\sin 70°}{12}$ | $\dfrac{\sin\theta}{12} = \dfrac{\sin 70°}{10}$ | $\dfrac{\sin\theta}{10} = \dfrac{\sin 70°}{12}$ |
| $\sin\theta = \dfrac{10\sin 70°}{12}$ | $\sin\theta = \dfrac{12\sin 70°}{10}$ | $\sin\theta = \dfrac{10\sin 70°}{12}$ |
| $\theta = \sin^{-1}\left(\dfrac{10\sin 70°}{12}\right)$ | $= 1.13$ | $\theta = \sin^{-1}\left(\dfrac{10\sin 70°}{12}\right)$ |
| $= 51.5°$ | $1.13 > 1$ | $= 51.5°$ |
| $180 - 51.5 = 128.5$ | So, there are no solutions. | $180 - 51.5 = 128.5$ |
| So, $\theta = 51.5°$ | | So, $\theta = 51.5°$ or $128.5°$ |

## Exercise 18B

For questions 1 to 4, you are given that $\sin\theta = 0.6$. Use the method demonstrated in Worked Example 18.6 to find the value of the following.

1  a  $\sin(\theta - \pi)$
   b  $\sin(\theta + \pi)$

2  a  $\sin(2\pi + \theta)$
   b  $\sin(2\pi - \theta)$

3  a  $\cos\left(\theta + \dfrac{\pi}{2}\right)$
   b  $\cos\left(\theta - \dfrac{\pi}{2}\right)$

4  a  $\cos\left(\dfrac{\pi}{2} - \theta\right)$
   b  $\cos\left(\dfrac{3\pi}{2} - \theta\right)$

For questions 5 to 8, you are given that $\cos 160° = -0.940$. Use the method demonstrated in Worked Example 18.7 to find the value of the following.

5  a  $\cos 520°$
   b  $\cos(-200°)$

6  a  $\cos(-160°)$
   b  $\cos 200°$

7  a  $\cos 340°$
   b  $\cos 20°$

8  a  $\sin 250°$
   b  $\sin 70°$

For questions 9 to 11, use the method demonstrated in Worked Example 18.8 to show that the following relationships hold.

9  a  $\tan(\theta - \pi) = \tan\theta$
   b  $\tan(2\pi - \theta) = -\tan\theta$

10  a  $\tan(-\theta) = -\tan\theta$
    b  $\tan(\pi - \theta) = -\tan\theta$

11  a  $\tan\left(\dfrac{\pi}{2} - \theta\right) = \dfrac{1}{\tan\theta}$
    b  $\tan\left(\dfrac{\pi}{2} + \theta\right) = -\dfrac{1}{\tan\theta}$

For questions 12 to 14, you are given that $\tan\left(\dfrac{\pi}{5}\right) = 0.73$. Use the method demonstrated in Worked Example 18.9 to find the value of the following.

12  a  $\tan\left(-\dfrac{\pi}{5}\right)$
    b  $\tan\left(\dfrac{9\pi}{5}\right)$

13  a  $\tan\left(\dfrac{6\pi}{5}\right)$
    b  $\tan\left(-\dfrac{9\pi}{5}\right)$

14  a  $\tan\left(\dfrac{4\pi}{5}\right)$
    b  $\tan\left(\dfrac{14\pi}{5}\right)$

For questions 15 to 20, use the method demonstrated in Worked Example 18.10 to find the exact value of the following.

15  a  $\sin\left(\dfrac{5\pi}{6}\right)$
    b  $\sin\left(\dfrac{7\pi}{6}\right)$

16  a  $\cos\left(\dfrac{7\pi}{4}\right)$
    b  $\cos\left(\dfrac{3\pi}{4}\right)$

17  a  $\tan\left(\dfrac{2\pi}{3}\right)$
    b  $\tan\left(\dfrac{5\pi}{3}\right)$

18  a  $\sin 120°$
    b  $\sin 300°$

19  a  $\cos 225°$
    b  $\cos 135°$

20  a  $\tan 150°$
    b  $\tan 210°$

For questions 21 to 24, use the method demonstrated in Worked Example 18.11 to find the possible size(s) of the angle $\theta$ in each triangle. Give your answers correct to the nearest degree.

21  a

b

22  a

b

23  a

b

24  a

b

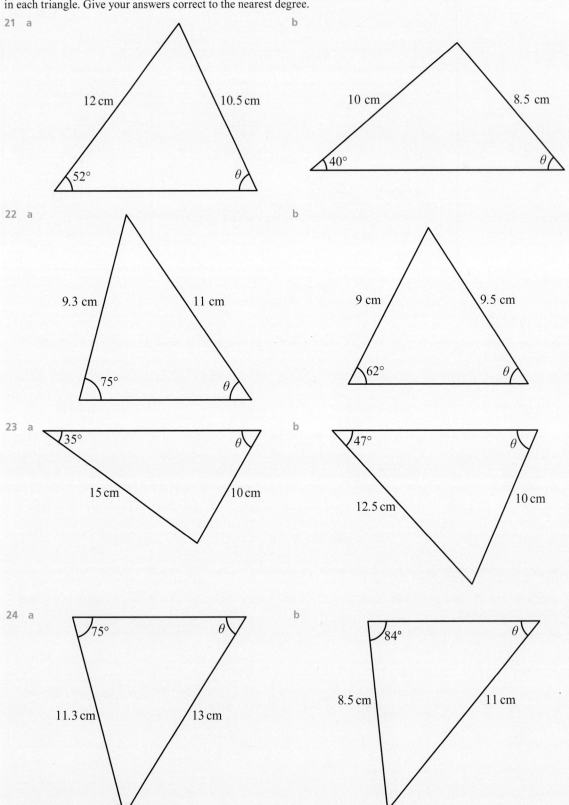

25 Evaluate $\sin x + \sin(2\pi - x)$.

26 Show that $\sin 60° \cos 30° + \cos 60° \sin 30° = \sin 90°$.

27 Find the exact value of $\left(\tan\left(\dfrac{\pi}{3}\right) - 1\right)^2$.

28 Write $\tan\left(\dfrac{\pi}{6}\right) + \cos\left(\dfrac{\pi}{6}\right)$ in the form $k\sqrt{3}$, where $k$ is a rational number.

 29 Simplify $\cos x + \cos\left(x + \dfrac{\pi}{2}\right) + \cos(x + \pi) + \cos\left(x + \dfrac{3\pi}{2}\right) + \cos(x + 2\pi)$.

30 In triangle $ABC$, $AB = 9$ cm, $AC = 4$ cm and angle $BA = 60°$. Find the exact length of $BC$.

31 In triangle $ABC$, $AB = 8$ cm, $AC = 11$ cm and angle $BAC = \theta°$. The area of the triangle is $35\,\text{cm}^2$. Find the possible values of $\theta$.

32 In triangle $ABC$, $AB = 7.5$ cm, $BC = 5.3$ cm and angle $BAC = 44°$. Find the two possible values of the length $AC$.

33 Triangle $KLM$ has $KL = 12$ cm, $LM = 15$ cm and angle $MKL = 55°$. Show that there is only one possible value for the length of the side $KM$ and find this value.

34 Find the exact length marked $c$ in this triangle.

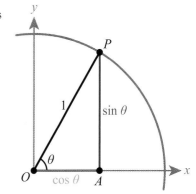

35 Find the equation of the line through the origin which makes an angle of $30°$ with the positive $x$-axis.

36 a Show that the equation of the line through the origin which makes an angle of $\theta$ with the positive $x$-axis is $y = x\tan\theta$.

  b Hence prove that $\tan\theta = \tan(180° + \theta)$.

  c Simplify $\tan\theta \times \tan(90° + \theta)$.

# 18C Trigonometric identities

There are many identities involving trigonometric functions but at this stage there are a few important ones you need to know.

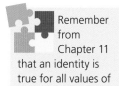

Remember from Chapter 11 that an identity is true for all values of the variable.

## ▉ The Pythagorean identity $\cos^2\theta + \sin^2\theta \equiv 1$

This identity, which relates cosine and sine, follows directly from their definitions on the unit circle:

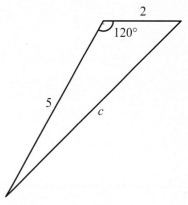

Using Pythagoras in the right-angled triangle $OAP$ gives the identity.

> **KEY POINT 18.11**
>
> ● $\cos^2\theta + \sin^2\theta \equiv 1$

You will meet the double angle formula for tan if you study Analysis and approaches HL.

**Tip**

The second and third versions of the $\cos 2\theta$ identity follow from the first using $\cos^2\theta + \sin^2\theta \equiv 1$ to substitute for the function you do not want.

## Double angle identities for sine and cosine

Double angle identities relate the value of $\sin 2\theta$ or $\cos 2\theta$ to values of $\sin\theta$ and/or $\cos\theta$.

> **KEY POINT 18.12**
>
> ● $\sin 2\theta \equiv 2\sin\theta\cos\theta$
>
> ● $\cos 2\theta \equiv \begin{cases} \cos^2\theta - \sin^2\theta \\ 2\cos^2\theta - 1 \\ 1 - 2\sin^2\theta \end{cases}$

**Proof 18.2**

Prove that $\sin 2\theta = 2\sin\theta\cos\theta$.

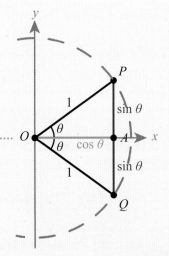

Bisect an angle of $2\theta$ in a triangle to give triangles with angle $\theta$

Use the unit circle in order to involve the definitions of sin and cos

Use $A = \dfrac{1}{2}ab\sin C$ to find an expression for the area of triangle $OPQ$ ┈┈┈ Area of triangle $OPQ = \dfrac{1}{2} \times 1 \times 1 \times \sin 2\theta$

$$= \dfrac{1}{2}\sin 2\theta$$

Use $A = \dfrac{1}{2}bh$ to find the area of the right-angled triangles $OAP$ and $OAQ$, and note that by the definition of sin and cos, $b = \cos\theta$ and $h = \sin\theta$ ┈┈┈ Area of triangle $OAP = \dfrac{1}{2}\sin\theta\cos\theta$

Area of triangle $OAQ = \dfrac{1}{2}\sin\theta\cos\theta$

Area of $OPQ =$ Area of $OAP +$ Area of $OAQ$ ┈┈┈ So, $\dfrac{1}{2}\sin 2\theta \equiv \dfrac{1}{2}\sin\theta\cos\theta + \dfrac{1}{2}\sin\theta\cos\theta$

$$\sin 2\theta \equiv 2\sin\theta\cos\theta$$

## ◼ The relationship between trigonometric ratios

One use of identities is to find the exact value of a particular trigonometric function when given the exact value of another trigonometric function.

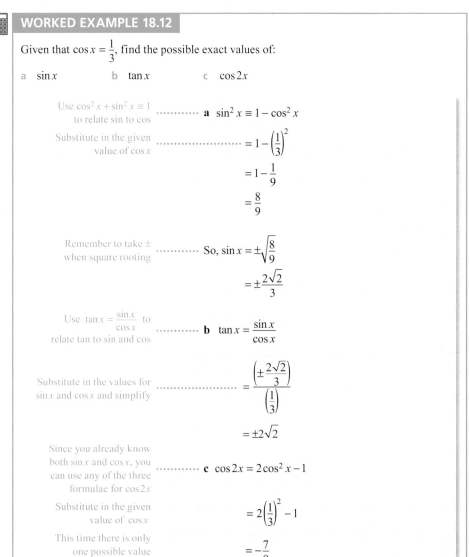

**WORKED EXAMPLE 18.12**

Given that $\cos x = \dfrac{1}{3}$, find the possible exact values of:

a  $\sin x$            b  $\tan x$            c  $\cos 2x$

| | |
|---|---|
| Use $\cos^2 x + \sin^2 x \equiv 1$ to relate sin to cos | **a** $\sin^2 x \equiv 1 - \cos^2 x$ |
| Substitute in the given value of $\cos x$ | $= 1 - \left(\dfrac{1}{3}\right)^2$ |
| | $= 1 - \dfrac{1}{9}$ |
| | $= \dfrac{8}{9}$ |
| Remember to take $\pm$ when square rooting | So, $\sin x = \pm\sqrt{\dfrac{8}{9}}$ |
| | $= \pm\dfrac{2\sqrt{2}}{3}$ |
| Use $\tan x = \dfrac{\sin x}{\cos x}$ to relate tan to sin and cos | **b** $\tan x = \dfrac{\sin x}{\cos x}$ |
| Substitute in the values for $\sin x$ and $\cos x$ and simplify | $= \dfrac{\left(\pm\dfrac{2\sqrt{2}}{3}\right)}{\left(\dfrac{1}{3}\right)}$ |
| | $= \pm 2\sqrt{2}$ |
| Since you already know both $\sin x$ and $\cos x$, you can use any of the three formulae for $\cos 2x$ | **c** $\cos 2x = 2\cos^2 x - 1$ |
| Substitute in the given value of $\cos x$ | $= 2\left(\dfrac{1}{3}\right)^2 - 1$ |
| This time there is only one possible value | $= -\dfrac{7}{9}$ |

Notice that in Worked Example 18.12 there were two possible answers for $\sin x$ and $\tan x$ due to the $\pm$ when square rooting.

If $x$ is restricted in the question, you will need to decide which of the two possible answers to give.

**WORKED EXAMPLE 18.13**

Given that $\sin x = \dfrac{3}{4}$ and $\dfrac{\pi}{2} < x < \pi$, find the exact value of $\sin 2x$.

$\sin 2x = 2\sin x \cos x$ so first you need to find the value of $\cos x$  ⋯⋯⋯⋯⋯⋯⋯⋯⋯  $\cos^2 x \equiv 1 - \sin^2 x$

To do this, use $\cos^2 x + \sin^2 x \equiv 1$  ⋯⋯⋯⋯⋯⋯⋯⋯⋯  $= 1 - \left(\dfrac{3}{4}\right)^2$

$= 1 - \dfrac{9}{16}$

$= \dfrac{7}{16}$

In the given interval, $\cos x < 0$ (you can see this from the unit circle)  ⋯⋯⋯⋯⋯⋯  Since $\dfrac{\pi}{2} < x < \pi$,

$\cos x = -\sqrt{\dfrac{7}{16}}$

$= -\dfrac{\sqrt{7}}{4}$

Now use $\sin 2x \equiv 2\sin x \cos x$  ⋯⋯⋯⋯⋯⋯⋯⋯  So, $\sin 2x \equiv 2\sin x \cos x$

$= 2\left(\dfrac{3}{4}\right)\left(-\dfrac{\sqrt{7}}{4}\right)$

$= -\dfrac{3\sqrt{7}}{8}$

One common application of double angle identities is finding exact values of half angles.

**WORKED EXAMPLE 18.14**

Use the exact value of $\cos 30°$ to show that $\sin 15° = \sqrt{\dfrac{2 - \sqrt{3}}{4}}$.

You need to relate $\sin 15°$ to the exact value of $\cos 30°$. The version of the cos double angle identity involving sin will do this  ⋯⋯⋯⋯⋯  Using $\cos 2\theta \equiv 1 - 2\sin^2 \theta$,

$\cos 30° = 1 - 2\sin^2 15°$

$\sin^2 15° = \dfrac{1 - \cos 30°}{2}$

Substitute in $\cos 30° = \dfrac{\sqrt{3}}{2}$  ⋯⋯⋯⋯⋯  $= \dfrac{1 - \dfrac{\sqrt{3}}{2}}{2}$

Multiply through by 2  ⋯⋯⋯⋯⋯  $= \dfrac{2 - \sqrt{3}}{4}$

You know that $\sin 15° > 0$ so take the positive square root  ⋯⋯⋯  Since $\sin 15° > 0$,

$\sin 15° = \sqrt{\dfrac{2 - \sqrt{3}}{4}}$

In Chapter 11, you learned how to prove identities by starting with one side and transforming it until you get the other side.

You can use the identities you met in this section to prove other identities.

**WORKED EXAMPLE 18.15**

Prove that $\dfrac{\cos x - \sin x}{\cos x + \sin x} \equiv \dfrac{1 - \sin 2x}{\cos 2x}$.

Start from the left-hand side (LHS). ············ $\text{LHS} = \dfrac{\cos x - \sin x}{\cos x + \sin x}$

You want to change to a fraction
with the denominator $\cos 2x$, which
equals $\cos^2 x - \sin^2 x$. So multiply
top and bottom by $(\cos x - \sin x)$

$\equiv \dfrac{(\cos x - \sin x)(\cos x - \sin x)}{(\cos x + \sin x)(\cos x - \sin x)}$

Expand the brackets and simplify ··········· $\equiv \dfrac{\cos^2 x + \sin^2 x - 2\sin x \cos x}{\cos^2 x - \sin^2 x}$

Use the identities:
$\cos^2 x + \sin^2 x \equiv 1$
$2\sin x \cos x \equiv \sin 2x$ ············· $\equiv \dfrac{1 - \sin 2x}{\cos 2x} = \text{LHS}$
$\cos^2 x - \sin^2 x \equiv \cos 2x$

## Exercise 18C

For questions 1 to 6, use the method demonstrated in Worked Example 18.12 to find possible value(s) of the trigonometric ratios.

1   a   Find $\sin x$ given that $\cos x = \dfrac{1}{4}$.

    b   Find $\sin x$ given that $\cos x = \dfrac{2}{3}$.

2   a   Find $\cos x$ given that $\sin x = -\dfrac{1}{3}$.

    b   Find $\cos x$ given that $\sin x = -\dfrac{3}{4}$.

3   a   Find $\cos 2x$ given that $\cos x = \dfrac{1}{2}$.

    b   Find $\cos 2x$ given that $\cos x = \dfrac{1}{3}$.

4   a   Find $\cos 2x$ given that $\sin x = -\dfrac{2}{3}$.

    b   Find $\cos 2x$ given that $\sin x = -\dfrac{1}{4}$.

5   a   Find $\cos x$ and $\tan x$ given that $\sin x = \dfrac{2}{5}$.

    b   Find $\cos x$ and $\tan x$ given that $\sin x = \dfrac{3}{4}$.

6   a   Find $\sin x$ and $\sin 2x$ given that $\cos x = -\dfrac{1}{4}$.

    b   Find $\sin x$ and $\sin 2x$ given that $\cos x = -\dfrac{1}{5}$.

For questions 7 to 10, use the method demonstrated in Worked Example 18.13 to find the exact values of trigonometric ratios.

7   a   Given that $\cos x = \dfrac{2}{3}$ and $0 < x < \dfrac{\pi}{2}$, find $\sin 2x$.

    b   Given that $\cos x = \dfrac{3}{4}$ and $0 < x < \dfrac{\pi}{2}$, find $\sin 2x$.

8   a   Given that $\sin x = -\dfrac{2}{5}$ and $\dfrac{3\pi}{2} < x < 2\pi$, find $\tan x$.

    b   Given that $\sin x = -\dfrac{1}{3}$ and $\dfrac{3\pi}{2} < x < 2\pi$, find $\tan x$.

9   a   Given that $\sin x = -\dfrac{1}{4}$ and $\pi < x < \dfrac{3\pi}{2}$, find $\sin 2x$.

    b   Given that $\sin x = -\dfrac{1}{3}$ and $\pi < x < \dfrac{3\pi}{2}$, find $\sin 2x$.

10   a   Given that $\cos x = -\dfrac{2}{3}$ and $\dfrac{\pi}{2} < x < \pi$, find $\tan x$.

    b   Given that $\cos x = -\dfrac{1}{2}$ and $\dfrac{\pi}{2} < x < \pi$, find $\tan x$.

For questions 11 to 14, use the method demonstrated in Worked Example 18.14, together with the exact values of trigonometric functions from Key Point 18.9, to find the exact values of the following.

11   a   $\sin 22.5°$

    b   $\cos 22.5°$

12   a   $\cos 75°$

    b   $\sin 75°$

13   a   $\sin \dfrac{\pi}{12}$

    b   $\cos \dfrac{\pi}{12}$

14   a   $\cos \dfrac{3\pi}{8}$

    b   $\sin \dfrac{3\pi}{8}$

15   Given that $\cos\theta = \dfrac{2}{5}$ and $\pi < \theta < 2\pi$, find the exact value of

    a   $\sin\theta$

    b   $\sin 2\theta$

16   Given that $\sin x = \dfrac{3}{7}$ and $\dfrac{\pi}{2} < x < \pi$, find the exact value of

    a   $\cos x$

    b   $\tan x$

**17** Given that $\sin\theta = -\dfrac{4}{9}$,

 a find the exact value of $\cos 2\theta$.          b find the possible values of $\cos\theta$.

**18** Express $3\sin^2 x + 7\cos^2 x$ in terms of $\sin x$ only.

**19** Express $4\cos^2 x - 5\sin^2 x$ in terms of $\cos x$ only.

**20** Prove that $1 + \tan^2\theta \equiv \dfrac{1}{\cos^2\theta}$.

**21** Prove the identity $(\sin x + \cos x)^2 + (\sin x - \cos x)^2 \equiv 2$.

**22** Show that $\cos 15° = \sqrt{\dfrac{2+\sqrt{3}}{4}}$.

**23** a Write down the exact value of $\cos\dfrac{\pi}{4}$.

 b Show that $\sin\dfrac{\pi}{8} = \sqrt{\dfrac{2-\sqrt{2}}{4}}$.

**24** Given that $\cos 2x = \dfrac{5}{6}$ and that $0 < x < \dfrac{\pi}{2}$, find the exact value of $\cos x$.

**25** Given that $\cos 2x = \dfrac{1}{3}$ and that $0 < x < \dfrac{\pi}{2}$, find the exact value of $\sin x$.

**26** Given that $\cos x = \dfrac{2}{3}$, find the exact value of $\cos 4x$.

**27** a Write down an expression for $\cos 2x$ in terms of $\sin x$.

 b Show that $\cos 4x \equiv 1 - 8\sin^2 x + 8\sin^4 x$.

**28** Prove that $\dfrac{1}{\sin^2 A} - \dfrac{1}{\tan^2 A} \equiv 1$.

**29** Prove that $\dfrac{1 - \cos 2\theta}{1 + \cos 2\theta} \equiv \tan^2\theta$.

**30** Prove that $\dfrac{\sin 2x}{\tan x} \equiv 2\cos^2 x$.

**31** Prove the identity $\dfrac{1}{\cos\theta} - \cos\theta \equiv \sin\theta\tan\theta$.

**32** Given that $0 < x < \dfrac{\pi}{2}$ is such that $2\sin 4x = 3\sin 2x$, find the exact value of $\cos x$.

**33** a Show that $\sin\dfrac{3\pi}{8} = \sqrt{\dfrac{2+\sqrt{2}}{4}}$ and find the exact value of $\cos\dfrac{3\pi}{8}$.

 b Hence show that $\tan\dfrac{3\pi}{8} = 1 + \sqrt{2}$.

# 18D Graphs of trigonometric functions

## ■ The circular functions $\sin x$, $\cos x$ and $\tan x$

Using the definition of sine from the unit circle, we can draw its graph:

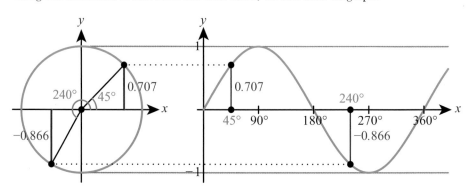

Sine repeats every $2\pi$ radians, so we say it is a periodic function with **period** $2\pi$ radians.

It has a minimum value of $-1$ and a maximum value of $+1$, so we say it has an **amplitude** of 1.

**KEY POINT 18.13**

The graph of $y = \sin x$:

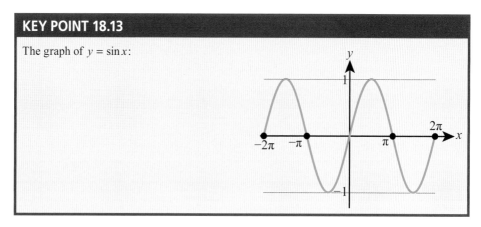

We can construct the cosine graph from the unit circle definition in the same way:

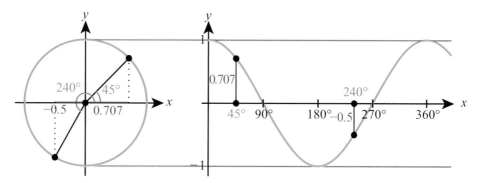

Like sine, cosine has a period of $2\pi$ radians and an amplitude of 1.

**KEY POINT 18.14**

The graph of $y = \cos x$:

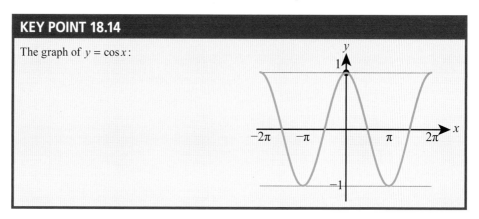

From the definition $\tan x = \dfrac{\sin x}{\cos x}$, we know that $y = \tan x$ will have vertical asymptotes whenever $\cos x = 0$, that is, at $x = \pm\dfrac{\pi}{2}, \pm\dfrac{3\pi}{2}$, and so on. We also know that $\tan x = 0$ whenever $\sin x = 0$, that is, at $x = 0, \pm\pi, \pm 2\pi$, and so on.

Tan is periodic, but this time the period is $\pi$ radians.

**KEY POINT 18.15**

The graph of $y = \tan x$:

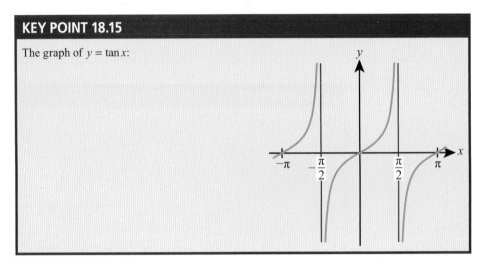

## Transformations of trigonometric functions

We can now apply the transformations you met in Section 16A to trigonometric functions.

**WORKED EXAMPLE 18.16**

a  Sketch the graph of $y = \tan\left(x - \dfrac{\pi}{4}\right)$ for $-\dfrac{3\pi}{2} \leqslant x \leqslant \dfrac{3\pi}{2}$, labelling all axis intercepts.

b  State the equations of the asymptotes.

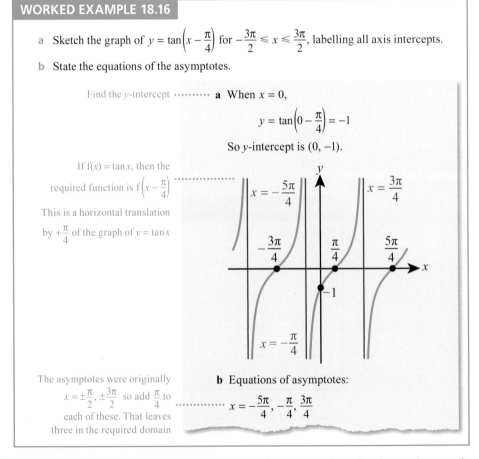

Find the $y$-intercept ·········· **a**  When $x = 0$,

$$y = \tan\left(0 - \frac{\pi}{4}\right) = -1$$

So $y$-intercept is $(0, -1)$.

If $f(x) = \tan x$, then the required function is $f\left(x - \dfrac{\pi}{4}\right)$

This is a horizontal translation by $+\dfrac{\pi}{4}$ of the graph of $y = \tan x$

The asymptotes were originally $x = \pm\dfrac{\pi}{2}, \pm\dfrac{3\pi}{2}$ so add $\dfrac{\pi}{4}$ to each of these. That leaves three in the required domain

**b**  Equations of asymptotes:

$$x = -\frac{5\pi}{4}, -\frac{\pi}{4}, \frac{3\pi}{4}$$

You can apply this process each time you meet a function such as this, but with sin and cos it is useful to be able to recognize straight away what effect a transformation will have on the amplitude and period.

 You know from Section 16A that $a$ will cause a vertical stretch with scale factor $a$ and $b$ will cause a horizontal stretch with scale factor $\frac{1}{b}$.

**KEY POINT 18.16**

The functions $a\sin(b(x+c))+d$ and $a\cos(b(x+c))+d$ have:

● amplitude $a$  ● period $\frac{2\pi}{b}$

The constant $d$ causes a vertical translation: $d$ will be halfway between the maximum and minimum values of the function.

The constant $c$ causes a horizontal translation, but care is needed here as you have not met a combination of two $x$-transformations (here $b$ and $c$ both affect $x$-coordinates). The following Worked Example shows how to deal with this.

**WORKED EXAMPLE 18.17**

For the function $f(x) = 2\cos\left(3\left(x+\frac{\pi}{6}\right)\right)+4,\ 0 \leqslant x \leqslant \pi$:

a  state the amplitude

b  state the period

c  find the coordinates of the maximum point

d  hence, sketch the graph of $y = f(x)$.

Compare to $a\cos(b(x+c))+d$, which has amplitude $a$ ············ **a** Amplitude $= 2$

Compare to $\cos(b(x+c))+d$, which has period $\frac{2\pi}{b}$ ············ **b** Period $= \frac{2\pi}{3}$

The max points of $\cos x$ occur when $x = 0, 2\pi, 4\pi$, and so on, but here $x$ has been replaced by $3\left(x+\frac{\pi}{6}\right)$ ············ **c** For max points,

$$3\left(x+\frac{\pi}{6}\right) = 0, 2\pi, 4\pi$$

$$x = -\frac{\pi}{6}, \frac{\pi}{2}, \frac{7\pi}{6}$$

The function is restricted to $0 \leqslant x \leqslant \pi$ ············ So, $x = \frac{\pi}{2}$

The function has moved up by 4 and has amplitude 2 ············ $y = 4+2 = 6$

So, max point is $\left(\frac{\pi}{2}, 6\right)$

**d**

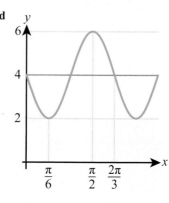

You can also work backwards from a given graph to find its equation.

**WORKED EXAMPLE 18.18**

The graph here has equation $y = a\sin(bx) + d$.

Find the values of $a$, $b$ and $d$.

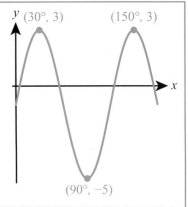

| The amplitude, $a$, is half of the difference between the max and min $y$-values | $a = \dfrac{3 - (-5)}{2} = 4$ |
|---|---|
| Find the period... | Period $= 150 - 30 = 120°$ |
| ... and then use the fact that $\text{period} = \dfrac{360°}{b}$ (note degrees, here) | $120 = \dfrac{360}{b}$ <br> $b = 3$ |
| $d$ is halfway between the max and min values | $d = \dfrac{3 + (-5)}{2} = -1$ |

## ▦ Real-life contexts

Sine and cosine functions are very useful in modelling periodic motion. This might be an object moving round a circular path, such as a Ferris wheel at a fairground, or an object oscillating on the end of a spring or the motion of the tide.

**WORKED EXAMPLE 18.19**

On one particular day, the depth of water in a harbour is 11 m at high tide and 5 m at low tide. High tide is at midnight and low tide at 12:00.

Find a model for the depth of water in the harbour, $d$, $t$ hours after midnight in the form $d = k\cos pt + q$, assuming that high tide occurs 24 hours after the previous high tide.

| The amplitude ($k$) is half of the difference between the highest and lowest points | $k = \dfrac{11 - 5}{2} = 3$ |
|---|---|
| There are 24 hours between high tides, so the period is 24 | Period $= 24$ hours |
| Use $\text{period} = \dfrac{2\pi}{p}$ | $24 = \dfrac{2\pi}{p}$ <br> $p = \dfrac{\pi}{12}$ |
| $q$ is halfway between the highest and lowest points | $q = \dfrac{11 + 5}{2} = 8$ |
| | So, $d = 3\cos\left(\dfrac{\pi}{12}t\right) + 8$ |

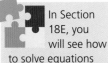 In Section 18E, you will see how to solve equations involving functions like this so that you can find the times when a tide is at a certain height.

## Exercise 18D

For questions 1 to 4, use the method demonstrated in Worked Example 18.16 to sketch the graph, labelling the axis intercepts and equations of the asymptotes.

1   a   $y = \tan\left(x - \dfrac{\pi}{6}\right)$ for $-\dfrac{\pi}{3} \leqslant x \leqslant \dfrac{5\pi}{3}$

   b   $y = \tan\left(x - \dfrac{\pi}{3}\right)$ for $-\dfrac{\pi}{6} \leqslant x \leqslant \dfrac{11\pi}{6}$

2   a   $y = \tan\left(x + \dfrac{\pi}{2}\right)$ for $0 < x < 2\pi$

   b   $y = \tan\left(x + \dfrac{\pi}{4}\right)$ for $-\dfrac{3\pi}{2} < x < \dfrac{3\pi}{2}$

3   a   $y = \tan 2x$ for $0 \leqslant x \leqslant 2\pi$

   b   $y = \tan 3x$ for $0 \leqslant x \leqslant \pi$

4   a   $y = \tan\left(\dfrac{x}{3}\right)$ for $0 \leqslant x \leqslant \pi$

   b   $y = \tan\left(\dfrac{x}{2}\right)$ for $-\pi \leqslant x \leqslant \pi$

For questions 5 to 8, use the method demonstrated in Worked Example 18.17 to find the period and the amplitude of each function, and the coordinates of the maximum point. Hence sketch the graph of $y = f(x)$ for $0 \leqslant x \leqslant 2\pi$.

5   a   $f(x) = 3\cos\left(2\left(x + \dfrac{\pi}{2}\right)\right) + 1$

   b   $f(x) = 2\cos\left(3\left(x + \dfrac{\pi}{3}\right)\right) - 1$

6   a   $f(x) = 2\sin\left(3\left(x - \dfrac{\pi}{3}\right)\right) + 2$

   b   $f(x) = 3\sin\left(2\left(x - \dfrac{\pi}{2}\right)\right) - 2$

7   a   $f(x) = \sin\left(\dfrac{1}{2}\left(x + \dfrac{\pi}{2}\right)\right) - 3$

   b   $f(x) = \cos\left(\dfrac{1}{2}\left(x - \dfrac{\pi}{2}\right)\right) + 3$

8   a   $f(x) = 2 - 3\sin(x + \pi)$

   b   $f(x) = 3 - 2\sin(x - \pi)$

For questions 9 to 12, you are given a function and its graph. Use the method demonstrated in Worked Example 18.18 to find the values of the constants in the equation.

9   a   $y = a\sin(bx) + d$

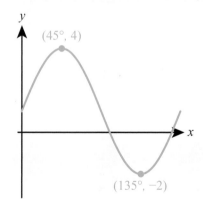

   b   $y = a\sin(bx) - d$

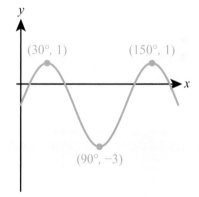

10   a   $y = a\sin(x - c) + d$

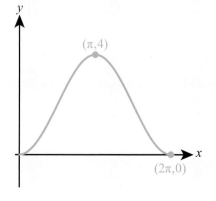

   b   $y = a\sin(x - c) + d$

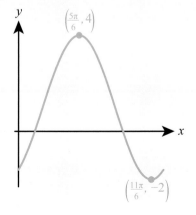

11    a    $y = a\cos(bx) + d$                                    b    $y = a\cos(bx) + d$

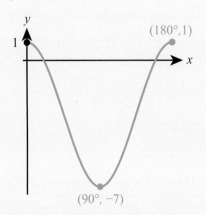

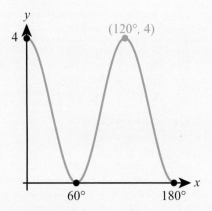

12    a    $y = a\cos(x + c) + d$                                 b    $y = a\cos(x + c) + d$

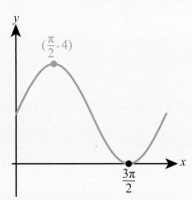

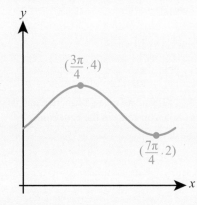

13    Sketch the graph of $y = \tan(x - 45°)$ for $0 \leqslant x \leqslant 180°$, showing the equations of its asymptotes.

14    Find the exact period of the function $f(x) = 3\sin(4x)$, where $x$ is in radians.

15    Find the minimum and maximum values of the function $g(x) = 10\cos(2x) - 7$.

16    The height, $h$ m, of a swing above ground is given by $h = 0.8 - 0.6\cos(\pi t)$, where $t$ is the time in seconds. Find the greatest height of the swing above ground.

17    The depth of water in a harbour varies according to the equation $d = 5 + 1.6\sin\left(\dfrac{\pi}{12}t\right)$, where $d$ is measured in metres and $t$ is the time, in hours, after midnight.

      a    Find the depth of water at high tide.

      b    Find the first time after midnight when the high tide occurs.

18    The height of a seat on a Ferris wheel above ground is modelled by the equation $h = 6.2 - 4.8\cos\left(\dfrac{\pi}{4}t\right)$, where $h$ is measured in metres and $t$ is the time, in minutes, since the start of the ride.

      a    How long does the wheel take to complete one revolution?

      b    Find the maximum height of the seat above ground.

      c    Find the height of the seat above ground 2 minutes and 40 seconds after the start of the ride.

**19** The diagram shows the graph with equation $y = a\sin(bx)$.
Find the values of $a$ and $b$.

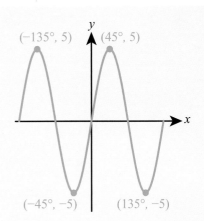

**20** The graph of $y = 4\tan(bx)$ is shown on the right.
Find the value of $b$.

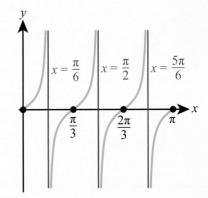

**21** a Sketch the graph of $y = \cos 2x$ and $y = 3\sin x$, for $0 \leqslant x \leqslant 2\pi$.

   b State the number of solutions of the equation $\cos 2x = 3\sin x$ for $0 \leqslant x \leqslant 2\pi$.

   c Find the number of solutions of the equation $\cos 2x = 3\sin x$ for $-2\pi \leqslant x \leqslant 4\pi$.

**22** A ball is attached to one end of a spring and hangs vertically. The ball is then pulled down and released. In subsequent motion, the height of the ball, $h$ metres, above ground at time $t$ seconds is given by $1.4 - 0.2\cos(15t)$.

   a Find the greatest height of the ball above ground.

   b How many full oscillations does the ball perform during the first 3 seconds?

   c Find the second time when the ball is 1.5 m above ground.

**23** Find the period of the function $f(x) = 2\sin 2x + 3\sin 6x$, where $x$ is in radians.

**24** The function $f(x)$ is defined by $f(x) = \dfrac{3}{7 + 3\sin(2x)}$.

   a Does the graph of $f(x)$ have any vertical asymptotes?

   b Find the maximum value of $f(x)$, and the first positive value of $x$ for which this maximum occurs.

In the language of functions from Section 3A, trigonometric functions are many-to-one.

# 18E Trigonometric equations

You can see from the graphs of trigonometric functions that, in general, there will be more than one value of $x$ for a given value of the function.

For example, if $\sin x = 0.5$, then $x$ could take two possible values between 0 and $\pi$: $x = \dfrac{\pi}{6}$ or $x = \pi - \dfrac{\pi}{6} = \dfrac{5\pi}{6}$. If the interval is larger there could be more solutions.

This is an example of a trigonometric equation. You need to be able to solve a range of different trigonometric equations both graphically and analytically.

 ■ **Solving trigonometric equations in a finite interval graphically**

See Section 17B for examples of this technique that do not involve trigonometric functions.

You can solve a trigonometric equation graphically in the same way that you have solved other equations with your GDC.

---
**WORKED EXAMPLE 18.20**

Using the function $d = 3\cos\left(\dfrac{\pi}{12}t\right) + 8$ from Worked Example 18.19, find the times at which the depth of water in the harbour is 9 m.

Use your GDC to find the intersection of $y = 3\cos\left(\dfrac{\pi}{12}x\right) + 8$ and $y = 9$ ............... $t = 4.702, 19.298$

```
Y1=3cos (πX÷12)+8
Y2=9



                              ISECT
X=4.701918624   Y=9
```

Convert 0.702 and 0.298 to minutes ............... So at 04:42 and 19:18
by multiplying by 60

---

■ **Solving trigonometric equations in a finite interval analytically**

You need to able to solve some trigonometric equations without a calculator – in these cases the standard angles from Key Point 18.9 will be involved.

The method varies slightly depending on the trigonometric function involved.

---
**KEY POINT 18.17**

To solve $\sin x = a$
- Find a solution: $x_1 = \sin^{-1} a$
- Find a second solution: $x_2 = \pi - x_1$
- Find other solutions by adding/subtracting $2\pi$ to/from $x_1$ and $x_2$.
---

**WORKED EXAMPLE 18.21**

Solve the equation $\sin x = \dfrac{\sqrt{2}}{2}$, for $0 < x < \dfrac{5\pi}{2}$.

You know that $\sin \dfrac{\pi}{4} = \dfrac{\sqrt{2}}{2}$ .................................... $x_1 = \sin^{-1} \dfrac{\sqrt{2}}{2} = \dfrac{\pi}{4}$

$x_2 = \pi - x_1$ .................................... $x_2 = \pi - \dfrac{\pi}{4} = \dfrac{3\pi}{4}$

Sketch the graph to see how many solutions there are in the required interval. It is easy to miss solutions, otherwise .............

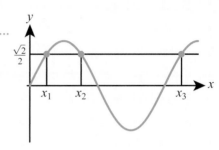

Now find the one remaining solution: ································ $x_3 = \dfrac{\pi}{4} + 2\pi = \dfrac{9\pi}{4}$

$x_3 = x_1 + 2\pi$

So, $x = \dfrac{\pi}{4}, \dfrac{3\pi}{4}, \dfrac{9\pi}{4}$

**KEY POINT 18.18**

To solve $\cos x = a$
- Find a solution: $x_1 = \cos^{-1} a$
- Find a second solution: $x_2 = 2\pi - x_1$
- Find other solutions by adding/subtracting $2\pi$ to/from $x_1$ and $x_2$.

**WORKED EXAMPLE 18.22**

Solve the equation $\cos\theta = 0.5$, for $-360° < \theta < 360°$.

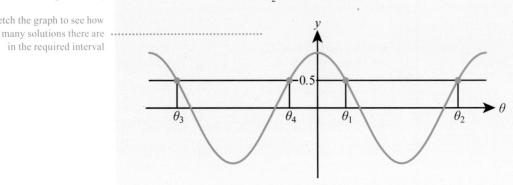

You know that $\cos 60° = 0.5$ ............................................ $\theta_1 = \cos^{-1} 0.5 = 60°$

$\theta_2 = 360 - \theta_1$ ............................................ $\theta_2 = 360 - 60 = 300°$

Sketch the graph to see how
many solutions there are ...............
in the required interval

Now find the two
remaining solutions: ............................................ $\theta_3 = 60 - 360 = -300°$
$\theta_3 = \theta_1 - 360$

$\theta_2 = \theta_2 - 360$ ............................................ $\theta_4 = 300 - 360 = -60°$

So, $\theta = -300°, -60°, 60°, 300°$

**KEY POINT 18.19**

To solve $\tan x = a$
● Find a solution: $x_1 = \tan^{-1} a$
● Find other solutions by adding/subtracting $\pi$ to/from $x_1$.

**WORKED EXAMPLE 18.23**

Solve the equation $\tan\theta = -\sqrt{3}$ for $0 < \theta < 2\pi$.

You know that $\tan\sqrt{3} = \dfrac{\pi}{3}$ and
also that $\tan(-\theta) = -\tan\theta$. ............................................ $\theta_1 = \tan^{-1}\left(-\sqrt{3}\right) = -\dfrac{\pi}{3}$

Therefore, $\tan\left(-\sqrt{3}\right) = -\dfrac{\pi}{3}$

Sketch the graph to see how
many solutions there are in
the required interval. Note ...............
that $\theta_1$ is not in the interval
but it can still be used to
find the solutions that are

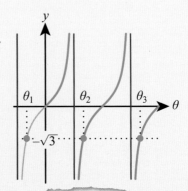

Now find the two
remaining solutions: .................................... $\theta_2 = -\frac{\pi}{3} + \pi = \frac{2\pi}{3}$
$\theta_2 = \theta_1 + \pi$

$\theta_3 = \theta_1 + 2\pi$ ........................... $\theta_3 = -\frac{\pi}{3} + 2\pi = \frac{5\pi}{3}$

So, $\theta = \frac{2\pi}{3}, \frac{5\pi}{3}$

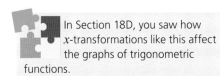 In Section 18D, you saw how $x$-transformations like this affect the graphs of trigonometric functions.

If you have an equation such as $\tan 2\theta = -\sqrt{3}$ to solve, you will need to adapt the process. One way of doing this is to make a substitution ($A = 2\theta$). Remember to change the interval from $\theta$ to the new variable $A$.

**WORKED EXAMPLE 18.24**

Solve the equation $\cos(x + 40°) = \frac{\sqrt{3}}{2}$, for $0° < x < 360°$.

$$\cos(x + 40°) = \frac{\sqrt{3}}{2}$$

Make the substitution
$A = x + 40$ .......................... Let $A = x + 40$:

This changes the interval to
$0 + 40 < x + 40 < 360 + 40$ .......................... $\cos A = \frac{\sqrt{3}}{2}$, for $40° < A < 400°$

Now solve the equation for $A$
using the standard method .......................... $A_1 = \cos^{-1}\frac{\sqrt{3}}{2} = 30°$

$A_2 = 360 - \theta_1$ .......................... $A_2 = 360 - 30 = 330°$

Sketch the graph to see how
many solutions there are in ..........................
the required interval. Note
that $A_1$ is not in the interval
but it can still be used to
find the solutions that are

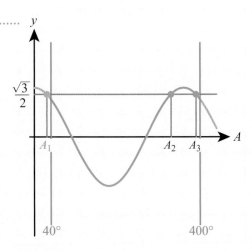

$A_3 = A_1 + 360$ .......................... $A_3 = 30 + 360 = 390°$

So, $A = 330°, 390°$

Convert back to $x$
using $x = A - 40$ .......................... $x = 290°, 350°$

If you have an equation containing more than one trigonometric function, you can often use an identity to convert it into an equation in just one function.

You met double angle identities in Section 18C.

**WORKED EXAMPLE 18.25**

Solve the equation $4\sin\theta\cos\theta = 1$ for $-\pi < \theta < \pi$.

The expression $\sin\theta\cos\theta$ appears in a double angle identity: $\sin 2\theta \equiv 2\sin\theta\cos\theta$

$$4\sin\theta\cos\theta = 1$$

$$2\sin\theta\cos\theta = \frac{1}{2}$$

You now have an equation in just the one trigonometric function

$$\sin 2\theta = \frac{1}{2}$$

Now solve the equation using the standard method

Let $A = 2\theta$:

This changes the interval to $-2\pi < 2\theta < 2\pi$

$\sin A = \frac{1}{2}$, for $-2\pi < A < 2\pi$

You know that $\sin\left(\frac{\pi}{6}\right) = \frac{1}{2}$

$A_1 = \sin^{-1}\frac{1}{2} = \frac{\pi}{6}$

$A_2 = \pi - A_1$

$A_2 = \pi - \frac{\pi}{6} = \frac{5\pi}{6}$

Sketch the graph to see how many solutions there are in the required interval

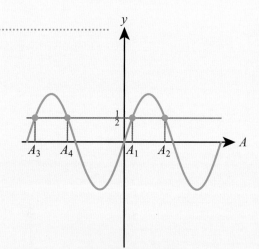

$A_3 = A_1 - 2\pi$

$A_3 = \frac{\pi}{6} - 2\pi = -\frac{11\pi}{6}$

$A_4 = A_2 - 2\pi$

$A_4 = \frac{5\pi}{6} - 2\pi = -\frac{7\pi}{6}$

So, $A = -\frac{11\pi}{6}, -\frac{7\pi}{6}, \frac{\pi}{6}, \frac{5\pi}{6}$

Convert back to $\theta$ using $= \frac{A}{2}$

$\theta = -\frac{11\pi}{12}, -\frac{7\pi}{12}, \frac{\pi}{12}, \frac{5\pi}{12}$

**Tip**

Another useful identity is $\frac{\sin x}{\cos x} \equiv \tan x$. For example, the equation $2\sin x = 3\cos x$ can be written as $\tan x = \frac{2}{3}$.

 You met disguised quadratics in Section 17A.

## ■ Equations leading to quadratic equations in $\sin x$, $\cos x$, $\tan x$

You will often meet trigonometric equations that need to be solved as quadratics.

 **WORKED EXAMPLE 18.26**

Solve the equation $\sin^2 x - 3\sin x + 2 = 0$, for $-\pi < x < \pi$.

Either make the substitution $y = \sin x$ or just note that this is a quadratic in $\sin x$ and factorize

$$\sin^2 x - 3\sin x + 2 = 0$$
$$(\sin x - 1)(\sin x - 2) = 0$$
$$\sin x = 1 \quad \text{or} \quad \sin x = 2$$

$\sin x$ can only take values between $-1$ and $1$

$\sin x = 2$ is impossible, so $\sin x = 1$

Sketch the graph to see that there is only one solution in the required interval

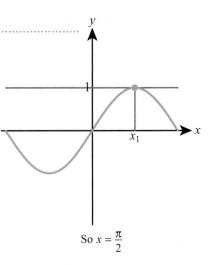

So $x = \dfrac{\pi}{2}$

**Tip**

Remember, you always want to aim for an equation in one trig function. To achieve this, you will often need to use an identity.

You may need to use an identity to turn the equation into a quadratic in one trigonometric function.

 **WORKED EXAMPLE 18.27**

Solve the equation $2\sin^2\theta = \cos\theta + 1$, for $0 < \theta < 360°$.

$$2\sin^2\theta = \cos\theta + 1$$

Use $\sin^2\theta \equiv 1 - \cos^2\theta$

$$2(1 - \cos^2\theta) = \cos\theta + 1$$

This is a quadratic in $\cos\theta$. Rearrange into the standard quadratic form

$$2 - 2\cos^2\theta = \cos\theta + 1$$
$$2\cos^2\theta + \cos\theta - 1 = 0$$

Factorise and solve (using the substitution $y = \cos\theta$ if necessary)

$$(2\cos\theta - 1)(\cos\theta + 1) = 0$$

$$\cos\theta = \frac{1}{2} \quad \text{or} \quad -1$$

A sketch shows there are three solutions in the required interval, two for $\cos\theta = \dfrac{1}{2}$ and one for $\cos\theta = -1$

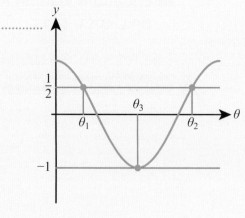

Solve each equation for $\cos\theta$ separately using the standard method

$$\cos\theta = \frac{1}{2}$$

$$\theta = \cos^{-1}\frac{1}{2} = 60°$$

$$\theta_1 = 60°$$

$$\theta_2 = 360 - 60 = 300°$$

$$\cos\theta = -1$$

$$\theta = \cos^{-1}(-1) = 180°$$

$$\theta_3 = 180°$$

So, $\theta = 60°,\ 180°,\ 300°$

## Exercise 18E

For questions 1 to 3, use the method demonstrated in Worked Example 18.20 to find all the solutions of the equation in the given interval. Make sure your GDC is in degrees or radians, as appropriate.

1  a  $2+5\cos(3x)=4$ for $0 \le x \le \pi$

 b  $3+4\cos(2x)=2$ for $0 \le x \le \pi$

3  a  $4\tan(\pi(x-2))=3$ for $-1 < x < 1$

 b  $3\tan\left(\dfrac{x-\pi}{2}\right)=5$ for $0 < x < 4\pi$

2  a  $4-2\sin(2x)=1$ for $0 \le x \le 360°$

 b  $3\sin(4x)-5=-1$ for $0 \le x \le 180°$

For questions 4 to 7, use the method demonstrated in Worked Example 18.21 to find all the solutions in the given interval, giving your answers in terms of $\pi$.

4  a  $\sin x = \dfrac{1}{2}$ for $0 \le x \le 2\pi$

 b  $\sin x = \dfrac{\sqrt{3}}{2}$ for $0 \le x \le 2\pi$

6  a  $\sin x = -\dfrac{\sqrt{3}}{2}$ for $0 < x < 2\pi$

 b  $\sin x = -\dfrac{1}{\sqrt{2}}$ for $0 < x < 2\pi$

5  a  $\sin x = \dfrac{\sqrt{2}}{2}$ for $-2\pi < x < 3\pi$

 b  $\sin x = 1$ for $-2\pi < x < 3\pi$

7  a  $\sin x = -1$ for $-3\pi < x < \pi$

 b  $\sin x = -\dfrac{1}{2}$ for $0 < x < 3\pi$

For questions 8 to 11, use the method demonstrated in Worked Example 18.22 to find all the solutions in the given interval.

8   a   $\cos\theta = \dfrac{1}{2}$ for $0 < \theta < 360°$

    b   $\cos\theta = \dfrac{\sqrt{2}}{2}$ for $0 < \theta < 360°$

10  a   $\cos\theta = -\dfrac{\sqrt{2}}{2}$ for $-180° < \theta < 180°$

    b   $\cos\theta = -\dfrac{1}{2}$ for $-180° < \theta < 180°$

9   a   $\cos\theta = \dfrac{\sqrt{3}}{2}$ for $-360° < \theta < 180°$

    b   $\cos\theta = \dfrac{1}{2}$ for $-360° < \theta < 180°$

11  a   $\cos\theta = -1$ for $-360° \leqslant \theta \leqslant 360°$

    b   $\cos\theta = 1$ for $-360° \leqslant \theta \leqslant 360°$

For questions 12 to 15, use the method demonstrated in Worked Example 18.23 to find all the solutions in the given interval, giving your answers in terms of $\pi$.

12  a   $\tan\theta = \sqrt{3}$ for $0 < \theta < 2\pi$

    b   $\tan\theta = \dfrac{1}{\sqrt{3}}$ for $0 < \theta < 2\pi$

14  a   $\tan\theta = -\dfrac{1}{\sqrt{3}}$ for $0 < \theta < 3\pi$

    b   $\tan\theta = -1$ for $0 < \theta < 3\pi$

13  a   $\tan\theta = 1$ for $-2\pi < \theta < 2\pi$

    b   $\tan\theta = \sqrt{3}$ for $-2\pi < \theta < 2\pi$

15  a   $\tan\theta = -\sqrt{3}$ for $-3\pi < \theta < \pi$

    b   $\tan\theta = 0$ for $-3\pi < \theta < \pi$

For questions 16 to 21, use the method demonstrated in Worked Example 18.24 to solve the equation in the given interval.

16  a   $\cos(x + 70°) = \dfrac{1}{2}$ for $0° < x < 360°$

    b   $\cos(x + 50°) = \dfrac{\sqrt{2}}{2}$ for $0° < x < 360°$

18  a   $\tan\left(x - \dfrac{\pi}{4}\right) = -1$ for $-\pi \leqslant x \leqslant \pi$

    b   $\tan\left(x - \dfrac{\pi}{3}\right) = -\dfrac{1}{\sqrt{3}}$ for $-\pi \leqslant x \leqslant \pi$

20  a   $\sin(3x) = -\dfrac{\sqrt{2}}{2}$ for $0 < x < \pi$

    b   $\sin(4x) = -\dfrac{1}{2}$ for $0 < x < \pi$

17  a   $\sin(x - 20°) = \dfrac{\sqrt{2}}{2}$ for $0° < x < 360°$

    b   $\sin(x - 30°) = \dfrac{\sqrt{3}}{2}$ for $0° < x < 360°$

19  a   $\cos(3x) = \dfrac{\sqrt{3}}{2}$ for $-\pi \leqslant x \leqslant \pi$

    b   $\cos(2x) = \dfrac{1}{2}$ for $-\pi \leqslant x \leqslant \pi$

21  a   $\tan(2x) = 1$ for $0 < x < 180°$

    b   $\tan(3x) = \sqrt{3}$ for $0 < x < 180°$

For questions 22 to 25, use the method demonstrated in Worked Example 18.25 to solve the equation in the given interval. You will need to use the identity $\sin 2x = 2\sin x\cos x$ or $\tan x = \dfrac{\sin x}{\cos x}$.

22  a   $4\sin x\cos x = \sqrt{3}$ for $0 \leqslant x \leqslant 2\pi$

    b   $4\sin x\cos x = \sqrt{2}$ for $0 \leqslant x \leqslant 2\pi$

24  a   $\sin x = \sqrt{3}\cos x$ for $0 < x < 2\pi$

    b   $\sqrt{3}\sin x = \cos x$ for $0 < x < 2\pi$

23  a   $\sin x\cos x = -\dfrac{1}{4}$ for $0° \leqslant x \leqslant 360°$

    b   $\sin x\cos x = -\dfrac{\sqrt{3}}{4}$ for $0° \leqslant x \leqslant 360°$

25  a   $\sin x - \cos x = 0$ for $-180° < x < 180°$

    b   $\sin x + \cos x = 0$ for $-180° < x < 180°$

For questions 26 to 28, use the method demonstrated in Worked Example 18.26 to find all the solutions of the equation in the given interval.

26  a   $2\sin^2 x - 3\sin x + 1 = 0$ for $0 \leqslant x \leqslant 2\pi$

    b   $2\sin^2 x + \sin x - 1 = 0$ for $0 \leqslant x \leqslant 2\pi$

28  a   $2\cos^2 x - 3\cos x - 2 = 0$ for $0 < x < 2\pi$

    b   $2\cos^2 x + 3\cos x - 2 = 0$ for $0 < x < 2\pi$

27  a   $\tan^2 x - 1 = 0$ for $0 \leqslant x \leqslant 180°$

    b   $\tan^2 x + \tan x = 0$ for $0 \leqslant x \leqslant 180°$

For questions 29 to 32, use the method demonstrated in Worked Example 18.27 to find the exact solutions of each equation in the interval $0 \leqslant x \leqslant 2\pi$.

29    a    $2\cos^2 x = 3 - 3\sin x$

       b    $2\cos^2 x = 1 + \sin x$

31    a    $\cos^2 x + 2 = 2\sin x$

       b    $\sin^2 x - 3 = 3\cos x$

30    a    $\cos^2 x - \cos x = \sin^2 x$

       b    $\cos^2 x + \cos x = \sin^2 x$

32    a    $2\cos^2 x = 2 - 3\sin x$

       b    $3\sin^2 x = 3 + 4\cos x$

33   Solve the equation $2\sin x = 3\cos x - 1$ for $0 < x < 2\pi$.

34   Solve the equation $2\sin x - 1 = 0$ for $0 < x < 360°$.

35   Solve the equation $\tan^2 x = 1$ for $-180 < x < 180°$.

36   Solve the equation $2\cos(x - 10°) + 1 = 0$ for $0 < x < 360°$.

37   Given that $\cos 2x = \cos x - 1$:

    a   Show that $2\cos^2 x - \cos x = 0$.

    b   Hence, solve the equation $\cos 2x = \cos x - 1$ for $0° < x < 360°$.

38   Solve $\sin x - \cos x = 0$ for $0 < x < 360°$.

39   Solve the equation $\sin 2x = \sin x$ for $0 \leqslant x \leqslant \pi$.

40   Solve the equation $2\cos^2 \theta - \sin \theta = 1$ for $0 \leqslant \theta \leqslant 360°$.

41   Given that $\cos^2 x + 3\cos x = 3\sin^2 x$ find the exact value of $\cos x$.

42   Solve the equation $\cos 2x = \cos x$ for $-\pi \leqslant x \leqslant \pi$.

43   For what values of $k$ does $k\sin^2 x + \sin x + 1 = 0$ have a solution?

## Checklist

- You should be able to convert between degrees and radians
  $360° = 2\pi$ radians

- You should be able to find the length of an arc of a circle
  $s = r\theta$
  where $r$ is the radius of the circle and $\theta$ is the angle subtended at the centre measured in radians.

- You should be able to find the area of a sector of a circle

  $A = \dfrac{1}{2}r^2\theta$

  where $r$ is the radius of the circle and $\theta$ is the angle subtended at the centre measured in radians.

- You should be able to define the sine and cosine functions in terms of the unit circle.

  For a point $P$ on the unit circle,
  - ☐   $\sin\theta$ is the $y$-coordinate of the point $P$
  - ☐   $\cos\theta$ is the $x$-coordinate of the point $P$.

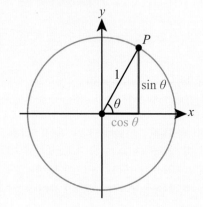

- You should be able to define the tangent function

  $\tan\theta = \dfrac{\sin\theta}{\cos\theta}$

- You should know some key relationships for the sine function:
  - $\sin(-\theta) = -\sin\theta$
  - $\sin(\pi - \theta) = \sin\theta$
  - $\sin(\theta + \pi) = -\sin\theta$
  - $\sin(\theta + 2\pi) = \sin\theta$

- You should know some key relationships for the cosine function:
  - $\cos(-\theta) = \cos\theta$
  - $\cos(\pi - \theta) = -\cos\theta$
  - $\cos(\theta + \pi) = -\cos\theta$
  - $\cos(\theta + 2\pi) = \cos\theta$

- You should know some key relationships for the tangent function:
  - $\tan(-\theta) = -\tan\theta$
  - $\tan(\theta + \pi) = \tan\theta$

- You should know the exact values of trigonometric ratios of certain angles:

| Radians | $0$ | $\dfrac{\pi}{6}$ | $\dfrac{\pi}{4}$ | $\dfrac{\pi}{3}$ | $\dfrac{\pi}{2}$ | $\pi$ |
|---|---|---|---|---|---|---|
| **Degrees** | $0$ | $30$ | $45$ | $60$ | $90$ | $180$ |
| **sin $\theta$** | $0$ | $\dfrac{1}{2}$ | $\dfrac{\sqrt{2}}{2}$ | $\dfrac{\sqrt{3}}{2}$ | $1$ | $0$ |
| **cos $\theta$** | $1$ | $\dfrac{\sqrt{3}}{2}$ | $\dfrac{\sqrt{2}}{2}$ | $\dfrac{1}{2}$ | $0$ | $-1$ |
| **tan $\theta$** | $0$ | $\dfrac{1}{\sqrt{3}}$ | $1$ | $\sqrt{3}$ | Not defined | $0$ |

- You should know about the ambiguous case of the sine rule:

  when using the sine rule to find an angle, there may be two possible solutions: $\theta$ and $180 - \theta$.

- You should know the Pythagorean identity $\cos^2\theta + \sin^2\theta \equiv 1$.

- You should know the double angle identities for sine and cosine:
  - $\sin 2\theta \equiv 2\sin\theta\cos\theta$

  - $\cos 2\theta \equiv \begin{cases} \cos^2\theta - \sin^2\theta \\ 2\cos^2\theta - 1 \\ 1 - 2\sin^2\theta \end{cases}$

- You should be able to sketch the graphs of trigonometric functions:

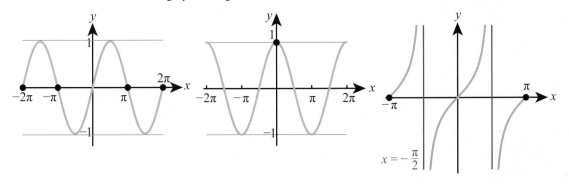

■ You should be able to apply transformations to the graphs of trigonometric functions.

The functions $a\sin(b(x+c))+d$ and $a\cos(b(x+c))+d$ have:
   □ amplitude $a$
   □ period $\dfrac{2\pi}{b}$

■ You should be able to apply trigonometric functions to real-life contexts.

■ You should be able to solve trigonometric equations graphically.

■ You should be able to solve trigonometric equations analytically.
   □ To solve $\sin x = a$
      — find a solution: $x_1 = \sin^{-1} a$
      — find a second solution: $x_2 = \pi - x_1$
      — find other solutions by adding/subtracting $2\pi$ to/from $x_1$ and $x_2$.
   □ To solve $\cos x = a$
      — find a solution: $x_1 = \cos^{-1} a$
      — find a second solution: $x_2 = 2\pi - x_1$
      — find other solutions by adding/subtracting $2\pi$ to/from $x_1$ and $x_2$.
   □ To solve $\tan x = a$
      — find a solution: $x_1 = \tan^{-1} a$
      — find other solutions by adding/subtracting $\pi$ to/from $x_1$.

■ You should be able to solve trigonometric equations that lead to quadratic equations.

## ■ Mixed Practice

**1** The height of a wave ($h$ m) at a distance ($x$ m) from the shore is modelled by the equation $h = 1.3\sin(2.5x)$.
   **a** Write down the amplitude of the wave.
   **b** Find the distance between consecutive peaks of the wave.

**2** Show that $\cos\left(\dfrac{\pi}{6}\right) - \tan\left(\dfrac{\pi}{6}\right) = \dfrac{1}{6}\tan\left(\dfrac{\pi}{3}\right)$.

**3** Solve $2\sin x = \sqrt{3}$ for $0 < x < 360°$.

**4** For the triangle shown in the diagram, find the two possible values of $\theta$.

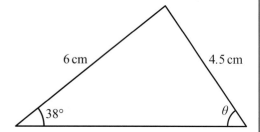

**5** Solve $\sin x + 2\cos 2x = 1$ for $0 < x < 2\pi$.

**6** Solve $2\cos\left(2x + \dfrac{\pi}{3}\right) = 1$ for $0 \leqslant x \leqslant 2\pi$.

**7** The obtuse angle $A$ has $\sin A = \dfrac{5}{13}$.

Find the value of:
   **a** $\cos A$
   **b** $\tan A$
   **c** $\sin 2A$.

**8** The diagram shows a Ferris wheel that moves with constant speed and completes a rotation every 40 seconds. The wheel has a radius of 12 m and its lowest point is 2 m above the ground.

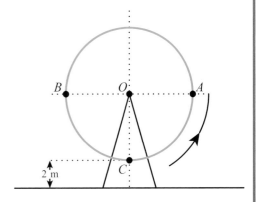

Initially, a seat $C$ is vertically below the centre of the wheel, $O$. It then rotates in an anticlockwise (counterclockwise) direction.

**a** Write down
  **i** the height of $O$ above the ground;
  **ii** the maximum height above the ground reached by $C$.

In a revolution, $C$ reaches points $A$ and $B$, which are at the same height above the ground as the centre of the wheel.

**b** Write down the number of seconds taken for $C$ to first reach $A$ and then $B$.

The sketch on the right shows the graph of function, $h(t)$, for the height above ground of $C$, where $h$ is measured in metres and $t$ is the time in seconds, $0 \leqslant t \leqslant 40$.

**c** **Copy** the sketch and show the results of part **a** and part **b** on your diagram. Label the points clearly with their coordinates.

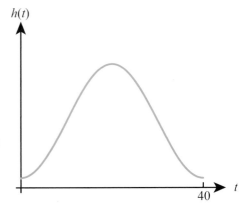

The height of $C$ above ground can be modelled by the function, $h(t) = a\cos(bt) + c$, where $bt$ is measured in degrees and $t$ is the time in seconds.

**d** Find the value of
  **i** $a$
  **ii** $b$
  **iii** $c$.

$C$ **first** reaches a height of 20 m above the ground after $T$ seconds.

**e** **i** Sketch a clearly labelled diagram of the wheel to show the position of $C$.
  **ii** Find the angle that $C$ has rotated through to reach this position.
  **iii** Find the value of $T$.

Mathematical Studies SL November 2011 Paper 2 Q5

**9** Given that $\sin x = \dfrac{3}{4}$, where $x$ is an obtuse angle, find the value of
**a** $\cos x$;
**b** $\cos 2x$.

Mathematics SL May 2015 Paper 1 TZ1 Q5

**10** Solve $3\tan^2 x - 1 = 0$ for $-\pi < x < \pi$.

**11** Prove that $\dfrac{\sin\theta}{\cos\theta} + \dfrac{\cos\theta}{\sin\theta} \equiv \dfrac{1}{\sin 2\theta}$.

**12** In triangle $ABC$, $AC = 10\,\text{cm}$, $BC = 15\,\text{cm}$, $AB = x\,\text{cm}$ and angle $BAC = 120°$. Find the exact value of $x$.

**13** In triangle $ABC$, $AB = 10\,\text{cm}$, $BC = 7\,\text{cm}$ and angle $BAC = 40°$. Find the difference in areas between two possible triangles $ABC$.

**14 a** Prove that $\tan 2x \equiv \dfrac{2\tan x}{1 - \tan^2 x}$.

**b** Hence solve $\tan 2x + \tan x = 0$ for $0 \leqslant x \leqslant \pi$.

**15** Prove the identity $\dfrac{1}{1 + \cos x} + \dfrac{1}{1 - \cos x} \equiv \dfrac{2}{\sin^2 x}$.

**16** The graph of $y = a\tan(x + c)$ is shown below. Find the values of $a$ and $c$, where $0 < c < \pi$

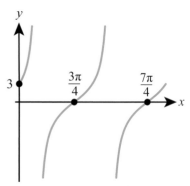

**17** Let $f(x) = \sin\left(x + \dfrac{\pi}{4}\right) + k$. The graph of f passes through the point $\left(\dfrac{\pi}{4}, 6\right)$.

**a** Find the value of $k$.

**b** Find the minimum value of $f(x)$.

Let $g(x) = \sin x$. The graph of g is translated to the graph of f by $p$ units in the horizontal direction and $q$ units in the vertical direction.

**c** Write down the value of $p$ and of $q$.

Mathematics SL November 2013 Paper 1 Q5

**18** Solve $2\cos^2 x - 5\cos x + 2 = 0$ for $0 < x < 2\pi$.

**19** A rectangle is drawn around a sector of a circle as shown. If the angle of the sector is 1 radian and the area of the sector is $7\,\text{cm}^2$, find the dimensions of the rectangle, giving your answers to the nearest millimetre.

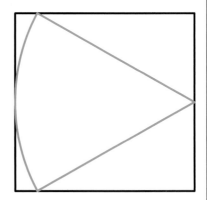

Mathematics HL May 2013 Paper 2 TZ1 Q5

**20** The following diagram shows two intersecting circles of radii 4 cm and 3 cm. The centre $C$ of the smaller circle lies on the circumference of the bigger circle. $O$ is the centre of the bigger circle and the two circles intersect at points $A$ and $B$.
Find:

**a** $B\hat{O}C$;

**b** the area of the shaded region.

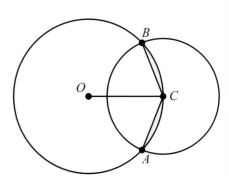

Mathematics HL May 2014 Paper 2 TZ2 Q4

**21** Prove that $\sin 4x \equiv 4\sin x\cos^3 x - 4\sin^3 x\cos x$.

**22** Prove the identity $\dfrac{1}{2}\sin 2x \equiv \dfrac{\tan x}{1+\tan^2 x}$ .

**23** Find the period of the function $f(x) = 3\sin\left(\dfrac{x}{2}\right) - 4\cos\left(\dfrac{x}{5}\right)$, where $x$ is in radians.

**24** Find the range of the function $f(x) = (4\cos(x - \pi) - 1)^2$.

**25** The function $I(t) = a\sin(bt)$ is known to model the current, in amps, flowing through a resistor at time, $t$, after the circuit is connected.

When $t = 1$ the current is 0.2 amps.

When $t = 2$ the current is $-0.1$ amps.

Find the maximum current flowing through the circuit.

**26** If $0 < k < 1$, find the sum of the solutions to $\sin x = k$ for $-\pi < x < 3\pi$.

**27 a** Sketch $y = x^2 - x$.
  **b** Hence find the values of $k$ for which $\sin^2 x - \sin x - k = 0$ has solutions.

**28** A bicycle chain is modelled by the arcs of two circles connected by two straight lines which are tangent to both circles.

The radius of the larger circle is 10 cm and the radius of the smaller circle is 6 cm. The distance between the centres $O_1$ and $O_2$ of the circles is 30 cm.
  **a** Find angle $AO_1O_2$.
  **b** Hence find the length of the bicycle chain, giving your answer to the nearest cm.

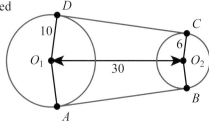

**29** Given that $\sin x + \cos x = \dfrac{2}{3}$, find $\cos 4x$.

Mathematics HL May 2014 Paper 1 TZ1 Q10

# 19 Analysis and approaches: Statistics and probability

## ESSENTIAL UNDERSTANDINGS

- Statistics is concerned with the collection, analysis and interpretation of data, and the theory of probability can be used to estimate parameters and predict the occurrence of events.
- Probability theory allows us to make informed choices, to evaluate risk and to make predictions about seemingly random events.

## In this chapter you will learn...

- how to use your GDC to find the regression line of $x$ on $y$
- how to use this regression line to predict values of $x$ not in the data set
- a formula for finding conditional probabilities
- how to test whether two events are independent
- how to standardize normal variables
- how to use inverse normal calculations to find an unknown mean and variance.

## CONCEPTS

The following concepts will be addressed in this chapter:
- Different statistical techniques require justification and the identification of their limitations and **validity**.
- Some techniques of statistical analysis, such as regression and standardization of formulae, can be applied in a practical context to apply to **general** cases.
- Modelling through statistics can be **reliable**, but may have limitations.

## PRIOR KNOWLEDGE

Before starting this chapter, you should already be able to complete the following:

1  a  Using technology, find the regression line of $y$ on $x$ for the following data:

| $x$ | 10 | 11 | 11 | 12 | 14 | 14 | 15 | 17 |
|---|---|---|---|---|---|---|---|---|
| $y$ | 7.2 | 6.5 | 6.8 | 6.4 | 5.9 | 5.4 | 5.6 | 4.1 |

   b  Find the value of $y$ when $x = 13$.

■ **Figure 19.1** Can we use the past to predict the future?

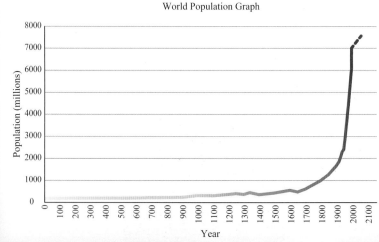

2 There are five red counters and three blue counters in a bag. Two counters are chosen at random without replacement. Find the probability that the second is blue given that the first was red.

3 The probability of a randomly chosen student being female is 0.48 and the probability that the student plays a musical instrument is 0.26. If these two factors are independent, find the probability that a randomly chosen student is female and plays a musical instrument.

4 Given $X \sim N(10, 4)$, find

   a   $P(X > 12.5)$

   b   the value of $x$ if $P(X < x) = 0.2$.

Often in real life, you gather data about a variable and suspect it comes from a certain distribution. However, you do not know the population parameters of that distribution – such as its mean or variance – so you cannot immediately do any calculations with it.

One of the main applications of statistics is to determine these unknown population parameters, given some experimental evidence about the distribution.

## TOK Links

Is it possible to have knowledge of the future?

## Starter Activity

Look at the pictures in Figure 19.1. What are the trends being shown? How confident are you in the predictions (implicitly) being made?

**Now look at this problem:**

An athlete keeps track of his personal best time for the 100 m at different ages:

| Age (years) | 11 | 13 | 15 | 17 | 19 |
|---|---|---|---|---|---|
| Time (s) | 13.1 | 12.6 | 11.9 | 11.4 | 11.1 |

Based on the data predict his personal best time when he is

   a   16 years old

   b   50 years old.

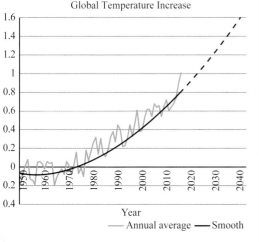

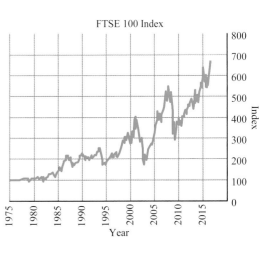

## 19A Linear regression

In Section 6D, you found the regression line of $y$ on $x$, that is, you found the equation of the line of best fit for bivariate data in the form $y = ax + b$. This then allowed you to predict values of $y$ from given values of $x$.

If, instead, you wanted to predict a value of $x$ from a given value of $y$, you might think it would just be a case of rearranging the equation you already have. However, in general, this will not give you an accurate prediction.

You need to recalculate the regression line to form a new equation in the form $x = ay + b$. This is the regression line of $x$ on $y$.

### ■ Equation of the regression line of $x$ on $y$

You do not need to know the formula for the regression line, but it useful to know that both the $y$-on-$x$ and $x$-on-$y$ lines pass through the mean point: $(\bar{x}, \bar{y})$.

With most calculators, to find the regression line of $x$ on $y$ you will need to enter the $x$-data in the $y$-column and the $y$-data in the $x$-column.

**WORKED EXAMPLE 19.1**

For the data below from Worked Example 6.24, find the equation of the regression line of $x$ on $y$ using technology.

| $x$ | 1 | 2 | 3 | 4 | 4 | 5 | 6 | 6 | 7 | 9 |
|-----|---|---|---|---|---|---|---|----|---|----|
| $y$ | 0 | 2 | 4 | 4 | 5 | 6 | 7 | 10 | 9 | 10 |

Enter the $x$-data in the $y$-column · · · · · · · · · · · · · · · · · · · · · · · Regression line is
and the $y$-data in the $x$-column                                           $x = 0.677y + 0.842$

Just remember that the output
is in the form $x = ay = b$
(and not $y = ax = b$)

```
LinearReg
  a =0.67678746
  b =0.84231145
  r =0.94742821
  r²=0.89762022
  MSe=0.66674828
y=ax+b
                    COPY
```

Notice that if you had rearranged the $y$-on-$x$ regression line (calculated in Worked Example 6.24), you would have got:

$$y = 1.33x - 0.543$$
$$1.33x = y + 0.543$$
$$x = 0.754y + 0.402$$

This is similar to, but certainly not the same as, the regression line found in Worked Example 19.1.

**TOOLKIT:** Problem Solving

Create or find an arbitrary set of data. Find the gradients of the *y*-on-*x* and *x*-on-*y* lines of best fit. Which line is closer to the *x*-axis? Will this always be the case?

Find the Pearson's correlation coefficient, *r*, and also $r^2$. Can you hypothesize any link between the gradients of the two lines and $r^2$?

To prove your hypotheses you might need to research the equations for *r* and the gradients of the regression lines.

 You know from Section 6D that when you interpolate the result can be considered reliable, but when you extrapolate this is no longer the case.

## Use of the equation for prediction purposes

With an *x*-on-*y* regression line, you can now predict *x*-values from given *y*-values. However, the same warning applies about extrapolation.

### CONCEPTS – VALIDITY

The justification of the equation of the *x*-on-*y* regression line assumes that *y* is known perfectly and *x* is uncertain. This explains why, in the absence of other information, we normally use it to predict *x* for a known *y*. However, in some data sets you might have a reasonable belief that one variable is dependent on the other. For example, you might believe that crop yield (*y*) depends on rainfall (*x*). In this situation it would not be **valid** to use an *x*-on-*y* regression line.

### WORKED EXAMPLE 19.2

Based on data with *y*-values between 5 and 15 and a correlation coefficient of −0.91, a regression line of *x* on *y* is formed:

$$x = -2.74y + 81.6$$

a   Use this line to predict the value of *x* when:

  i   *y* = 12

  ii  *y* = 20.

b   Comment on the reliability of your answers in **a**.

*Substitute each value of y into the regression line* ········· **a  i**  $x = -2.74(12) + 81.6 = 48.7$

  **ii**  $x = -2.74(20) + 81.6 = 26.8$

**b**  The prediction when *y* = 12 can be considered reliable as 12 is within the range of known *y*-values and the correlation coefficient is reasonably close to −1.

The prediction when *y* = 20, however, cannot be considered reliable as the relationship needed to be extrapolated significantly beyond the range of given data to make this prediction.

CONCEPTS – RELIABILITY

If you were to take a different sample, then you could get a slightly different regression line and therefore a different prediction. This is an issue of **reliability**. In an effort to overcome this, in more advanced work, a single value is not given for the prediction, but rather a range called a confidence interval. The way that the $x$-on-$y$ line is calculated means that it can produce a confidence interval for $x$-values but not $y$-values. These intervals are narrowest at the mean point and get wider as you get further from this point, reflecting increasing uncertainty.

## Exercise 19A

For questions 1 and 2, use the method demonstrated in Worked Example 19.1 to find the equation of the regression line of $x$ on $y$.

1 a

| $x$ | 3 | 4 | 6 | 7 | 9 | 10 | 12 | 13 |
|---|---|---|---|---|---|---|---|---|
| $y$ | 2 | 3 | 5 | 4 | 7 | 6 | 5 | 7 |

  b

| $x$ | 1 | 2 | 2 | 3 | 5 | 7 | 8 | 8 |
|---|---|---|---|---|---|---|---|---|
| $y$ | −3 | 0 | 1 | −2 | −1 | 2 | 4 | 3 |

2 a

| $x$ | 1 | 1 | 2 | 3 | 5 | 6 | 6 | 8 | 9 |
|---|---|---|---|---|---|---|---|---|---|
| $y$ | 6 | 7 | 5 | 4 | 5 | 4 | 2 | 1 | 2 |

  b

| $x$ | 14 | 13 | 15 | 19 | 19 | 22 | 23 | 24 | 22 |
|---|---|---|---|---|---|---|---|---|---|
| $y$ | 14 | 12 | 12 | 11 | 9 | 9 | 8 | 6 | 5 |

For questions 3 and 4, you are given the minimum and maximum $y$ data values, the correlation coefficient and the equation of the regression line of $x$ on $y$. Use the method demonstrated in Worked Example 19.2 to:

  i  predict the value of $x$ for the given value of $y$

  ii comment on the reliability of your prediction.

|   |   | Minimum value | Maximum value | $r$ | Regression line | $y$ |
|---|---|---|---|---|---|---|
| 3 | a | 10 | 20 | 0.973 | $x = 1.62y − 7.31$ | 18 |
|   | b | 1 | 7 | −0.875 | $x = −0.625y + 1.37$ | 9 |
| 4 | a | 20 | 50 | 0.154 | $x = 2.71y + 0.325$ | 27 |
|   | b | 12 | 27 | −0.054 | $x = 4.12y − 2.75$ | 22 |

5  The table shows the distance from the nearest train station and the average house price for seven villages. The correlation coefficient between the two variables is −0.789.

| $x$ = distance (km) | 0.8 | 1.2 | 2.5 | 3.7 | 4.1 | 5.5 | 7.4 |
|---|---|---|---|---|---|---|---|
| $y$ = average house price (000's £) | 240 | 185 | 220 | 196 | 187 | 156 | 162 |

Adam wants to estimate the distance from the train station for a village where the average house price is £205 000.

  a  Find the equation of the appropriate regression line.

  b  Use your regression line to calculate Adam's estimate.

  c  Comment on the reliability of this estimate.

**6** Elena collected some data on the number of people visiting a park in a day ($x$) and maximum daily temperature ($y°C$) over a period of time. She calculated the equations of two regression lines:

$y = 0.0928x - 4.26$ and $x = 12.5y + 68.2$

Use the appropriate equation to predict the number of people visiting the park on a day when the maximum temperature is 23.6°C.

**7** The table shows the data for height and arm length for a sample of 10 15-year-olds.

| Height (cm) | 154 | 148 | 151 | 165 | 154 | 147 | 172 | 156 | 168 | 152 |
|---|---|---|---|---|---|---|---|---|---|---|
| Arm length (cm) | 65 | 63 | 58 | 71 | 59 | 65 | 75 | 62 | 61 | 61 |

Use the appropriate regression line to estimate:

a  arm length of a 15-year-old with a height of 161 cm

b  height of a 15-year-old whose arm length is 56 cm.

**8** The table shows the marks on two mathematics papers achieved by eight students in a class.

| Paper 1 | 71 | 63 | 87 | 51 | 72 | 68 | 65 | 91 |
|---|---|---|---|---|---|---|---|---|
| Paper 2 | 62 | 57 | 72 | 55 | 60 | 58 | 47 | 71 |

Paolo scored 64 marks in paper 2 but missed paper 1. Use an appropriate regression line to estimate what mark he would have got in paper 1.

**9** A class of eight students took a mathematics test and a chemistry test. The table shows the marks on the two tests.

| Mathematics mark ($x$) | 72 | 47 | 82 | 65 | 71 | 83 | 81 | 57 |
|---|---|---|---|---|---|---|---|---|
| Chemistry mark ($y$) | 50 | 38 | 57 | 50 | 32 | 65 | 57 | 45 |

The correlation coefficient between the two sets of marks is 0.698 and the equation of the regression line of $y$ on $x$ is $y = 0.583x + 8.57$.

a  Explain why this regression line should not be used to predict the mathematics mark for a student who scored 55 marks in chemistry.

b  Use an appropriate regression line to predict the mathematics mark for this student.

c  Could your regression line be reliably used to predict the mathematics mark for a student who scored 72 marks in chemistry? Justify your answer.

**10** A sales manager recorded the monthly sales ($\$x000$) and the amount spent on advertising in the previous month ($\$y00$) over a period of time. The equation of the regression line of $y$ on $x$ is $y = 0.745x - 0.290$ and the equation of the regression line of $x$ on $y$ is $x = 1.01y + 1.32$.

a  Estimate the monthly profit the month after $250 was spent on advertising.

b  Find the mean monthly profit and the mean amount spent on advertising.

**11** Antonia inputted some data into a spreadsheet. The spreadsheet reported that the $y$-on-$x$ regression line is $y = -0.5x + 4$ and $r^2 = 1$.

a  State, with justification, the value of $r$.

b  Find the equation of the $x$-on-$y$ regression line in the form $x = ay + b$.

**12** Phillip has some data on the distance travelled in a straight line by a snail $y$ (in cm) after a time $t$ (in minutes). He finds that the equation of the regression line is $y = 6t + 5$ and the correlation coefficient is 0.6.

a  Based on this regression line, estimate the speed of the snail. Explain, based on the regression line, why your answer is only an estimate.

The distance travelled in metres is denoted $Y$ and the time taken in hours is denoted $T$.

b  Find the equation of the $Y$-on-$T$ for Phillip's data.

c  Find the correlation coefficient of $Y$ and $T$.

# 19B Conditional probability

## ■ A formula for conditional probability

In Section 7B, you used Venn diagrams to find conditional probabilities, but it was also mentioned that this leads to a formula.

---
**KEY POINT 19.1**

$$P(A\mid B) = \frac{P(A\cap B)}{P(B)}$$
---

Sometimes it is much more convenient to use the formula than to use a Venn diagram.

---
**WORKED EXAMPLE 19.3**

$P(A) = 0.4$, $P(B) = 0.7$ and $P(A\mid B) = 0.2$.

Find:

a   $P(A \cap B)$

b   $P(B\mid A)$.

Rearrange the formula ·········· **a**   $P(A \cap B) = P(A\mid B)\,P(B)$

$$P(A\mid B) = \frac{P(A\cap B)}{P(B)}$$

$$= 0.2 \times 0.7$$

$$= 0.14$$

Interchange $A$ and $B$ in the

formula $P(A\mid B) = \dfrac{P(A\cap B)}{P(B)}$ ·········· **b**   $P(B\mid A) = \dfrac{P(B\cap A)}{P(A)}$

Note that $P(A \cap B)$ and

$P(B \cap A)$ are the same thing

$$= \frac{0.14}{0.4}$$

$$= 0.35$$
---

## ■ Testing for independence

You also have met the idea that if you know the events $A$ and $B$ are independent, then $P(A \cap B) = P(A)\,P(B)$. This can also be used to determine whether or not events are independent.

---
**KEY POINT 19.2**

The events $A$ and $B$ are independent if

$P(A\mid B) = P(A)$

or equivalently $P(A \cap B) = P(A)\,P(B)$
---

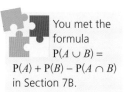

You met the formula
$$P(A \cup B) =$$
$$P(A) + P(B) - P(A \cap B)$$
in Section 7B.

---

**WORKED EXAMPLE 19.4**

Two events $A$ and $B$ are such that $P(A) = 0.5$, $P(B) = 0.6$ and $P(A \cup B) = 0.8$.

Determine whether or not $A$ and $B$ are independent.

Find $(A \cap B)$ using the formula ········· $0.8 = 0.5 + 0.6 - P(A \cap B)$
$P(A \cup B) = P(A) + P(B) - P(A \cap B)$ $P(A \cap B) = 0.3$

Calculate $P(A) P(B)$... ········· $P(A) P(B) = 0.5 \times 0.6 = 0.3$

... and compare that to $P(A \cap B)$ ········· So, $P(A \cap B) = P(A) P(B)$
and therefore $A$ and $B$ are independent

---

**WORKED EXAMPLE 19.5**

The table below shows the probabilities of a particular train service being on time or late when it does or does not rain.

|  | Rain | Does not rain |
|---|---|---|
| **On time** | 0.15 | 0.55 |
| **Late** | 0.2 | 0.1 |

Determine whether or not the train being late and it raining are independent.

$$P(\text{rain}) = 0.15 + 0.2 = 0.35$$

$P(\text{late}|\text{rain}) = \dfrac{P(\text{late} \cap \text{rain})}{P(\text{rain})}$ ········· $P(\text{late}|\text{rain}) = \dfrac{0.2}{0.35} = 0.571$

Compare ········· So, $P(\text{late}|\text{rain}) \neq P(\text{late})$
$P(\text{late}|\text{rain})$ to $P(\text{rain})$ and therefore the train being late and it raining are not independent

---

**Tip**

In Worked Example 19.5, instead of finding $P(\text{late}|\text{rain})$, you could find $P(\text{late})$ and then check whether or not $P(\text{late} \cap \text{rain}) = P(\text{late}) P(\text{rain})$.

---

# Exercise 19B

For questions 1 to 6, use the method demonstrated in Worked Example 19.3 to find the required probability.

1 a $P(A \cap B) = 0.3$, $P(B) = 0.6$; find $P(A|B)$
  b $P(A \cap B) = 0.2$, $P(B) = 0.8$; find $P(A|B)$

2 a $P(A \cap B) = 0.4$, $P(A) = 0.5$; find $P(B|A)$
  b $P(A \cap B) = 0.1$, $P(A) = 0.4$; find $P(B|A)$

3 a $P(B) = 0.5$, $P(A|B) = 0.8$; find $P(A \cap B)$
  b $P(B) = 0.7$, $P(A|B) = 0.9$; find $P(A \cap B)$

4 a $P(A) = 0.7$, $P(B|A) = 0.3$; find $P(A \cap B)$
  b $P(A) = 0.3$, $P(B|A) = 0.2$; find $P(A \cap B)$

5 a $P(A \cap B) = 0.2$, $P(B|A) = 0.8$; find $P(A)$
  b $P(A \cap B) = 0.3$, $P(B|A) = 0.6$; find $P(A)$

6 a $P(A \cap B) = 0.5$, $P(A|B) = 0.8$; find $P(B)$
  b $P(A \cap B) = 0.4$, $P(A|B) = 0.8$; find $P(B)$

For questions 7 to 12, use the method demonstrated in Worked Example 19.4 to determine whether or not events $A$ and $B$ are independent.

7 a $P(A) = 0.6$, $P(B) = 0.5$, $P(A \cap B) = 0.4$
  b $P(A) = 0.5$, $P(B) = 0.4$, $P(A \cap B) = 0.2$

8 a $P(A) = 0.7$, $P(B) = 0.3$, $P(A \cap B) = 0.21$
  b $P(A) = 0.4$, $P(B) = 0.6$, $P(A \cap B) = 0.2$

9 a $P(A) = 0.6$, $P(B) = 0.5$, $P(A \cup B) = 0.8$
  b $P(A) = 0.8$, $P(B) = 0.3$, $P(A \cup B) = 0.86$

10 a $P(A) = 0.5$, $P(B) = 0.4$, $P(A \cup B) = 0.52$
  b $P(A) = 0.6$, $P(B) = 0.2$, $P(A \cup B) = 0.71$

11 a $P(A) = 0.7$, $P(A \cap B) = 0.4$, $P(A \cup B) = 0.9$
  b $P(A) = 0.5$, $P(A \cap B) = 0.1$, $P(A \cup B) = 0.6$

12 a $P(A \cap B) = 0.12$, $P(B) = 0.3$, $P(A \cup B) = 0.38$
  b $P(A \cap B) = 0.35$, $P(B) = 0.6$, $P(A \cup B) = 0.8$

For questions 13 to 15, the table shows some information about the probabilities of events $A$ and $B$. Use the method demonstrated in Worked Example 19.5 to determine whether $A$ and $B$ are independent.

**13 a**

|       | $A$  | Not $A$ |
|-------|------|---------|
| $B$   | 0.25 | 0.15    |
| Not $B$ | 0.30 | 0.30  |

**b**

|       | $A$  | Not $A$ |
|-------|------|---------|
| $B$   | 0.3  | 0.2     |
| Not $B$ | 0.3 | 0.2   |

**14 a**

|       | $A$  | Not $A$ |
|-------|------|---------|
| $B$   | 0.12 | 0.28    |
| Not $B$ | 0.18 | 0.42  |

**b**

|       | $A$  | Not $A$ |
|-------|------|---------|
| $B$   | 0.25 | 0.25    |
| Not $B$ | 0.15 | 0.35  |

**15 a**

|       | $A$  | Not $A$ |
|-------|------|---------|
| $B$   | 0.25 | 0.25    |
| Not $B$ | 0.25 | 0.25  |

**b**

|       | $A$  | Not $A$ |
|-------|------|---------|
| $B$   | 0.4  | 0       |
| Not $B$ | 0   | 0.6   |

**16** Events $A$ and $B$ satisfy $P(A) = 0.4$, $P(B) = 0.3$ and $P(A \cap B) = 0.1$. Find
a $P(A|B)$                b $P(B|A)$.

**17** Given that $P(A) = 0.3$, $P(B) = 0.5$ and $P(B|A) = 0.5$, find
a $P(A \cap B)$           b $P(A|B)$.

**18** Two events $A$ and $B$ are such that $P(A) = 0.5$, $P(B) = 0.4$ and $P(A \cup B) = 0.8$.
a Find $P(A \cap B)$.
b Show that $A$ and $B$ are not independent.
c Find $P(A|B)$.

**19** Given that $P(A) = 0.3$, $P(B) = 0.5$ and $P(A|B) = 0.3$,
a explain how you can tell that $A$ and $B$ are independent
b write down the value of $P(A \cap B)$
c find $P(A \cup B)$.

**20** The table shows the probabilities of a child having a sibling and a child having a pet.

|        | Sibling | No sibling |
|--------|---------|------------|
| Pet    | 0.28    | 0.12       |
| No pet | 0.42    | 0.18       |

Determine whether the events 'having a sibling' and 'having a pet' are independent.

**21** The table shows some probabilities associated with events $A$ and $B$.

|      | $A$  | $A'$ |
|------|------|------|
| $B$  | $x$  | 0.26 |
| $B'$ | 0.21 | $y$  |

Given that $A$ and $B$ are independent, find the possible values of $x$ and $y$.

**22** Given that $P(A) = \frac{1}{3}$, $P(B|A') = \frac{2}{5}$ and $P(A|B) = \frac{2}{11}$ find $P(B|A)$.

**23** Alessia has two boxes of pens. Box 1 contains five blue pens and four red pens, and Box 2 contains three blue pens and seven red pens. She rolls a fair six-sided dice. If the dice shows a '1' she selects a pen at random from Box 1; otherwise she selects a pen at random from Box 2. Given that the pen is red, find the probability that it came from Box 1.

**24** Events $A$ and $B$ satisfy $P(A) = 0.6$, $P(B) = 0.4$, $P(B|A) = 0.5$. Find $P(A|B)$.

**25** $P(A) = \frac{1}{3}$, $P(A|B) = \frac{1}{4}$, $P(A \cup B) = \frac{3}{4}$. Find $P(B)$.

**26** Given that events $A$ and $B$ are mutually exclusive (and neither has probability 0), prove that $A$ and $B$ are not independent.

**27** Prove that if events $A$ and $B$ are independent, then events $A'$ and $B'$ are also independent.

**28** Prove that $P(A \cup B) - P(A \cap B) = P(A)(1 - P(B|A)) + P(B)(1 - P(A|B))$.

# 19C Normal distribution

## Standardization of normal variables

You know from Section 8C what proportion of the data in a normal distribution is one standard deviation from the mean, or two standard deviations from the mean, but you might also want to know how many standard deviations from the mean a particular value lies. This is called the $z$-value.

---

**KEY POINT 19.3**

If $X \sim \mathrm{N}(\mu, \sigma^2)$, the $z$-value is

$$z = \frac{x - \mu}{\sigma}$$

It measures the number of standard deviations from the mean.

---

**WORKED EXAMPLE 19.6**

Given that $X \sim \mathrm{N}(30, 7.84)$,

a   find how many standard deviations $x = 35$ is away from the mean

b   find the value of $X$ which is 0.75 standard deviations below the mean.

You will need $\sigma$ so square root the variance $\sigma^2$ ......... **a**   $\sigma = \sqrt{7.84} = 2.8$

Use $z = \frac{x - \mu}{\sigma}$ to find the $z$-score. This is the number of standard deviations from the mean ............. $z = \frac{35 - 30}{2.8} = 1.79$

35 is 1.79 standard deviations from the mean

A value below the mean has a negative $z$-value ......... **b**   $-0.75 = \frac{x - 30}{2.8}$

$$x - 30 = -2.1$$
$$x = 27.9$$

---

Converting the values of a random variable $X \sim \mathrm{N}(\mu, \sigma^2)$ to $z$-values gives a new random variable $Z$. Whatever the mean and variance of $X$, the new random variable $Z$ will always be normally distributed with mean 0 and variance 1.

---

**KEY POINT 19.4**

If $X \sim \mathrm{N}(\mu, \sigma^2)$ and $Z = \frac{X - \mu}{\sigma}$, then $X \sim \mathrm{N}(0, 1)$.

---

The random variable $Z$ is said to have the **standard normal distribution**, and the process of converting from $X$ to $Z$ is referred to as standardization.

---

**CONCEPTS – GENERALIZATION**

If $X$ has mean $\mu$ and standard deviation $\sigma$ then you can reason that $\frac{X - \mu}{\sigma}$ will have mean 0 and standard deviation 1. However, it is not at all obvious that this quantity will follow the same distribution as $X$. The fact that it does is what allows the many different situations which can be modelled by a normal distribution to be **generalized**. Historically, this meant that only tables of the N(0, 1) distribution needed to be created and all probabilities for any normal distribution could be found.

**WORKED EXAMPLE 19.7**

Given $X \sim N(4, 0.5^2)$, write in terms of $Z$,

a   $P(X < 4.3)$

b   $P(3 < X < 5)$.

Standardize by finding ·········· **a**   $P(X < 4.3) = P\left(Z < \dfrac{4.3 - 4}{0.5}\right)$
the $z$-score of 4.3

$$= P(Z < 0.6)$$

Standardize again by finding ·········· **b**   $P(3 < X < 5) = P\left(\dfrac{3 - 4}{0.5} < Z < \dfrac{5 - 4}{0.5}\right)$
the $z$-scores of 3 and 5

$$= P(-2 < Z < 2)$$

## ■ Inverse normal calculations where mean and standard deviation are unknown

If you do not know both the mean and variance of a particular normal distribution, then you cannot directly work with it. Instead you will need to convert it to the standard normal distribution and use this to help you calculate the mean and variance of the original distribution.

**WORKED EXAMPLE 19.8**

The random variable $X$ is normally distributed with standard deviation $\sigma = 12$. Experimental data suggest that $P(X > 70) = 0.15$.

Estimate the mean of $X$.

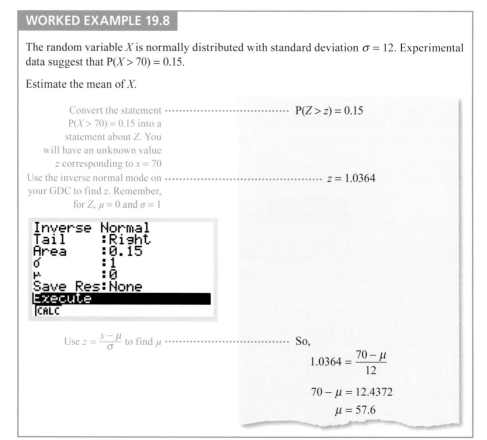

Convert the statement ································· $P(Z > z) = 0.15$
$P(X > 70) = 0.15$ into a
statement about $Z$. You
will have an unknown value
$z$ corresponding to $x = 70$

Use the inverse normal mode on ······························· $z = 1.0364$
your GDC to find $z$. Remember,
for $Z$, $\mu = 0$ and $\sigma = 1$

```
Inverse Normal
Tail     :Right
Area     :0.15
σ        :1
μ        :0
Save Res:None
Execute
|CALC
```

Use $z = \dfrac{x - \mu}{\sigma}$ to find $\mu$ ······························ So,

$$1.0364 = \frac{70 - \mu}{12}$$

$$70 - \mu = 12.4372$$

$$\mu = 57.6$$

**WORKED EXAMPLE 19.9**

The weights of adult cats are thought to be normally distributed. Data shows that 20% of adult cats weigh more than 5.8 kg and 30% weigh less than 4.5 kg.

Estimate the mean and variance of the weight of adult cats.

Define the random variable ⋯⋯⋯⋯⋯⋯⋯ Let $X$ be the weight of an adult cat.
$X$ and state its distribution

Then $X \sim N(\mu, \sigma^2)$

You know that 20% of adult ⋯⋯⋯⋯⋯⋯ $P(X > 5.8) = 0.2$
cats weigh more than 5.8 kg

Use the inverse normal mode ⋯⋯⋯⋯⋯ $P(Z > z_1) = 0.2$
on your GDC to find $z_1$
$\qquad z_1 = 0.84162$

```
Inverse Normal
Tail    :Right
Area    :0.2
σ       :1
μ       :0
Save Res:None
Execute
|CALC
```

Use $z = \frac{x - \mu}{\sigma}$ to form an ⋯⋯⋯⋯⋯⋯⋯ So,
equation in $\mu$ and $\sigma$

$$0.84162 = \frac{5.8 - \mu}{\sigma}$$

$$5.8 - \mu = 0.84162\sigma \qquad (1)$$

You know that 30% of adult ⋯⋯⋯⋯⋯⋯ $P(X < 4.5) = 0.3$
cats weigh less than 4.5 kg

Use the inverse normal mode ⋯⋯⋯⋯⋯ $P(Z < z_2) = 0.3$
on your GDC to find $z_2$
$\qquad z_2 = -0.52440$

```
Inverse Normal
Tail    :Left
Area    :0.3
σ       :1
μ       :0
Save Res:None
Execute
|CALC
```

Use $z = \frac{x - \mu}{\sigma}$ to form a ⋯⋯⋯⋯⋯⋯ So,
second equation in $\mu$ and $\sigma$

$$-0.52440 = \frac{4.5 - \mu}{\sigma}$$

$$4.5 - \mu = -0.52440\sigma \qquad (2)$$

Use your GDC to solve these ⋯⋯⋯⋯⋯ Solving (1) and (2) simultaneously:
simultaneous equations
$\qquad \mu = 5.00, \sigma = 0.952$

```
anX+bnY=Cn
        a       b        c
1[      1   0.8412    5.8]
2[      1  -0.524     4.5]

                 4.5
SOLV DEL CLR EDIT
```

So,
$\qquad$ mean = 5.00 kg
$\qquad$ variance = 0.906 kg$^2$

# Exercise 19C

For questions 1 to 4, you are given a normal distribution and a value of $x$. Use the method demonstrated in Worked Example 19.6 to find how many standard deviations the given $x$ value is away from the mean.

1   a   $X \sim N(25, 3^2)$, $x = 29$

    b   $X \sim N(13, 4^2)$, $x = 14$

2   a   $X \sim N(6.5, 1.2^2)$, $x = 4.7$

    b   $X \sim N(8.1, 0.8^2)$, $x = 6.2$

3   a   $X \sim N(258, 169)$, $x = 245$

    b   $X \sim N(48, 16)$, $x = 35$

4   a   $X \sim N(-7.2, 1.44)$, $x = -6.2$

    b   $X \sim N(-10.5, 4.41)$, $x = -11.5$

For questions 5 to 8, use the method demonstrated in Worked Example 19.6 to find the required value of $x$.

5   a   $X \sim N(62, 7.5^2)$ and $x$ is 1.2 standard deviations above the mean

    b   $X \sim N(138, 4.5^2)$ and $x$ is 1.2 standard deviations above the mean

6   a   $X \sim N(4.5, 1.3^2)$ and $x$ is 0.7 standard deviations below the mean

    b   $X \sim N(0.67, 0.8^2)$ and $x$ is 0.7 standard deviations below the mean

7   a   $X \sim N(-45, 400)$ and $x$ is 2.5 standard deviations above the mean

    b   $X \sim N(-11, 64)$ and $x$ is 2.5 standard deviations above the mean

8   a   $X \sim N(135, 2.56)$ and $x$ is 1.3 standard deviations below the mean

    b   $X \sim N(12, 6.25)$ and $x$ is 1.3 standard deviations below the mean

For questions 9 to 12, you are given a normal distribution for $X$. Use the method demonstrated in Worked Example 19.7 to write the given probability expressions in terms of probabilities of the standard normal variable $Z$ (do not evaluate the probabilities).

9   a   $X \sim N(12, 4^2)$, $P(X < 18)$

    b   $X \sim N(23, 5^2)$, $P(X < 30)$

10   a   $X \sim N(6.5, 1.5^2)$, $P(4.2 < X < 6.8)$

    b   $X \sim N(8.2, 2.4^2)$, $P(4.2 < X < 5.5)$

11   a   $X \sim N(57, 64)$, $P(X > 62)$

    b   $X \sim N(168, 144)$, $P(X > 185)$

12   a   $X \sim N(12, 1.96)$, $P(11.6 < X < 12.5)$

    b   $X \sim N(21, 6.25)$, $P(23.5 < X < 24.2)$

For questions 13 to 16, the random variable $X$ follows a normal distribution with an unknown mean and given standard deviation ($\sigma$). You are given one probability statement about $X$. Use the method demonstrated in Worked Example 19.8 to find the mean of $X$ correct to two significant figures.

13   a   $\sigma = 6.3$, $P(X < 16) = 0.73$

    b   $\sigma = 12$, $P(X < 30) = 0.75$

14   a   $\sigma = 5.5$, $P(X < 11.5) = 0.21$

    b   $\sigma = 1.2$, $P(X < 7.2) = 0.12$

15   a   $\sigma = 5$, $P(X > 77) = 0.036$

    b   $\sigma = 4.5$, $P(X > 25) = 0.060$

16   a   $\sigma = 1.5$, $P(X > 7.8) = 0.14$

    b   $\sigma = 2.1$, $P(X > 10.2) = 0.10$

For questions 17 to 20, the random variable $X$ follows a normal distribution with an unknown mean ($\mu$) and standard deviation ($\sigma$). You are given two probability statements about $X$. Use the method demonstrated in Worked Example 19.9 to find $\mu$ and $\sigma$ correct to two significant figures.

17   a   $P(X < 19.5) = 0.274$, $P(X > 22.6) = 0.261$

     b   $P(X < 12.2) = 0.309$, $P(X > 13.8) = 0.309$

18   a   $P(X < 4.2) = 0.0564$, $P(X > 6.1) = 0.232$

     b   $P(X < 4.8) = 0.0703$, $P(X > 7.2) = 0.146$

19   a   $P(X < 20) = 0.106$, $P(X > 40) = 0.338$

     b   $P(X < 20) = 0.0912$, $P(X > 65) = 0.0478$

20   a   $P(X < 15) = 0.0631$, $P(X > 42) = 0.0498$

     b   $P(X < 16) = 0.137$, $P(X > 36) = 0.0501$

**21** Random variable $X$ follows a normal distribution. What is the probability that the value of $X$ is between 1 and 1.5 standard deviations above the mean?

**22** A manufacturer knows that his machines produce bolts that have diameters that follow a normal distribution with standard deviation 0.065 cm. In a random sample of bolts, 12% had a diameter larger than 2.5 cm. Estimate the mean diameter of the bolts.

**23** Random variable $X$ has a normal distribution with mean 21.4. It is known that $P(X < 20) = 0.065$. Find the standard deviation of $X$.

**24** The times taken to complete a test can be modelled by a normal distribution with mean 36 minutes. Experience suggests that 15% of students take less than 30 minutes to complete the test. Estimate the standard deviation of the times.

**25** Random variable $X$ has the distribution $N(25, \sigma^2)$. It is known that $P(10 < x < 25) = 0.48$. Find the value of $\sigma$.

**26** The weights of apples from a particular orchard can be modelled by a normal distribution with mean 130 g. It is found that 99% of the apples weigh between 110 g and 150 g. Estimate the standard deviation of the weights.

**27** An examiner believes that the scores on a test can be modelled by a normal distribution. She finds that 7% of the scores are below 35 and 5% of the scores are above 70. Estimate the mean and variance of the scores.

**28** The wingspan of a species of pigeon is modelled by a normal distribution with mean 60 cm. It is known that 60% of pigeons have the wingspan between 55 cm and 65 cm. Estimate the percentage of pigeons whose wingspan is less than 50 cm.

**29** It is claimed that the range of a distribution can be estimated as six times the standard deviation. For a normally distributed variable, what percentage of the values lies within this range?

**30** A normal distribution has the interquartile range equal to 12.8. Find the variance of the distribution.

## Checklist

■ You should be able to use your GDC to find the regression line of $x$ on $y$, and use this regression line to predict values of $x$ not in the data set.

■ You should be able to find conditional probabilities using the formula

$$P(A|B) = \frac{P(A \cap B)}{P(B)}$$

■ You should know how to test whether two events are independent.

Events $A$ and $B$ are independent if $P(A|B) = P(A)$

or equivalently if $P(A \cap B) = P(A) P(B)$

■ You should know how to standardize normal variables

If $X \sim N(\mu, \sigma^2)$, the $z$-value is

$$z = \frac{x - \mu}{\sigma}$$

It measures the number of standard deviations from the mean.

■ You should know how to use inverse normal calculations to find an unknown mean and variance.

If $X \sim N(\mu, \sigma^2)$ and $Z = \frac{X - \mu}{\sigma}$, then

$Z \sim N(0, 1)$

---

## ■ Mixed Practice

**1** The table shows the marks scored by seven students on two different tests.

| Test 1 | 67 | 52 | 71 | 65 | 48 | 52 | 61 |
|--------|----|----|----|----|----|----|----|
| Test 2 | 60 | 50 | 69 | 62 | 51 | 49 | 58 |

**a** Find the product-moment correlation coefficient between the two marks.
**b** Use the appropriate regression line to predict the mark on the first test for a student who scored 53 marks on the second test.
**c** Comment on the reliability of your estimate.

**2** In each part, explain why the given regression line cannot be reliably used to estimate the value of $x$ when $y = 23$.
**a** The correlation coefficient between $x$ and $y$ is 0.215 and the equation of the regression line of $x$ on $y$ is $x = 1.12y + 32.5$.
**b** The correlation coefficient between $x$ and $y$ is 0.872 and the equation of the regression line of $y$ on $x$ is $y = 1.12x + 12.7$.
**c** The maximum value of $y$ in the sample is 19, the correlation coefficient between $x$ and $y$ is 0.872 and the equation of the regression line of $x$ on $y$ is $x = 1.12y + 32.5$.

**3** Events $A$ and $B$ are such that $P(A) = \frac{2}{3}$, $P(B) = \frac{1}{5}$ and $P(B|A) = \frac{3}{20}$. Find
**a** $P(A \cap B)$
**b** $P(A|B)$

**4** The probability that a child has green eyes is 0.6 and the probability that they have brown hair is 0.4. The probability that they have neither green eyes nor brown hair is 0.2.

  **a** Find the probability that a child has both green eyes and brown hair.

  **b** Hence determine whether the events 'having green eyes' and 'having brown hair' are independent.

**5** Asher's 200 m running times can be modelled by a normal distribution with mean $\mu$ and variance $\sigma^2$. The probability that his time is more than 26.2 seconds is 0.12 and the probability that his time is below 22.5 seconds is 0.05.

  **a** Show that $\mu + 1.1745\sigma = 26.2$ and find another similar equation for $\mu$ and $\sigma$.

  **b** Hence find the mean and the standard deviation of the times.

**6** Events $A$ and $B$ are such that $P(A) = \frac{2}{5}$, $P(B) = \frac{11}{20}$ and $P(A|B) = \frac{2}{11}$.

  **a** Find $P(A \cap B)$.

  **b** Find $P(A \cup B)$.

  **c** State with a reason whether or not events $A$ and $B$ are independent.

Mathematics HL May 2014 Paper 2 TZ2 Q1

**7** Each day a factory recorded the number ($x$) of boxes it produces and the total production cost ($y$) dollars. The results for nine days are shown in the following table.

| $x$ | 26 | 44 | 65 | 43 | 50 | 31 | 68 | 46 | 57 |
|---|---|---|---|---|---|---|---|---|---|
| $y$ | 400 | 582 | 784 | 625 | 699 | 448 | 870 | 537 | 724 |

  **a** Write down the equation of the regression line of $y$ on $x$.

  Use your regression line as a model to answer the following.

  **b** Interpret the meaning of

   **i** the gradient

   **ii** the $y$-intercept.

  **c** Estimate the cost of producing 60 boxes.

  **d** The factory sells the boxes for $19.99 each. Find the least number of boxes that the factory should produce in one day in order to make a profit.

  **e** Comment on the appropriateness of using your model to

   **i** estimate the cost of producing 5000 boxes

   **ii** estimate the number of boxes produced when the total production cost is $540.

Mathematics SL 2012 Specimen Paper 2 Q8

**8** Events $A$ and $B$ satisfy $P(A) = 0.6$, $P(B) = 0.4$ and $P(A \cup B) = 0.76$. Show that $A$ and $B$ are independent.

**9** A machine produces cylindrical rods with diameters that are normally distributed with standard deviation 17 mm. In a random sample of 60 rods, it is found that 10 of them have diameters smaller than 4.5 cm. Estimate the mean diameter of the rods.

**10** Random variable $X$ follows the distribution $N(\mu, \sigma^2)$. It is known that $P(X < 12) = 0.1$ and $P(X < 25) = 0.85$. Find the values of $\mu$ and $\sigma$.

**11** The mass of dogs can be modelled by a normal distribution. It is known that 30% have mass less than 2.4 kg and 7% of dogs have mass greater than 10.7 kg. Find the percentage of dogs that have mass greater than 8 kg.

**12** A random variable follows a normal distribution. Given that its interquartile range is 17, find its variance.

**13** The times taken for a student to complete a test can be modelled by a normal distribution with mean 45 minutes and standard deviation 10 minutes.
  **a** What is the probability that a randomly chosen student takes longer than 50 minutes to complete a test?
  **b** Given that a student takes more than 50 minutes to complete a test, find the probability that they take more than 60 minutes.

**14** The masses of apples are normally distributed with mean 126 g and variance 144 g$^2$. An apple is classified as 'large' if it weighs more than 150 g.
  **a** What percentage of apples are classified as 'large'?
  **b** What percentage of large apples weigh more than 160 g?

**15** A student sits a national test and is told that the marks follow a normal distribution with mean 100. The student received a mark of 124 and is told that he is at the 68th percentile. Calculate the variance of the distribution.

Mathematics HL May 2014 Paper 2 TZ1 Q2

**16** The duration of direct flights from London to Singapore in a particular year followed a normal distribution with mean $\mu$ and standard deviation $\sigma$.

92% of flights took under 13 hours, while only 12% of flights took under 12 hours 35 minutes.

Find $\mu$ and $\sigma$ to the nearest minute.

Mathematics HL November 2013 Paper 2 Q4

**17** A factory makes lamps. The probability that a lamp is defective is 0.05. A random sample of 30 lamps is tested.
  **a** Find the probability that there is at least one defective lamp in the sample.
  **b** Given that there is at least one defective lamp in the sample, find the probability that there are at most two defective lamps.

Mathematics SL May 2012 Paper 2 TZ2 Q7

**18** Given that $P(A \cup B) = 1$, $P(A) = \frac{7}{8}$ and $P(A'|B) = \frac{1}{4}$, find $P(B)$.

**19** Two events $A$ and $B$ are such that $P(B|A) = \frac{3}{5}$, $P(A \cap B') = \frac{1}{3}$ and $P(A \cup B') = \frac{11}{12}$.
  **a** Find $P(A \cap B)$.
  **b** Determine whether $A$ and $B$ are independent.

**20** A machine cuts wire into pieces with lengths that can be modelled by a normal distribution with mean 30 cm and standard deviation 2.5 cm. Any pieces longer than 35 cm are discarded and the rest are packed in boxes of 10. Find the probability that, in a randomly selected box, at least one piece of wire is longer than 34 cm.

**21** The table shows six pairs of values $(x, y)$.

| $x$ | −6 | −2 | −4 | 3 | 5 | 6 |
|---|---|---|---|---|---|---|
| $y$ | 75 | 8 | 40 | 10 | 49 | 70 |

  **a** Calculate the correlation coefficient between $x$ and $y$.
  **b** Calculate $t$, the correlation coefficient between $x^2$ and $y$.
  **c** Use a suitable regression line to find the relationship between $x$ and $y$ in the form $y = kx^2 + c$.

**22** A machine manufactures a large number of nails. The length, $L$ mm, of a nail is normally distributed, where $L \sim N(50, \sigma^2)$.

**a** Find $P(50 - \sigma < L < 50 + 2\sigma)$.

**b** The probability that the length of a nail is less than 53.92 mm is 0.975.

Show that $\sigma = 2.00$ (correct to three significant figures).

All nails with length at least $t$ mm are classified as large nails.

**c** A nail is chosen at random. The probability that it is a large nail is 0.75.

Find the value of $t$.

**d i** A nail is chosen at random from the large nails. Find the probability that the length of the nail is less than 50.1 mm.

**ii** Ten nails are chosen at random from the large nails. Find the probability that at least two nails have a length that is less than 50.1 mm.

Mathematics SL May 2015 Paper 2 TZ2 Q9

# 20 Analysis and approaches: Differentiation

## ESSENTIAL UNDERSTANDINGS

- Calculus describes rates of change between two variables.
- Calculus helps us understand the behaviour of functions and allows us to interpret features of their graphs.

## In this chapter you will learn...

- how to differentiate functions such as $\sqrt{x}$, $\sin x$ and $\ln x$
- how to use the chain rule to differentiate composite functions
- how to differentiate products and quotients
- how to interpret the second derivative of a function
- how to find local minimum and maximum points on a graph
- how to use calculus to find the optimal solution to a problem
- how to identify points of inflection on a graph.

### CONCEPTS

The following concepts will be addressed in this chapter:

- The derivative may be **represented** physically as a rate of change and geometrically as the gradient or slope function.
- Mathematical modelling can provide effective solutions to real-life problems in optimization by maximizing or minimizing a **quantity**.

■ **Figure 20.1** What factors affect the profits of company, the yield of a harvest, our fitness and our happiness?

## PRIOR KNOWLEDGE

Before starting this chapter, you should already be able to complete the following:

1  If $f(x) = x^2 + 2$ find $f'(x)$.

2  Find the equation of the tangent to the curve $y = x^3$ at $x = 1$.

3  Solve $e^x = 5$.

You already know how to differentiate some simple functions, but in this chapter you will significantly extend the number of functions you can differentiate. You will also meet further applications of differentiation used to describe graphs and see how this can be translated into many real-world situations where optimal solutions are required.

## Starter Activity

Look at the pictures in Figure 20.1. Think about what they represent and discuss the following question: What are the variables you might consider changing if you wanted to achieve an optimal result in each of these situations? Sketch graphs to show how the outcomes might vary depending on the different variables you have considered.

**Now look at this problem:**

The gradient on the curve $y = f(x)$ when $x = 2$ is 10. The function $g(x)$ is defined as $f(2x)$. What is the gradient on the graph of $y = g(x)$ when $x = 1$?

## LEARNER PROFILE – Balanced

Is being a good mathematician the same as being good at sitting maths exams? Always remember that in the real world you are rarely asked to do things in silence, alone and without any reference materials. Although examinations are important, they need to be put into perspective and should not be seen as the only purpose of studying mathematics.

# 20A Extending differentiation

So far you have only seen how to differentiate functions of the form $f(x) = ax^n + bx^m + \ldots$, where $n, m \in \mathbb{Z}$. It would be useful, however, to be able to differentiate the many other types of function you have met, too.

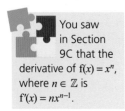 You saw in Section 9C that the derivative of $f(x) = x^n$, where $n \in \mathbb{Z}$ is $f'(x) = nx^{n-1}$.

## ■ Derivative of $x^n$, $n \in \mathbb{Q}$

The same result that you already know for differentiating integer powers of $x$ also applies when $n$ is any rational number.

### KEY POINT 20.1

If $f(x) = x^n$, where $n \in \mathbb{Q}$, then $f'(x) = nx^{n-1}$.

### WORKED EXAMPLE 20.1

Find $f'(x)$ for $f(x) = \sqrt[3]{x}$.

Write $f(x)$ in the form $x^n$ ............................ $f(x) = \sqrt[3]{x}$

$$= x^{\frac{1}{3}}$$

Use $f'(x) = nx^{n-1}$ ............................ $f'(x) = \dfrac{1}{3}x^{\frac{1}{3}-1}$

$$= \dfrac{1}{3}x^{-\frac{2}{3}}$$

As before, this result applies to multiples and sums of terms, and you might need to use the laws of exponents to get the expression in the correct form first.

### WORKED EXAMPLE 20.2

Find $\dfrac{dy}{dx}$ for $y = \dfrac{3x^2 - 4}{\sqrt{x}}$.

Use the laws of exponents to get the expression into the form $ax^n + bx^m$ ............................ $y = \dfrac{3x^2 - 4}{\sqrt{x}}$

$$= \dfrac{3x^2}{x^{\frac{1}{2}}} - \dfrac{4}{x^{\frac{1}{2}}}$$

$$= 3x^{\frac{3}{2}} - 4x^{-\frac{1}{2}}$$

Then differentiate term by term ............................ $\dfrac{dy}{dx} = 3 \times \dfrac{3}{2}x^{\frac{3}{2}-1} - 4\left(-\dfrac{1}{2}\right)x^{-\frac{1}{2}-1}$

$$= \dfrac{9}{2}x^{\frac{1}{2}} + 2x^{-\frac{3}{2}}$$

## ■ Derivatives of $\sin x$ and $\cos x$

A different class of function that is very familiar, but that you have not yet learned how to differentiate, is trigonometric functions. It turns out that the results given are only true if the angle $x$ is measured in radians.

If you are studying analysis and approaches HL you will see how to prove these results.

### KEY POINT 20.2

If $f(x) = \sin x$, then $f'(x) = \cos x$.

### KEY POINT 20.3

If $f(x) = \cos x$, then $f'(x) = -\sin x$.

In Section 20C, you will see how to differentiate $\tan x$ although you only need to remember that derivative if you are studying analysis and approaches HL.

Again, these results can be applied to multiples and sums of terms.

### WORKED EXAMPLE 20.3

Differentiate $f(x) = 2\sin x - 5\cos x$.

Differentiate term by term ······················· $f'(x) = 2\cos x - 5(-\sin x)$
$$= 2\cos x + 5\sin x$$

## ■ Derivatives of $e^x$ and $\ln x$

Two other commonly occurring functions you have met are the exponential and natural logarithm functions.

The exponential function has the important property that its derivative is itself.

### KEY POINT 20.4

If $f(x) = e^x$, then $f'(x) = e^x$.

The derivative of the natural logarithm function follows from the fact that it is the inverse of the exponential function.

### KEY POINT 20.5

If $f(x) = \ln x$, then $f'(x) = \frac{1}{x}$.

**TOK Links**

Does Key Point 20.4 reflect a property of e, or is it the definition of e? Who decides the starting points (called axioms) of mathematics? What criteria do you use to determine if one fact is more fundamental than another?

---

**Proof 20.1**

Given that for $y = e^x$, $y' = e^x$, prove that for $y = \ln x$, $y' = \dfrac{1}{x}$.

| | |
|---|---|
| Pick any point on the curve $y = e^x$ | Let the point P lie on $y = e^x$ and have $x$-coordinate $p$<br>So P is the point $(p, e^p)$ |
| Reflecting a point in the line $y = x$ swaps the $x$- and $y$-coordinates | Let the point $Q$ be the reflection of P in the line $y = x$<br>So $Q$ is the point $(e^p, p)$ |
| The graphs of a function and its inverse are reflections in the line $y = x$ | But since $y = e^x$ and $y = \ln x$ are inverse functions, $Q$ lies on $y = \ln x$ |
| | The gradient of $y = e^x$ at P is $e^p$ |
| The gradient of the tangent to $y = e^x$ at P is $e^p$, so reflecting this tangent in the line $y = x$ gives the gradient of the tangent to $y = \ln x$ at $Q$ | So the gradient of $y = \ln x$ at $Q$ is $\dfrac{1}{e^p}$ |

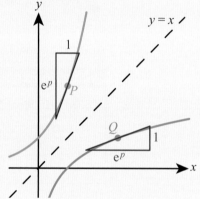

But $e^p$ is the $x$-coordinate of $Q$ so the gradient of $y = \ln x$ is $\dfrac{1}{x}$

---

You can use the laws of logarithms to get the expression into the correct form for differentiating.

**WORKED EXAMPLE 20.4**

Given $y = \ln x^3$, find $y'$.

| | |
|---|---|
| Use $\ln x^m = m \ln x$ | $y = \ln x^3$<br>$= 3\ln x$ |
| Now differentiate | $y' = 3\left(\dfrac{1}{x}\right)$<br>$= \dfrac{3}{x}$ |

In Section 20B, you will see how you can differentiate an expression such as $\ln x^3$ without needing to use the laws of logarithms first.

You can use these results to find the equations of tangents and normals.

> **WORKED EXAMPLE 20.5**
>
> Find the equation of the tangent to the curve $y = \cos x + 2e^x$ at the point $(0, 3)$.
>
> Differentiate term by term .............................. $y' = -\sin x + 2e^x$
>
> Find the value of the ·································· When $x = 0$,
> gradient ($y'$) at $x = 0$
> $$y' = -\sin 0 + 2e^0$$
> $$= 0 + 2$$
> $$= 2$$
>
> Substitute into the equation ···················· $y - 3 = 2(x - 0)$
> of a straight line:
> $$y = 2x + 3$$
> $$y - y_1 = m(x - x_1)$$

 You saw how to find the equations of tangents and normals in Section 9D.

## Exercise 20A

For questions 1 to 6, use the method demonstrated in Worked Example 20.1 to find $f'(x)$ for the following functions.

1  a  $f(x) = x^{\frac{2}{3}}$

  b  $f(x) = x^{\frac{3}{4}}$

2  a  $f(x) = x^{-\frac{1}{2}}$

  b  $f(x) = x^{-\frac{4}{3}}$

3  a  $f(x) = 6x^{\frac{3}{2}}$

  b  $f(x) = 8x^{\frac{1}{4}}$

4  a  $f(x) = 9x^{-\frac{2}{3}}$

  b  $f(x) = 10x^{-\frac{2}{5}}$

5  a  $f(x) = 2\sqrt{x}$

  b  $f(x) = 6\sqrt[4]{x}$

6  a  $f(x) = \dfrac{-12}{\sqrt[3]{x}}$

  b  $f(x) = \dfrac{-5}{\sqrt[5]{x}}$

For questions 7 to 9, use the method demonstrated in Worked Example 20.2 to differentiate the following expressions.

7  a  $\dfrac{9x + 1}{x^{\frac{4}{3}}}$

  b  $\dfrac{2x - 1}{x^{\frac{3}{2}}}$

8  a  $\dfrac{12x^2 - x}{5x^{\frac{3}{4}}}$

  b  $\dfrac{6x^3 + x}{7x^{\frac{2}{3}}}$

9  a  $\dfrac{2 - 3x}{\sqrt[3]{x}}$

  b  $\dfrac{5 + 8x}{\sqrt[4]{x}}$

For questions 10 and 11, use the method demonstrated in Worked Example 20.3 to differentiate the following functions.

10  a  $f(x) = 3\sin x$

   b  $f(x) = 4\cos x$

11  a  $f(x) = \dfrac{1}{2}\cos x - 5\sin x$

   b  $f(x) = \dfrac{3}{4}\sin x - 2\cos x$

For questions 12 and 13, use the method demonstrated in Worked Example 20.4 to find $\dfrac{dy}{dx}$ for the following graphs.

12  a  $y = 3\ln x$

   b  $y = -4\ln x$

13  a  $y = \ln x^2$

   b  $y = \ln x^{-1}$

For questions 14 and 15, use the method demonstrated in differentiating $e^x$ in Worked Example 20.5 to find $y'$ for the following graphs.

14  a  $y = 5e^x$

   b  $y = -6e^x$

15  a  $y = -\dfrac{e^x}{2}$

   b  $y = \dfrac{3e^x}{4}$

**16** Find the gradient of the graph of $y = \sin x$ at $x = \dfrac{\pi}{3}$.

**17** The height of a falcon above the ground in metres ($h$) during a swooping attack $t$ seconds after it spots some prey is modelled by

$$h = \frac{(10t + 20)}{\sqrt{t}}$$

What is the rate of change of height when $t = 1$? What is the significance of the sign of your answer?

**18** Find the point of the curve $y = 2 + 5x - e^x$ which has gradient 4.

**19** Find the gradient of the graph $y = e^x + 3x + 4$ when $x = \ln 2$.

**20** Find the equation of the tangent to $y = 3\cos x, 0 < x < \dfrac{\pi}{2}$ that is parallel to $2y + 3x + 3 = 0$.

**21** Find the equation of the tangent to $y = e^x - 2x$ that is perpendicular to $y = -x + 4$.

**22** Find the point on the curve $y = \sqrt{x}$ that has a tangent of $y = \dfrac{x}{6} + \dfrac{3}{2}$.

**23** a Find the equation of the normal to $y = 2x - e^x$ at $x = \ln\dfrac{5}{2}$.

   b Show that this normal does not meet the original curve at any other point.

**24** The number of bacteria in millions, $N$, after $t$ hours of being introduced to an agar dish is modelled by $N = (1 + \sqrt{t})^2$.

   a What is the initial population introduced?

   b How long does it take for the population to reach 9 million?

   c Find the rate of increase of bacteria when $t = 4$.

   d State one limitation of this model.

**25** Find the interval in which $\ln x - 2x$ is a decreasing function.

**26** The tangent to the curve $y = \sqrt{x}$ at $x = k$ is perpendicular to the tangent to the curve $y = \dfrac{1}{\sqrt{x}}$, also at $x = k$. Find the value of $k$.

**27** a The curve $y = \cos x$ for $-\pi < x < \pi$ meets the $x$-axis at two points, labelled $P$ and $Q$. The tangents at $P$ and $Q$ meet at $R$. Find the area of the triangle $PQR$.

   b The curve $y = \sin x$ for $0 \leqslant x < 2\pi$ meets the $x$-axis at two points, labelled $A$ and $B$. The tangents at $A$ and $B$ meet at $C$. Find the area of the triangle $ABC$.

**28** Find the coordinates of the point on the curve $y = \sqrt{x}$ that has a tangent passing through the point $(0, 1)$.

**29** The line $2\sqrt{3}y - 3x + \dfrac{13\pi}{2} = 0$ is a tangent to the curve $y = \sin(x) - \dfrac{1}{2}$ at the point $P$. Find the coordinates of $P$.

**30** Find the $x$-coordinate of all points on the graph of $y = x^2 - 2\ln x$ where the tangent is parallel to $y + 3x = 9$.

# 20B The chain rule for composite functions

In Worked Example 20.4 we used one of the laws of logarithms to enable us to differentiate $\ln x^3$. However, there is no equivalent method to enable us to differentiate, say, $\sin x^3$ or $e^{x^3}$.

All of these are examples of composite functions, $f(g(x))$, where $g(x) = x^3$, so we need a method for differentiating composite functions. This is provided by the **chain rule**.

---

**KEY POINT 20.6**

The chain rule:

If $y = f(u)$, where $u = g(x)$ then $\dfrac{dy}{dx} = \dfrac{dy}{du} \times \dfrac{du}{dx}$.

---

---

**WORKED EXAMPLE 20.6**

Differentiate $y = (x^2 - 3x)^6$.

This is a composite function $\cdots\cdots\cdots\cdots\cdots\cdots$ $y = u^6$, where $u = x^2 - 3x$

Use $\dfrac{dy}{dx} = \dfrac{dy}{du} \times \dfrac{du}{dx}$, with $\dfrac{dy}{du} = 6u^5$ $\cdots\cdots\cdots\cdots$ $\dfrac{dy}{dx} = 6u^5 \times (2x - 3)$

and $\dfrac{du}{dx} = 2x - 3$

Replace $u$ with $x^2 - 3x$ $\cdots\cdots\cdots\cdots\cdots$ $= 6(x^2 - 3x)^5(2x - 3)$

---

**WORKED EXAMPLE 20.7**

Differentiate $y = \sin 4x$.

This is a composite function $\cdots\cdots\cdots\cdots\cdots\cdots$ $y = \sin u$, where $u = 4x$

Use $\dfrac{dy}{dx} = \dfrac{dy}{du} \times \dfrac{du}{dx}$, with $\dfrac{dy}{du} = \cos u$ $\cdots\cdots\cdots\cdots$ $\dfrac{dy}{dx} = \cos u \times 4$

and $\dfrac{du}{dx} = 4$

Replace $u$ with $4x$ $\cdots\cdots\cdots\cdots\cdots$ $= 4\cos 4x$

---

**WORKED EXAMPLE 20.8**

Differentiate $y = e^{3x^2}$.

This is a composite function $\cdots\cdots\cdots\cdots\cdots\cdots$ $y = e^u$, where $u = 3x^2$

Use $\dfrac{dy}{dx} = \dfrac{dy}{du} \times \dfrac{du}{dx}$, with $\dfrac{dy}{du} = e^u$ $\cdots\cdots\cdots\cdots$ $\dfrac{dy}{dx} = e^u \times 6x$

and $\dfrac{du}{dx} = 6x$

Replace $u$ with $3x^2$ $\cdots\cdots\cdots\cdots\cdots$ $= 6xe^{3x^2}$

---

If you have a composite of three functions, you will need an extra derivative in the chain rule.

---

**WORKED EXAMPLE 20.9**

Differentiate $y = (\ln 2x)^5$.

This is a composite of $\cdots\cdots\cdots\cdots\cdots$ $y = u^5$, where $u = \ln v$ and $v = 2x$
three functions

Use the chain rule $\cdots\cdots\cdots\cdots$ $\dfrac{dy}{dx} = 5u^4 \times \dfrac{1}{v} \times 2$
with three derivatives:

$\dfrac{dy}{dx} = \dfrac{dy}{du} \times \dfrac{du}{dv} \times \dfrac{dv}{dx}$, where

$\dfrac{dy}{du} = 5u^4$, $\dfrac{du}{dv} = \dfrac{1}{v}$ and $\dfrac{dv}{dx} = 2$

Replace $u$ with $\ln v$ $\cdots\cdots\cdots\cdots$ $= 5(\ln 2x)^4 \times \dfrac{1}{2x} \times 2$
and $v$ with $2x$

$= \dfrac{5}{x}(\ln 2x)^4$

---

## Exercise 20B

For questions 1 to 3, use the method demonstrated in Worked Example 20.6 to find $\dfrac{dy}{dx}$ for the graphs given.

1  a  $y = (3x + 2)^4$                2  a  $y = (x^2 + 3)^{\frac{3}{4}}$                3  a  $y = \sqrt{2x^3 + x}$

   b  $y = (2x - 7)^5$                   b  $y = (4 - x^2)^{\frac{2}{3}}$                   b  $y = \sqrt[3]{5x - x^3}$

For questions 4 to 7, use the method demonstrated in Worked Example 20.7 to find f'(x) for the functions given.

4  a  $f(x) = \sin 2x$          5  a  $f(x) = \cos \pi x$          6  a  $f(x) = \sin(3x + 1)$          7  a  $f(x) = \cos(2 - 3x)$

   b  $f(x) = \sin \dfrac{1}{3} x$          b  $f(x) = \cos 5x$          b  $f(x) = \sin(1 - 4x)$          b  $f(x) = \cos\left(\dfrac{1}{2}x + 4\right)$

For questions 8 to 11, use the method demonstrated in Worked Example 20.8 to differentiate the expressions given.

8  a  $e^{3x}$                  9  a  $e^{-x^3}$                  10  a  $\ln 4x$                  11  a  $\ln(x^2 + 1)$

   b  $e^{\frac{x}{2}}$                   b  $e^{4x^2}$                   b  $\ln \dfrac{x}{3}$                   b  $\ln(3 - 4x^2)$

For questions 12 to 14, use the method demonstrated in Worked Example 20.9 to find the derivative of the graphs given.

12  a  $y = \sin^2 3x$              13  a  $y = e^{\cos 2x}$              14  a  $y = (\ln 3x)^{\frac{1}{2}}$

    b  $y = \cos^2 4x$                 b  $y = e^{\sin 5x}$                 b  $y = (\ln 2x)^{\frac{1}{3}}$

15  If $f(x) = \dfrac{1}{e^x}$ find f'(x).

16  Find the gradient of the curve $y = (e^x)^2 + 5x$ when $x = \ln 3$.

17  Find the equation of the tangent to the curve $y = \sqrt{x^2 + 9}$ at $x = 4$.

18  Find the coordinates of the point on the graph of $y = \ln(2x - 5)$ that has its tangent parallel to $y = 2x$.

19  Find the tangent to $y = \cos\left(\dfrac{1}{x}\right)$ at $x = \dfrac{2}{\pi}$.

20  Find the tangent to the curve $y = \ln(x - 2)$ that is perpendicular to $y = -3x + 2$.

21  a  Find the derivative of $\sin^2 x$.

    b  Hence, find all points on the curve of $y = \sin^2 x$, $0 < x < 2\pi$ where the gradient of the tangent is zero.

22  a  Show that $(\cos x + \sin x)^2 = 1 + \sin 2x$.

    b  Use differentiation to deduce that $\cos 2x = \cos^2 x - \sin^2 x$.

23  Given that $\sin 3x = 3\sin x - 4\sin^3 x$ prove that $\cos 3x = 4\cos^3 x - 3\cos x$.

24  The tangent to the curve $y = e^{kx}$ at $x = \dfrac{1}{k}$ is called $L$. Prove that for all $k > 0$ the $y$-intercept of $L$ is independent of $k$.

25  Find the $x$-coordinate of any point on the curve $y = \ln(x^2 - 8)$ where the tangent is parallel to $y = 6x - 5$.

26  a  Show that if $y = e^{2x}$ then $\dfrac{dy}{dx} = 2y$.

    b  Is it true that if $\dfrac{dy}{dx} = 2y$ then $y = e^{2x}$? Justify your answer.

# 20C The product and quotient rules

You know that you cannot differentiate a product of functions term by term as you can with a sum. Sometimes you can get round this. For example, you can expand the product $x^2(2x + 1)$ to give $2x^3 + x^2$ and then differentiate.

However, clearly you cannot do this with a product such as $x^2 \ln x$, so we need a method for differentiating products of functions. This is provided by the **product rule**.

> **KEY POINT 20.7**
>
> The product rule:
> If $y = u(x)v(x)$ then $\dfrac{dy}{dx} = u\dfrac{dv}{dx} + v\dfrac{du}{dx}$.

> **WORKED EXAMPLE 20.10**
>
> Differentiate $y = x^2 \ln x$
>
> Define the functions $\cdots\cdots\cdots\cdots\cdots\cdots$ Let $u = x^2$ and $v = \ln x$
> $u(x)$ and $v(x)$
>
> Find their derivatives $\cdots\cdots\cdots\cdots\cdots$ Then $\dfrac{du}{dx} = 2x$ and $\dfrac{dv}{dx} = \dfrac{1}{x}$
>
> Substitute into $\dfrac{dy}{dx} = u\dfrac{dv}{dx} + v\dfrac{du}{dx}$ $\cdots\cdots\cdots\cdots$ $\dfrac{dy}{dx} = x^2\left(\dfrac{1}{x}\right) + (\ln x)(2x)$
>
> $\qquad\qquad\qquad\qquad\qquad = x + 2x\ln x$

You may need to use the product rule and the chain rule together.

> **WORKED EXAMPLE 20.11**
>
> Differentiate $f(x) = e^x(x^2 + 3)^4$.
>
> Define the functions $u(x)$ and $\cdots\cdots\cdots\cdots$ Let $u = e^x$ and $v = (x^2 + 3)^4$
> $v(x)$ for use in the product rule
>
> $\qquad\qquad\qquad\qquad\qquad$ Then $u' = e^x$
>
> For $v'$ you need the chain $\cdots\cdots\cdots\cdots$ and $v' = 4(x^2 + 3)^3 \times 2x = 8x(x^2 + 3)^3$
> rule. Remember to multiply
> by the derivative of $x^2 + 3$
>
> Substitute into $f'(x) = uv' + vu'$ $\cdots\cdots\cdots\cdots$ $f'(x) = e^x \times 8x(x^2 + 3)^3 + (x^2 + 3)^4 \times e^x$
>
> $\qquad\qquad\qquad\qquad\qquad\qquad = 8xe^x(x^2 + 3)^3 + e^x(x^2 + 3)^4$

# Be the Examiner 20.1

Differentiate $f(x) = \sin(x^2 - 5x)$.

Which is the correct solution? Identify the errors made in the incorrect solutions.

| Solution 1 | Solution 2 | Solution 3 |
|---|---|---|
| Let $u = \sin$ and $v = x^2 - 5x$ <br> Then $u' = \cos$ and $v' = 2x - 5$ <br> $f'(x) = \sin(2x - 5) + \cos(x^2 - 5x)$ | $f'(x) = (2x - 5)\cos(x^2 - 5x)$ | $f'(x) = \cos(2x - 5)$ |

Just as with products, you know that you cannot differentiate the top and bottom of a quotient separately. You could rewrite a quotient as a product and use the product rule to differentiate. For example:

$$\frac{x^2 + 6x - 2}{(x + 3)^2} = (x^2 + 6x - 2)(x + 3)^{-2}$$

However, it is often simpler to use a method that allows you to differentiate quotients directly. This is called the **quotient rule**.

> **KEY POINT 20.8**
>
> The quotient rule:
>
> If $y = \dfrac{u(x)}{v(x)}$, then $\dfrac{dy}{dx} = \dfrac{v\dfrac{du}{dx} - u\dfrac{dv}{dx}}{v^2}$.

**WORKED EXAMPLE 20.12**

Differentiate $y = \dfrac{x^2 + 6x - 2}{(x+3)^2}$, giving your answer in the form $\dfrac{a}{(x+3)^b}$ where $a, b \in \mathbb{Z}$.

Define the functions $u(x)$ and $v(x)$ ................ Let $u = x^2 + 6x - 2$ and $v = (x+3)^2$

Note that for $\dfrac{dv}{dx}$ you need ................ Then $\dfrac{du}{dx} = 2x + 6$ and $\dfrac{dv}{dx} = 2(x+3)$
the chain rule, but the derivative of $x = 1$ is just 1

Substitute into $\dfrac{dy}{dx} = \dfrac{v\frac{du}{dx} - u\frac{dv}{dx}}{v^2}$ ........ $\dfrac{dy}{dx} = \dfrac{(x+3)^2(2x+6) - (x^2+6x-2)2(x+3)}{((x+3)^2)^2}$

Cancel a factor of $(x-3)$ ............ $= \dfrac{(x+3)(2x+6) - 2(x^2+6x-2)}{(x+3)^3}$

Expand the numerator and simplify ............ $= \dfrac{2x^2 + 12x + 18 - 2x^2 - 12x + 4}{(x+3)^3}$

$= \dfrac{22}{(x+3)^3}$

**WORKED EXAMPLE 20.13**

Show that if $f(x) = \tan x$ then $f'(x) = \dfrac{1}{\cos^2 x}$.

Use $\tan x = \dfrac{\sin x}{\cos x}$ to write ........ $f(x) = \tan x = \dfrac{\sin x}{\cos x}$
$\tan x$ in terms of functions you can differentiate

Now use the quotient rule ........ Let $u = \sin x$ and $v = \cos x$

Then $u' = \cos x$ and $v' = -\sin x$

Substitute into ........ $f'(x) = \dfrac{\cos x \cos x - \sin x(-\sin x)}{\cos^2 x}$
$f'(x) = \dfrac{vu' - uv'}{v^2}$

$= \dfrac{\cos^2 x + \sin^2 x}{\cos^2 x}$

Use $\cos^2 x + \sin^2 x \equiv 1$ ........ $= \dfrac{1}{\cos^2 x}$

# Be the Examiner 20.2

Differentiate $f(x) = \dfrac{5}{e^{2x}}$.

Which is the correct solution? Identify the errors made in the incorrect solutions.

| Solution 1 | Solution 2 | Solution 3 |
|---|---|---|
| Let $u = 5$ and $v = e^{2x}$<br>Then $u' = 5$ and $v' = 2e^{2x}$<br>$f'(x) = \dfrac{e^{2x} \times 0 - 5 \times 2e^{2x}}{(e^{2x})^2}$<br>$= -\dfrac{10e^{2x}}{(e^{2x})^2} = -\dfrac{10}{e^{2x}}$ | $f(x) = \dfrac{5}{e^{2x}} = 5e^{-2x}$<br>Let $u = 5$ and $v = e^{-2x}$<br>Then $u' = 0$ and $v' = -2e^{-2x}$<br>$f'(x) = 5(-2e^{-2x}) + e^{-2x} \times 0$<br>$= -10e^{-2x}$ | $f(x) = \dfrac{5}{e^{2x}} = 5e^{-2x}$<br>$f'(x) = -10e^{-2x}$ |

## Exercise 20C

For questions 1 to 4, use the method demonstrated in Worked Example 20.10 to find $\dfrac{dy}{dx}$ for the graphs given.

1  a  $y = x \sin x$

   b  $y = x^{\frac{1}{2}} \sin x$

2  a  $y = x^2 \cos x$

   b  $y = x^{-1} \cos x$

3  a  $y = x^{-\frac{1}{2}} e^x$

   b  $y = x^3 e^x$

4  a  $y = x^{\frac{2}{3}} \ln x$

   b  $y = x^4 \ln x$

For questions 5 to 9, use the method demonstrated in Worked Example 20.11 to find f'($x$) for the functions given.

5  a  $f(x) = x(2x + 1)^{\frac{1}{2}}$

   b  $f(x) = x^2 (3x - 4)^{\frac{3}{2}}$

6  a  $f(x) = x^2 \sin 3x$

   b  $f(x) = x^{\frac{3}{4}} \sin 2x$

7  a  $f(x) = x \cos(4x - 1)$

   b  $f(x) = x^{-1} \cos 5x$

8  a  $f(x) = x^{\frac{1}{2}} e^{4x+5}$

   b  $f(x) = x^{-2} e^{1-x}$

9  a  $f(x) = x^{-1} \ln(2x - 3)$

   b  $f(x) = x^3 \ln(5 - x)$

For questions 10 to 12, use the method demonstrated in Worked Example 20.12 to find the derivative of the graphs given.

10  a  $y = \dfrac{x^2}{x + 2}$

    b  $y = \dfrac{3x}{x - 4}$

11  a  $y = \dfrac{x - 2}{\sqrt{x + 1}}$

    b  $y = \dfrac{4 - x}{(x - 3)^2}$

12  a  $y = \dfrac{e^x}{(3x + 1)^2}$

    b  $y = \dfrac{e^x}{(2x - 1)^3}$

13  Find the equation of the tangent to the curve $y = \dfrac{\sin x}{x}$ at $x = \dfrac{\pi}{2}$.

14  Find the equation of the normal to the curve $y = x \cos 2x$ at $x = \dfrac{\pi}{4}$.

15  If $f(x) = x e^{2x}$ find f'(3).

16  Differentiate $x e^x \ln x$.

17  Differentiate $e^{x \sin x}$.

18  Find the point on the curve $y = e^{3x} - 11x$ where the tangent has gradient 13.

19  Find the coordinates of the points on the curve $y = x e^{-x}$ where the gradient is zero.

20  Find the coordinates of the points on the curve $y = x \ln x$ where the gradient is 2.

21  Find and simplify an expression for the derivative of $f(x) = \dfrac{x}{\sqrt{k + x^2}}$ where $x$ is a positive constant.

22  If $f(x) = x\sqrt{2 + x}$ show that $f'(x) = \dfrac{ax + b}{2\sqrt{2 + x}}$ where $a$ and $b$ are integers to be found.

23  Show that $f(x) = \dfrac{x^2}{1 + x}$ is an increasing function if $x > a$ or if $x < b$ where $a$ and $b$ are constants to be determined.

24  The population of rabbits on an island, $P$ thousands, $n$ years after they were introduced is modelled by

$$P = \dfrac{4}{1 + 3e^{-n}}$$

   a  Find an expression for the population growth rate, $\dfrac{dP}{dn}$, and explain why this predicts that the population is always growing.

   b  Find (i) the initial population and (ii) the initial population growth rate.

   c  What is the value of $P$ as $n$ gets very large?

   d  Hence sketch the population as a function of time.

25  a  Show that if $x > 0$ then $x^x = e^{x \ln x}$.

    b  Hence find the equation of the tangent to the curve $y = x^x$ at $x = 1$.

# 20D The second derivative

## ◼ Finding the second derivative

Differentiating the derivative f'(*x*) gives the second derivative f''(*x*), which measures the rate of change of the gradient.

If the notation $\dfrac{dy}{dx}$ is used for the derivative, then $\dfrac{d^2y}{dx^2}$ is used for the second derivative.

---

**WORKED EXAMPLE 20.14**

If f(*x*) = $x^3$ + 3*x* find f''(2).

First find f'(*x*) ········································· f'(*x*) = $3x^2$ + 3

Then differentiate again ························· f''(*x*) = 6*x*

Now substitute in *x* = 2 ····················· f''(2) = 12

---

## ◼ Graphical behaviour of functions, including the relationship between the graphs of f, f' and f''

In Section 9B, you saw how to sketch the graph of the derivative from the graph of the original function by thinking about the gradient at various different points on the function.

To sketch the graph of the second derivative, you first sketch the graph of the derivative as before, and then repeat the process.

---

**WORKED EXAMPLE 20.15**

Sketch the second derivative of this function.

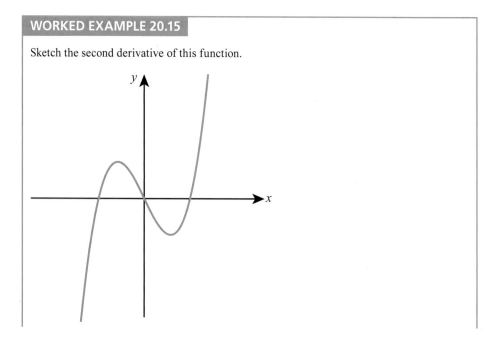

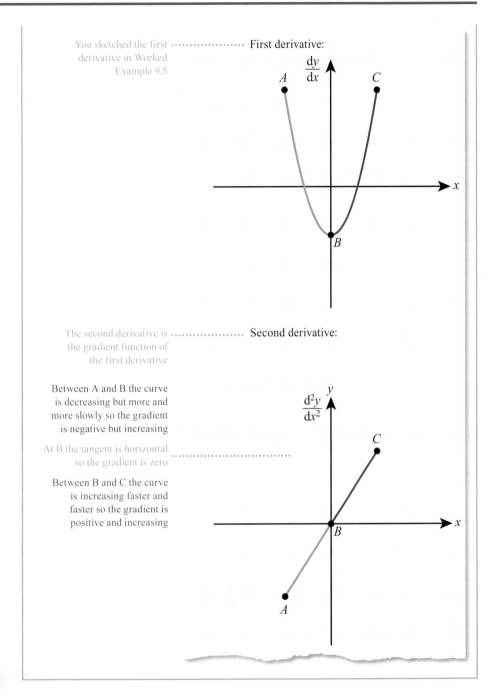

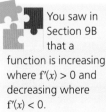 You saw in Section 9B that a function is increasing where f'(x) > 0 and decreasing where f'(x) < 0.

## Concave-up and concave-down

You already know that you can classify different parts of the graph of a function as being increasing or decreasing depending on whether the gradient (the derivative) is positive or negative.

There is another classification based on the gradient of the gradient (the second derivative). If you look at Worked Example 20.15, you can see that:

■ Where the graph of $y = f(x)$ was shaped like this

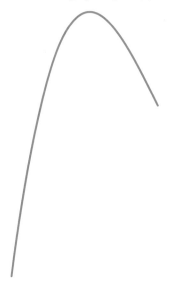

the graph of $y = f''(x)$ ended up being negative.

■ Where the graph of $y = f(x)$ was shaped like this

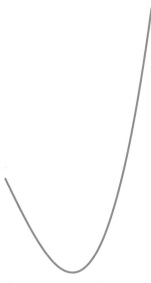

the graph of $y = f''(x)$ ended up being positive.

## Tip

A curve does not have to have a turning point to be described as concave-up or concave-down. Any part of the graphs above can be described as concave-up or concave-down. One useful trick is to think where you would naturally put a compass point (or your wrist) when sketching small sections of the curve. If it is above the curve it is concave-up. If it is below the curve it is concave-down.

**KEY POINT 20.9**

A function f(x) is:
- **concave-up** where f″(x) > 0
- **concave-down** where f″(x) < 0

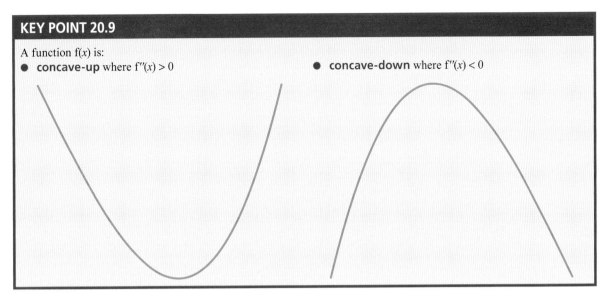

**WORKED EXAMPLE 20.16**

Using one or more of the words 'increasing', 'decreasing', 'concave-up', 'concave-down', describe the sections of the graph:

a   *A* to *B*

b   *C* to *E*

c   *F* to *G*.

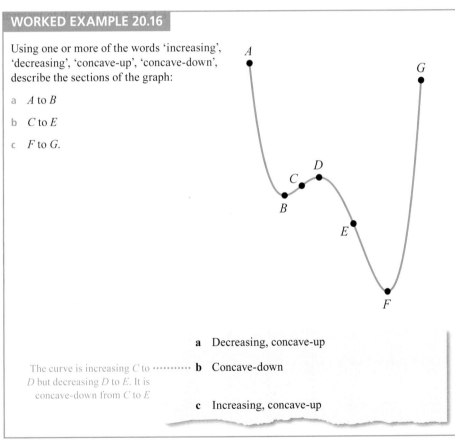

                                           **a**   Decreasing, concave-up

The curve is increasing *C* to ·········· **b**   Concave-down
*D* but decreasing *D* to *E*. It is
concave-down from *C* to *E*

                                            **c**   Increasing, concave-up

**You are the Researcher**

Just as the first derivative is related to the graphical concept of the gradient of a graph, the second derivative is related to the graphical concept of curvature. You might like to investigate how to find a formula for the radius of curvature of a graph. This is a core problem in an area of mathematics called differential geometry.

**WORKED EXAMPLE 20.17**

Find the values of $x$ for which the function $f(x) = x^4 + 2x^3 - 36x^2 + 5x - 4$ is concave-down.

Find the second derivative ·······································

$$f'(x) = 4x^3 + 6x^2 - 72x + 5$$
$$f''(x) = 12x^2 + 12x - 72$$

For concave-down $f''(x) < 0$ ·······················

$$f''(x) < 0$$
$$12x^2 + 12x - 72 < 0$$

Solve the quadratic inequality ·······················

$$x^2 + x - 6 < 0$$
$$(x + 3)(x - 2) < 0$$

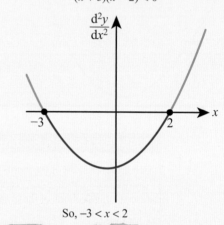

So, $-3 < x < 2$

## Exercise 20D

For questions 1 to 4, use the method demonstrated in Worked Example 20.14 to evaluate the second derivative of the functions given when $x = 2$.

1   a   $f(x) = 5x^2 + 2x + 1$
   b   $f(x) = 6x^2 + 3x + 5$

2   a   $f(x) = 5x^3 + x^2 + 3x + 7$
   b   $f(x) = x^3 - 2x^2 + 4x + 11$

3   a   $y = e^{-2x}$
   b   $y = 2e^{3x}$

4   a   $y = \ln x$
   b   $y = -4\ln(x) + 2$

For questions 5 and 6, use the method demonstrated in Worked Example 20.15 to sketch the second derivative of the functions shown.

5   a

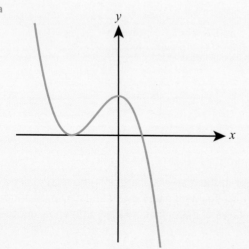

   b

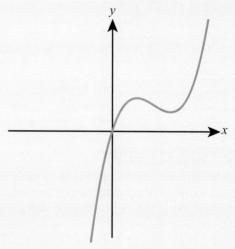

6  a

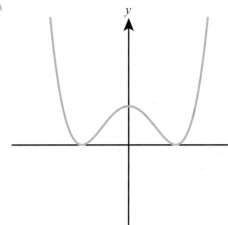

b

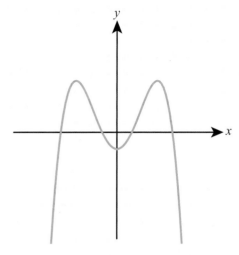

7  Use the method demonstrated in Worked Example 20.16 to give one or more of the words 'increasing', 'decreasing', 'concave-up', 'concave-down' to describe the given sections of each graph.

a

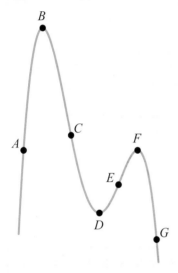

b

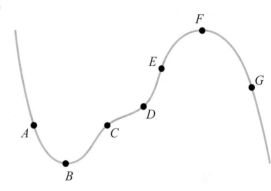

i   $A$ to $B$
ii  $C$ to $F$
iii $F$ to $G$

i   $A$ to $B$     ii   $C$ to $E$     iii   $E$ to $G$

For questions 8 and 9, use the method demonstrated in Worked Example 20.17 to find the values of $x$ for which the function:

8  a  $f(x) = x^3 - 4x^2 + 5x - 2$ is concave-down
   b  $f(x) = x^3 + 9x^2 - 4x + 1$ is concave-up

9  a  $f(x) = x^4 - 4x^3 - 18x^2 + 2x - 1$ is concave-up
   b  $f(x) = x^4 - 6x^3 + 2x - 7$ is concave-down

10  If $f(x) = x^3 + kx^2 + 3x + 1$ and $f''(1) = 10$ find the value of $k$.

11  If $f(x) = x^3 + ax^2 + bx + 1$ with $f''(-1) = -4$ and $f'(1) = 4$ find $a$ and $b$.

12  If $y = xe^x$ find $\dfrac{d^2 y}{dx^2}$.

13  Sketch a graph where for all $x$:

    $f(x) < 0$

    $f'(x) > 0$

    $f''(x) < 0$

14  Sketch a graph where for all positive $x$, $f'(x) > 0$ and $f''(x) > 0$ and for all negative $x$, $f'(x) > 0$ and $f''(x) < 0$.

15  The graph of $x^3 - kx^2 + 4x + 7$ is concave-up for all $x > 1$ and this is the only section which is concave-up. Find the value of $k$.

**16**   If $f(x) = x^n$ and $f''(1) = 12$ find the possible values of $n$.

**17**   Prove that the graph of $y = x^2 + bx + c$ is always concave-up.

**18**   If $y = \sin 3x + 2\cos 3x$ prove that $\dfrac{d^2 y}{dx^2} = -9y$.

**19**   If $y = \ln(a + x)$ find $\dfrac{d^2 y}{dx^2}$. Hence, prove that the graph of $y = \ln(a + x)$ is always concave-down.

**20**   Find the point on the graph $y = x^2 \ln x$ where the second derivative equals 1.

**21**   Find the interval in which the gradient of the graph $y = x^3 - 5x^2 + 4x - 2$ is decreasing.

**22**   The curve $y = x^3 + ax^2 + bx + 7$ is concave-up for all $x > 2$ and this is the only section which is concave-up.

     a   Find the value of $a$.

     b   Find a condition on $b$ for the curve to be strictly increasing.

**23**   The curve $y = ax^3 - bx^2$ (where $a$ and $b$ are positive) is only increasing for $x < 0$ and $x > 4$.

     a   Evaluate $\dfrac{b}{a}$.

     b   Prove that the curve is concave-up for $x > 2$.

**24**   If $y''(x) = 10$, $y'(0) = 10$ and $y(0) = 2$ find $y(x)$.

# 20E Local maximum and minimum points

When you sketched graphs of derivatives, you saw that at points where the function changes from increasing to decreasing (or vice versa) the gradient is zero. These points are called local maximum or local minimum points.

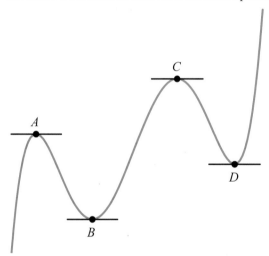

- $A$ and $C$ are local maximum points
- $B$ and $D$ are local minimum points.

## Tip

The term 'local' is used because there might be several 'maximum' (or 'minimum') points, or because the maximum (or minimum) of the function might actually be at the end point of the graph.

---

**KEY POINT 20.10**

At a local maximum or minimum point, $f'(x) = 0$.

## Tip

The term **stationary point** is often used to mean any point where $f'(x) = 0$.

 You will see in Section 20F that there is another possibility other than a maximum or minimum for a point where $f'(x) = 0$.

### WORKED EXAMPLE 20.18

Find the value of $x$ at any local maximum/minimum points of $f(x) = x^2 - 2x^{\frac{3}{2}} + x$.

Maximum/minimum points occur where $f'(x) = 0$ ········· $f'(x) = 0$
$$2x - 3x^{\frac{1}{2}} + 1 = 0$$

This is a disguised quadratic. Solve using the substitution $u = x^{\frac{1}{2}}$ (or factorizing directly as a quadratic in $x^{\frac{1}{2}}$) ········· Let $u = x^{\frac{1}{2}}$:
$$2u^2 - 3u + 1 = 0$$
$$(2u - 1)(u - 1) = 0$$
$$u = \frac{1}{2} \text{ or } 1$$

Convert back to $x$ ········· So,
$$x^{\frac{1}{2}} = \frac{1}{2} \text{ or } 1$$
$$x = \frac{1}{4} \text{ or } 1$$

## ■ Testing for maximum and minimum

In Worked Example 20.18 you know you have local maximum/minimum points at $x = \frac{1}{4}$ and $x = 1$, but you don't know which is which.

However, you already know how to determine whether the graph is making the shape of a local maximum or a local minimum – you look at the second derivative and see whether the function is concave-down or concave-up at that point.

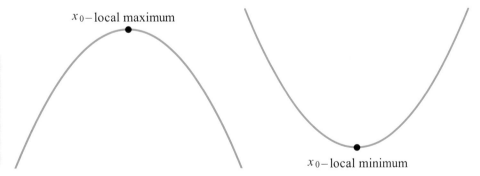

$x_0$–local maximum

$x_0$–local minimum

 You saw in Section 20D that a function is concave-down where $f''(x) < 0$ and concave-up where $f''(x) > 0$.

### KEY POINT 20.11

Given $f'(a) = 0$ then if
- $f''(a) < 0$ then there is a local maximum at $x = a$
- $f''(a) > 0$ then there is a local minimum at $x = a$
- $f''(a) = 0$ then no conclusion can be drawn.

**WORKED EXAMPLE 20.19**

Determine whether each of the points found in Worked Example 20.18 is a local maximum or minimum.

Find the second derivative ························ $f'(x) = 2x - 3x^{\frac{1}{2}} + 1$

$$f''(x) = 2 - \frac{3}{2}x^{-\frac{1}{2}}$$

Evaluate the second ··················· $f''\left(\frac{1}{4}\right) = 2 - \frac{3}{2}\left(\frac{1}{4}\right)^{-\frac{1}{2}}$
derivative at $x = \frac{1}{4}$

$$= 2 - 3$$
$$= -1 < 0$$

So, there is a local maximum at $x = \frac{1}{4}$

Evaluate the second ··················· $f''(1) = 2 - \frac{3}{2}(1)^{-\frac{1}{2}}$
derivative at $x = 1$

$$= 2 - \frac{3}{2}$$
$$= \frac{1}{2} > 0$$

So, there is a local minimum at $x = 1$

 ### ■ Optimization

You can now apply the ideas about local maximum and minimum points to maximize or minimize functions in real-life contexts.

**WORKED EXAMPLE 20.20**

A business produces $x$ thousand units of a particular product each month, where $x > 0$. The profit P (in thousands of dollars) is given by $P = 16x^{\frac{3}{2}} - x^3 - 25$.
a   Find the number of units the business should produce each month to maximize its profit, showing that this is indeed a maximum.
b   Find this maximum monthly profit.

Find $\frac{dP}{dx}$ ·········· **a**   $\frac{dP}{dx} = 24x^{\frac{1}{2}} - 3x^2$

For a maximum ···················· $\frac{dP}{dx} = 0$
(or minimum) $\frac{dP}{dx} = 0$

$$24x^{\frac{1}{2}} - 3x^2 = 0$$

Solve by factorizing ············· $3x^{\frac{1}{2}}\left(8 - x^{\frac{3}{2}}\right) = 0$

$$8 - x^{\frac{3}{2}} = 0 \quad (\text{since } x \neq 0)$$
$$x^{\frac{3}{2}} = 8$$

Take power $\frac{2}{3}$ of both sides ················· $x = 8^{\frac{2}{3}} = 4$

Find the value of $\dfrac{d^2P}{dx^2}$ at $x = 4$ to ············ $\dfrac{d^2P}{dx^2} = 12x^{-\frac{1}{2}} - 6x$

show that this is a maximum

When $x = 4$,

$$\dfrac{d^2P}{dx^2} = 12\left(4^{-\frac{1}{2}}\right) - 6(4)$$

$$= 6 - 24$$

$$= -18 < 0$$

So, this is a maximum

Remember $x$ is in thousands ············· So, the business should produce 4000 items per month

Substitute $x = 4$ into P ··········· **b**  $P = 16\left(4^{\frac{3}{2}}\right) - 4^3 - 25$

$$= 16(8) - 4^3 - 25$$

$$= 64(2 - 1) - 25$$

$$= 39$$

Maximum value of P is $\$39\,000$

## CONCEPTS – QUANTITY

The calculus part of optimization is only one part of the process – and often the easiest part! Deciding what **quantity** you actually want to optimize and what quantities might have an impact on it is not as easy as you might think. In the 1920s, an experiment was conducted at the Hawthorne Factory in Illinois in the USA to see if raising light levels increased productivity. The workers' productivity increased with increasing light levels, but then returned to previous levels after the study finished despite the light remaining at the increased level. The modern interpretation of this study, called the Hawthorne effect, is that the factor leading to increased productivity was actually just the experience of being observed, and not the increased light levels at all.

Predicting the link between quantities is not always obvious. In an attempt to minimize the number of cobras in India, the colonial British government offered a fee for cobra heads. This resulted in people breeding cobras and the number of cobras growing massively. Goodhart's law states that the effect of making a measure of success a target influences people's behaviour with sometimes perverse effects. For example, in some health systems it is commonly observed that targets for minimizing the time taken to treat an illness simply results in doctors not officially diagnosing the illness until access to treatment is available.

Finally, not every objective can be easily quantified. For example, the UN's World Happiness report tries to measure the happiness of a country using measures such as GDP per capita and life expectancy. Is this a valid measure of happiness? Is your result in a pure maths examination the best measure of your ability to have a successful career?

**TOOLKIT:** Modelling

Revisit the Starter Activity at the beginning of this chapter, in which you considered the effects of different variables on the following:

■ the profits of a company
■ a yield of crops
■ your exercise regime
■ your happiness.

Are there any potential unexpected outcomes that might stem from increasing or decreasing different variables that you didn't originally consider? How would these impact upon the conditions required to achieve an optimal result?

## Exercise 20E

For questions 1 to 5, use the method demonstrated in Worked Examples 20.18 and 20.19 to find any local maximum/minimum points of the following functions. In each case determine whether each point is a maximum or minimum.

1  a  $f(x) = x^3 - 3x^2 - 9x + 17$

   b  $f(x) = 2x^3 - 9x^2 + 12x + 5$

2  a  $f(x) = 2x^{\frac{1}{2}} - 3x + 1$

   b  $f(x) = x^{-\frac{1}{2}} + 4x - 2$

3  a  $f(x) = \sin x - \frac{x}{2}, \ 0 < x < 2\pi$

   b  $f(x) = \cos x + \frac{x}{2}, \ 0 < x < 2\pi$

4  a  $f(x) = e^x - 5x$

   b  $f(x) = \frac{x}{2} - e^x$

5  a  $f(x) = 3\ln x - 2x$

   b  $f(x) = x^{\frac{1}{2}} - \ln x$

6   The curve $y = x^2 + bx + c$ has a minimum value at $(1, 2)$. Find the values of $b$ and $c$.

7   The curve $y = x^4 + bx + c$ has a minimum value at $(1, -2)$. Find the values of $b$ and $c$.

8   The profit $\$P$ million made by a business which invests $\$x$ million in advertising is given by
$P = 10x^2 - 10x^4$ for $x \geqslant 0$.
Find the maximum profit the company can make, based on this model.

9   A rectangle has width $x$ cm and length $20 - x$ cm.
   a  Find the perimeter of the rectangle.
   b  Find the maximum area of the rectangle.

10  Find the stationary points on the curve $y = \dfrac{x^2}{1 + x}$.

11  Find and classify the stationary points on the curve $y = x - 2\sqrt{x}$.

12  Find the stationary point on the curve $y = \sin x - 0.5x$ for $0 < x < \pi$ and justify that it is a local maximum.

13  A cuboid is formed with a square base of side length $x$ cm. The other side of the cuboid is of length $9 - x$ cm. Find the maximum volume of the cuboid.

14  The function f is defined by $f(x) = \dfrac{\ln x}{x}$ for $x > 0$.
   a  Find $f'(x)$.
   b  Find the coordinates of the stationary point on the curve $y = f(x)$.
   c  Show that $f''(x) = \dfrac{2\ln x - 3}{x^3}$.
   d  Hence classify the stationary point.

15  Find and classify the stationary points on the curve $y = e^{\sin x}$ for $0 < x < 2\pi$.

16  The number of live bacteria, $N$ million, $t$ minutes after being introduced into an agar solution is modelled by $N = e^t - 2t + 5$.
   a  Find the initial number of live bacteria.
   b  Find the minimum value of $N$.

17  a  A manufacturer of widgets believes that if he sets the price of widgets at $\$x$ he will make a profit of $\$x - 4$ per widget and sell $1000 - 100x$ widgets each day. Based on this model, what price should he charge to maximize his profit?
   b  An alternative model suggests that the number of widgets sold will be $\dfrac{1000}{(x+1)^2}$. Explain one advantage of this model compared to the one described in a.
   c  Based on the alternative model, find the price the manufacturer should charge to maximize his profit.

18  A drone of mass $m$ kg (which is larger than 0.25) can travel at a constant speed of $4 - \dfrac{1}{m}$ metres per second for a time of $\dfrac{200}{m}$ seconds.
   a  Find an expression for the distance travelled by a drone of mass $m$.
   b  What mass should the manufacturer select to maximize the distance the drone can travel?

19  Find the range of the function $f(x) = e^x - 3x + 7$.

20  Find the range of the function $f(x) = 3x^4 - 4x^3 - 12x^2 + 2$.

21  Find and classify the stationary points on the curve $y = e^{2x} - 6e^x + 4x + 8$.

22  The height, $h$ metres, of a point on a hanging chain at a given $x$-coordinate, is modelled by $h(x) = e^x + e^{-2x}$.
Find the minimum height of the chain.

# 20F Points of inflection with zero and non-zero gradients

When a function changes from concave-up to concave-down (or vice versa) the second derivative changes from positive to negative (or negative to positive), so there must be a point where the second derivative is zero. This is called a **point of inflection**.

Points of inflection may either have zero or non-zero gradient, and the function may be either increasing or decreasing either side of the point.

Notice that the tangent will cross the curve at a point of inflection.

> **KEY POINT 20.12**
>
> At a point of inflection
> - $f''(x) = 0$ and
> - the concavity of the function changes.

**WORKED EXAMPLE 20.21**

The function $f(x) = e^x - 3x^2$ has a point of inflection.

a   Find the value of $x$ at which this occurs.

b   Determine whether this is a stationary or non-stationary point of inflection.

| | |
|---|---|
| Find the second derivative ··········· **a** | $f'(x) = e^x - 6x$ |
| | $f''(x) = e^x - 6$ |
| At a point of inflection $f''(x) = 0$ ················ | $f''(x) = 0$ |
| | $e^x - 6 = 0$ |
| There is no need to check ····················· that this is indeed a point of inflection as it is the only candidate and the question asks for the point of inflection | $e^x = 6$ |
| | $x = \ln 6$ |
| At a stationary point, $f'(x) = 0$, ··········· **b** so check whether this is the case for the inflection point at $x = \ln 6$ | $f'(\ln 6) = e^{\ln 6} - 6\ln 6$ |
| | $= 6 - 6\ln 6 \neq 0$ |
| | So, $x = \ln 6$ is a non-stationary point of inflection |

Just because $f''(x) = 0$, it does not mean that you definitely have a point of inflection – you do also need to check that the concavity changes.

**WORKED EXAMPLE 20.22**

Find any inflection points of the function $f(x) = 3x^5 + 5x^4$.

Find the second derivative
$$f'(x) = 15x^4 + 20x^3$$
$$f''(x) = 60x^3 + 60x^2$$

At a point of inflection $f''(x) = 0$
$$f''(x) = 0$$
$$60x^3 + 60x^2 = 0$$

There are two candidates
$$60x^2(x+1) = 0$$
$$x = 0 \text{ or } -1$$

Find the sign of the second derivative either side of $x = -1$. (Note, don't use $x = 0$ as you know that $f''(0) = 0$.) The function is **concave-down** for $x$ just less than $-1$ and **concave-up** for $x$ just greater than $-1$
$$f''(-2) = 60(-2)^2(-2+1)$$
$$= -240 < 0$$
$$f''\left(-\frac{1}{2}\right) = 60\left(-\frac{1}{2}\right)^2\left(-\frac{1}{2}+1\right)$$
$$= \frac{15}{2} > 0$$

Concavity changes either side of $x = -1$, so this is a point of inflection.

You have already found that the function is concave-up for $x$ just less than 0 (using $x = -\frac{1}{2}$)
$$f''(x) = 60(1)^2(1+1)$$
$$= 120 > 0$$

Function is concave-up either side of $x = 0$ so this is not a point of inflection.

---

**Tip**

Remember a stationary point is any point for which $f'(x) = 0$. It may be a local maximum, local minimum or point of inflection.

---

 **WORKED EXAMPLE 20.23**

Find the coordinates of the stationary point on the curve $y = x^4$ and determine its nature.

Stationary points occur where $y' = 0$
$$y' = 0$$
$$4x^3 = 0$$
$$x = 0$$

Find the $y$-coordinate
When $x = 0$,
$$y = 0^4 = 0$$
So the only stationary point is $(0, 0)$

Find the value of the second derivative at $x = 0$
$$y'' = 12x^2$$

$y'' = 0$ does not necessarily mean a point of inflection. No conclusion can be drawn about the nature of the stationary point from this
When $x = 0$,
$$y'' = 12(0)^4$$
$$= 0$$
So inconclusive

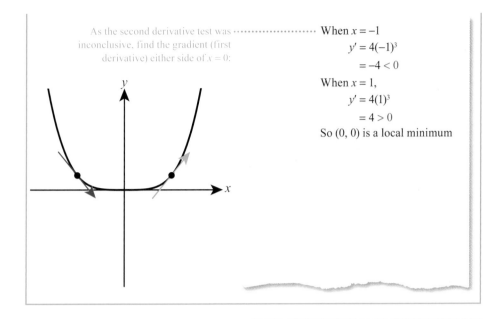

As the second derivative test was inconclusive, find the gradient (first derivative) either side of $x = 0$:

When $x = -1$
$$y' = 4(-1)^3$$
$$= -4 < 0$$
When $x = 1$,
$$y' = 4(1)^3$$
$$= 4 > 0$$
So $(0, 0)$ is a local minimum

**TOOLKIT: Problem Solving**

$\dfrac{d^2y}{dx^2} = 0$ is always the case at a point of inflection, but not all points with $\dfrac{d^2y}{dx^2} = 0$ are points of inflection. Mathematicians describe this by saying $\dfrac{d^2y}{dx^2} = 0$ is a necessary but not sufficient condition for a point to be a point of inflection. Can you think of any other conditions in other areas of mathematics which are necessary but not sufficient?

A sufficient condition for a point to be a point of inflection is that $\dfrac{d^2y}{dx^2} = 0$ and $\dfrac{d^3y}{dx^3} \neq 0$. Can you justify this?

# Exercise 20F

For questions 1 to 5, use the method demonstrated in Worked Example 20.21 to find the value(s) of $x$ at which the given functions have the given number of points of inflection. For each point of inflection, state whether it is stationary or non-stationary.

1 a $f(x) = x^3 - 6x^2 - 4x + 3$: 1 point of inflection

  b $f(x) = x^3 - 12x^2 + x + 5$: 1 point of inflection

2 a $f(x) = x^4 - 6x^3$: 2 points of inflection

  b $f(x) = x^4 - 4x^3 + 16x - 16$: 2 points of inflection

3 a $f(x) = \sin x - x$, $x \in [0, 2\pi]$: 3 points of inflection

  b $f(x) = \cos x - x$, $x \in [0, 2\pi]$: 2 points of inflection

4 a $f(x) = 4e^x - x^2$: 1 point of inflection

  b $f(x) = e^x - \dfrac{1}{2}x^2$: 1 point of inflection

5 Find the coordinates of the point of inflection on the curve $y = x^3 + 9x^2 + x - 1$.

6 Find the point of inflection on the curve $y = x^5 - 80x^2$.

7 The curve $y = x^3 + bx^2 + c$ has a point of inflection at $(1, 3)$. Find the values of $b$ and $c$.

8 The curve $y = 2x^3 - ax^2 + b$ has a point of inflection at $(1, -7)$.
Find the values of $a$ and $b$.

9 The curve $y = x^3 + ax^2 + bx + c$ has a stationary point of inflection at $(1, 3)$. Find the values of $a$, $b$ and $c$.

10 a Find the point of inflection of the function $f(x) = 8 \ln x + x^2$, where $x > 0$.

  b Determine whether the point of inflection is stationary or non-stationary.

11 a Show that the graph $y = 2 \ln x + x^2$ has no stationary points.

  b Find the coordinates of the point of inflection on the graph.

**12** The graph below shows $y = f'(x)$. Make a sketch of the graph.

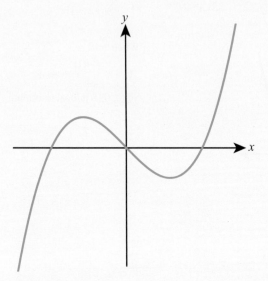

    **a** Mark all points corresponding to a local minimum of $f(x)$ with a $P$.

    **b** Mark all points corresponding to a local maximum of $f(x)$ with a $Q$.

    **c** Mark all points corresponding to a point of inflection of $f(x)$ with a $R$.

    **d** Are any of the points of inflection stationary points of inflection? Explain your answer.

**13**  **a** Find the $x$-coordinates of the stationary points of the function $f(x) = x^4 - 6x^2 + 8x - 3$.

    **b** Determine the nature of these stationary points.

**14**  **a** Find the $x$-coordinates of the stationary points of the function $f(x) = 5x^4 - x^5$.

    **b** Determine the nature of these stationary points.

**15** Find the exact coordinates of the points of inflection on the curve $y = e^{-x^2}$, fully justifying your answer.

**16** Find the point of inflection on the curve $y = x^2 \ln x$.

**17** Find the coordinates of any inflection points on the curve $y = 3x^5 - 10x^4 + 10x^3 + 2$.

**18** The function $f(x) = x^3 + ax^2 + bx + c$ has a stationary point of inflection.

    Show that $a^2 = 3b$.

**TOOLKIT: Modelling**

When a new business starts up, it will often use market research to set pricing. Try doing this for a hypothetical new business in your school – for example, selling fruit. See if you can create a model for the cost of purchasing $n$ items (remember, that you can often get a discount when buying larger quantities). Use a survey or some other method to estimate the probability of people purchasing your items at different prices and use this to form a model of the profit as a function of $n$. How valid do you think your model is? How could you improve your model?

## Checklist

■ You should be able to differentiate functions such as $\sqrt{x}$, $\sin x$ and $\ln x$.

    ☐ If $f(x) = x^n$, where $n \in \mathbb{Q}$, then $f'(x) = nx^{n-1}$

    ☐ If $f(x) = \sin x$, then $f'(x) = \cos x$

    ☐ If $f(x) = \cos x$, then $f'(x) = -\sin x$

    ☐ If $f(x) = e^x$, then $f'(x) = e^x$

    ☐ If $f(x) = \ln x$, then $f'(x) = \dfrac{1}{x}$

▥ How to use the chain rule to differentiate composite functions.
  ☐ If $y = f(u)$, where $u = g(x)$, then

$$\frac{dy}{dx} = \frac{dy}{du} \times \frac{du}{dx}$$

▥ How to differentiate products and quotients.
  ☐ If $y = u(x)v(x)$ then

$$\frac{dy}{dx} = u\frac{dv}{dx} + v\frac{du}{dx}$$

  ☐ If $y = \dfrac{u(x)}{v(x)}$ then

$$\frac{dy}{dx} = \frac{v\dfrac{du}{dx} - u\dfrac{dv}{dx}}{v^2}$$

▥ How to interpret the second derivative of a function.
  ☐ A function $f(x)$ is:
    — concave-up where $f''(x) > 0$

$y = f(x)$

    — concave-down where $f''(x) < 0$

$y = f(x)$

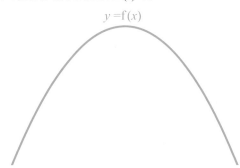

▥ How to find local minimum and maximum points on a graph.
  ☐ At a local maximum or minimum $f'(x) = 0$
  ☐ Given $f'(a) = 0$ then if
    — $f''(a) < 0$ then there is a local maximum at $x = a$
    — $f''(a) > 0$ then there is a local minimum at $x = a$
    — $f''(a) = 0$ then no conclusion can be drawn.

▥ How to use calculus to find the optimal solution to a problem.

▥ How to identify points of inflection on a graph.
  ☐ At a point of inflection
    — $f''(x) = 0$ and
    — the concavity of the function changes.

## ■ Mixed Practice

**1** Consider the function $f(x) = ax^3 - 3x + 5$, where $a \neq 0$.
   **a** Find $f'(x)$.
   **b** Write down the value of $f'(0)$.
   The function has a local maximum at $x = -2$.
   **c** Calculate the value of $a$.

Mathematical Studies SL May 2013 Paper 1 TZ1 Q15

**2** Find the coordinates of the point of inflection on the curve $y = x^3 + 6x^2 - 4x + 1$.

**3** The curve $y = x^2 + bx + c$ has a minimum value at $(2, 3)$. Find the values of $b$ and $c$.

**4** The curve $y = x^3 + bx^2 + cx + 5$ has a point of inflection at $(2, -3)$. Find the values of $b$ and $c$.

**5** Find the gradient of the tangent to the curve $y = \sqrt{x-1}$ at $x = 5$.

**6** Find the equation of the tangent to the curve $y = \ln x$ at $x = 1$.

**7** Find the equation of the normal to the curve $y = e^{2x}$ at $x = 0$.

**8** The diagram shows the graph of the function defined by $y = x(\ln x)^2$ for $x > 0$.

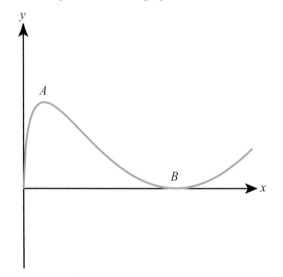

The function has a local maximum at the point $A$ and a local minimum at the point $B$.
   **a** Find the coordinates of the points $A$ and $B$.
   **b** Given that the graph of the function has exactly one point of inflection, find its coordinates.

Mathematics HL November 2012 Paper 1 Q4

**9** A tranquillizer is injected into a muscle from which it enters the bloodstream. The concentration $C$ in mg$\,$l$^{-1}$, of tranquillizer in the bloodstream can be modelled by the function $C(t) = \dfrac{2t}{3+t^2}$, $t \geqslant 0$ where $t$ is the number of minutes after the injection.

Find the maximum concentration of tranquillizer in the bloodstream.

Mathematics HL November 2014 Paper 1 Q5

**10** Two cyclists are at the same road intersection. One cyclist travels north at $20\,\text{kmh}^{-1}$. The other cyclist travels west at $15\,\text{kmh}^{-1}$.

Use calculus to show that the rate at which the distance between the two cyclists changes is independent of time.

Mathematics HL November 2014 Paper 2 Q4

**11** Consider $f(x) = \ln(x^4 + 1)$.
  **a** Find the value of $f(0)$.
  **b** Find the set of values of $x$ for which f is increasing.
  The second derivative is given by $f''(x) = \dfrac{4x^2(3 - x^4)}{(x^4 + 1)^2}$.
  The equation $f''(x) = 0$ has only three solutions, when $x = 0$, $\pm\sqrt[4]{3}$ ($\pm 1.316\ldots$)
  **c i** Find $f''(1)$.
  **ii** **Hence**, show that there is no point of inflection on the graph of f at $x = 0$.
  **d** There is a point of inflection on the graph of f at $x = \sqrt[4]{3}$ ($1.316\ldots$).
  Sketch the graph of f, for $x \geqslant 0$.

Mathematics SL May 2013 Paper 1 TZ1 Q10

**12** Let $f(x) = \dfrac{x}{-2x^2 + 5x - 2}$ for $2 \leqslant x \leqslant 4$, $x \neq \dfrac{1}{2}$. The graph of f is given below.

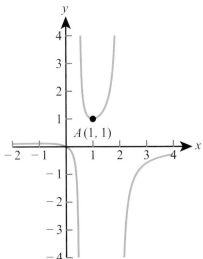

The graph of f has a local minimum at $A(1, 1)$ and a local maximum at $B$.

  **a** Use the quotient rule to show that $f'(x) = \dfrac{2x^2 - 2}{(-2x^2 + 5x - 2)^2}$.
  **b** Hence find the coordinates of $B$.
  **c** Given that the line $y = k$ does not meet the graph of f, find the possible values of $k$.

Mathematics SL May 2012 Paper 1 TZ2 Q10

**13** Find the equation of the tangent to the curve $y = x \cos x$ at $x = 0$.
**14** If $y = x^2 \, e^{2x}$, find $\dfrac{d^2 y}{dx^2}$.

**15** The two shorter sides of a right-angled triangle are $x$ cm and $6 - x$ cm.
  **a** Find the maximum area of the triangle.
  **b** Find the minimum perimeter of the triangle.

**16** If $f(x) = xe^{kx}$ and $f''(0) = 10$ find $f'(1)$.

**17** The graph of $x^3 - kx^2 + 8x + 2$ is concave-up for all $x > 4$ and this is the only section which is concave-up. Find the value of $k$.

**18** Differentiate $xe^x \sin x$.

**19** If $f(x) = \dfrac{x}{\sqrt{4 + x}}$ show that $f'(x) = \dfrac{ax + b}{2(4 + x)^c}$ where $a$, $b$ and $c$ are constants to be found.

**20** **a** Find the set of values for which $f(x) = xe^{-x}$ is concave-down.
  **b** Hence, find the point of inflection on the curve $y = f(x)$.

**21** Find the coordinates of the stationary point on the curve $y = e^{-x} \sin x$ for $0 < x < \pi$.

**22** The volume of water, $V$ million cubic metres, in a lake $t$ hours after a storm is modelled by $V = 2te^{-t} + 5$.
  **a** What is the initial volume of the lake?
  **b** What is the maximum volume of the lake?
  **c** When is the lake emptying fastest?

**23** Prove that the equation of the tangent to the curve $y = \ln x + kx$ at $x = 1$ always passes through the point $(0, -1)$ independent of the value of $k$.

**24** Let $f(x) = \dfrac{100}{(1 + 50e^{-0.2x})}$. Part of the graph of f is shown below.

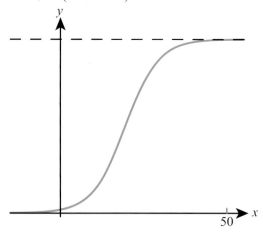

  **a** Write down $f(0)$.
  **b** Solve $f(x) = 95$.
  **c** Find the range of f.
  **d** Show that $f'(x) = \dfrac{1000e^{-0.2x}}{\left(1 + 50e^{-0.2x}\right)^2}$.
  **e** Find the maximum rate of change of f.

Mathematics SL May 2013 Paper 2 TZ1 Q9

**25** The function $f(x) = \dfrac{x-a}{x-b}$ is defined for $x \neq b$ and $a > b$.

    **a**  Show that $f'(x) > 0$ for all values of $x$ in the domain.

    **b**  Does this mean that if $p > q$ then $f(p) > f(q)$? Justify your answer.

**26**  **a**  State the largest possible domain for the function $y = \sqrt{4 - x^2}$.

    **b**  Prove that for any point on the graph of $y = \sqrt{4 - x^2}$, the normal passes through the origin.

**27** The graph $y = kx^n$ has the property that at every point the product of the gradient and the $y$-value equals 1.

    **a**  Find the value of $n$.

    **b**  Find the possible values of $k$.

# 21 Analysis and approaches: Integration

## ESSENTIAL UNDERSTANDINGS

- Calculus describes rates of change between two variables and the accumulation of limiting areas.
- Understanding these rates of change and accumulations allows us to model, interpret and analyse real-world problems and situations.
- Calculus helps us understand the behaviour of functions and allows us to interpret features of their graphs.

## In this chapter you will learn…

- how to integrate functions such as $\sqrt{x}$, $\sin x$ and $e^x$
- how to integrate by inspection
- how to find definite integrals
- how to link definite integrals to areas between a curve and the x-axis, and also between two curves
- how to apply calculus to kinematics, linking displacement, velocity, acceleration and distance travelled.

### CONCEPTS

The following concept will be addressed in this chapter:
- Derivatives and integrals describe real-world kinematics problems in two and three-dimensional **space** by examining displacement, velocity and acceleration.

### PRIOR KNOWLEDGE

Before starting this chapter, you should already be able to complete the following:

1. Find $y$ if $\dfrac{dy}{dx} = x^2 + 2$ and $y = 4$ when $x = 0$.

2. Use technology to evaluate the area between the curve $y = x^4$ and the x-axis between $x = 1$ and $x = 3$.

3. If $y = e^{2x} + \sin x + 2$, find $\dfrac{dy}{dx}$.

4. If $f(x) = \sqrt{x^2 + 3}$, find $f'(x)$.

■ **Figure 21.1** How is integration applied in studying the motion of bodies in the real world?

In Chapter 10, you saw that the operation which reverses differentiation is called integration, but that surprisingly this has applications in evaluating areas and accumulations. Now that you know how to differentiate more functions, it is natural to extend your knowledge of integration to reverse at least some of these new derivatives. We can then apply these new integrals to more complex situations, including one of the most important topics in physics: the study of motion.

## Starter Activity

Look at the photographs in Figure 21.1. Using what you learned about integration in Chapter 10, think about how it could be used to describe the motion of these different entities.

**Now look at this problem:**

If you know that

$$\int_0^a f(x) \, dx = A$$

Which of the following can you evaluate?

$$\int_0^a f(x-1) \, dx$$

$$\int_1^{a+1} f(x-1) \, dx$$

$$\int_0^a f(2x) \, dx$$

$$\int_0^{\frac{a}{2}} f(2x) \, dx$$

# 21A Further indefinite integration

## ■ Extending the list of integrals

In Chapter 20, you saw that the derivative of $x^n$ with respect to $x$ is $nx^{n-1}$ for any $n \in \mathbb{Q}$. This means that $x^{n+1}$ differentiates to $(n+1)x^n$, so $\dfrac{1}{n+1}x^{n+1}$ integrates to $x^n$ as long as $n \neq -1$ (to avoid dividing by zero). Since integration is the reverse of differentiation this means that

 You saw a similar rule in Chapter 10, but we have now extended it to all rational values of $n$ other than –1. Remember that we always need a '+c' when integrating since this will disappear when it is differentiated.

> **KEY POINT 21.1**
>
> $$\int x^n \, dx = \frac{x^{n+1}}{n+1} + c \text{ for } n \neq -1$$

> **WORKED EXAMPLE 21.1**
>
> Find $\displaystyle\int 3\sqrt{x} \, dx$.
>
> | | |
> |---|---|
> | Rewrite the integral to make it explicitly include $x^n$ | $\displaystyle\int 3x^{\frac{1}{2}}dx$ |
> | The constant factor can be taken out of the integral | $= 3\displaystyle\int x^{\frac{1}{2}}dx$ |
> | Apply Key Point 21.1 | $= \dfrac{3x^{\left(\frac{1}{2}+1\right)}}{\frac{1}{2}+1} + c$ |
> | Simplify | $= \dfrac{3x^{\frac{3}{2}}}{\frac{3}{2}} + c$ |
> | | $= 2x^{\frac{3}{2}} + c$ |

In Chapter 20, we also saw that $\sin x$ differentiates to $\cos x$ when $x$ is measured in radians. This can be reversed to turn this statement into an integral.

**Tip**

Because the rules of differentiating and integrating trigonometric functions are so similar, they are often confused. Make sure you remember which way round they work.

> **KEY POINT 21.2**
>
> $$\int \cos x dx = \sin x + c$$

Since $\cos x$ differentiates to $-\sin x$ it follows that $-\cos x$ differentiates to $\sin x$. Reversing this result gives an integral.

> **KEY POINT 21.3**
>
> $$\int \sin x dx = -\cos x + c$$

You also saw in Chapter 20 that $e^x$ differentiates to $e^x$. This statement can also be reversed.

**KEY POINT 21.4**

$$\int e^x \, dx = e^x + c$$

You know that $\ln x$ differentiates to $\dfrac{1}{x}$. This suggests that $\int \dfrac{1}{x} dx = \ln x + c$. However, there is a small complication that you need to consider. This integral can be linked to the area between the curve and the $x$-axis and this area exists for negative values of $x$ even though $\ln x$ is not defined in this region:

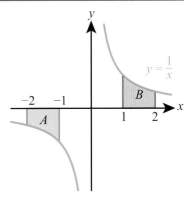

We can use the symmetry of the graph of $y = \dfrac{1}{x}$ to see that that area $A$ and area $B$ are equal. This leads to the following interpretation.

This fills in the hole in Key Point 21.1 – integrating $x^n$ where $n = -1$.

**KEY POINT 21.5**

$$\int \dfrac{1}{x} \, dx = \ln|x| + c$$

**WORKED EXAMPLE 21.2**

If $\dfrac{dy}{dx} = 3e^x + \cos x + 2$ find a general expression for $y$.

| We reverse differentiation using integration | $y = \int 3e^x + \cos x + 2 \, dx$ |
| We can split up the sum into separate integrals and take constant factors out | $= 3\int e^x \, dx + \int \cos x \, dx + \int 2 \, dx$ |
| The integrals can be evaluated using the key points above | $= 3e^x + \sin x + 2x + c$ |

## ■ Integrating f$(ax + b)$

From the chain rule you know that when differentiating f$(ax + b)$ you get f$'(ax + b)$ In words, this says that you can differentiate the function as normal and then multiply by $a$. When integrating we can reverse this logic to say that when integrating f$(ax + b)$ we can integrate as normal then divide by $a$.

**KEY POINT 21.6**

If $\int f(x) \, dx = F(x)$, then $\int f(ax + b) \, dx = \dfrac{1}{a} F(ax + b)$

## Tip

In Worked Example 21.3, you might have worried whether you needed to divide the constant by 5, too. However, $\frac{c}{5}$ is just another constant so it actually does not matter.

**WORKED EXAMPLE 21.3**

Find $\int \sin(5x + 2)\mathrm{d}x$.

We can identify $f(x) = \sin x$.
This integrates to $-\cos x$

$$\int \sin(5x + 2)\mathrm{d}x = -\frac{1}{5}\cos(5x + 2) + c$$

## ■ Integration by inspection

The chain rule says that the derivative of a composite function $f(g(x))$ is $g'(x)f'(g(x))$. This is a very common type of expression to look out for – a composite function multiplied by the derivative of the inner function – because it can easily be integrated.

**KEY POINT 21.7**

$$\int g'(x)f'(g(x))\ \mathrm{d}x = f(g(x)) + c$$

 Key Point 21.6 is a special case of the above result (slightly rearranged) with $g(x) = ax + b$. You will meet a further generalization, called integration by substitution, if you study analysis and approaches HL.

**WORKED EXAMPLE 21.4**

Find $\int 2x\ \mathrm{e}^{(x^2)}\ \mathrm{d}x$.

$f'(x) = \mathrm{e}^x$ and $g(x) = x^2$. The composite function is multiplied by $2x$ which is $g'(x)$ so we can apply Key Point 21.7 directly

$$\int 2x\ \mathrm{e}^{(x^2)}\ \mathrm{d}x = \mathrm{e}^{(x^2)} + c$$

Sometimes, the expression multiplying the composite function is not exactly the derivative of the inner function but is within a constant factor of it. In this situation we can introduce the required factor in the integral, but to keep the value unchanged we need to divide by the same factor too.

**WORKED EXAMPLE 21.5**

Find $\int \dfrac{x}{3x^2 - 4}\mathrm{d}x$.

## Tip

Worked Example 21.5 makes use of the very common integral:

$$\int \frac{f'(x)}{f(x)}\,\mathrm{d}x = \ln\big|f(x)\big| + c$$

This integral occurs so often you might like to just learn this result.

We can identify $f'(x) = \frac{1}{x}$ and $g(x) = 3x^2 - 4$

therefore $g'(x) = 6x$. To get this factor we need to introduce an extra factor of 6, which must be balanced by $\frac{1}{6}$

$$\int \frac{x}{3x^2 - 4}\mathrm{d}x = \frac{1}{6}\int \frac{6x}{3x^2 - 4}\mathrm{d}x$$

The integral is now of the required form to apply Key Point 21.7. We can use the fact that the integral of $\frac{1}{x}$ is $\ln|x| + c$

$$= \frac{1}{6}\ln\big|3x^2 - 4\big| + c$$

**TOOLKIT:** Problem Solving

Consider $\int \dfrac{1}{2x}\,dx$. Using the method of Worked Example 21.5, we can make the top of the fraction the derivative of the bottom:

$$\int \frac{1}{2x}\,dx = \frac{1}{2}\int \frac{2}{2x}\,dx = \frac{1}{2}\ln|2x| + c$$

Alternatively, we can use algebra to take out a factor of $\dfrac{1}{2}$ from the integral.

$$\int \frac{1}{2x}\,dx = \frac{1}{2}\int \frac{1}{x}\,dx = \frac{1}{2}\ln|x| + c$$

Which of these two solutions is correct?

## Be the Examiner 21.1

Find $\int \sin x \cos x\,dx$.

Which is the correct solution? Identify the errors in the incorrect solutions.

| Solution 1 | Solution 2 | Solution 3 |
|---|---|---|
| Since $\dfrac{d}{dx}(\sin x) = \cos x$ $\int (\sin x)^1 \cos x\,dx$ $= \dfrac{\sin^2 x}{2} + c$ | $\int \sin x \cos x\,dx$ $= \dfrac{1}{2}\int \sin 2x\,dx$ $= -\dfrac{1}{4}\cos 2x + c$ | Since $\dfrac{d}{dx}(\cos x) = -\sin x$ $-\int -\sin x(\cos x)^1\,dx$ $= -\dfrac{\cos^2 x}{2} + c$ |

## Exercise 21A

For questions 1 to 6, use the method demonstrated in Worked Example 21.1 to find the given integrals.

1  a  $\int x^{\frac{2}{3}}\,dx$

   b  $\int x^{\frac{3}{4}}\,dx$

2  a  $\int x^{-\frac{1}{2}}\,dx$

   b  $\int x^{-\frac{4}{3}}\,dx$

3  a  $\int 10x^{\frac{3}{2}}\,dx$

   b  $\int 5x^{\frac{1}{4}}\,dx$

4  a  $\int 4x^{-\frac{2}{3}}\,dx$

   b  $\int 3x^{-\frac{2}{5}}\,dx$

5  a  $\int \frac{1}{2}\sqrt[3]{x}\,dx$

   b  $\int \frac{1}{3}\sqrt[5]{x}\,dx$

6  a  $\int \frac{6}{\sqrt[4]{x}}\,dx$

   b  $\int \frac{7}{\sqrt{x}}\,dx$

For questions 7 to 10, use the method demonstrated in Worked Example 21.2 to find an expression for $y$.

7  a  $\dfrac{dy}{dx} = 3\cos x$

   b  $\dfrac{dy}{dx} = -\cos x$

8  a  $\dfrac{dy}{dx} = -2\sin x$

   b  $\dfrac{dy}{dx} = \dfrac{1}{2}\sin x$

9  a  $\dfrac{dy}{dx} = 5e^x$

   b  $\dfrac{dy}{dx} = -\dfrac{4}{3}e^x$

10  a  $\dfrac{dy}{dx} = \dfrac{2}{x}$

   b  $\dfrac{dy}{dx} = \dfrac{1}{2x}$

For questions 11 to 16, use the method demonstrated in Worked Example 21.3 to find the following integrals.

11  a  $\int \sqrt{2x+1}\,dx$

    b  $\int \sqrt[3]{1-2x}\,dx$

12  a  $\int \dfrac{1}{\sqrt{3-5x}}\,dx$

    b  $\int \dfrac{1}{\sqrt[4]{2x-7}}\,dx$

13  a  $\int \cos(2-3x)\,dx$

    b  $\int \cos(4x+3)\,dx$

14  a  $\int \sin\!\left(\dfrac{1}{2}x - 5\right)dx$

    b  $\int \sin(5-2x)\,dx$

15  a  $\int e^{5x+2}\,dx$

    b  $\int e^{1-3x}\,dx$

16  a  $\int \dfrac{1}{4x-5}\,dx$

    b  $\int \dfrac{1}{3-2x}\,dx$

For questions 17 to 19, use the method demonstrated in Worked Example 21.4 to find the given integrals.

17 a $\int 2x(x^2+4)^3\,dx$

b $\int 4x^3(x^4-2)^{\frac{3}{2}}\,dx$

18 a $\int \cos x \sin^3 x\,dx$

b $\int -\sin x \cos^2 x\,dx$

19 a $\int \dfrac{3x^2+2}{x^3+2x}\,dx$

b $\int \dfrac{4x^3-5}{x^4-5x}\,dx$

For questions 20 to 22, use the method demonstrated in Worked Example 21.5 to find the given integrals.

20 a $\int x^2\sqrt{x^3+4}\,dx$

b $\int x\sqrt[3]{1-x^2}\,dx$

21 a $\int 4xe^{-x^2}\,dx$

b $\int 9x^2e^{x^3}\,dx$

22 a $\int \dfrac{x^2}{x^3+5}\,dx$

b $\int \dfrac{x^3}{x^4-3}\,dx$

23 Given that $\dfrac{dy}{dx}=3\sqrt{x}$ and that $y=12$ when $x=9$, find an expression for $y$ in terms of $x$.

24 Function f satisfies $f'(x)=2\cos x-3\sin x$ and $f(0)=5$. Find an expression for $f(x)$.

25 Given that $\dfrac{dy}{dx}=2e^x-\dfrac{5}{x}$, and that $y=0$ when $x=1$, find an expression for $y$ in terms of $x$.

26 Find $\int \dfrac{4x^2+3}{2x}\,dx$.

27 Given that $\dfrac{dV}{dt}=t-\dfrac{1}{2}\sin t$, and that $V=2$ when $t=0$, find an expression for $V$ in terms of $t$.

28 It is given that $x(0)=5$ and that $\dfrac{dx}{dt}=5-2e^t$. Find an expression for $x$ in terms of $t$.

29 Find $\int (5-2x)^6\,dx$.

30 Find $\int (3\sin(2x)-2\cos(3x))\,dx$.

31 Find $\int 2x(x^2+1)^5\,dx$.

32 Given that $\dfrac{dy}{dx}=x\sin(3x^2)$ find an expression for $y$ in terms of $x$.

33 Find $\int \dfrac{3x}{\sqrt{x^2+2}}\,dx$.

34 Given that $f'(x)=\dfrac{1}{(2x+3)^2}$, and that the graph of $y=f(x)$ passes through the point $(-1, 1)$, find an expression for $f(x)$.

35 The curve $y=f(x)$ passes through the point $(1, 1)$ and has gradient $f'(x)=\dfrac{2\ln x}{3x}$. Find an expression for $f(x)$.

36 Given that $f'(x)=\dfrac{2}{x}$ and that $f(-1)=5$, find the exact value of $f(-3)$.

37 Find $\int 4\cos^3 x\sin x+kx^2+1\,dx$ where $k$ is a constant.

38 Find $\int \dfrac{e^{2x}-e^{-2x}}{e^x}\,dx$.

39 Find $\int \sqrt{x^4+x^2}\,dx$ for $x>0$.

40 a Starting from a double angle identity for cosine, show that $\sin^2 x\equiv\dfrac{1}{2}(1-\cos 2x)$

b Hence find $\int \sin^2 x\,dx$

41 a Express $\cos 2x$ in terms of $\sin x$.

b Hence find $\int \sin^2 x\,dx$.

42 Find $\int \tan x\,dx$, showing all your working.

43 Find $\int \sin^3 x\,dx$.

44 Find $\int \dfrac{1}{x\ln x}\,dx$.

# 21B Further links between area and integrals

You saw in Chapter 10 that we can represent the area under a curve $y = f(x)$ between $x = a$ and $x = b$ (called the **limits** of the integral) using $\int_a^b f(x)dx$ and you evaluated this using technology. In this section, you will see how this can be evaluated analytically.

It turns out the definite integral is found by evaluating the integral at the value of the upper limit and subtracting the value of the integral at the lower limit. This is often expressed using 'square bracket' notation.

> **KEY POINT 21.8**
>
> $$\int_a^b g'(x)\, dx = \left[g(x)\right]_a^b = g(b) - g(a)$$

It is not at all obvious that the process of reversing differentiation has anything to do with the area under a curve. This is something called the 'Fundamental Theorem of Calculus' and it was formalized by English mathematician and physicist Isaac Newton and German mathematician and philosopher Gottfried Leibniz in the late seventeenth century. Although, today, most historians agree that the two made their discoveries about calculus independently, at the time there was a great controversy over the question of who developed these ideas first, and whether one had stolen them from the other.

## Proof 21.1

Prove that the area under the curve $y = f(x)$ is $\int_a^b f(x)dx$.

Consider a general representation of $y = f(x)$.
The increase in the area when $b$ is changed is
approximately a rectangle of height $f(b)$ .................

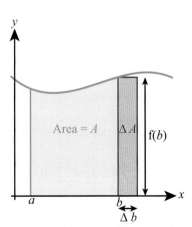

The change in area, $\Delta A$, is approximately the red rectangle, so $\Delta A = f(b)\, \Delta b$.

So $\dfrac{\Delta A}{\Delta b} = f(b)$. In the limit as $\Delta b$ gets very small this expression becomes $\dfrac{dA}{db}$.

To find $A$ we need to undo the differentiation – this is integration. We can use the standard notation that the indefinite integral of $f(x)$ is $F(x)$. This is the area from some unspecified starting point up to $b$ (hence it still contains an unknown constant) ·········· Therefore $A = F(b) + c$

The area between $a$ and $b$ can be found as the difference between the area ·········· up to $b$ and the area up to $a$

So area between $a$ and $b$ is

$$(F(b) + c) - (F(a) + c) = F(b - F(a)$$

$$= \int_a^b f(x)\,dx$$

---

## WORKED EXAMPLE 21.6

Evaluate $\int_1^2 x^3\ dx$.

First of all, find the indefinite integral. For a definite integral you do not need the $+c$ although it does not matter if you put it in – it will end up cancelling out in the next line ·········· $\int_1^2 x^3\ dx = \left[\dfrac{x^4}{4}\right]_1^2$

Evaluate the expression at the upper limit ·········· and subtract the value at the lower limit

$$= \left(\dfrac{2^4}{4}\right) - \left(\dfrac{1^4}{4}\right)$$

$$= 4 - \dfrac{1}{4}$$

$$= 3.75$$

---

Not every definite integral can be evaluated analytically. When technology is available it is often better to use it to evaluate definite integrals.

## WORKED EXAMPLE 21.7

Evaluate $\int_0^1 e^{-x^2}dx$, giving your answer to three significant figures.

We can do this using a function on the ·········· calculator or using the graph

From GDC: $\int_0^1 e^{-x^2}dx \approx 0.747$

```
∫(e^-X²,0,1)
          0.7468241328




Solve d/dz d²/dx ∫dx        ▷
```

**TOK Links**

$e^{-x^2}$ is an example of a function which you might be told cannot be integrated. This really means that the indefinite integral cannot be written in terms of other standard functions, but who decides what is a 'standard function'? The area under the curve does exist, so we could (and indeed some mathematical communities do) define a new function that is effectively the area under this curve from 0 up to $x$ (if you are interested it is $\frac{\sqrt{\pi}}{2}\,\mathrm{erf}(x)$). Does just giving the answer a name increase your knowledge?

## ■ Area between a curve and the *x*-axis

Definite integrals give the area between a curve and the *x*-axis when the function between the limits is entirely above the *x*-axis. If the curve is below the *x*-axis then the integral will give a negative value. The modulus of this value is the area.

**WORKED EXAMPLE 21.8**

Find the area enclosed by $x = -1$, $x = 1$, $y = 3x^2 - 4$ and the *x*-axis.

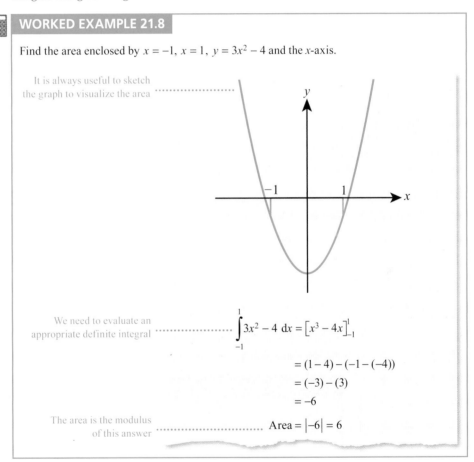

It is always useful to sketch the graph to visualize the area

We need to evaluate an appropriate definite integral

$$\int_{-1}^{1} 3x^2 - 4 \ \mathrm{d}x = \left[x^3 - 4x\right]_{-1}^{1}$$

$$= (1 - 4) - (-1 - (-4))$$
$$= (-3) - (3)$$
$$= -6$$

The area is the modulus of this answer

$$\text{Area} = |{-6}| = 6$$

**Tip**

If we are not told any units for $x$ and $y$ then we do not give units in the answer.

Sometimes areas are partly above and partly below the $x$-axis. In that case, you have to find each part separately.

**WORKED EXAMPLE 21.9**

Find the shaded area on the curve of $y = x^2 - 2x$.

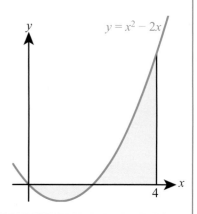

<table>
<tr><td>First find where it goes from above the axis to below the axis</td><td>···· It crosses the axis when $$x^2 - 2x = 0$$ $$x(x-2) = 0$$ So it crosses when $x = 0$ and $x = 2$</td></tr>
<tr><td>Split the integral into two parts. Consider first the region below the $x$ axis</td><td>$$\int_0^2 x^2 - 2x\,dx = \left[\frac{x^3}{3} - x^2\right]_0^2 = -\frac{4}{3} - 0$$ So, the first area is $\left|-\frac{4}{3}\right| = \frac{4}{3}$</td></tr>
<tr><td>Then the second region</td><td>$$\int_2^4 x^2 - 2x\,dx = \left[\frac{x^3}{3} - x^2\right]_2^4 = \frac{16}{3} - \left(-\frac{4}{3}\right) = \frac{20}{3}$$</td></tr>
<tr><td>Combine the two areas</td><td>···· Total area is $\frac{4}{3} + \frac{20}{3} = 8$</td></tr>
</table>

**Tip**

A common mistake in problems like that in Worked Example 21.9 is to think that the area is $\int_0^4 x^2 - 2x\,dx$, which is $\frac{16}{3}$. This is the 'net area' – how much more is above the $x$-axis than below – and this is sometimes very useful, but it is not the answer to the given question.

With a calculator you can find the total area between $x = a$, $x = b$, the curve $y = f(x)$ and the $x$-axis using the following formula.

**KEY POINT 21.9**

$$\text{Area} = \int_a^b |f(x)|\ dx$$

**WORKED EXAMPLE 21.10**

Find the area between the curve $y = x^3 - x$ and the $x$ axis between $x = -1$ and $x = 1$

Write the required area in terms of a ························ Area $= \int_{-1}^{1} |x^3 - x| \, dx$
definite integral from Key Point 21.9

Evaluate using technology ························ $= 0.5$ from GDC

```
∫(Abs (X^3-X),-1,1)
                0.5
```

```
Abs  Int  Frac  Rnd  Intg  RndFi
```

**TOOLKIT:** Proof

Justify each of the following rules for definite integrals:

$$\int_{a}^{b} f(x)dx = -\int_{b}^{a} f(x)dx$$

$$\int_{a}^{b} f(x)dx + \int_{b}^{c} f(x)dx = \int_{a}^{c} f(x)dx$$

$$\int_{a}^{b} f(x)dx + \int_{a}^{b} g(x)dx = \int_{a}^{b} f(x) + g(x) \, dx$$

## Area between curves

If we want to find the area, $A$, between two curves, it can be determined by finding the area under the top curve and subtracting the area under the bottom curve.

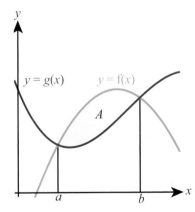

**Tip**

Sometimes people worry about what happens if one of the graphs goes below the $x$-axis. As long as the function $f(x)$ is equal or above $g(x)$ between $a$ and $b$, then Key Point 21.10 will still work.

**KEY POINT 21.10**

$$\text{Area} = \int_{a}^{b} f(x) \, dx - \int_{a}^{b} g(x) \, dx = \int_{a}^{b} f(x) - g(x) \, dx$$

**WORKED EXAMPLE 21.11**

Find the area enclosed by $y = x^2 - 3x$ and $y = 7x - x^2$.

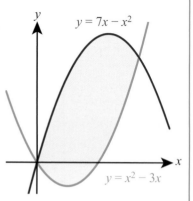

First you must find the points of intersection ............... Graphs meet when $x^2 - 3x = 7x - x^2$

So
$$2x^2 - 10x = 0$$
$$2x(x - 5) = 0$$
$$x = 0 \text{ or } x = 5$$

Use Key Point 21.10, noting that between 0 and 5 the curve $y = 7x - x^2$ is above the curve $y = x^2 - 3x$

$$\text{Area} = \int_0^5 (7x - x^2) - (x^2 - 3x) \, dx$$

$$= \int_0^5 10x - 2x^2 \, dx$$

Evaluate the definite integral ...............

$$= \left[ 5x^2 - \frac{2x^3}{3} \right]_0^5$$

$$= \frac{125}{3}$$

## Exercise 21B

For questions 1 to 6, use the method demonstrated in Worked Example 21.6 to find the given definite integrals.

1 a $\int_1^4 x^{\frac{1}{2}} \, dx$

   b $\int_0^8 x^{\frac{1}{3}} \, dx$

2 a $\int_{-2}^1 x^4 \, dx$

   b $\int_{-1}^0 x^5 \, dx$

3 a $\int_{-3}^{-1} x^2 \, dx$

   b $\int_{-4}^{-2} x^3 \, dx$

4 a $\int_{\frac{\pi}{3}}^{\frac{\pi}{2}} \sin x \, dx$

   b $\int_0^{\frac{\pi}{2}} \cos x \, dx$

5 a $\int_0^{\ln 3} e^x \, dx$

   b $\int_{\ln 2}^{\ln 7} e^x \, dx$

6 a $\int_1^5 \frac{1}{x} \, dx$

   b $\int_2^3 \frac{1}{x} \, dx$

For questions 7 and 8, use the method demonstrated in Worked Example 21.7 to find the given definite integrals.

7 a $\int_0^3 \sin(x^2) \, dx$

   b $\int_1^6 \cos(x^3) \, dx$

8 a $\int_2^3 \frac{1}{\ln x} \, dx$

   b $\int_4^9 \ln(\ln x) \, dx$

For questions 9 and 10, use the method demonstrated in Worked Examples 21.8 and 21.9 to find the shaded areas.

9  a

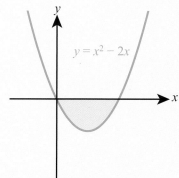

b

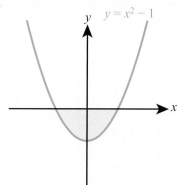

10  a

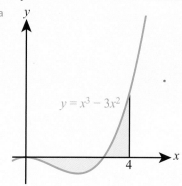

b

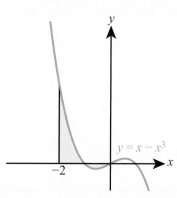

For questions 11 and 12, use the method demonstrated in Worked Example 21.10 to find the shaded areas:

11  a

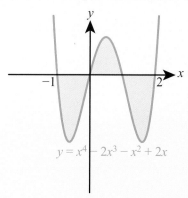

b

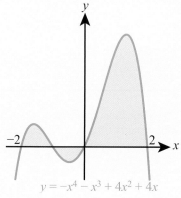

12  a

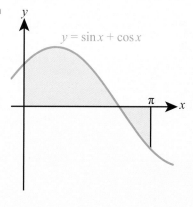

b

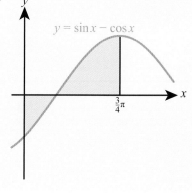

For questions 13 to 15, use the method demonstrated in Worked Example 21.11 to find the shaded areas.

**13  a**

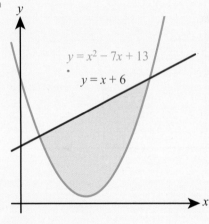

$y = x^2 - 7x + 13$

$y = x + 6$

**14  a**

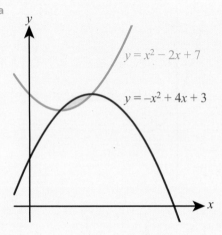

$y = x^2 - 2x + 7$

$y = -x^2 + 4x + 3$

**b**

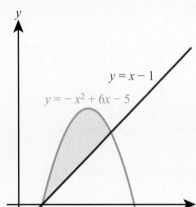

$y = x - 1$

$y = -x^2 + 6x - 5$

**b**

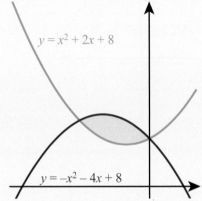

$y = x^2 + 2x + 8$

$y = -x^2 - 4x + 8$

**15  a**

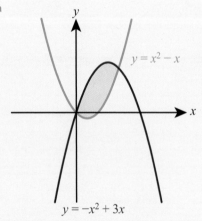

$y = x^2 - x$

$y = -x^2 + 3x$

**b**

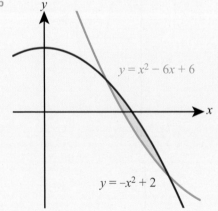

$y = x^2 - 6x + 6$

$y = -x^2 + 2$

**16** Evaluate $\displaystyle\int_{1}^{5} \sqrt{3x + 1}\,dx$.

**17** The curve with equation $y = 3\sin 2x$ crosses the $x$-axis at $x = 0$ and $x = \dfrac{\pi}{2}$. Find the exact area enclosed by this part of the curve and the $x$-axis.

**18** Find the area enclosed by the graph of $y = \sqrt{\ln x}$, the $x$-axis and the lines $x = 2$ and $x = 5$.

**19** Find the value of $a$ such that $\displaystyle\int_{1}^{a} \dfrac{3}{\sqrt{x}}\,dx = 24$.

**20** a Find the coordinates of the points where the curve $y = 9 - x^2$ crosses the $x$-axis.

b Find the area enclosed by the curve and the $x$-axis.

**21** The graph of $y = x^3 - 2x^2 - x + 2$ is shown in the diagram.

a Evaluate $\displaystyle\int_{-1}^{1} y\,dx$ and $\displaystyle\int_{1}^{2} y\,dx$.

b Hence find the total area of the shaded region.

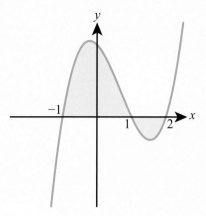

**22** The graph of $y = \ln(10 - x^2)$, shown in the diagram, crosses the $x$-axis at the points $(-3, 0)$ and $(3, 0)$. Find the shaded area.

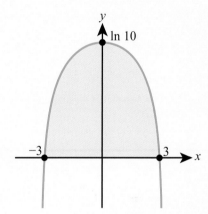

**23** Find the exact value of $\displaystyle\int_{0}^{\ln 3} 2e^{-3x}\,dx$.

**24** Find the exact value of $\displaystyle\int_{-9}^{-3} \frac{5}{x}\,dx$.

**25** Show that the value of the integral $\displaystyle\int_{k}^{2k} \frac{1}{x}\,dx$ is independent of $k$.

**26** The diagram shows the graphs of $y = \sin x$ and $y = \cos x$ for $0 \le x \le \dfrac{\pi}{2}$.

a Write down the coordinates of the point of intersection of the two graphs.

b Find the exact area of the shaded region.

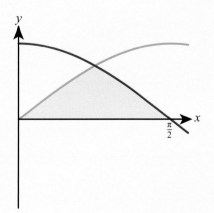

**27** The curves with equations $y = e^{-x^2}$ and $y = e^{-x}$ are shown in the diagram.

   a  Find the coordinates of the points $A$ and $B$.

   b  Find the area enclosed by the two curves.

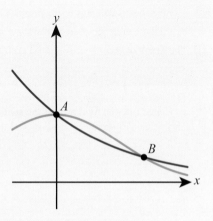

**28** a  On the same axes, sketch the graphs of $y = \cos(x^2)$ and $y = 2 - 2\sin(x^2)$ for $0 \leqslant x \leqslant \dfrac{\pi}{2}$.

   b  Find the area enclosed by the two graphs.

**29** Find the area enclosed between the graphs of $y = x^3 - 4x$ and $y = 2x - x^2$ for $x \geqslant 0$.

**30** Find the area enclosed by the graph of $y = \sin 2x$ and the $x$-axis for $0 \leqslant x \leqslant \pi$.

**31** Find the exact value of $\displaystyle\int_0^{\sqrt{3}} x\sqrt{x^2 + 1}\, dx$.

**32** The curves shown in the diagram have equations $y = e^x$ and $y = e^{-2x}$.

   Find the area of the shaded region.

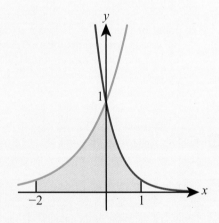

**33** Find the exact value of $\displaystyle\int_1^4 \frac{1}{2x + 3}\, dx$.

**34** Evaluate $\displaystyle\int_2^3 \frac{1}{x - 5}\, dx$.

**35** Given that $\displaystyle\int_2^5 f(x)dx = 10$, evaluate $\displaystyle\int_2^5 (3f(x) - 1)dx$.

**36** Given that $\displaystyle\int_0^6 f(x)dx = 7$, evaluate $\displaystyle\int_0^3 5f(2x)dx$.

**37** a  Differentiate $x \ln x$.

   b  Hence evaluate $\displaystyle\int_1^e \ln x\, dx$.

**38** The diagram shows the graph of $y = f(x)$.

Given that $\displaystyle\int_{b}^{d} f(x)\,dx = 1$, $\displaystyle\int_{a}^{d} f(x)\,dx = 5$ and $\displaystyle\int_{a}^{d} |f(x)|\,dx = 17$, find $\displaystyle\int_{a}^{c} f(x)\,dx$.

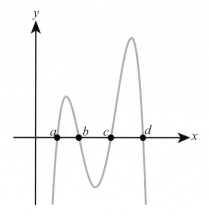

## 21C Kinematics

### Derivatives in kinematics

Kinematics is the study of motion over time. The basic quantity studied in kinematics is displacement. Displacement is the position of an object relative to a reference point and it is conventionally given the symbol *s*. Time is given the symbol *t*.

 The letter s is used because it comes from the German word 'Strecke', which translates as 'route'. For much of the nineteenth and early twentieth centuries, Germany was at the heart of the development of mathematics and physics, so many terms studied in that era have names originating from German.

The rate of change of displacement with respect to time is called velocity. It is given the symbol *v*. Based on this definition we can write the following.

**KEY POINT 21.11**

$$v = \frac{ds}{dt}$$

The rate of change of velocity with respect to time is called acceleration. It is given the symbol *a*.

**KEY POINT 21.12**

$$a = \frac{dv}{dt} = \frac{d^2s}{dt^2}$$

**You are the Researcher**

The derivative of acceleration with respect to time is called jerk. The derivative of jerk with respect to time is called jounce. You might like to find out more about how these quantities are used in physical applications, such as in the design of roller coasters.

**You are the Researcher**

How do computers generate random numbers? How do they generate random numbers from a normal distribution?

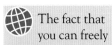

The fact that you can freely choose $t = 0$ and $s = 0$ turns out to be remarkably powerful. The German mathematician Emily Noether proved in 1915 that these observations actually lead to the conservation of momentum and energy.

## CONCEPTS – SPACE

The quantity of displacement is often confused with distance. Imagine a footballer running from one end of the pitch to the other and back. Her total distance travelled is approximately 200 m, but her final displacement in **space** is zero. Her speed while running might always be 8 m/s but her velocity one way will be 8 m/s and the other way will be –8 m/s.

Another vital idea when dealing with these quantities is realizing that you have freedom to choose many starting points. You can often pick $t = 0$ to be any convenient time.

## WORKED EXAMPLE 21.12

If $s = \sin 3t$ find an expression for $a$.

First find $v$ by using the rule for differentiating composite functions (the chain rule) ............ $v = \dfrac{\mathrm{d}s}{\mathrm{d}t} = 3\cos 3t$

Differentiate again to find $a$ ............ $a = \dfrac{\mathrm{d}v}{\mathrm{d}t} = -9\sin 3t$

The process can also be reversed, using integration.

## KEY POINT 21.13

Change in velocity from $t = a$ to $t = b$ is given by $\displaystyle\int_a^b a\ \mathrm{d}t$.

Change in displacement from $t = a$ to $t = b$ is given by $\displaystyle\int_a^b v\ \mathrm{d}t$.

## WORKED EXAMPLE 21.13

If when $t = 4$ seconds, $s = 1$ metre and for the next 5 seconds the velocity is given by $\dfrac{1}{\sqrt{t}}$ m/s find the displacement when $t = 9$.

We can use a definite integral from Key Point 21.13 to find the ............ Difference in displacment $= \displaystyle\int_4^9 v\ \mathrm{d}t$
difference in displacement

We substitute in what we know about the velocity, putting it in a form which is useful for ............ $= \displaystyle\int_4^9 t^{-0.5}\mathrm{d}t$
applying the rules of integration

We can find the indefinite integral ............ $= \left[\dfrac{t^{0.5}}{0.5}\right]_4^9$

Before substituting in limits, it is worth simplifying ............ $= \left[2\sqrt{t}\right]_4^9$

$= 2\sqrt{9} - 2\sqrt{4}$

$= 2$

Remember to take into account the initial displacement ............ So, the final displacement is $1 + 2 = 3$ m

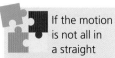

## Tip

Remember that, if you have technology available, definite integrals are often best done on the calculator.

If the motion is not all in a straight line then we need to represent displacement, velocity and acceleration using vectors. You will learn more about these if you study analysis and approaches HL.

The net displacement between $t = a$ and $t = b$ can include some forward and backward motion cancelling each other out. If we want to just find the distance travelled, we have to ignore the fact that some of the motion is forward and some is backward. We can do this by just considering the modulus of the velocity.

---

### KEY POINT 21.14

The distance travelled between $t = a$ and $t = b$ is given by

$$\int_a^b |v| \, dt$$

---

### WORKED EXAMPLE 21.14

The motion of a particle is described by $v = e^{-t} \sin t$. Find the distance travelled in the first $2\pi$ seconds of the motion.

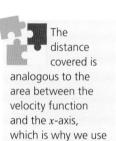 The distance covered is analogous to the area between the velocity function and the *x*-axis, which is why we use a similar method to Key Point 21.9.

Write down an expression for the distance travelled using Key Point 21.14 ⋯⋯⋯ $\text{Distance} = \displaystyle\int_0^{2\pi} \left| e^{-t} \sin t \right| \, dt$

Evaluate this integral on your GDC ⋯⋯⋯ $= 0.544$ (3 s.f.)

```
∫(Abs (e^-Xsin X),0,2
π)
            0.5441476396
```
`Abs` `Int` `Frac` `Rnd` `Intg` `RndFi`

---

The speed is the given by the modulus of the velocity.

### WORKED EXAMPLE 21.15

The velocity of a particle is given by $t(t-2)e^{-t}$ for $0 < t < 4$. Find its maximum speed.

Use the fact that speed is $|v|$ ⋯⋯⋯ $\text{Speed} = \left| t(t-2)e^{-t} \right|$

Use your calculator to sketch this graph and find the maximum. You should show ⋯⋯⋯ a sketch of this graph in your working

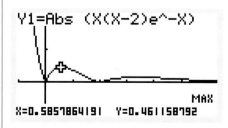
```
Y1=Abs (X(X-2)e^-X)


                          MAX
X=0.5857864191  Y=0.461158792
```

*(0.586, 0.461)*

$v = \left| t(t-2)e^{-t} \right|$

So the maximum speed is 0.461

## Exercise 21C

Note: In this exercise all displacements are in metres, times are in seconds and velocities are in metres per second.

For questions 1 to 4, use the method demonstrated in Worked Example 21.12 to find an expression for the velocity and acceleration in the cases given.

1 a $s = t^3 + 3t^2 + 1$

  b $s = t^4 - 5t + 2$

2 a $s = \cos 2t$

  b $s = \sin \dfrac{t}{2}$

3 a $s = e^{3t}$

  b $s = e^{-4t}$

4 a $s = \ln \dfrac{t}{3}$

  b $s = \ln 2t$

For questions 5 and 6, use the method demonstrated in Worked Example 21.13 to find the change in displacement in the cases given.

5 a $v = 4\sqrt[3]{t}$ between $t = 1$ and $t = 8$

  b $v = 5\sqrt[4]{t}$ between $t = 1$ and $t = 16$

6 a $v = e^{\frac{t}{2}}$ between $t = 0$ and $t = \ln 9$

  b $v = e^{-t}$ between $t = 0$ and $t = \ln 5$

For questions 7 and 8, use the method demonstrated in Worked Example 21.13 to find the change in velocity in the cases given.

7 a $a = 2\cos \dfrac{t}{3}$ between $t = 0$ and $t = \pi$

  b $a = 3\sin 2t$ between $t = 0$ and $t = \dfrac{\pi}{2}$

8 a $a = \dfrac{1}{2t+1}$ between $t = 4$ and $t = 12$

  b $a = \dfrac{1}{3t-2}$ between $t = 1$ and $t = 6$

For questions 9 to 11, use the method demonstrated in Worked Example 21.14 to find the distance travelled in the cases given.

9 a $v = \dfrac{\sin t}{t}$ between $t = 1$ and $t = 5$

  b $v = \dfrac{\cos t}{t}$ between $t = 1$ and $t = 6$

11 a $v = 5(\ln t)^3$ between $t = 0.5$ and $t = 2$

  b $v = \dfrac{3\ln t}{t}$ between $t = 0.5$ and $t = 3$

10 a $v = 2e^{-t^2} - 1$ between $t = 0$ and $t = 2$

  b $v = 2e^{-t^3} - 1$ between $t = 0$ and $t = 1.5$

For questions 12 and 13, use the method demonstrated in Worked Example 21.15 to find the maximum speed in the cases given.

12 a $v = 3e^{-t}\sin 2t$ between $t = 0$ and $t = 5$

  b $v = 4e^{-t}\cos t$ between $t = 1$ and $t = 4$

13 a $v = \sqrt{t}(t-1)(t-2)$ between $t = 0$ and $t = 2$

  b $v = \sqrt{t}(t-1)(t-3)$ between $t = 0$ and $t = 3$

**14** The displacement of a particle, $s$, at time $t$ is modelled by $s = 10t - t^2$.

  a Find the displacement after 2 seconds.

  b Find the velocity after 3 seconds.

  c Find the acceleration after 4 seconds.

**15** The velocity of a particle is given by $v = e^{-0.5t}$. Find the displacement of the particle after 3 seconds, relative to its initial position.

**16** The velocity of a particle is given by $v = 4 - t^2$.

  a Find the displacement after 6 seconds.

  b Find the distance travelled in the first 6 seconds.

**17** The acceleration of a particle is modelled by $2t + 1$. If the initial velocity is $3\,\text{m s}^{-1}$, find the velocity after 4 seconds.

**18** A bullet is fired through a viscous fluid. Its velocity is modelled by $v = 256 - t^4$ from $t = 0$ until it stops.

  a What is the initial speed of the bullet?

  b What is the acceleration after 2 seconds?

  c How long does it take to stop?

  d What is the distance travelled by the bullet?

**19** Find the distance travelled by a particle in its first 2 seconds of travel if the speed is modelled by $v = e^t - 2$.

**20** The displacement of a particle, $s$ metres, at time $t$ seconds is modelled by $s = 6t^2 - t^3$ for $0 \leqslant t \leqslant 6$.

  a Find the times at which the displacement is zero.

  b Find the times at which the velocity is zero.

  c Find the time at which the displacement is maximum.

**21** The displacement of a particle is modelled by $s = t\sin t$ for $0 \leqslant t \leqslant 2\pi$

  a  Find an expression for $v$.

  b  Hence find the initial velocity.

  c  Find an exact expression for the velocity when $t = \frac{\pi}{4}$.

  d  What is the initial acceleration?

  e  At what times is the velocity equal to 1?

  f  What is the total distance travelled by the particle?

**22** If a particle has a constant acceleration, $a$, initial velocity $u$ and initial displacement zero, show that:

  a  $v = u + at$          b  $s = ut + \frac{1}{2}at^2$.

**23** If $s = \sin \omega t$ where $\omega$ is a constant, show that $a = -\omega^2 s$.

**24** The velocity of a ball thrown off a cliff, $v$ metres per second, after a time $t$ seconds is modelled by $v = 5 - 10t$. The ball is initially 60 m above sea level. It is thrown upwards.

  a  What is the vertical acceleration of the ball?

  b  When does the ball reach its maximum height above sea level? What is that maximum height?

  c  When does the ball hit the sea?

  d  What is the vertical distance travelled by the ball?

  e  State one assumption being made in this model.

**25** The velocity, in metres per second, of a bicycle $t$ seconds after passing a time check point is modelled by $v = 18 - 2t^2$ for $0 < t < 6$.

  a  What is the initial speed of the bicycle?

  b  When is the velocity zero?

  c  What is the acceleration of the bicycle after 2 seconds?

  d  What is the distance travelled by the bicycle between $t = 0$ and $t = 6$?

  e  Find the displacement after 6 seconds.

  f  Without any further calculus, deduce the distance travelled in the first 3 seconds.

**26** The displacement of a particle is modelled by $s = -t^3 + 6t^2 - 2t$. Find the maximum velocity of the particle.

**27** A cyclist accelerates from rest. For the first 2 seconds his acceleration is modelled by $2t$. For the next 2 seconds his acceleration is modelled by $\frac{16}{t^2}$. Find the cyclists's velocity 4 seconds after starting.

**28** Given that a particle travels with velocity $\frac{1}{t} - 1$ find the distance travelled between $t = 0.5$ and $t = 2$.

**29** The displacement of a particle, $s$ metres after $t$ seconds, is given by $s(t) = t(10 - t)$.

  a  Find the displacement after 10 seconds.

  b  Find the distance travelled in the first 10 seconds.

**30** If $v = t(t-1)(t-4)$ for $0 \leqslant t \leqslant 4$ find:

  a  the maximum speed.          b  the minimum speed.          c  the average speed.

**31** a  If a particle has acceleration $a = At$, initial velocity $u > 0$ and initial displacement zero, find expressions for:

  i  velocity in terms of $A$, $t$ and $u$

  ii  displacement in terms of $A$, $t$ and $u$.

  b  Show that the average velocity up to time $t$ is given by $\frac{v + 2u}{3}$.

**32** In a race, Jane starts 42 m ahead of Aisla. Jane has velocity given by $t + 2$. Aisla has velocity given by $t^2$. How long does it take Aisla to overtake Jane?

**33** A juggler throws a ball vertically upward. The ball's speed is modelled by $v = 5 - 10t$.

Half a second later, the juggler throws a second ball vertically upward in exactly the same way from the same position. At what height above the release position do the two balls collide?

**34** By considering kinematic quantities, prove that if $a$, $b$, $c$ and $d$ are positive and $\frac{a}{b} < \frac{c}{d}$ then $\frac{a}{b} < \frac{a+c}{b+d} < \frac{c}{d}$.

**TOOLKIT:** Modelling

Try to create a model of your travel graph for a journey. How can you include uncertainty in your model? Use your model to try to answer real-world questions, such as when you should leave if you want to be sure of getting somewhere by a particular time.

## Checklist

■ You should be able to integrate functions such as $\sqrt{x}$, $\sin x$ and $e^x$.

☐ $\displaystyle\int x^n \ dx = \frac{x^{n+1}}{n+1} + c$ for $n \neq -1$

☐ $\displaystyle\int \cos x \, dx = \sin x + c$

☐ $\displaystyle\int \sin x \, dx = -\cos x + c$

☐ $\displaystyle\int e^x \ dx = e^x + c$

☐ $\displaystyle\int \frac{1}{x} \ dx = \ln|x| + c$

■ You should be able to integrate by inspection.

☐ If $\displaystyle\int f(x)dx = F(x)$ then $\displaystyle\int f(ax+b)dx = \frac{1}{a}F(ax+b)$

☐ $\displaystyle\int g'(x)f'(g(x)) \ dx = f(g(x)) + c$

■ You should be able to find definite integrals

☐ $\displaystyle\int_a^b g'(x) \ dx = \left[g(x)\right]_a^b = g(b) - g(a)$

■ You should be able to link definite integrals to areas between a curve and the $x$-axis, and also between two curves.

☐ Area $= \displaystyle\int_a^b |f(x)| \ dx$

☐ Area $= \displaystyle\int_a^b f(x)dx - \int_a^b g(x)dx = \int_a^b f(x) - g(x)dx$

■ You should be able to apply calculus to kinematics, linking displacement, velocity, acceleration and distance travelled.

☐ $v = \dfrac{ds}{dt}$

☐ $a = \dfrac{dv}{dt} = \dfrac{d^2s}{dt^2}$

☐ Change in velocity from $t = a$ to $t = b$ is given by $\displaystyle\int_a^b a \ dt$

☐ Change in displacement from $t = a$ to $t = b$ is given by $\displaystyle\int_a^b v \ dt$

☐ The distance travelled between $t = a$ and $t = b$ is given by $\displaystyle\int_a^b |v| dt$

## ■ Mixed Practice

**1** Evaluate $\displaystyle\int_{1}^{4} 3x^2 - 4 \, \mathrm{d}x$.

**2** Given that $\dfrac{\mathrm{d}y}{\mathrm{d}x} = \dfrac{1}{4\sqrt{x}}$ and that $y = 3$ when $x = 4$, find $y$ in terms of $x$.

**3** Find the area between the graph of $y = \cos x - \sin x$, the $x$-axis and the lines $x = 0$ and $x = \dfrac{\pi}{6}$.

**4** Find $\displaystyle\int \dfrac{3x - 2}{4x^2} \, \mathrm{d}x$.

**5** A particle moves in a straight line so that its displacement at time $t$ seconds is $s$ metres, where $s = 3t - 0.06t^3$. Find the velocity and acceleration of the particle after 8.6 seconds.

**6** Let $\mathrm{f}(x) = \displaystyle\int \dfrac{12}{2x - 5} \, \mathrm{d}x$, for $x \geqslant \dfrac{5}{2}$. The graph of f passes through $(4, 0)$. Find $\mathrm{f}(x)$.

Mathematics SL May 2013 Paper 1 TZ1 Q6

**7** A particle moves in a straight line with velocity $v = 12t - 2t^3 - 1$, for $t \geqslant 0$, where $v$ is in centimetres per second and $t$ is in seconds.
**a** Find the acceleration of the particle after 2.7 seconds.
**b** Find the displacement of the particle after 1.3 seconds.

Mathematics SL May 2012 Paper 2 TZ2 Q5

**8** Find $\displaystyle\int \dfrac{3x^2 - 2}{x\sqrt{x}} \, \mathrm{d}x$.

**9** A curve has gradient $3\cos x \sin^2 x$ and passes through the point $(\pi, 2)$. Find the equation of the curve.

**10 a** Sketch the graph of $y = \cos x$ for $0 \leqslant x \leqslant \dfrac{\pi}{2}$.
**b** The graph intersects the $y$-axis at the point $A$ and the $x$-axis at the point $B$. Write down the coordinates of $A$ and $B$.
**c** Find the equation of the straight line which passes through the points $A$ and $B$.
**d** Calculate the exact area enclosed by the line and the curve.

**11** Function f is defined on the domain $0 \leqslant x \leqslant 2$ by $\mathrm{f}(x) = x^2 - kx$. On the graph of $y = \mathrm{f}(x)$, the area below the $x$-axis equals the area above the $x$-axis. Find the value of $k$.

**12** The diagram shows the graphs of $y = x^2 + 1$ and $y = 7 - x$.
**a** Find the coordinates of the points $A$, $B$ and $C$.
**b** Find the area of the shaded region.

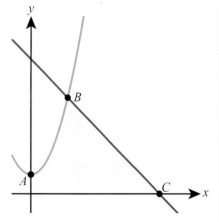

**13** A particle moves in a straight line with the velocity at time $t$ seconds $v = (9t - 3t^2)\,\mathrm{m\,s^{-1}}$.
**a** Find the acceleration of the particle when $t = 3$.
**b** The initial displacement of the particle from the origin is $5\,\mathrm{m}$. Find the displacement from the origin after 3 seconds.

**14 a** Sketch the graph of $y = 2\sin x + 1$ for $0 \leqslant x \leqslant 2\pi$.

A particle's velocity at time $t$ seconds is $v = (2\sin t + 1)\,\mathrm{m\,s^{-1}}$.

**b** Find the speed of the particle after 2.5 seconds.

**c** Find the displacement of the particle from the initial position after $2\pi$ seconds.

**d** Find the distance travelled by the particle in the first $2\pi$ seconds.

**15** The diagram shows the graph of $f(x) = \dfrac{x}{x^2+1}$, for $0 \leqslant x \leqslant 4$, and the line $x = 4$.

Let $R$ be the region enclosed by the graph of f, the $x$-axis and the line $x = 4$.

Find the area of $R$.

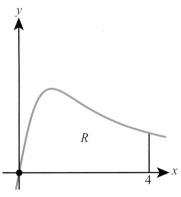

*Mathematics SL November 2014 Paper 1 Q6*

**16** Let $f(x) = \cos x$, for $0 \leqslant x \leqslant 2\pi$. The diagram shows the graph of f.

There are $x$-intercepts at $= \dfrac{\pi}{2}, \dfrac{3\pi}{2}$.

The shaded region $R$ is enclosed by the graph of f, the line $x = b$, where $b > \dfrac{3\pi}{2}$, and the $x$-axis. The area of $R$ is $\left(1 - \dfrac{\sqrt{3}}{2}\right)$. Find the value of $b$.

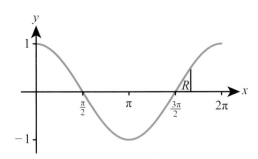

*Mathematics SL May 2015 Paper 1 TZ1 Q7*

**17** The diagram shows the graphs of the **displacement**, **velocity** and **acceleration** of a moving object as functions of time, $t$.

**a** Complete the following table by noting which graph A, B or C corresponds to each function.

| Function | Graph |
|---|---|
| displacement | |
| acceleration | |

**b** Write down the value of $t$ when the velocity is greatest.

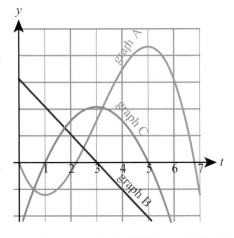

*Mathematics SL May 2009 Paper 1 TZ1 Q4*

**18** Given that $\int_{2}^{5} f(x)dx = 3$, find $\int_{5}^{8} 2f(x-3)dx$.

**19** The diagram shows the graph of the derivative, $y = f'(x)$, of a function f.

**a** Find the range of values of $x$ for which f is decreasing.

**b** Find the set of values of $x$ for which f is concave-up.

**c** The total area enclosed by the graph of $y = f'(x)$ and the $x$-axis is 20. Given that $f(a) = 8$ and $f(d) = 2$, find the value of $f(0)$.

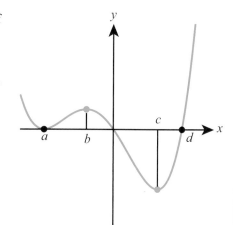

**20** The graphs of $f(x) = -x^2 + 2$ and $g(x) = x^3 - x^2 - bx + 2$, $b > 0$, intersect and create two closed regions. Show that these two regions have equal areas.

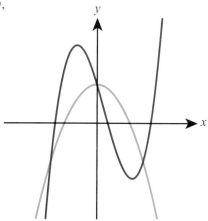

Mathematics HL November 2011 Paper 1 Q7

**21 a** By writing $\sin^2 x$ in terms of $\cos 2x$ find $\int \sin^2(3x)dx$

**b** **Deduce** from your previous result $\int \cos^2(3x)dx$

Analysis and approaches SL:

# Practice Paper 1

Non-calculator.
1 hour 30 minutes, maximum mark for the paper [80 marks].

## Section A

1   A biology student measured the lengths of 200 leaves collected from the local park. The results
    are displayed in the cumulative frequency diagram.

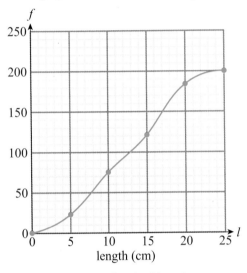

    a  Estimate the median leaf length.

    b  Estimate the number of leaves which were longer than 18 cm.

    c  What is the probability that a randomly selected leaf is longer than 10 cm?

                                                                                                              [6]

2   The line shown in the diagram has equation $y = -3x + c$. It crosses the coordinate axes at points $A$
    and $B$.

    a  Write down the value of $c$.

    b  Find the coordinates of $A$.

    c  Find the coordinates of the midpoint of $AB$.

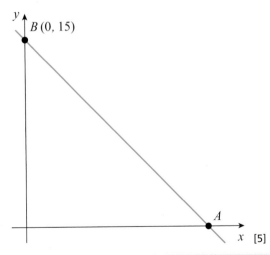

                                                                                                              [5]

3   The table shows the number of students from two schools studying Maths SL and Maths HL.

|            | School 1 | School 2 |
|------------|----------|----------|
| **Maths SL** | 20       | 30       |
| **Maths HL** | 10       | 20       |

   a   How many students are there in the two schools together?

   b   What is the probability that a randomly selected student is from School 1 and studies Maths HL?

   c   Given that a student is from School 2, what is the probability that they study Maths SL?

   [5]

4   a   Write down the value of $\log_2\left(\frac{1}{4}\right)$.

   b   Solve the equation $(\log_2 x)^2 - \log_2 x - 6 = 0$.

   [5]

5   The table shows some values of function f and its first and second derivatives.

| $x$       | 0  | 1  | 2  | 3  | 4  | 5 | 6 |
|-----------|----|----|----|----|----|---|---|
| $f(x)$    | 1  | 4  | 1  | 0  | −3 | 0 | 2 |
| $f'(x)$   | 2  | 0  | −3 | −4 | 0  | 1 | 3 |
| $f''(x)$  | −1 | −3 | 0  | 2  | 4  | 3 | 1 |

   a   Find $f \circ f(3)$.

   b   Is f increasing or decreasing at $x = 2$?

   c   Write down the coordinates of the stationary points of f. Determine whether each is a local maximum or local minimum.

   d   Write down the coordinates of the point of inflection of f.

   [6]

6   a   Evaluate $\dfrac{\tan 60°}{\sin 60°} - \sin 60° \tan 60°$.

   b   Prove the identity $\dfrac{\tan x}{\sin x} - \sin x \tan x \equiv \cos x$.

   [6]

7   The curve shown in the diagram has equation $y = (2x - 4)\, e^{-(x^2 - 4x)}$.

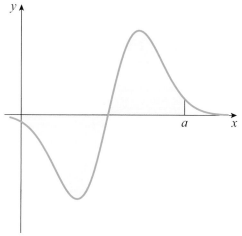

   The shaded area equals $2(e^4 - 1)$. Find the value of $a$.

   [7]

## Section B

**8**  *[13 marks]*

The diagram shows the parabola $P$ with equation $y = x^2 - 6x + 9$.

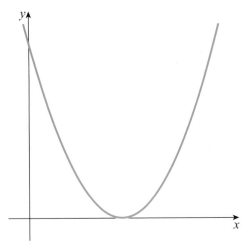

**a**  Write down the coordinates of the vertex of $P$.

[1]

**b**  Parabola $P$ is translated 2 units to the right and 4 units up to obtain parabola $Q$.

  **i**    Write down the coordinates of the vertex of $Q$.

  **ii**   Find the equation of $Q$ in the form $y = ax^2 + bx + c$.

[3]

**c**  Find the coordinates of the point of intersection of $P$ and $Q$.

[3]

**d**  Line $L$ has equation $y = x - k$.

  **i**    Show that the $x$-coordinate of any point of intersection of $P$ and $L$ satisfy the equation $x^2 - 7x + (9 + k) = 0$.

  **ii**   Given that $L$ is tangent to $P$, find the value of $k$.

[6]

**9**  *[13 marks]*

Daniel and Alessia take turns to take a shot at a goal, with Daniel going first. The probability that Daniel scores a goal is $p$ and the probability that Alessia scores is $\frac{1}{2}$ on each attempt, independently of all other attempts. The first person to score wins the game.

A part of the tree diagram for the games is shown below.

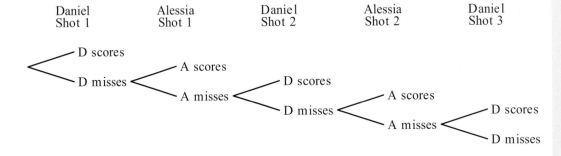

a  Find the probability that Alessia wins on her first shot.

[2]

b  The probability that Alessia wins on her second shot is $(1-p)^a \left(\dfrac{1}{2}\right)^b$. Write down the value of $a$ and $b$.

[2]

c  Write down an expression for the probability that Alessia wins:
  i   on her 3rd shot
  ii  on her $k$th shot.

[3]

d  The game continues until one person wins. The probability that Alessia wins the game is $\dfrac{1}{5}$. Find the value of $p$.

[6]

10  *[14 marks]*

Function f is defined by $f(x) = \ln(x^3 + 1)$ for $x > -1$.

You are given that the range of f is $\mathbb{R}$ and that $f'(x) = \dfrac{3x^2}{x^3+1}$.

a  Explain why f never decreases.

[1]

b  Show that $f''(x) = \dfrac{3x(2-x^3)}{(x^3+1)^2}$.

[4]

c  Explain how you can tell that $(0, 0)$ is a point of inflection on the graph of $y = f(x)$.

[2]

d  Find the coordinates of the second point of inflection on the graph of $y = f(x)$.

[3]

e  Show that f has an inverse function. Find an expression for $f^{-1}(x)$ and state its domain and range.

[4]

Analysis and approaches SL:

# Practice Paper 2

Calculator.

1 hour 30 minutes, maximum mark for the paper [80 marks].

## Section A

1  The diagram shows a triangle $ABC$ where $AB = 18\,\text{cm}$, $AC = 13\,\text{cm}$ and $B\hat{A}C = 0.75\,\text{radians}$.

   $CD$ is an arc of a circle with centre $A$.

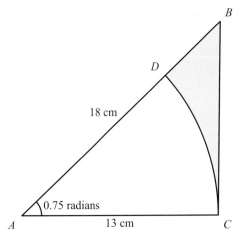

   a  Find the length of the arc $CD$.

   b  Find the shaded area.

   [5]

2  a  Differentiate the function $f(x) = (\ln x)^2$

   b  Show that the values of $x$ where the tangent to the curve $y = (\ln x)^2$ is parallel to the tangent of $y = x^3 - x^2 - 6x$ satisfy
   $$2\ln x = 3x^3 - 2x^2 - 6x$$

   c  Use technology to find these values of $x$.

   [6]

3  The triangle $ABC$ has $AB = 10\,\text{cm}$, $BC = 15\,\text{cm}$ and $A\hat{C}B = 28°$.

   Find the possible sizes of $B\hat{A}C$.

   [5]

4  The curve $C$ has equation $y = x + \dfrac{2}{x}$ where $x > 0$, and the line $l$ has equation $y = 5 - x$.

   a  On the same set of axes, sketch $C$ and $l$, labelling the coordinates of any axis intercepts.

   b  Find the area of the region enclosed between $C$ and $l$.

   [6]

5  For the function $f(x) = \dfrac{2x + 3}{x + 1}$, find

   a  the largest possible domain

   b  the range

   c  an expression for the inverse function, $f^{-1}(x)$.

   [6]

6   A particle moves with velocity $v = e^t \cos 2t$ m s$^{-1}$ for $t \geqslant 0$.

   a  Find an expression for the acceleration after time $t$ seconds.

   b  Find the distance moved in the first 2 seconds of motion.

[6]

7   A square-based pyramid has height 50 cm. The sloping edges make an angle of 75° with the base. Find:

   a  the length of each sloping edge

   b  the length of the diagonal of the square base

   c  the volume of the pyramid.

[7]

## Section B

8   *[11 marks]*

   The line $l_1$ has the equation $3x + 2y = 12$. It crosses the $y$-axis at the point $N$.

   a  Find:

   i    the gradient of $l_1$

   ii   the coordinates of $N$.

[3]

   The line $l_2$ passes through the point $P(3, -5)$ and is perpendicular to $l_1$.

   b  Find the equation of $l_2$ in the form $ax + by + c = 0$.

[3]

   The point of intersection of $l_1$ and $l_2$ is $Q$.

   c  Find the coordinates of $Q$.

   d  Calculate the distances:

   i    $NQ$

   ii   $PQ$.

   e  Find the area of triangle $NPQ$.

[5]

9   *[14 marks]*

   Waiting times at a doctor's surgery are normally distributed with mean $\mu$ and standard deviation $\sigma$. It is known that there is a 10% chance of waiting more than 19 minutes and a 5% chance of waiting less than 3 minutes.

   a  i    Show that $\mu + 1.2816\sigma = 19$.

   ii   Write down a second equation in $\mu$ and $\sigma$.

   iii  Hence find $\mu$ and $\sigma$.

[7]

   A wait is defined to be 'short' if it is less than $t$ minutes.

   b  The probability of a visit being short is 0.2. Find $t$.

[2]

   c  i    A patient is chosen at random. Find the probability that they wait more than 15 minutes.

   ii   From a randomly chosen group of 10 patients, find the probability that at least half wait more than 15 minutes.

[5]

**10** *[14 marks]*

The point $P$ with $x$-coordinate $p > 0$ lies on the curve $y = \dfrac{1}{x}$.

a  i   Find, in terms of $p$, the gradient of the tangent to the curve at $P$.

   ii   Show that the equation of the tangent to the curve at $P$ is $x + p^2y - 2p = 0$.

                                                                                       [5]

The tangent to the curve at $P$ intersects the $x$-axis at $Q$ and the $y$-axis at $R$.

b  i   Find the coordinates of $Q$ and $R$.

   ii   Find the area of the triangle $OQR$, where $O$ is the origin.

                                                                                       [4]

c  Show that the distance $QR$ is given by $2\sqrt{p^2 + \dfrac{1}{p^2}}$.

                                                                                       [3]

d  Find the minimum distance $QR$.

                                                                                       [2]

# Answers

## Toolkit: Using technology

Computer algebra systems (CAS)

a  $\sqrt{1-x^2}$

b  1

Spreadsheets

b  370 m, 20 m s$^{-1}$

Dynamic geometry packages

a  $AB = 2\,MC$

b  1:1

Programming

c  Welcome to Mathematics for the International Baccalaureate. We hope that you enjoy this book and learn lots of fun maths! (This is an example of a ROT13 Caesar Shift Cipher)

## Toolkit: Algebra Practice

1  a  $11x + 15y$  
   b  $10x^2 + 7xy$  
   c  $2x + 3$  
   d  $x + 4$  
   e  $xy^2z$  
   f  $63x^2y$  
   g  $8x^3$  
   h  $\dfrac{x}{6}$  
   i  $\dfrac{1+2x}{y}$  
   j  $\dfrac{y}{3z}$  
   k  $\dfrac{x}{2}$  
   l  $-1$  
   m  $\dfrac{x^2}{15}$  
   n  $\dfrac{2(x+1)}{5}$  
   o  $\dfrac{2x}{y}$

2  a  $2x^2 - 6x$  
   b  $x^2 + 6x + 9$  
   c  $x^2 + x - 20$  
   d  $x^3 + 3x^2 + 2x$  
   e  $x^2 + y^2 + 2xy - 1$  
   f  $x^3 + 9x^2 + 23x + 15$

3  a  $4(3 - 2y)$  
   b  $3(x - 2y)$  
   c  $7x(x - 2)$  
   d  $5xy(x + 2y)$  
   e  $(2z + 5)(x + y)$  
   f  $(3x - 1)(x - 2)$

4  a  $x = -\dfrac{7}{8}$  
   b  $x = 4$  
   c  $x = -8$  
   d  $x = 5$  
   e  $x < -\dfrac{13}{2}$  
   f  $x \geq \dfrac{5}{3}$

5  a  $x = 6,\ y = 4$  
   b  $x = 3,\ y = 1$  
   c  $x = -1,\ y = 2$

6  a  $-2$  
   b  9  
   c  12  
   d  2  
   e  12.5  
   f  5

7  a  $x = \dfrac{y-4}{2}$  
   b  $x = \dfrac{3y}{2-y}$  
   c  $x = \dfrac{1-2y}{y-1}$  
   d  $x = \dfrac{by}{y-a}$  
   e  $x = \pm\sqrt{a^2 + 4}$  
   f  $x = -\dfrac{1}{2y}$

8  a  $2 - \sqrt{2}$  
   b  $2\sqrt{15}$  
   c  4  
   d  $3\sqrt{3}$  
   e  $3 - 2\sqrt{2}$  
   f  $-1$

9  a  $\sqrt{2}$  
   b  $\sqrt{2} + 1$  
   c  $\dfrac{\sqrt{3}}{3}$

10  a  $x = -3 \text{ or} -2$  
    b  $x = \pm\dfrac{3}{2}$  
    c  $x = -1 \pm \sqrt{6}$

11  a  $\dfrac{b-a}{ab}$  
    b  $\dfrac{3x+5}{x^2}$  
    c  $\dfrac{x^2 - x + 13}{x - 1}$  
    d  $\dfrac{2}{(1-x)(1+x)}$  
    e  $\dfrac{25 - 2a^2}{a(5-a)}$  
    f  $\dfrac{2(x^2 - 2)}{(x-2)(x-1)}$

# Chapter 1 Prior Knowledge

**1 a** 81 **b** 40
**2 a** $3.4271 \times 10^2$ **b** $8.56 \times 10^{-3}$
**3** $2^6$

# Exercise 1A

**1 a** $x^6$ **b** $x^{12}$
**2 a** $y^6$ **b** $z^{10}$
**3 a** $a^7$ **b** $a^{11}$
**4 a** $5^{17}$ **b** $2^{24}$
**5 a** $x$ **b** $x^3$
**6 a** $y^4$ **b** $z^6$
**7 a** $b^6$ **b** $b^8$
**8 a** $11^8$ **b** $7^5$
**9 a** $x^{15}$ **b** $x^{32}$
**10 a** $y^{16}$ **b** $z^{25}$
**11 a** $c^{14}$ **b** $c^{14}$
**12 a** $3^{50}$ **b** $13^{28}$
**13 a** $48x^7$ **b** $15x^7$
**14 a** $3a^3$ **b** $5b^4$
**15 a** $20x^5y^3z$ **b** $12x^8yz^5$
**16 a** $2x^5$ **b** $3x^6$
**17 a** $\dfrac{1}{2x^3}$ **b** $\dfrac{1}{4x}$
**18 a** $\dfrac{5}{3x^2}$ **b** $\dfrac{3x^2}{4}$
**19 a** $7x^2y^3$ **b** $\dfrac{2x^3z}{y}$
**20 a** $\dfrac{1}{10}$ **b** $\dfrac{1}{7}$
**21 a** $\dfrac{1}{27}$ **b** $\dfrac{1}{25}$
**22 a** $\dfrac{4}{3}$ **b** $\dfrac{7}{5}$
**23 a** $\dfrac{9}{4}$ **b** $\dfrac{125}{8}$
**24 a** $\dfrac{7}{9}$ **b** $\dfrac{5}{16}$
**25 a** $\dfrac{6}{x}$ **b** $\dfrac{10}{x^4}$
**26 a** $\dfrac{8}{9}$ **b** $\dfrac{81}{64}$

**27 a** $27u^{-6}$ **b** $32v^{-15}$
**28 a** $\dfrac{1}{4}a^{10}$ **b** $\dfrac{1}{27}b^{21}$
**29 a** $25x^4y^6$ **b** $81a^8b^{-8}$
**30 a** $\dfrac{x^3}{27}$ **b** $\dfrac{25}{x^4}$
**31 a** $\dfrac{27x^6}{8y^9}$ **b** $\dfrac{25u^2v^6}{49b^8}$
**32 a** $\dfrac{16v^4}{9u^2}$ **b** $\dfrac{27b^6}{8a^9}$
**33 a** $2x - 7x^6$ **b** $3y + 5y^3$
**34 a** $5u^2 + 6uv^2$ **b** $5a^2b^5c - 4ac^2$
**35 a** $5p - 3p^{-1}q^2$ **b** $2s^2t^{-2} + 3s^3t^2$
**36 a** $3^8$ **b** $3^{24}$
**37 a** $2^9$ **b** $5^{10}$
**38 a** $2^5$ **b** $3^{11}$
**39 a** $2^{17}$ **b** $2^7$
**40 a** $x = 4$ **b** $x = 3$
**41 a** $x = -1$ **b** $x = 6$
**42 a** $x = \dfrac{7}{3}$ **b** $x = \dfrac{9}{2}$
**43 a** $x = -4$ **b** $x = -3$
**44** $2x + 4x^2$
**45** $x^5y^2$
**46** $\dfrac{b^6}{8a^3}$
**47 a** $n = \dfrac{1000}{\sqrt{D}}$ **b** $250\,000$
   **c** \$10 million
**48 a** $k_A = 8$, $k_B = 40$ **b** $\dfrac{n}{5}$
   **c** method B
**49** $x = 2$
**50** $x = 5$
**51** $x = -1$
**52** $x = -1$
**53** $x = -2$
**54 a** $R = 3.2T^2$ **b** $250\,K$
**55 a** $\dfrac{3}{v}$ **b** $1.5v$
   **c** $40\,km$ per hour
**56** $x = 1$, $y = -3$
**57** $x = 4$
**58** $x = 6$
**59** $x = 1$, 3 or $-5$
**60** $2^{7000}$
**61** 7

# Exercise 1B

1 a $32\,000$    b $6\,920\,000$
2 a $0.048$    b $0.000\,985$
3 a $6.1207 \times 10^2$    b $3.07691 \times 10^3$
4 a $3.0617 \times 10^{-3}$    b $2.219 \times 10^{-2}$
5 a $6.8 \times 10^7$    b $9.6 \times 10^{11}$
6 a $1 \times 10^0$    b $1.2 \times 10^{-5}$
7 a $2.5 \times 10^{21}$    b $3.6 \times 10^{13}$
8 a $2 \times 10^{-2}$    b $4 \times 10^{-4}$
9 a $5 \times 10^0$    b $2.5 \times 10^2$
10 a $2.1 \times 10^{11}$    b $3.1 \times 10^9$
11 a $2.1 \times 10^5$    b $3.02 \times 10^8$
12 a $7.6 \times 10^5$    b $8.91 \times 10^{14}$
13 a $4.01 \times 10^4$    b $6.13 \times 10^{13}$
20 a $1.22 \times 10^8$    b $400$
   c $4 \times 10^2$
21 $6 \times 10^{-57}$
22 $1.99 \times 10^{-23}$ g
23 a $1.5 \times 10^{-14}$ m    b $1.77 \times 10^{-42}$ m$^3$
24 a $2.98$    b $7.41 \times 10^8$
   c Europe
25 a $1.5$    b $a + b + 1$
26 a $4$    b $a - b - 1$
27 $r = p + q + 1$

# Exercise 1C

1 a $1$    b $2$
2 a $5$    b $6$
3 a $0$    b $-1$
4 a $-2$    b $-4$
5 a $1$    b $2$
6 a $3$    b $4$
7 a $0$    b $-1$
8 a $-2$    b $-3$
9 a $100$    b $1000$
10 a $50\,002$    b $332$
11 a $1.55$    b $-1.99$
12 a $e^2$    b $e^5$
13 a $e^{y+1}$    b $e^{y^2}$
14 a $\frac{1}{2}e^{y-3} - 2$    b $2e^{2y+1} + 12$
15 a $5\log x$    b $5\log x$

16 a $2\log 3x$    b $\dfrac{1}{\log 2x}$
17 a $1$    b $\dfrac{1}{\ln x}$
18 a $-15$    b $17$
19 a $4.5$    b $-1.5$
20 a $13$    b $7$
21 a $3$    b $7$
22 a $8$    b $9$
23 a $25$    b $81$
24 a $\dfrac{1}{3}$    b $\dfrac{1}{6}$
25 a $\dfrac{1}{32}$    b $\dfrac{1}{64}$
26 a $2.10$    b $3.15$
27 a $5.50$    b $2.88$
28 a $-0.301$    b $-1.40$
29 a $\log 5$    b $\log 7$
30 a $\log 0.2$    b $\log 0.06$
31 a $\ln 3$    b $\ln 7$
32 a $2 + \log 7$    b $\log 13 - 4$
33 a $2 + \log 70$    b $\log 13 - 2$
34 a $\ln(k - 2) - 2$    b $\ln(2k + 1) - 2$
35 a $0.349$    b $-0.398$
36 a $1.26$    b $0.847$
37 $10\,000$
38 $332$
39 $a + b$
40 $0.699$
41 $0.824$
42 $0.845$
43 $\log 1.6$
44 a $7.60$    b $0.0126$ moles per litre
45 a $10$    b $6.93$ days
46 a i $1000$    b $10\ln 3 = 11.0$ hours
    ii $1221$
47 $4.62$ years
48 a $57.0$ decibels    b $67.0$ decibels
   c Increases the noise level by 10 decibels.
   d $10^{-3}$ W m$^{-2}$
49 a $e^{\ln 20}$    b $\dfrac{\ln 7}{\ln 20}$
50 $x = 100,\ y = 10$ or $x = -100,\ y = -10$
51 $1$

# Chapter 1 Mixed Practice

**1** a $9.3 \times 10^3$ cm   b $5\,280\,000$ cm$^2$

**2** $\dfrac{9y}{x}$

**3** $\dfrac{y^6}{9x^4}$

**4** a $4\,000\,000$   b $1\,000\,000$   c $16$

**5** $x = 3$

**9** $500$ s

**10** $x = 99$

**11** $x = 0.5e^3$

**12** $x = 0.693$

**13** $x = 0.531$

**14** $x = \ln\left(\dfrac{y+1}{5}\right)$

**15** a $m = 3, n = 4$   b $x = 7.5$

**16** a $2.8$   b $a + b + 1$

**17** a $1.2$   b $a - b$

**18** $0.0259$ to $0.0326$

**19** a $9.42\,\mathrm{m\,s^{-1}}$   b Yes

**20** a $3$
   b Strength increases by one.
   c $3160$ m (It would have had to be measured using a special damped seismograph.)

**21** $x = \log\left(\dfrac{2}{3}\right) \approx -0.176$

**22** $x = 1, y = -2$

**23** $x = 10, y = 0.1$

# Chapter 2 Prior Knowledge

**1** $x = 4, y = -1$

**2** a $80$   b $36$

**3** $x = \pm 2$

# Exercise 2A

**1** a $49$   b $68$

**2** a $-22$   b $-40$

**3** a $2$   b $7$

**4** a $-0.5$   b $-2.5$

**5** a $u_1 = -37, d = 7$   b $u_1 = -43, d = 9$

**6** a $8 - 3(n-1)$   b $10 + 4(n-1)$

**7** a $3 + 2(n-1)$   b $10 + 4(n-1)$

**8** a $1 + 5(n-1)$   b $20 - 3(n-1)$

**9** a $34$   b $24$

**10** a $13$   b $29$

**11** a $372$   b $720$

**12** a $-11$   b $-236$

**13** a $352$   b $636$

**14** a $184$   b $390$

**15** a $-45$   b $0$

**16** a $32$   b $99$

**17** a $105$   b $205$

**18** a $806$   b $354$

**19** a $13$   b $10$

**20** a $16$   b $16$

**21** a $216$   b $2230$

**22** a $4$   b $31$   c $465$

**23** a $8$   b $305$

**24** a $5$   b $112$

**25** a £$312$   b £$420$

**26** £$300$

**27** $14$

**28** a $387$   b $731$

**29** $140$

**30** a $21$   b $798$

**31** a $u_1 = -7, d = 7$   b $945$

**32** a $296$   b $390$

**33** $632$

**34** $x = 2$

**35** $42$

**36** $42$

**37** $960$

**38** a $876$   b $1679$   c $480$

**39** $408.5$

**40** $53$

**41** $-20$

**42** a Day $41$   b $3997.5$ minutes

**43 a** 19 days    **b** 45 days

**44 a** 3, 7    **b** 199

**45** $2n + 3$

**46** 0

**47** 70 336

**48** 735

**49** 0.2 m

**50** −1

**52 b** 150

# Exercise 2B

**1 a** 20 971 520    **b** 1458

**2 a** −15 625    **b** −1 441 792

**3 a** $\dfrac{1}{32}$    **b** $-\dfrac{2}{729} \approx 0.00274$

**4 a** $u_1 = 7, r = 2$    **b** $u_1 = \dfrac{4}{3}, r = 3$

**5 a** $u_1 = 3, r = \pm 2$    **b** $u_1 = \dfrac{5}{9}, r = \pm 3$

**6 a** $u_1 = -3, r = 2$ or $u_1 = 3, r = -2$

  **b** $u_1 = 7168, r = \dfrac{1}{2}$ or $u_1 = -7168, r = -\dfrac{1}{2}$

**7 a** 6    **b** 8

**8 a** 13    **b** 8

**9 a** 5465    **b** 4095

**10 a** 190.5    **b** 242

**11 a** 153.75    **b** $\dfrac{1456}{9}$

**12 a** 363    **b** 2800

**13 a** 976 560    **b** 324 753

**14 a** 68 796    **b** 488 280 000

**15 a** $\dfrac{189}{16}$    **b** $\dfrac{671}{27}$

**16 a** 1    **b** 255

**17 a** 2    **b** 96    **c** 3069

**18 a** 1.5    **b** 182.25

**19** 1920 cm$^2$

**20** 0.0375 mg ml$^{-1}$

**21** 15.5

**22** 0.671 seconds

**23 a** 3580 m$^3$    **b** 11 days

**24** 255

**25 a** $9.22 \times 10^{18}$    **b** $1.84 \times 10^{19}$

  **c** 2450 years

**26** $x^{2-n} y^{n+1}$

**27** 3069

**28** −1

**29** 88 572

**30 a** 0.246 m    **b** 3.19 m

  **c** The height is so small that measurement error and other inaccuracies would be overpowering.

# Exercise 2C

**1 a** $2382.03    **b** $6077.45

**2 a** $580.65    **b** $141.48

**3 a** $6416.79    **b** $1115.87

**4 a** 48 years    **b** 5 years

**5 a** 173 months    **b** 77 months

**6 a** 7.18%    **b** 4.81%

**7 a** 14.9%    **b** 7.18%

**8 a** $418.41    **b** $128.85

**9 a** £13 311.16    **b** £7119.14

**10 a** $97    **b** $294

**11 a** £737.42    **b** £2993.68

**12 a** $103    **b** $5130

**13 a** £94    **b** £2450

**14 a** €1051.14    **b** €579.64

**15 a** $598.74    **b** $4253.82

**16** £900.41

**17** £14 071.00

**18** $8839.90

**19** €16 360.02

**20** £8870

**21 a** monthly    **b** £28.55

**22** £6814.36

**23**

| Year | Start-year value ($) | Depreciation expense ($) | End-year value ($) |
|---|---|---|---|
| 1 | 20 000 | 6 000 | 14 000 |
| 2 | 14 000 | 4 200 | 9 800 |
| 3 | 9 800 | 2 940 | 6 860 |
| 4 | 6 860 | 2 058 | 4 802 |
| 5 | 4 802 | 1 441 | 3 361 |
| 6 | 3 361 | 1 008 | 2 353 |
| 7 | 2 353 | 706 | 1 647 |
| 8 | 1 647 | 147 | 1 500 |

**24** 15.9%

**25** 12.7%

**26** a  2.44%   b  12.8%

**27** a  250 billion marks
    b  0.0024 marks     c  0.0288 marks

## Chapter 2 Mixed Practice

**1** a  €6847.26    b  7.18%

**2** a  i  $d_n$    b  i  $\frac{1}{2}$

    ii  $b_n$       ii  $-\frac{3}{256}$

**3** a  2    b  −4    c  300

**4** a  4    b  512    c  43690

**5** 8

**6** 10

**7** a  48.8 cm    b  9

**8** a  €563.50    b  6.25 years

**9** a  8.04 billion    b  2033

**10** a  i  20    b  7290

    ii  10

**11** a  i  7    ii  16

    c  2    d  25

    e  $n = 498$    f  $n = 36$

**12** $x = 4$

**13** b  74

**14** 8190

**15** 16 months

**16** $a^{2n} b^{3-n}$

**17** $11.95 million

**18** $10450

**19** 8.63%

**20** $1212.27

**21** 22 years

**22** a  $a_1 = 6$

    b  i  $a_2 = 8$    ii  $a_3 = 10$

    c  2

    d  i  22    ii  594 m

    e  28

    f  16 m

**23** b  32

**24** a  B – 12th day    b  A – 6th day

**25** a  £68 500    b  £1 402 500    c  £43 800

**26** 3

## Chapter 3 Prior Knowledge

**1**
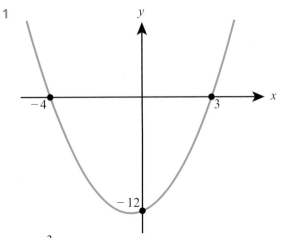

**2**  $x \geq \frac{3}{2}$

**3**  0

**4**  $x = -2.30$

**5**
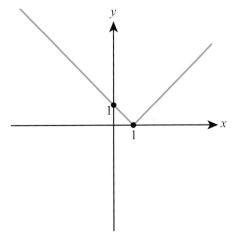

## Exercise 3A

**1** a  9    b  14

**2** a  7    b  47

**3** a  −17    b  19

**4** a  −15    b  −44

**5** a  Yes    b  Yes

**6** a  No    b  No

**7** a  Yes    b  No

**8** a  No    b  Yes

**9** a  Yes    b  Yes

10 a  $\mathbb{R}$          b  $\mathbb{R}$

11 a  $x \neq 0$          b  $x \neq 0$

12 a  $x \geqslant -5$          b  $x \geqslant 3.5$

13 a  $x > -0.6$          b  $x > 4$

14 a  $x \neq 2.5$          b  $x \neq -3$

15 a  $f(x) \geqslant -2$          b  $g(x) \geqslant 7$

16 a  $f(x) \leqslant 18$          b  $g(x) < 4$

17 a  $f(x) \geqslant 0$          b  $g(x) > 0$

18 a  $f(x) \geqslant 2$          b  $g(x) < 3$

19 a  40          b  $-7$

20 a  1          b  2

21 a  $-3$          b  $-1$

22 a

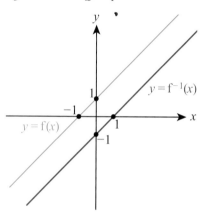

b

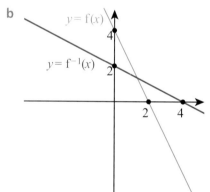

23 a  No inverse   b  No inverse

24 a

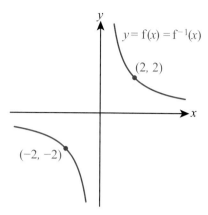

b

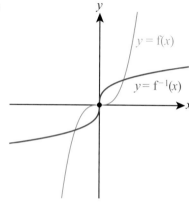

25 a

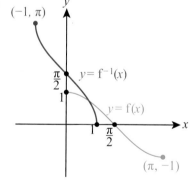

b

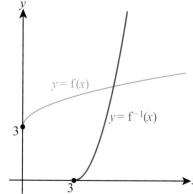

**26**   a   $-13$         b   $x = 3$

**27**   $x = 41$

**28**   a   $5.7\,\mathrm{m\,s^{-1}}$

     b   No, the car cannot accelerate uniformly for that long / The model predicts an unreasonable speed of $114\,\mathrm{m\,s^{-1}}$.

**29**   a   $x \neq 5$         b   $\dfrac{1}{3}$

**30**   a   $-\dfrac{5}{4}$         b   $q(x) \geqslant -2$

     c   $x = 4$ or $-4$

**31**   a   4.95 billion

     b   E.g. smartphones are likely to get replaced by another technology

**32**   a   $f(x) = 1.3x$

     b   Amount in pounds if $x$ is the amount in dollars.

**33**   a   1         b   $x = 2$

**34**   a   $x \geqslant 2.5$     b   $f(x) \geqslant 0$     c   $x = 7$

**35**   a   60         b   129.9; 130

     c   It predicts a non-integer number of fish.

**36**   a   14         b   4

**37**

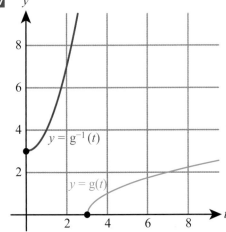

**38**   a   $x > 0$         b   4         c   $\dfrac{1}{9}$

**39**   $x > 5$

**40**   a   $x < \dfrac{7}{3}$         b   $x = -31$

**41**   a   19         b   $f(x) \geqslant 4$

     c   1 is not in the range

**42**   a   9                 b   2.5

     c

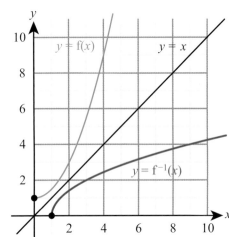

**43**   a   0.182         b   0.892

**44**   a   $x \geqslant 1, x \neq 17$     b   $f(x) \geqslant \dfrac{3}{4}$ or $f(x) < 0$

**45**   a and b

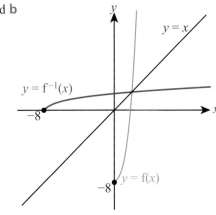

     c   $x = 2$

# Exercise 3B

**1**   a

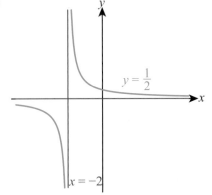

b

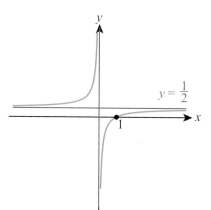

2 a

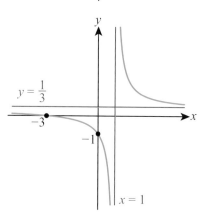

b

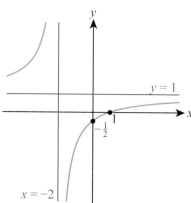

3 a

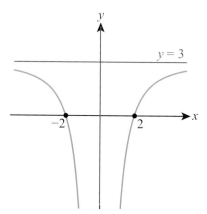

b

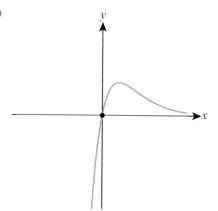

4 a $f(x) \geqslant \dfrac{11}{3}$      b $g(x) \leqslant \dfrac{81}{8}$

5 a $f(x) \geqslant -0.288$      b $g(x) \geqslant 1.50$

6 a $-1.17 \leqslant f(x) \leqslant 1.17$      b $-1.57 \leqslant g(x) \leqslant 2.72$

7 a $(0.7, 0.55); x = 0.7$

   b $\left(-\dfrac{2}{7}, \dfrac{39}{7}\right); x = -\dfrac{2}{7}$

8 a $(-0.293, 1.75), (-1.71\ 1.75), (-1,2); x = -1$

   b $(0,-1), (1,-1), \left(\dfrac{1}{2}, -\dfrac{3}{4}\right); x = \dfrac{1}{2}$

9 a

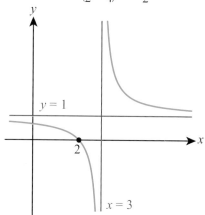

b

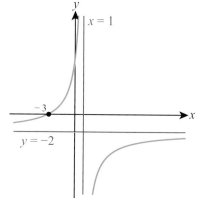

**10 a**

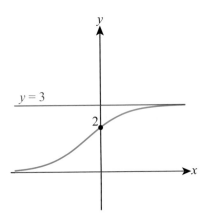

**b**

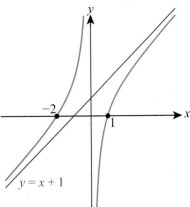

**11 a**

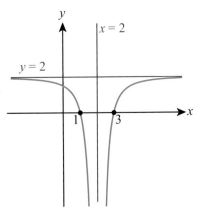

**b**

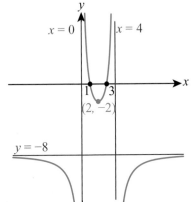

**12 a** $(-1.49, -4.78)$      **b** $(-0.661, 0.421)$

**13 a** $(-2, 7), (-1, 5), (1, 7)$

     **b** $(-2, 23), (1, 5), (2, 3), (3, 3)$

**14 a** $\left(-\sqrt{2}, -2\sqrt{2} - 2\right), \left(\sqrt{2}, 2\sqrt{2} - 2\right)$

     **b** $\left(1 - \sqrt{2}, 7 - \sqrt{2}\right), \left(1 + \sqrt{2}, 7 + \sqrt{2}\right)$

**15 a** $x = 0.040, 1.78$      **b** $x = 0.213, 1.632$

**16 a** $x = 1, 6.71$      **b** $x = -3.48, 2.48$

**17 a** $x = 0.063, 1.59$      **b** $x = 0.288, 49.0$

**18 a** $x = -2.18, 0.580$      **b** $x = -1.67, 0.977$

**19**

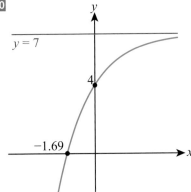

**20**

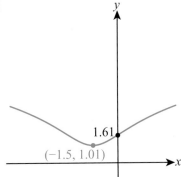

$(-1.69, 0), (0, 4), y = 7$

**21 a** $x \neq -2$

     **b**

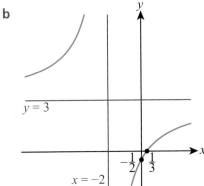

$x = -2, y = 3$

**22**

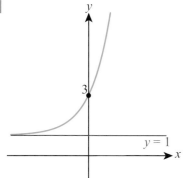

**23** (1.84, 3.16)

**24** (−1, 2.5)

**25** 235

**26** $x = -1$

**27**

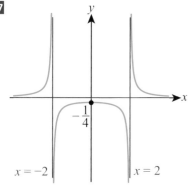

$x = -2$, $x = 2$, $y = 0$

**28**

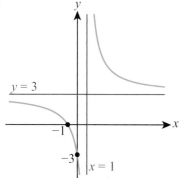

**29**

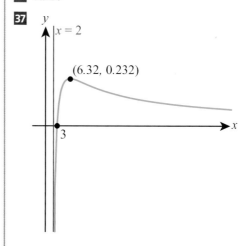

**30**

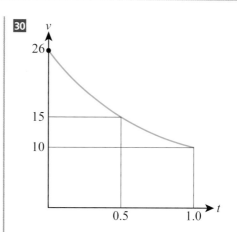

**31**

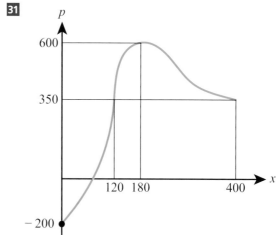

**32** $x = 0.755$

**33** $x = -2.20, -0.714, 1.91$

**34** $-2.41, 0.414, 2$

**35** $-1.41, 1.41$

**36** $0.920$

**37**

(6.32, 0.232)

**38** $x \geqslant 2$

# Chapter 3 Mixed Practice

**1** G

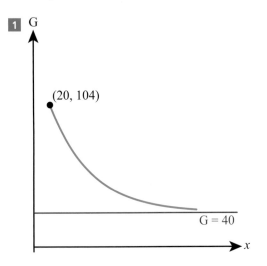

(20, 104)

$G = 40$

**b** $3538.09

**2** $x = 1.86, 4.54$

**3** **a** $x \geqslant -5$      **b** $x = -2.38$

**4** $(-1.61, 0.399)$ and $(0.361, 2.87)$

**5**

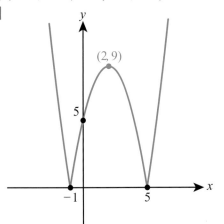

(2, 9)

5

$-1$      5

**6** **a**

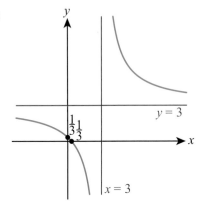

$y = 3$

$\frac{1}{3}\frac{1}{3}$

$x = 3$

**b** $x \neq 3, \ \mathrm{f}(x) \neq 3$

**7** **a** $20.0\,\mathrm{m\,s^{-1}}$      **b** $0.630, 17.1\,\mathrm{s}$      **c** $5\,\mathrm{s}$

**8** **b** $1.30\,\mathrm{s}$

**9** **a** $100°\mathrm{C}$      **b** $95.3°\mathrm{C}$      **c** $8.82\,\mathrm{km}$

**10** **a** $\mathrm{f}(x) = 1.8x + 32$

   **b** Temperature in Celsius if $x$ is temperature in Fahrenheit.

**11** **a** $p = -2, q = 4$

   **b** **i** $x \neq 2$

      **ii** $\mathrm{g}(x) > 0$

      **iii** $x = 2$

**12** **a** $\mathrm{f}(x) \geqslant 20$      **b** $12$

**13** $x = -5.24, 3.24$

**14** $x = 1, 2.41$

**15** $x \neq \pm 3, \ \mathrm{h}(x) \leqslant -2$ or $\mathrm{h}(x) > 0$

**16** $-9.25 \leqslant \mathrm{f}(x) < 3.$

**17** **a** $4.89$      **b** $x = 28.9$

**18** **a** $-19 < \mathrm{g}(x) \leqslant 8$      **b** $x = -1, x = 1$

   **c** $-20$ is not in the range of g

**19**

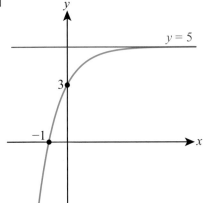

$y = 5$

3

$-1$

**20**

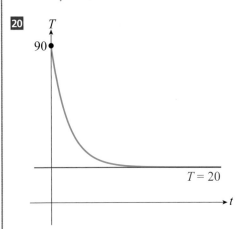

$T$

90

$T = 20$

$t$

**21**

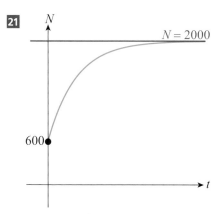

**22 a** $1.10 \, \text{ms}^{-1}$

   **b** e.g. The car will stop by then

**23** $x = -2.50, -1.51, 0.440$

**24 a**

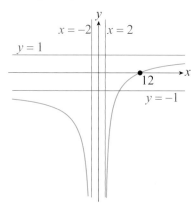

   **b i** $(12, 0)$

   **ii** $x = \pm 2, y = \pm 1$

**25** $x = 2.27, 4.47$     **26** $-1.34$

**27 a** $x \neq 0, 4$

   **b** $f(x) \leq -5.15$ or $f(x) \geq 3.40$

**28 a** $x > 0, x \neq e^{-3}$

   **b** $g(x) < 0$ or $g(x) \geq 0.271$

**29 a** $x > 2, g(x) \in \mathbb{R}$

   **b**

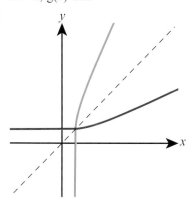

   **c** $x = 2.12$

# Chapter 4 Prior Knowledge

**1** $24 \, \text{cm}$

**2 a** $10$          **b** $(-1, 1)$

# Exercise 4A

**1 a** $1$                    **b** $2$

**2 a** $-3$                   **b** $-1$

**3 a** $\dfrac{1}{2}$                   **b** $-\dfrac{1}{2}$

**4 a** $(2, 0), (0, -6)$          **b** $(4, 0), (0, 8)$

**5 a** $(5, 0), (0, 2.5)$          **b** $(-3, 0), (0, 2)$

**6 a** $(3, 0), (0, 4)$          **b** $(4.5, 0), (0, -3)$

**7 a** $2x - y + 3 = 0$          **b** $5x - y + 1 = 0$

**8 a** $2x + y - 4 = 0$          **b** $3x + y - 7 = 0$

**9 a** $x - 2y + 7 = 0$          **b** $x + 3y - 9 = 0$

**10 a** $y = 2x + 5; 2, 5$          **b** $y = 3x + 4; 3, 4$

**11 a** $y = -\dfrac{2}{3}x - 2; -\dfrac{2}{3}, -2$     **b** $y = -\dfrac{5}{2}x + 5; -\dfrac{5}{2}, 5$

**12 a** $y = -0.6x + 1.4; -0.6, 1.4$

   **b** $y = -5.5x - 2.5; -5.5, -2.5$

**13 a** $y - 4 = 2(x - 1)$          **b** $y - 2 = 3(x - 5)$

**14 a** $y - 3 = -5(x + 1)$          **b** $y + 1 = -2(x - 2)$

**15 a** $y + 1 = \dfrac{2}{3}(x - 1)$          **b** $y - 1 = -\dfrac{3}{4}(x - 3)$

**16 a** $2x - y - 5 = 0$          **b** $2x - y + 7 = 0$

**17 a** $x + y - 10 = 0$          **b** $3x + y + 2 = 0$

**18 a** $x + 2y - 7 = 0$          **b** $x + 2y - 4 = 0$

**19 a** $3x + 4y - 13 = 0$          **b** $8x + 5y - 19 = 0$

**20 a** $y = 3x + 4$          **b** $y = -x + 9$

**21 a** $y - 5 = 1.5(x - 1)$          **b** $y + 3 = 0.5(x + 1)$

**22 a** $x - y - 6 = 0$          **b** $2x - y - 7 = 0$

**23 a** $x + 3y - 4 = 0$          **b** $x + 5y + 8 = 0$

**24 a** $2x - 5y = 0$          **b** $3x + 4y + 2 = 0$

**25 a** $y = -x + 5$          **b** $y = -\dfrac{1}{3}x + 8$

**26 a** $y - 5 = -4(x - 1)$          **b** $y - 4 = 3(x - 2)$

**27 a** $5x + y + 3 = 0$          **b** $2x + y - 7 = 0$

**28 a** $x - 2y + 5 = 0$          **b** $x - 3y - 10 = 0$

**29 a** $2x - 5y + 37 = 0$          **b** $2x + 3y + 3 = 0$

**30 a** $(1, 2)$          **b** $(5, 1)$

**31 a** $(1, 3)$          **b** $(2, 5)$

**32 a** $(2, 5)$          **b** $(-3, -4)$

33 a $\left(\frac{2}{5}, \frac{11}{5}\right)$    b $\left(\frac{11}{5}, \frac{2}{5}\right)$

34 a $\frac{7}{4}$    b $y = -\frac{4}{7}x + \frac{53}{7}$

35 a $-\frac{4}{3}$    b $4x + 3y = 36$

36 a $-\frac{5}{7}$    b $\frac{17}{5}$

37 a $\frac{1}{5}$    b $x - 5y = -8$

38 a $-\frac{7}{2}$    b No

c $y = \frac{2}{7}x - \frac{51}{7}$

39 0.733

40 a (8, 11)    b 8

c $x = 8$    d 56

41 a $y = -\frac{3}{2}x - 6$    b $x + 4y = -1$

c (−4.6, 0.9)

42 6.10 m

43 a $0.5t + 0.1$    b 9.8 s

44 a N/m    b 0.018 N

c larger    d 0.467 m

45 a $C = 0.01m + 5$    b $6.80

c 500 minutes

46 a $P = 10n - 2000$    b 350

c $P = 8n - 1200$    d 400, fewer

48 1390 m

# Exercise 4B

1 a 5    b 13

2 a 13    b 10

3 a $\sqrt{29}$    b $\sqrt{85}$

4 a 3    b 11

5 a 6    b 9

6 a $\sqrt{98}$    b $\sqrt{65}$

7 a (4, −1)    b (4, 6)

8 a (3, −4, 3)    b (−1, 2, −1)

9 a (5.5, 2, −3)    b (−5.5, 3.5, 7)

10 a (1.5, 0.5, 5.5)    b $\sqrt{171}$

11 (6, 3, −8)

12 $a = 3, b = 20, c = 4.5$

13 $\sqrt{196.5}$

14 $4.52\,\text{m s}^{-1}$

15 $\sqrt{30}$

16 $k = \pm 2$

17 $a = 2.97$

18 a $a = -2.2, b = -8.6$    b $y + 5 = -\frac{1}{16}(x - 4)$

19 a $\frac{4}{7}$    b $7x + 4y = -20$

c (0, −5)    d $\sqrt{65}$

20 a 1.8 m    b 3.33 m

21 a (3, 7)    b (6, 5) and (0, 9)

22 a $y + x = 9$    b $B(4, 5)\ D(5, 4)$

c 5

23 a $\sqrt{50} \approx 7.07\,\text{m}$    b $\sqrt{74} \approx 8.60\,\text{m}$

# Chapter 4 Mixed Practice

1 a i (2, −2)    b $k = -\frac{2}{3}$

ii $\frac{3}{2}$

iii $-\frac{2}{3}$

2 a $s = 6, t = -2$    b $4x + 5y = 23$

3 b $\frac{3}{2}$    c $-\frac{2}{3}$    d 4

4 a −1    b $y = x + 3$    c 4.5

5 a (2.5, −1)    b 9.22

c $y = \frac{7}{6}x - \frac{47}{12}$

6 a (2.5, −1, 4)    b 9.43

7 a $P(0, 3)\ Q(6, 0)$    b $\sqrt{45}$

c (2, 2)

8 a $-\frac{7}{4}$    b $k = 2$    c $d = 1$

11 ±20

12 a $p = 1, q = -18$    b 27.5

13 1.5

14 4.37

15 b $y = -2x + 5$    c $S(1, 3)$

16 10 m

17 a

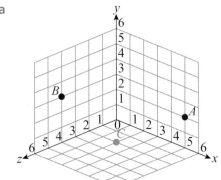

b (2.5, 2.5, 2)    c 6.48

**18** a $\dfrac{1}{2}$　　　　b　$(-1, 5)$

　　c　$2x + y = 3$　　e　4.47

**19** a　42　　　　b　15.2　　　　c　5.51

**20** 4.32

**21** a　$y = 3x - 13$　　　　b　$(6, 5)$ and $(4, -1)$

**23** 135 m

**24** 4 m

# Chapter 5 Prior Knowledge

**1** 95°

**2** 12.4 cm

**3** 60

**4** $\alpha = 140°, \beta = 85°$

**5** 050°

**6** Volume = 1570 cm$^3$, Surface area = 785 cm$^2$

# Exercise 5A

**1** a　40.7 cm$^2$, 24.4 cm$^3$　　b　1580 m$^2$, 5880 m$^3$

**2** a　8.04 m$^2$, 2.14 m$^3$　　b　0.126m$^2$, 0.004 19m$^3$

**3** a　170 m$^2$, 294 m$^3$　　b　101 km$^2$, 134 km$^3$

**4** a　47.9 cm$^2$, 38.6 cm$^3$

　　b　4.40 mm$^2$, 0.293 mm$^3$

**5** a　16$\pi$ cm$^2$, $\dfrac{32\pi}{3}$ cm$^3$　　b　36$\pi$ m$^2$, 36$\pi$ m$^3$

**6** a　64$\pi$ m$^2$, $\dfrac{256\pi}{3}$ m$^3$　　b　144$\pi$ m$^2$, 288$\pi$ m$^3$

**7** a　$\dfrac{3\pi}{4}$ m$^2$, $\dfrac{\pi}{12}$ m$^3$　　b　$\dfrac{4\pi}{3}$ mm$^2$, $\dfrac{16\pi}{81}$ mm$^3$

**8** a　90$\pi$ cm$^2$, 100$\pi$ cm$^3$　　b　36$\pi$ cm$^2$, 16$\pi$ cm$^3$

**9** a　6 cm$^3$　　b　16 cm$^3$

**10** a　20 cm$^3$　　b　18 cm$^3$

**11** a　26.7 cm$^3$　　b　2.67 cm$^3$

**12** a　64 mm$^3$, 144 mm$^2$　　b　336 cm$^3$, 365 cm$^2$

**13** a　2 mm$^3$, 13.8 mm$^2$　　b　320 cm$^3$, 319 cm$^2$

**14** a　5.33 cm$^3$, 33.9 cm$^2$　　b　816 cm$^3$, 577 cm$^2$

**15** a　75.4 cm$^3$, 109 cm$^2$　　b　20.9 cm$^3$, 46.4 cm$^2$

**16** 707 cm$^2$

**17** 96.5 cm$^2$

**18** 8.31 m$^3$

**19** 134 cm$^3$

**20** 3210 cm$^3$

**21** 3400 cm$^2$, 8120 cm$^3$

**22** $4.21 \times 10^4$ cm$^3$

**23** a　12.0 cm　　b　1100 cm$^2$

**24** a　3.61 cm　　b　36.9 cm

**25** a　192 cm$^3$　　b　3.58 cm

**26** a　15 cm　　b　1980 cm$^2$

**27** a　82.5 cm$^2$　　b　474 cm$^2$

　　c　594 cm$^3$

**28** 55.4 cm$^2$, 23.0 cm$^3$

**29** 1140 mm$^2$, 3170 mm$^3$

**30** 1880 mm$^3$, 889 mm$^2$

**31** a　13 cm　　b　815 cm$^2$

**32** 242 m$^2$

**33** a　151 mm$^2$, 134 mm$^3$　　b　13.1 mm

**34** a　9.35 m$^3$　　b　0.190 m$^3$

# Exercise 5B

**1** a　29.7°　　b　58.0°

**2** a　53.1°　　b　39.5°

**3** a　40.1°　　b　53.1°

**4** a　2.5　　b　3.86

**5** a　7.83　　b　11.5

**6** a　9.40　　b　14.1

**7** a　17.4　　b　26.3

**8** a　2.68　　b　25.7

**9** a　344　　b　4.23

**10** a　71.6°　　b　78.7°

**11** a　18.4°　　b　15.9°

**12** a　21.8°　　b　12.5°

**13** a　7.13°　　b　10.6°

**14** a　29.7°　　b　45°

**15** a　10.5°　　b　14.5°

**16** a　8.49 cm　　b　6.53 mm

**17** a　3.42 cm　　b　3.29 mm

**18** a　3.31 cm　　b　4.10 mm

**19** a　37.9°　　b　50.4°

**20** a　39.9°　　b　32.9°

**21** a　20.7°　　b　51.2°

**22** a　$\sqrt{37}$ cm　　b　7 mm

**23** a　5.06 cm　　b　6.40 mm

**24** a　5.23 cm　　b　6.43 mm

**25** a　39.4°　　b　41.6°

**26** a　42.6°　　b　62.8°

**27 a** 53.8°        **b** 97.3°

**28 a** 28.5 cm²     **b** 10.5 mm²

**29 a** 127 cm²      **b** 1.73 mm²

**30 a** 631 cm²      **b** 1710 mm²

**31 a** 52.2°        **b** 61.8°

**32 a** 8.30 cm      **b** 5.23 mm

**33 a** 7.04 cm      **b** 8.36 mm

**34** 4.16 cm

**35** 32.5°

**36 a** (0, 3), (−6, 0)     **b** 26.6°

**37 a**

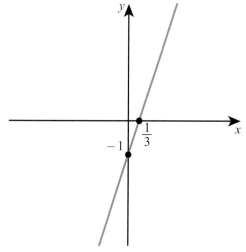

    **b** 71.6°

**38** 6.98 cm

**39** 58.5°, 78.5° (or 121.5° and 15.5°)

**40** 87.4°

**41 a** 18.6°        **b** 32.0

**42** 19.0

**43** $x = 11.9$ cm, $\theta = 15.4°$

**44** 38.7°

**45 b** 49.4°

**46** 82.2°

**47** 16.2 cm

**48** 5.69 cm

**49** 29.0°

**50** 47.0°

**51** 21.0°

**52** 5.63 cm

**53** 15.2

**54** $\theta = 52.0°$, $AB = 8.70$

**55** 9.98

**56** $h = \dfrac{d \tan 40}{\tan 50 - \tan 40}$

**57 a** $x = \dfrac{h}{\tan 30}$, $y = \dfrac{4 - h}{\tan 10}$

    **b** 3.73

# Exercise 5C

**1 a** 35.5°        **b** 59.0°

**2 a** 16.7°        **b** 51.3°

**3 a** 55.6°        **b** 18.9°

**4 a** 48.0°        **b** 82.6°

**5 a** 68.9°        **b** 59.8°

**6 a** 67.4°        **b** 41.6°

**7 a** 49.3°        **b** 48.2°

**8 a** 69.3°        **b** 46.6°

**9 a** 2.73 cm      **b** 8.22 m

**10 a** 36.9°       **b** 42.8°

**11 a** 36.6°       **b** 46.5°

**12 a** 31.2°       **b** 15.0°

**13 a** 45.3°       **b** 45.3°

**14 a** 50.6°       **b** 61.5°

**15 a** 17.8°       **b** 67.4°

**16 a** 54.2°       **b** 43.4°

**17 a** 1.16 km, 243°     **b** 2.71 km, 321°

**18 a** 4.00 km, 267°     **b** 7.62 km, 168°

**19 a** 3.97 km, 317°     **b** 16.5 km, 331°

**20** 57.1 m

**21 a** 15.8        **b** 18.4°

**22 a**

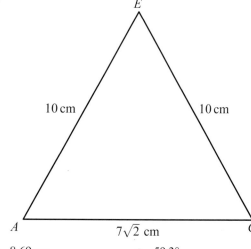

    **b** 8.69 cm     **c** 59.3°

**23** 56.3°

**24 a**

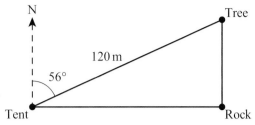

**b** 67.1 m

**25 a**

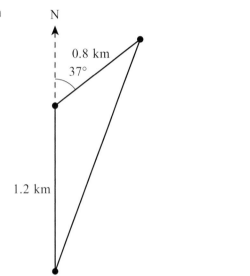

**b** 1.90 km

**26** 13.8 m

**27** 42.3 m

**28** 3.18 m

**29** 55.6 m

**30 a** 12.1 cm  **b** 35.4°  **c** 41.9°

**31 a** 13.02 m  **b** 3.11°

**32 a** 7.81 cm  **b** 45.2°

**33** 191 m, 273°

**34** 2.92 km, 008.6°

**35** 146 m

**36 a** 9.51 m  **b** 10.9 m  **c** 48.3°

**37** 3.96 m

**38 a** 9 – assumes lengths given are internal or thin glass or no large objects in the space

  **b** 359 000 W

  **c** yes – maximum angle is 51.8°

**39 a** 2.45 m  **b** 42.9 m$^3$

  **c** 75.5%  **d** 94.0%

**40** 38.9 cm$^2$

**41** 39.9°

**42** 92.3 m

# Chapter 5 Mixed Practice

**1** 35.0 m

**2 a** $AC = 22.6$ cm,  $AG = 27.7$ cm

  **b**

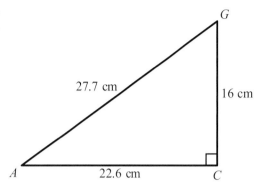

  **c** 35.3°

**3 a** 32.5 cm  **b** 24.1 cm  **c** 29.1 cm

**4** 45.2°

**5 a** 6  **b** $-\dfrac{1}{2}$  **c** 26.6°

**6 a** 12.9 cm

  **b** 80.5°

**7** 12.3 cm

**8** 3.07 cm

**9** 21.8°

**10** 32.4°

**11** 37.4°

**12** 18.3 cm

**13** 55.4 m

**14** 17.9 m

**15 a**

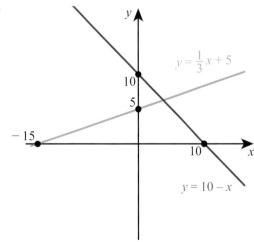

  **b** (3.75, 6.25)

  **c** 63.4°

**16** 23 900 cm³

**17** a  65.9°          b  21.0°

**18** 5.46

**19** 21

**20** 17.6 m

**21** a  131 cm³        b  313 cm³

**22** a  5.12 cm        b  9.84 cm          c  31.4°

**23** a  5 cm

  b  9.43 cm

  c  58.0°

  a  720 m

  b  72 300 m²

  c  88.3°

**24** a  i  22.5 m

    ii

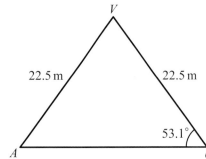

  c  27.0 m

  e  41 600 m³

  f  44 900 kg

**25** 22.5

# Chapter 6 Prior Knowledge

**1** a  4.75      b  5.5      c  6        d  9

**2** a  1.5      b  (0, 7)

# Exercise 6A

**1** a  9 lions, 21 tigers

  b  9 strawberry, 11 chocolate

**2** a  24 boys, 16 girls

  b  18 HL, 27 SL

**3** a  10 football, 14 hockey, 16 basketball

  b  6 cod, 9 haddock, 5 mackerel

**4** a  24 chairs, 9 tables, 4 beds

  b  8 oak, 7 willow, 4 chestnut

**5** a  all households in Germany

  b  convenience sampling

  c  e.g. Households in the city may have fewer pets than in the countryside.

**6** a  the number/proportion of pupils in each year group

  b  quota sampling

  c  i  keep

    ii  discard

**7** a  convenience sampling

  b  all residents of the village

  c  e.g. People using public transport may have different views from those who drive.

  d  would need access to all the residents

**8** Not necessarily correct, it could just be an extreme sample.

**9** a  continuous

  b  They would all be destroyed.

  c  list in serial number order, select every 20th

**10** a  quota

  b  more representative of the scarves sold

  c  12 red, 12 green, 10 blue, 6 white

**11** a  quota

  b  more representative of the population

  c  difficult to compile a list of all the animals

**12** a  continuous

  b  Basketball team are likely to be taller than average.

  c  systematic sampling

  d  Some samples not possible, e.g. it is not possible to select two students adjacent on the list.

**13** 5 cats, 9 dogs and 6 fish

**14** a

| Gender/age | 12 | 13 | 14 |
|---|---|---|---|
| **Boys** | 4 | 5 | 5 |
| **Girls** | 0 | 4 | 2 |

  b  no

**15** a  discrete

  b  no

  c  convenience

  d  e.g. Different species may live in different parts of the field.

**16** a  i  possible

    ii  not necessarily true

    iii  possible/quite likely

   b  iii  would be unlikely

   c  discard –32, keep 155

**17** a  20%

   b  i  84

      ii  65%

**18** a  100 million

   b  e.g. No fish die or are born; captured cod mix thoroughly with all other cod.

# Exercise 6B

**1** a  $16.5 \leqslant t < 18.5$; $20.5 \leqslant t \leqslant 22.5$; $22.5 \leqslant t < 24.5$; 3, 12, 3, 1, 1
      20 observations

   b  $14.5 \leqslant t < 17.5$; $20.5 \leqslant t < 23.5$; 10, 8, 2
      20 observations

**2** a  $300 \leqslant n \leqslant 499$; $500 \leqslant n \leqslant 699$; $700 \leqslant n \leqslant 899$; 11
      40 observations

   b  $200 \leqslant n \leqslant 399$; $400 \leqslant n \leqslant 599$; $600 \leqslant n \leqslant 899$; 20
      40 observations

**3** a  mean = 3.86, med = 3, mode = 1

   b  mean = 5.57, med = 6, mode = 6

**4** a  mean = 4.7, med = 4.5, mode = 3

   b  mean = 31.2, med = 24.5, mode = 24

**5** a  $x = 10$          b  $x = 7$

**6** a  $x = \dfrac{31}{9}$       b  $x = 6.1$

**7** a  $y = 10$          b  $y = 52$

**8** a  $y = 4.6$        b  $y = 2.2$

**9** a  i  13          b  i  20.5

      ii  $0 \leqslant x < 10$      ii  $25 \leqslant x < 35$

**10** a  i  13.5         b  i  10.2

      ii  $12.5 \leqslant x < 14.5$   ii  $11.5 \leqslant x < 15.5$

**11** a  10.4           b  4.96

**12** a  5.24           b  32.8

**13** a  $x < 5$ or $x > 125$   b  $x < 10$ or $x > 26$

**14** a  $x < 4.5$, $Q_3 > 12.5$  b  $x < 15.5$ or $x > 63.5$

**15** a  mean = 46, sd = 8

    b  mean = 62, sd = 18

**16** a  median = 25, IQR = 13

    b  median = 0.5, IQR = 2.1

**17** a  med = 360, IQR = 180

    b  med = 5, IQR = 2.6

**18** a  mean = 8, sd = 3

    b  mean = 41, sd = 3

**19** a  48         b  19         c  6.17, 1.31

**20** a  31         b  $1.4 \leqslant m < 1.6$  c  1.34

**21** a  28

    b  $15.5 \leqslant t < 17.5$

    c  17.3 s; used midpoints, actual times not available

**22** a  4.25

    b  1.5

    c  The second artist's songs are longer on average and their lengths are more consistent.

**23** a  4         b  3         c  yes (11)

**24** a  67

    b  23 cm

    c  24.7 cm

    d

| $20.5 \leqslant l < 23.5$ | $23.5 \leqslant l < 26.5$ | $26.5 \leqslant l < 29.5$ |
|---|---|---|
| 23 | 30 | 14 |

    e  44

    f  no

**25** a  $a = 12.5$         b  13.4

**26** a  $x = 5.5$         b  30.9

**27** 63.8

**28** 62

**29** a  $\dfrac{161 + 24a}{50}$      b  $a = 10$

**30** a  $x = 9.5$

    b  $Q_1 = 5$, $Q_2 = 5$, $Q_3 = 6.5$

    c  yes

**31** a  £440, £268     b  median

    c  $576, $351

**32** a  mean = 4, sd = 2.94

    b  mean = 2012, sd = 8.83

**33** 10.6°C, 2°C

**34** mean = 9.18 km, var = 11.9 km$^2$

**35** med = –75, IQR = 42

**36** a  med = 42, IQR = 8    b  $m = 58$

**37** 74%

**38** a  $p = 12$         b  0.968

**39** $p = 7$, $q = 4$

**40** $c = 30$

**41** 10

**42** (4, 8) or (5, 7)

# Exercise 6C

You might find that different GDCs or programs give slightly different values for the quartiles, resulting in slightly different answers from those given here. In an exam, all feasible answers would be allowed.

1  a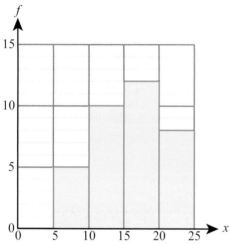

b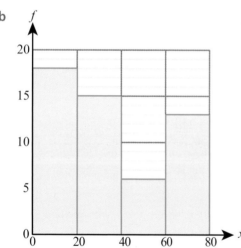

2  a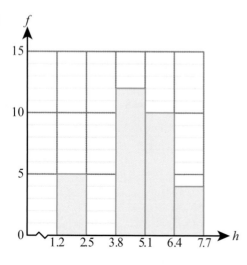

b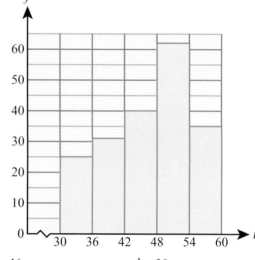

3  a  16            b  30

4  a  23            b  4

5  a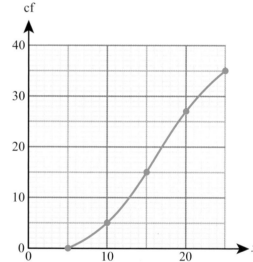

b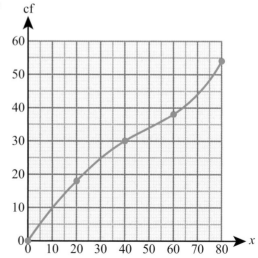

**6  a**

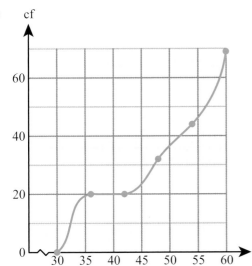

**b**

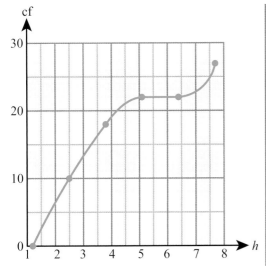

**7  a  i**  13

  **ii**  20

  **iii**  6

  **iv**  21

  **b  i**  4.6

  **ii**  6.5

  **iii**  2.3

  **iv**  6.6

**8  a**

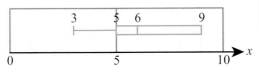

  **b**

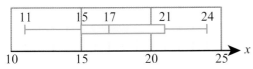

**9  a**

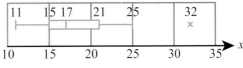

  **b**

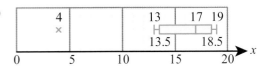

**10  a**

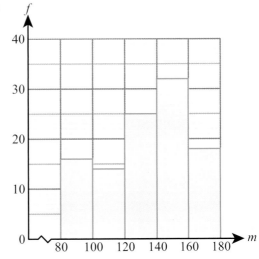

  **b**  25

**11  a**  105

  **b**

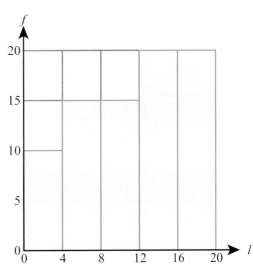

  **c**  46%

**12 a**

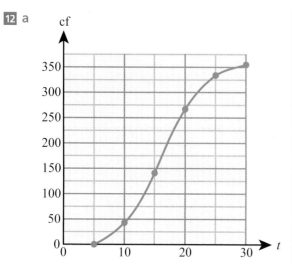

**b i** around 17°C

**ii** around 7°C

**13 a** 45

**b** 22.2%

**c**

| Time (min) | $5 \leqslant t < 10$ | $10 \leqslant t < 15$ | $15 \leqslant t < 20$ | $20 \leqslant t < 25$ | $25 \leqslant t < 30$ |
|---|---|---|---|---|---|
| Freq | 7 | 9 | 15 | 10 | 4 |

**d** 16.9 minutes

**14 a**

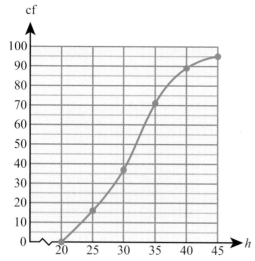

**b** median ≈ 31 cm, IQR ≈ 9 cm

**c**

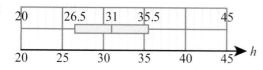

**15 a**

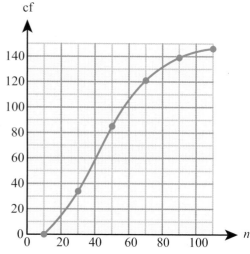

**b** 40

**c** 45, 32

**d**

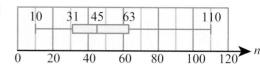

**e** Overall, fewer candidates take History SL. History has a larger spread of numbers than maths (based on the IQR).

**16** 159 cm

**17 a** 160          **b** 90%

**c** 75          **d** 72, 18

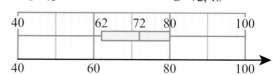

**e** The second school has a higher median score, but more variation in the scores.

**18 a** $Q_1 = 1, Q_2 = 2, Q_3 = 3$

**b** 2

**c** 7 is an outlier

**d**

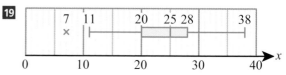

**19**

**20 a**

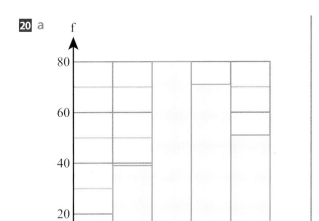

**b** 59.9

**20** A2, B3, C1

# Exercise 6D

**1 a**

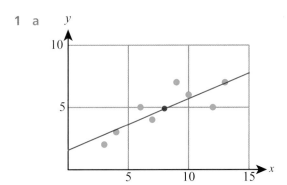

**b**

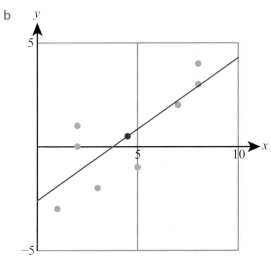

**2 a**

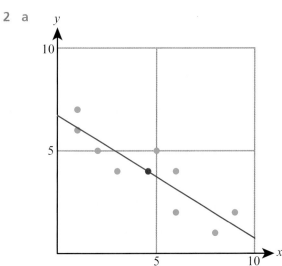

**b**

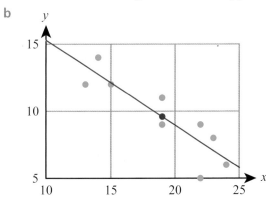

**3 a**

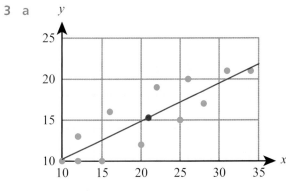

**b**

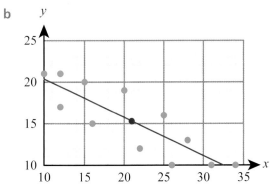

**4 a** weak positive      **b** strong positive

**5 a** strong negative      **b** weak negative

**6 a** no linear correlation (circle)

   **b** no linear correlation (V-shape)

**7 a** no correlation (slightly scattered around a vertical line)

   **b** no correlation (slightly scattered around a horizontal line)

**8 a i** $r = 0.828$      **b i** $r = -0.886$

     **ii** strong positive      **ii** strong negative

     **iii** yes            **iii** yes

**9 a i** $r = 0.542$

     **ii** weak/moderate positive

     **iii** no

   **b i** $r = -0.595$

     **ii** weak/moderate negative

     **iii** yes

**10 a** $y = 0.413x + 1.57$    **b** $y = 0.690x - 2.60$

**11 a** $y = -0.598x + 6.72$    **b** $y = -0.632x + 21.6$

**12 a i** $y = 21.9$         **ii** reliable

   **b i** $y = -4.26$

     **ii** not reliable (extrapolation)

**13 a i** $y = 73.5$

     **ii** not reliable (no correlation)

   **b i** $y = 87.9$

     **ii** not reliable (no correlation)

**14 a, c, d**

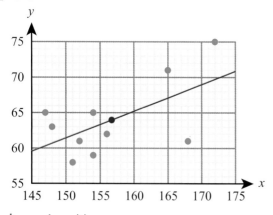

   **b** weak positive

   **c** $156.7\,\text{cm}, 64\,\text{cm}$

   **e** arm length $\approx 61.5\,\text{cm}$

   **f i** appropriate

     **ii** not appropriate (extrapolation)

     **iii** not appropriate (different age from sample)

**15 a, d**

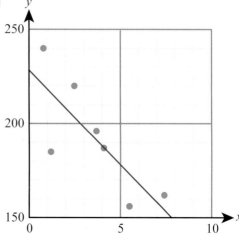

   **b** weak negative

   **c** $3.6\,\text{km}, \$192\,000$

   **e** average price $\approx \$161\,000$

**16** A1, B2, C3, D4

**17 a** 0.688            **b** $y = 0.418x + 18.1$

   **c** 46.5 (or 47)

   **d** no (correlation does not imply causation)

**18 a** 0.745

   **b** The larger the spend on advertising, the larger the profit.

   **c** $y = 10.8x + 188$

   **d i** \$1270      **ii** \$2350

   **e** the first one (no extrapolation required)

**19 a** $-0.0619$

   **b** $y = -0.037x + 51.6$

   **c** no (no correlation)

**20 a**

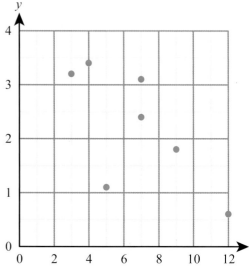

   **b** $-0.695$

c  It shows statistically significant negative correlation.

**21 a**

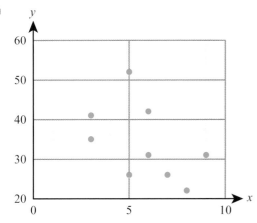

b  weak negative       c  −0.480

d  not significant

**22 a**  moderate positive correlation

b  yes (there is correlation and value is within range of data)

c  40.8 cm

**23 a**  0.820. There is a moderate positive correlation between time spent practising and test mark.

b  $m = 0.631t + 5.30$

c  For every extra minute practice he can expect 0.631 extra marks. With no practice he can expect around 5 marks.

**24 a**  Positive correlation: the larger the advertising, budget, the larger the profit.

b  no (correlation does not imply cause)

c  i  For each €1000 euros spent on advertising, the profit increases by €3250.

ii  With no advertising the profit would be €13 800.

**25 a**

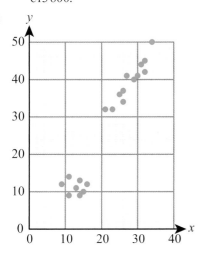

b  summer and winter

c  positive correlation in the summer, no correlation in winter

d  39.5

**26 a**

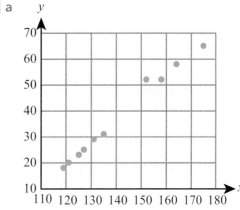

b  children and adults

c  children (126, 24.3), adults (162, 56.8)

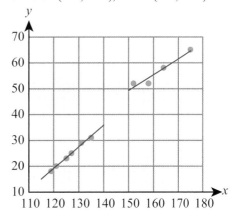

d  19.0 kg

**27 a**  $LQ(x) = 12$, $UQ(x) = 24$, $LQ(y) = 11$, $UQ(y) = 19$

c, d, e

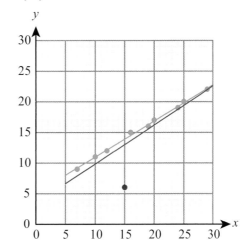

# Chapter 6 Mixed Practice

**1** **a** **i** systematic sampling

**ii** e.g. Students may take books out on the same day each week.

**b** **i** Each possible sample of 10 days has an equal chance of being selected.

**ii** Representative of the population of all days.

**c** **i** 11

**ii** 17.4

**iii** 3.17

**2** **a**

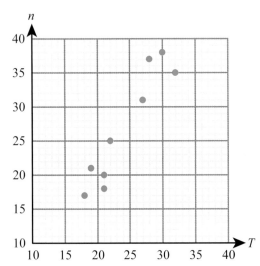

**b** strong positive correlation

**c** $n = 1.56T - 10.9$

**d** 29.7

**3** **a** 2.38 kg

**b**

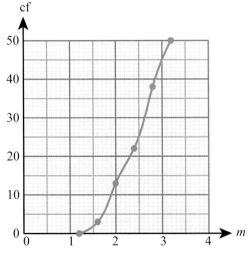

**c** 2.5 kg, 0.9 kg

**d**

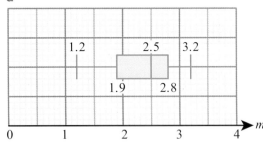

**4** **a** discrete

**b** 0

**c** **i** 1.47

**ii** 1.5

**iii** 1.25

**5** **a** 4

**b**

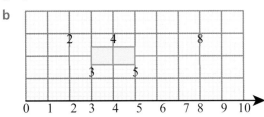

Number of books read

**c** 10

**6** **a** 0.996

**b** $a = 3.15$, $b = -15.4$

**c** 66.5

**7** **a**

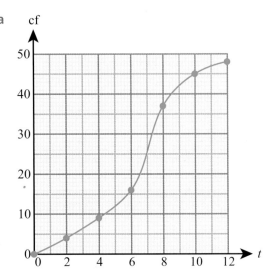

**b** **i** 6.9 minutes

**ii** 2.9 minutes

**iii** 9.3 minutes

c

| Time | Freq |
|---|---|
| $0 \leqslant t < 2$ | 4 |
| $2 \leqslant t < 4$ | 5 |
| $4 \leqslant t < 6$ | 7 |
| $6 \leqslant t < 8$ | 21 |
| $8 \leqslant t < 10$ | 8 |
| $10 \leqslant t < 12$ | 3 |

d 6.38 minutes

**8** a med = 46, $Q_1 = 33.5$, $Q_3 = 56$

b yes (93)

c

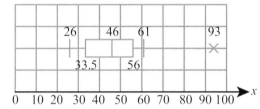

**9** a 121 cm, 22.9 cm$^2$

b 156 cm, 22.9 cm$^2$

**10** mean = $55.02, sd = $43.13

**11** $x = 6$

**12** a

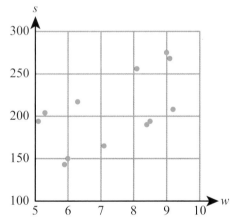

b $w = 19.1s + 99.0$

c 0.994 – strong positive correlation

d 252 g

e 137 g to 175 g; extrapolating from the data so not reliable.

**13** a i Athletes generally do better after the programme.

ii Better athletes improve more.

b 11.6 miles

c i 0.84

ii $Y = 1.2X + 3.2$

# Chapter 7 Prior Knowledge

**1** a $\{2, 3, 4, 5, 6, 7, 8, 9\}$

b $\{5, 8\}$

**2**

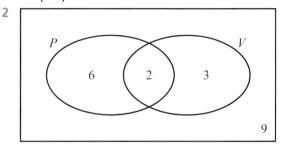

**3**

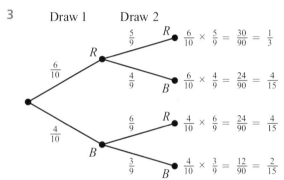

# Exercise 7A

**1** a $\dfrac{3}{10}$     b $\dfrac{2}{15}$

**2** a $\dfrac{1}{5}$     b $\dfrac{3}{5}$

**3** a $\dfrac{1}{2}$     b $\dfrac{1}{3}$

**4** a $\dfrac{1}{4}$     b $\dfrac{1}{13}$

**5** a $\dfrac{1}{26}$     b $\dfrac{3}{26}$

**6** a 0.94     b 0.55

**7** a $\dfrac{11}{20}$     b $\dfrac{23}{40}$

**8** a 0.85     b 0.13

**9** a $\dfrac{47}{120}$     b $\dfrac{41}{48}$

**10** a 0.73     b 0.66

**11** a 0.44     b 0.11

**12** a 4     b 27

**13** a 12     b 6

**14** a 4.8     b 7.5

**15 a**  1.6                    **b**  1.5
**16 a**  0.0743                 **b**  66.9
**17**  7.5
**18**  15
**19**  8
**20**  0.75
**21**  $1162.50
**22 a**  1.5                    **b**  0.8
**23**  $\dfrac{1}{16}$
**24**  $\dfrac{78}{25}$

# Exercise 7B

**1**

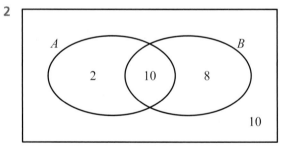

**a**  $\dfrac{3}{25}$                    **b**  $\dfrac{4}{25}$

**2**

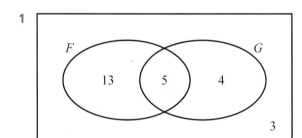

**a**  $\dfrac{4}{15}$                    **b**  $\dfrac{1}{3}$

**3**

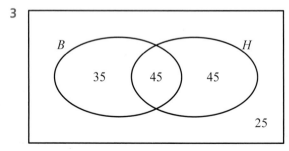

**a**  $\dfrac{7}{30}$                    **b**  $\dfrac{3}{10}$

**4**

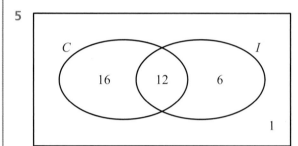

**a**  $\dfrac{37}{44}$                    **b**  $\dfrac{7}{11}$

**5**

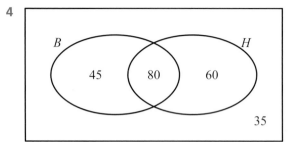

**a**  $\dfrac{6}{35}$                    **b**  $\dfrac{16}{35}$

**6**

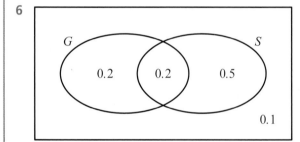

**a**  0.1                    **b**  0.5

**7**

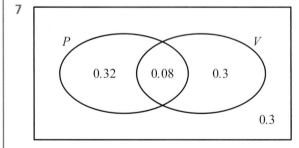

**a**  0.3                    **b**  0.32

**8**

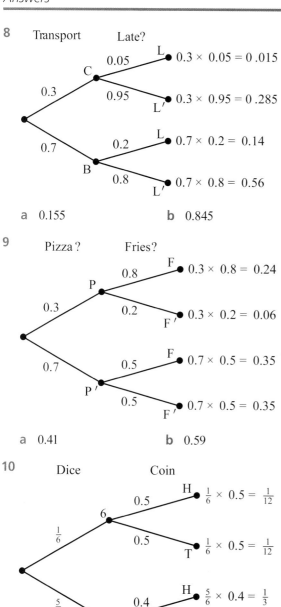

Transport   Late?

C   
0.3   
0.05 → L • $0.3 \times 0.05 = 0.015$   
0.95 → L′ • $0.3 \times 0.95 = 0.285$

0.7   
B   
0.2 → L • $0.7 \times 0.2 = 0.14$   
0.8 → L′ • $0.7 \times 0.8 = 0.56$

a  0.155        b  0.845

**9**

Pizza?   Fries?

P   
0.3   
0.8 → F • $0.3 \times 0.8 = 0.24$   
0.2 → F′ • $0.3 \times 0.2 = 0.06$

0.7   
P′   
0.5 → F • $0.7 \times 0.5 = 0.35$   
0.5 → F′ • $0.7 \times 0.5 = 0.35$

a  0.41        b  0.59

**10**

Dice   Coin

$\frac{1}{6}$   
6   
0.5 → H • $\frac{1}{6} \times 0.5 = \frac{1}{12}$   
0.5 → T • $\frac{1}{6} \times 0.5 = \frac{1}{12}$

$\frac{5}{6}$   
6′   
0.4 → H • $\frac{5}{6} \times 0.4 = \frac{1}{3}$   
0.6 → T • $\frac{5}{6} \times 0.6 = \frac{1}{2}$

a  $\frac{5}{12}$        b  $\frac{7}{12}$

**11**

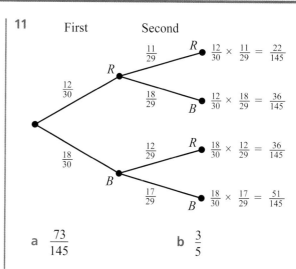

First   Second

$\frac{12}{30}$   
R   
$\frac{11}{29}$ → R • $\frac{12}{30} \times \frac{11}{29} = \frac{22}{145}$   
$\frac{18}{29}$ → B • $\frac{12}{30} \times \frac{18}{29} = \frac{36}{145}$

$\frac{18}{30}$   
B   
$\frac{12}{29}$ → R • $\frac{18}{30} \times \frac{12}{29} = \frac{36}{145}$   
$\frac{17}{29}$ → B • $\frac{18}{30} \times \frac{17}{29} = \frac{51}{145}$

a  $\frac{73}{145}$        b  $\frac{3}{5}$

**12**

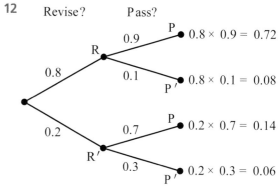

Revise?   Pass?

0.8   
R   
0.9 → P • $0.8 \times 0.9 = 0.72$   
0.1 → P′ • $0.8 \times 0.1 = 0.08$

0.2   
R′   
0.7 → P • $0.2 \times 0.7 = 0.14$   
0.3 → P′ • $0.2 \times 0.3 = 0.06$

a  0.72        b  0.14

**13**

|  |  | First Dice | | | | | |
|---|---|---|---|---|---|---|---|
|  |  | **1** | **2** | **3** | **4** | **5** | **6** |
| **Second Dice** | **1** | 2 | 3 | 4 | 5 | 6 | 7 |
|  | **2** | 3 | 4 | 5 | 6 | 7 | 8 |
|  | **3** | 4 | 5 | 6 | 7 | 8 | 9 |
|  | **4** | 5 | 6 | 7 | 8 | 9 | 10 |
|  | **5** | 6 | 7 | 8 | 9 | 10 | 11 |
|  | **6** | 7 | 8 | 9 | 10 | 11 | 12 |

a  $\frac{5}{36}$        b  $\frac{1}{6}$

**14**

| | | First Dice | | | | | |
|---|---|---|---|---|---|---|---|
| | | **1** | **2** | **3** | **4** | **5** | **6** |
| Second Dice | **1** | = | > | > | > | > | > |
| | **2** | < | = | > | > | > | > |
| | **3** | < | < | = | > | > | > |
| | **4** | < | < | < | = | > | > |
| | **5** | < | < | < | < | = | > |
| | **6** | < | < | < | < | < | = |

a $\dfrac{1}{6}$  b $\dfrac{5}{12}$

**15**

| | H | T |
|---|---|---|
| **H** | H, H | T, H |
| **T** | H, T | T, T |

a $\dfrac{1}{4}$  b $\dfrac{1}{2}$

**16**

| | B | G |
|---|---|---|
| **B** | B, B | G, B |
| **G** | B, G | G, G |

a $\dfrac{1}{2}$  b $\dfrac{1}{4}$

**17**

| *AILT* | *IALT* | *LAIT* | *TAIL* |
|---|---|---|---|
| *AITL* | *IATL* | LATI | TALI |
| ALIT | ILAT | *LIAT* | *TIAL* |
| ALTI | ILTA | LITA | TILA |
| ATIL | ITAL | *LTAI* | *TLAI* |
| ATLI | ITLA | *LTIA* | *TLIA* |

a $\dfrac{1}{6}$  b $\dfrac{1}{2}$

**18 a** $\dfrac{38}{109}$  **b** $\dfrac{55}{109}$

**19 a** $\dfrac{3}{19}$  **b** $\dfrac{25}{57}$

**20 a** $\dfrac{18}{77}$  **b** $\dfrac{16}{77}$

**21 a** $\dfrac{23}{65}$  **b** $\dfrac{13}{23}$

**22 a** 0.2  **b** 0.3

**23 a** 0.8  **b** 0.9

**24 a** 0.5  **b** 0.7

**25 a** 0.2  **b** 0.1

**26 a** $\dfrac{11}{15}$  **b** $\dfrac{17}{20}$

**27 a** $\dfrac{19}{49}$  **b** $\dfrac{30}{49}$

**28 a** $\dfrac{25}{51}$  **b** $\dfrac{1}{17}$

**29 a** $\dfrac{3}{31}$  **b** $\dfrac{18}{53}$

**30 a** $\dfrac{16}{39}$  **b** $\dfrac{4}{9}$

**31 a** $\dfrac{5}{18}$  **b** $\dfrac{5}{9}$

**32 a** $\dfrac{5}{6}$  **b** $\dfrac{5}{9}$

**33 a** $\dfrac{7}{12}$  **b** $\dfrac{9}{14}$

**34 a** $\dfrac{25}{37}$  **b** $\dfrac{28}{37}$

**35 a** $\dfrac{1}{7}$  **b** $\dfrac{1}{17}$

**36 a** 0.5  **b** $\dfrac{2}{7}$

**37 a** $\dfrac{19}{35}$  **b** $\dfrac{4}{7}$

**38 a** 0.14  **b** 0.24

**39 a** 0.3  **b** 0.4

**40 a**

| 1 | 2 | 3 | 4 | 5 | 6 |
|---|---|---|---|---|---|
| 2 | 4 | 6 | 8 | 10 | 12 |
| 3 | 6 | 9 | 12 | 15 | 18 |
| 4 | 8 | 12 | 16 | 20 | 24 |
| 5 | 10 | 15 | 20 | 25 | 30 |
| 6 | 12 | 18 | 24 | 30 | 36 |

**b** 40

**41** $\dfrac{13}{16}$

**42 a**

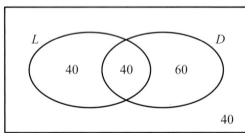

**b** $\dfrac{2}{9}$  **c** $\dfrac{1}{2}$

**43** 0.497

**44 a i** $\dfrac{1}{8}$  **ii** $\dfrac{1}{8}$

**b** yes

**45** a

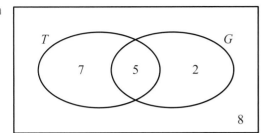

b $\dfrac{7}{22}$   c $\dfrac{5}{7}$

**46** a 47   b $\dfrac{83}{130}$   c $\dfrac{33}{71}$

**47** 0.582

**48** 0.75

**49** a $\dfrac{1}{15}$   b 0.88

**50** $\dfrac{1}{36}$

**51** a $\dfrac{7}{17}$   b $\dfrac{79}{187}$   c $\dfrac{26}{187}$

**52** 0.5

**53** 0.5

**54** a $\dfrac{1}{5}$   b $\dfrac{11}{20}$

**55** a $\dfrac{13}{23}$   b $\dfrac{52}{161}$   c same

**56** a 0.226   b 0.001 73

**57** a

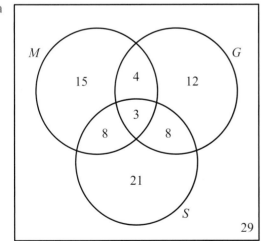

b 0.21
c $\dfrac{11}{40}$

**58** a and b

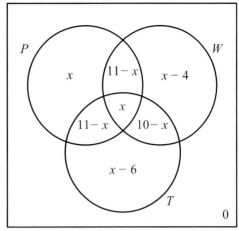

c $\dfrac{4}{15}$   d $\dfrac{11}{15}$   e $\dfrac{1}{4}$

**59** a $\dfrac{1}{3}$   b $\dfrac{1}{2}$

# Chapter 7 Mixed Practice

**1** 416

**2** a $\dfrac{56}{115}$   b $\dfrac{21}{115}$   c $\dfrac{32}{59}$   d $\dfrac{21}{32}$

**3** a $\dfrac{2}{15}$   b $\dfrac{37}{60}$

**4** a

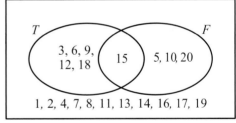

b i $\dfrac{1}{5}$   ii $\dfrac{3}{14}$

**5** a 0.2   b $\dfrac{2}{3}$

**6** a 0.18   b yes

**7** 0.92

**8** a

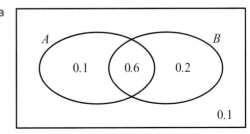

b 0.75   c 0.667

**9** a   i   $\dfrac{2}{9}$      ii   $\dfrac{1}{18}$      b   $\dfrac{5}{18}$

**10** a   0.143      b   0.111      c   0.238

**11** a

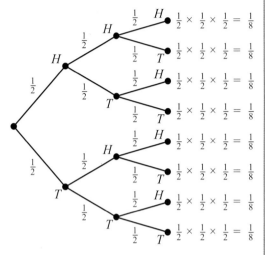

Toss 1    Toss 2    Toss 3

b   $\dfrac{1}{8}$      c   $\dfrac{7}{8}$      d   $\dfrac{3}{8}$

**12** Asher $\left(\dfrac{48}{91} > \dfrac{24}{49}\right)$

**13** 0.310

**14** a   $\dfrac{8}{23}$      b   $\dfrac{13}{23}$

**15** a   $\dfrac{1}{4}$      b   $\dfrac{7}{66}$      c   $\dfrac{25}{66}$

**16** a

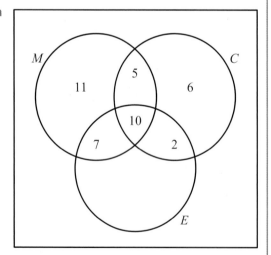

b   16

c   i   3      ii   56

d   i   0.22      ii   0.05

    iii   0.62      iv   $\dfrac{31}{39}$

**17** b   6 or 15

**18** a   15%      b   60%

    c   i   0.442      d   0.642

**19** a, b

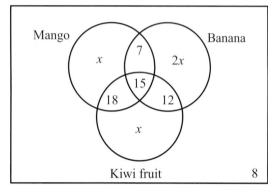

c   $x = 10$

d   i   50      ii   82

e   i   0.08      ii   0.37      iii   $\dfrac{15}{22}$

f   $\dfrac{14}{2475}$

# Chapter 8 Prior Knowledge

**1** a   $\dfrac{1}{3}$      b   $\dfrac{8}{15}$

**2**

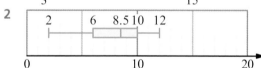

# Exercise 8A

**1** a

| $x$ | 0 | 1 | 2 |
|---|---|---|---|
| $P(X = x)$ | $\dfrac{2}{7}$ | $\dfrac{4}{7}$ | $\dfrac{1}{7}$ |

b

| $x$ | 0 | 1 | 2 |
|---|---|---|---|
| $P(X = x)$ | $\dfrac{4}{15}$ | $\dfrac{8}{15}$ | $\dfrac{1}{5}$ |

**2** a

| $x$ | 0 | 1 | 2 |
|---|---|---|---|
| $P(X = x)$ | $\dfrac{1}{4}$ | $\dfrac{1}{2}$ | $\dfrac{1}{4}$ |

b

| $x$ | 0 | 1 | 2 |
|---|---|---|---|
| $P(X = x)$ | $\dfrac{1}{4}$ | $\dfrac{1}{2}$ | $\dfrac{1}{4}$ |

**3 a**

| $x$ | 0 | 1 | 2 | 3 |
|---|---|---|---|---|
| P($X = x$) | $\frac{1}{8}$ | $\frac{3}{8}$ | $\frac{3}{8}$ | $\frac{1}{8}$ |

**b**

| $x$ | 0 | 1 | 2 | 3 |
|---|---|---|---|---|
| P($X = x$) | $\frac{1}{8}$ | $\frac{3}{8}$ | $\frac{3}{8}$ | $\frac{1}{8}$ |

**4 a** P($X = x$) $= \frac{1}{6}$ for $x = 1, 2, ..., 6$

**b** P($X = x$) $= \frac{1}{8}$ for $x = 1, 2, ..., 8$

**5 a**

| $x$ | 0 | 1 | 2 | 3 |
|---|---|---|---|---|
| P($X = x$) | $\frac{125}{216}$ | $\frac{25}{72}$ | $\frac{5}{72}$ | $\frac{1}{216}$ |

**b**

| $x$ | 0 | 1 | 2 | 3 |
|---|---|---|---|---|
| P($X = x$) | $\frac{343}{512}$ | $\frac{147}{512}$ | $\frac{21}{512}$ | $\frac{1}{512}$ |

**6 a**

| $x$ | 0 | 1 | 2 |
|---|---|---|---|
| P($X = x$) | 0.64 | 0.32 | 0.04 |

**b**

| $x$ | 0 | 1 | 2 |
|---|---|---|---|
| P($X = x$) | 0.09 | 0.42 | 0.49 |

**7 a** $k = 0.17$
   **i** 0.77     **ii** 0.416
  **b** $k = 0.26$
   **i** 0.49     **ii** 0.245

**8 a** $k = 0.2$
   **i** 0.4     **ii** 0.75
  **b** $k = 0.36$
   **i** 0.43     **ii** 0.814

**9 a** $k = 0.125$
   **i** 0.6     **ii** 0.833
  **b** $k = 0.2$
   **i** 0.9     **ii** 0.667

**10 a** 2.1    **b** 1.9

**11 a** 2.1    **b** 2.24

**12 a** 5.5    **b** 9.2

**13 a** $k = 0.4$    **b** 0.7    **c** 2.9

**14 a** $k = 0.2$    **b** 0.4    **c** 6.2

**15 a**

| $x$ | 0 | 1 | 2 |
|---|---|---|---|
| $p$ | $\frac{4}{13}$ | $\frac{48}{91}$ | $\frac{15}{91}$ |

  **b** 0.857

**16 a** $\frac{1}{17}$

  **b**

| $h$ | 0 | 1 | 2 |
|---|---|---|---|
| P($H = h$) | $\frac{19}{34}$ | $\frac{13}{34}$ | $\frac{1}{17}$ |

  **c** 0.5

**17** no

**18** $n = 1$

**19** \$1.50

**20 a**    Toss 1    Toss 2    Toss 3

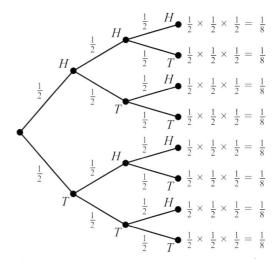

  **b** $\frac{3}{8}$

  **c**

| $x$ | 0 | 1 | 2 | 3 |
|---|---|---|---|---|
| P($X = x$) | $\frac{1}{8}$ | $\frac{3}{8}$ | $\frac{3}{8}$ | $\frac{1}{8}$ |

  **d** 1.5

**21 a**

| $x$ | 2 | 3 | 4 | 5 | 6 | 7 | 8 |
|---|---|---|---|---|---|---|---|
| P($X = x$) | $\frac{1}{16}$ | $\frac{1}{8}$ | $\frac{3}{16}$ | $\frac{1}{4}$ | $\frac{3}{16}$ | $\frac{1}{8}$ | $\frac{1}{16}$ |

  **b** 5

**22 a** 0.4    **b** 2.13

**23 a** $k = \frac{1}{12}$    **b** 0.75    **c** 5.17

**24 a** $c = 0.48$    **b** $\frac{3}{11}$    **c** 1.92

**25** $a = 0.3, b = 0.2$

# Exercise 8B

**1** **a** yes, $X \sim B\left(30,\frac{1}{2}\right)$

   **b** yes, $X \sim B\left(45,\frac{1}{6}\right)$

**2** **a** no; number of trials not constant
   **b** no; number of trials not constant

**3** **a** no; probability not constant
   **b** no; probability not constant

**4** **a** yes, $X \sim B(50, 0.12)$
   **b** yes, $X \sim B(40, 0.23)$

**5** **a** no; trials not independent
   **b** no; trials not independent

**6** **a** 0.160     **b** 0.180

**7** **a** 0.584     **b** 0.874

**8** **a** 0.596     **b** 0.250

**9** **a** 0.173     **b** 0.136

**10 a** 0.661     **b** 0.127

**11 a** 0.381     **b** 0.280

**12 a** 0.571     **b** 0.280

**13 a** 0.0792     **b** 0.231

**14 a** 0.882     **b** 0.961

**15 a** 0.276     **b** 0.001 69

**16 a** 6.25, 2.17     **b** 10, 2.58

**17 a** 5, 1.58     **b** 10, 2.24

**18 a** 5, 2.04     **b** 3.33, 1.67

**19 a** $B\left(10,\frac{1}{6}\right)$     **b** 0.291     **c** 0.225

**20 a** 0.292     **b** 0.736     **c** 10.2

**21 a** 0.160     **b** 0.872
   **c** 0.121     **d** 4.8

**22 a** 0.983     **b** 5     **c** 0.383

**23 a** All have same probability; employees independent of each other.
   **b** e.g. May not be independent, as could infect each other.
   **c** 0.0560
   **d** 0.0159

**24** 0.433

**25 a** 0.310     **b** 0.976     **c** 0.644

**26 a** 0.650     **b** 0.764

**27** 0.104

**28** 0.132

**29 a** 0.0296     **b** 0.000 876
   **c** e.g. A source of faulty components, such as a defective machine, would affect several components.

# Exercise 8C

**1** **a** 0.159     **b** 0.345

**2** **a** 0.726     **b** 0.274

**3** **a** 0.260     **b** 0.525

**4** **a** 0.523     **b** 0.244

**5** **a** 0.389     **b** 0.552

**6** **a** 0.246     **b** 0.252

**7** **a** 0.189     **b** 0.792

**8** **a** 0.133     **b** 0.132

**9** **a** 13.4     **b** 6.78

**10 a** 8.07     **b** 15.9

**11 a** 31.0     **b** 46.0

**12 a** 28.1     **b** 46.7

**13** no; not symmetrical

**14 a**

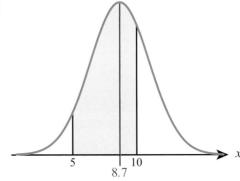

   **b** 0.660

**15 a** 0.465     **b** 0.0228

**16 a** 0.308     **b** 0.328

**17 a** 0.274     **b** 10.4     **c** 0.282

**18 a** 0.0478     **b** 6.42 minutes

**19 a** 0.0831     **b** 6.34 hours     **c** 62.3

**20 a** 4.53     **b** 6.47 m

**21** 12.8 s

**22 a** 15.4     **b** 6.74

**23** 4.61

**24** 20.9

**25** 15.2

**26** predicts 4% get a negative score

**27 a** symmetrical

b

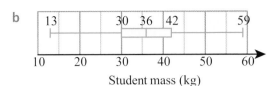

Student mass (kg)

**28** a   0.0228 e.g. anticipating the start gun

    b   0.159          c   0.487

    d   The same distribution is true in all races.
       Unlikely to be true.

**29** $728

# Chapter 8 Mixed Practice

**1** a   $k = 0.5$          b   0.6          c   2.9

**2** a   0.296          b   0.323

**3** a   0.347          b   0.227

**4** 215

**5** 0.346

**6** a   $\dfrac{1}{2}, \dfrac{1}{3}, \dfrac{1}{6}$

    b   no; expected outcome is not 0

**7** $N = 7$

**8** a   0.933

    b   i

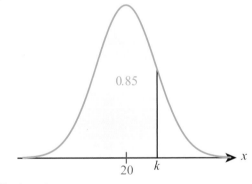

      ii   $k = 23.1$

**9** a   $p = \dfrac{1}{3}$          b   0.153          c   0.0435

**10** a   0.206          b   0.360

**11** a   0.356

     b   a box of six containing exactly one VL egg

**12** a   0.354          b   0.740

**13** 0.227

**14** 0.232

**15** a   3.52          b   0.06          c   0.456

**16** $a = \dfrac{1}{6}$

**17** a   10.8 cm          b   0.698%

**18** 2.28% of times would be negative.

**19** 320 ml

**20** a   $\dfrac{1}{9}$          b   0.0426

**21** a   0.925          b   $k = 20.4$

**22** a   0.835          b   $k = 1006.58$          c   $a = 6.58$

**23** a   i   1          ii   0.0579

    b   ii   $a = 0.05$, $b = 0.02$

    c   Bill (0.19)

**24** a   i   0.845          ii   1.69

    b   That Josie's second throw has the same
      distribution as the first and is independent.
      These seem unlikely.

**25** $n = 9$

**26** 10

**27** a   0.919          b   0.0561

**28** 277

**29** b   6

# Chapter 9 Prior Knowledge

**1**   $\dfrac{1}{4}x^{-1} - \dfrac{1}{2}x^{-2}$

**2**   $y = 3x + 11$

**3**   −2

# Exercise 9A

**1** a   0.6          b   0

**2** a   1.5          b   3

**3** a   2          b   3

**4** a   1          b   1

**5** a   0.693

     In further study you might learn how to show
     that this is actually ln 2

    b   1.10

**6** a   0.2          b   0

**7** a   0.5          b   0.5

**8** a   1          b   10

**9** a   −0.5          b   0

**10** a   0.805          b   0.347

**11** a   0          b   4

**12** a   6          b   0

**13** a   0.5          b   0.0833

14 a  −1                  b  −2

15 a  1                   b  0.693

16 a  $\dfrac{dz}{dv}$    b  $\dfrac{da}{db}$

17 a  $\dfrac{dp}{dt}$    b  $\dfrac{db}{dx}$

18 a  $\dfrac{dy}{dn}$    b  $\dfrac{dt}{df}$

19 a  $\dfrac{dh}{dt}$    b  $\dfrac{dw}{dv}$

20 a  $\dfrac{dw}{du}$    b  $\dfrac{dR}{dT}$

21 a  $\dfrac{dy}{dx} = y$   b  $\dfrac{dy}{dx} = \dfrac{x}{2}$

22 a  $\dfrac{ds}{dt} = kt^2$   b  $\dfrac{dq}{dp} = k\sqrt{q}$

23 a  $\dfrac{dP}{dt} = kL(P)$   b  $\dfrac{dP}{dt} = kA(P)$

24 a  $\dfrac{dh}{dx} = \dfrac{1}{h}$   b  $\dfrac{dr}{d\theta} = k$

25 a  $s'(t) = 7$   b  $q'(x) = 7x$

26 a  $\dfrac{d(pH)}{dT} = k$

   b  $\dfrac{dC}{dn} = k(1000 - n)$

   Technically, the number of items produced is not a continuous quantity so we should not formally use derivatives. However, as long as $n$ is large, as is usually the case in economic models, then approximating it as continuous is reasonable.

27 a  $\dfrac{dV}{dt} = V + 1$   b  5

28  57.3

29  1

30 a  $\dfrac{x^2 - 1}{x - 1}$

   b  2, the gradient of the curve $y = x^2$ at $x = 1$

31  3

32  $\dfrac{1}{6}$

# Exercise 9B

1 a  $x > 1$                    b  $x < 0$

2 a  $-1 < x < 1$               b  $x < -1$ or $x > 6$

3 a  $0 < x < 90$ or $270 < x < 360$

   b  $180 < x < 360$

4 a

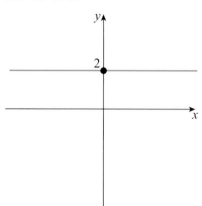

   b

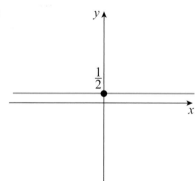

5 a

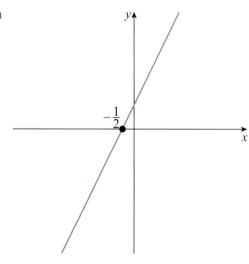

**b**

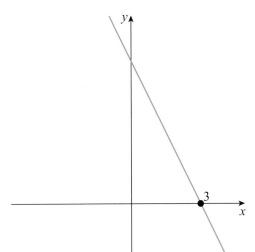

**6  a**

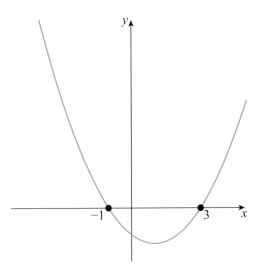

**b**

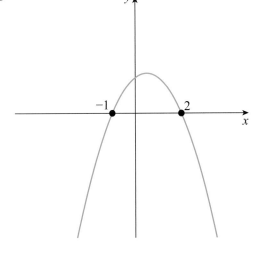

**7  a**

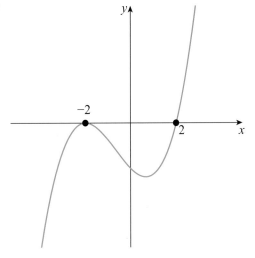

**b**

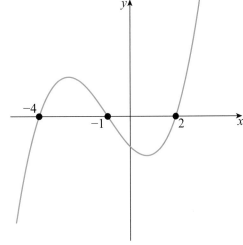

**8  a**

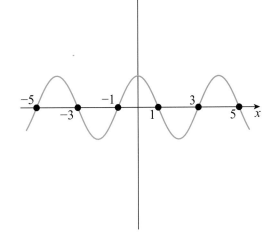

b

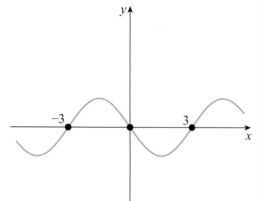

9  a

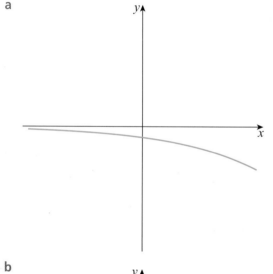

b

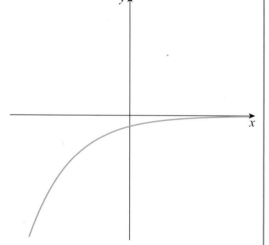

10 a   $x > -1$              b   $x < 3$
11 a   $x < -3$ or $x > 3$    b   $-4 < x < 0$

12 a   $x < -1$ or $x > 0$        b   $x < -\frac{1}{2}$ or $x > \frac{1}{2}$

13 a

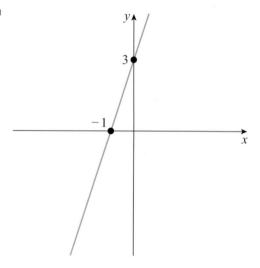

b

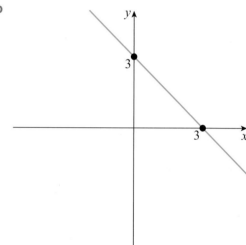

14 a

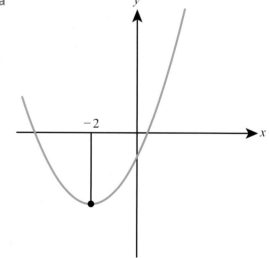

b

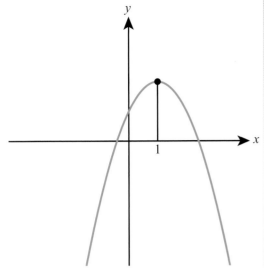

**15 a**

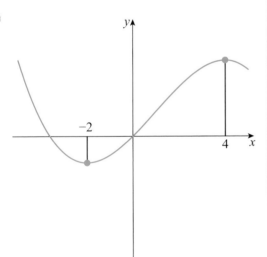

b

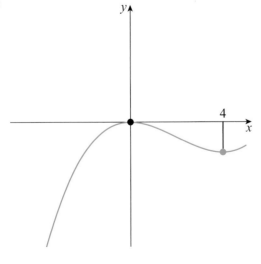

**16 a**

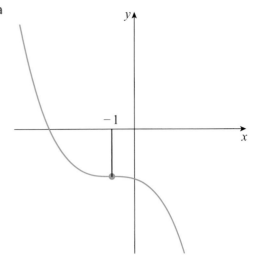

b

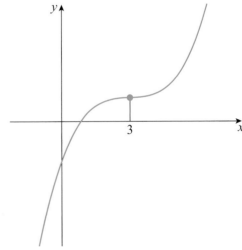

**17 a**

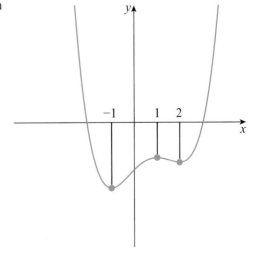

**b**

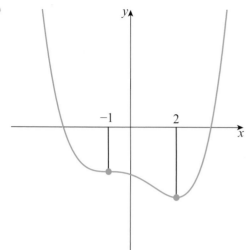

**18 a**

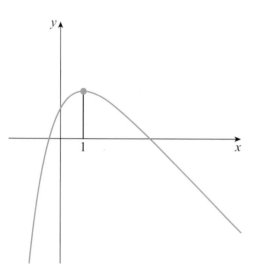

**b**

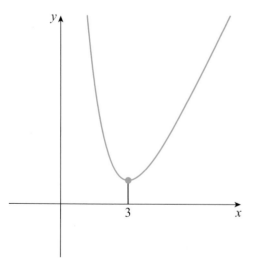

**19**

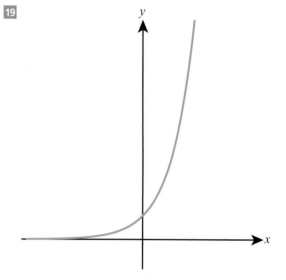

**20**

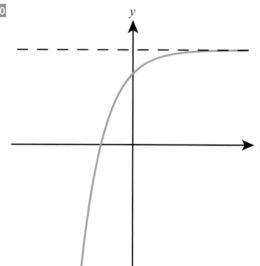

**21 a**

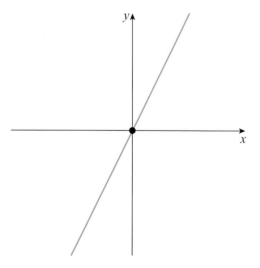

**b**

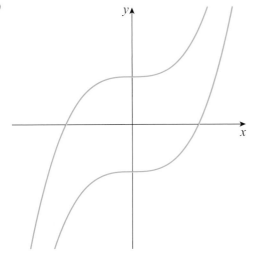

**22 a**

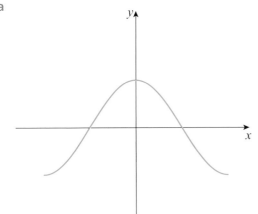

**b**

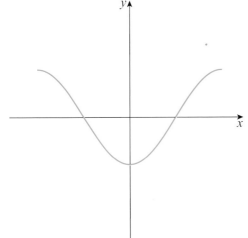

**c** Gradient specifies shape but not vertical position.

**23 a**

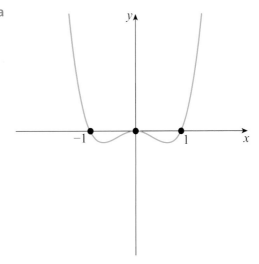

**b** $x < -1$ or $x > 1$

**c** $x < -0.707$ or $0 < x < 0.707$

**24** **a** and **e**, **b** and **d**, **f** and **c**

**25** **a** and **c**, **f** and **b**, **d** and **e**

# Exercise 9C

**1 a** $f'(x) = 4x^3$      **b** $g'(x) = 6x^5$

**2 a** $h'(u) = -u^{-2}$      **b** $z'(t) = -4t^{-5}$

**3 a** $\dfrac{dy}{dx} = 8x^7$      **b** $\dfrac{dp}{dq} = 1$

**4 a** $\dfrac{dz}{dt} = -5t^{-6}$      **b** $\dfrac{ds}{dr} = -10r^{-11}$

**5 a** $f'(x) = -4$      **b** $g'(x) = 14x$

**6 a** $\dfrac{dy}{dx} = 6$      **b** $\dfrac{dy}{dx} = 15x^4$

**7 a** $\dfrac{dy}{dx} = 0$      **b** $\dfrac{dy}{dx} = 0$

**8 a** $g'(x) = -x^{-2}$      **b** $h'(x) = -3x^{-4}$

**9 a** $\dfrac{dz}{dx} = -6x^{-3}$      **b** $\dfrac{dy}{dt} = 50t^{-6}$

**10 a** $\dfrac{dy}{dx} = -x^{-5}$      **b** $\dfrac{dy}{dx} = -x^{-6}$

**11 a** $f'(x) = \dfrac{3}{2x^2}$      **b** $f'(x) = \dfrac{5}{x^4}$

**12 a** $f'(x) = 2x - 4$      **b** $g'(x) = 4x - 5$

**13 a** $\dfrac{dy}{dx} = 9x^2 - 10x + 7$      **b** $\dfrac{dy}{dx} = -4x^3 + 12x - 2$

**14 a** $\dfrac{dy}{dx} = 2x^3$    **b** $\dfrac{dy}{dx} = -\dfrac{9}{2}x^5$

**15 a** $\dfrac{dy}{dx} = 3 - \dfrac{3}{4}x^2$    **b** $\dfrac{dy}{dx} = -6x^2 + x^3$

**16 a** $f'(x) = 8x^3 - 15x^2$

  **b** $g'(x) = 3x^2 + 6x - 9$

**17 a** $g'(x) = 2x + 2$    **b** $f'(x) = 2x - 1$

**18 a** $h'(x) = -\dfrac{2}{x^2} - \dfrac{2}{x^3}$

  **b** $g'(x) = 8x - \dfrac{18}{x^3}$

**19 a** $a$    **b** $2ax + b$
**20 a** $2ax + 3 - a$    **b** $3x^2 + b^2$
**21 a** $2x$    **b** $3a^2x^2$
**22 a** $2ax^{2a-1}$    **b** $-ax^{-a-1} + bx^{-b-1}$
**23 a** $7a + \dfrac{6b}{x^3}$    **b** $10b^2x - \dfrac{3a}{cx^2}$
**24 a** $2a^2x$    **b** $18a^2x$
**25 a** $2x + a + b$    **b** $2abx + b^2 + a^2$
**26 a** $-\dfrac{10}{x^2}$    **b** $\dfrac{5}{x^2}$
**27 a** $-\dfrac{4}{x^3} - \dfrac{1}{x^2}$    **b** $-\dfrac{3}{x^4} - \dfrac{4}{x^3}$
**28 a** $0.5 + 3x$    **b** $7x^3 + 4x^7$
**29 a** $x < 1$    **b** $x < -4$
**30 a** $x > 6$    **b** $x > 0$
**31** $6 + \dfrac{4}{t^2}$
**32** $1 - 2m^{-2}$
**33** $\dfrac{3}{2}k$
**34** $x > \dfrac{1}{2}$
**35** $x > -\dfrac{b}{2}$
**36** $-1$
**37** $1 - 3x^2$
**38 a** $-\dfrac{k}{r^2}$    **b** $-\dfrac{V^2}{k}$    **c** $4$
**39 a** $\dfrac{dA}{dL} = q + 2qL$
  **b** $q > 0$

# Exercise 9D

**1 a** 16    **b** 20
**2 a** $\dfrac{1}{8}$    **b** 16.75
**3 a** 32    **b** $-7$
**4 a** $\dfrac{23}{9}$    **b** 100.4
**5 a** 108    **b** 6
**6 a** 24    **b** $-\dfrac{3}{16}$
**7 a** 1    **b** $\pm 2$
**8 a** $\pm 2$    **b** $\pm \dfrac{1}{2}$
**9 a** $-1, -3$    **b** $\pm \dfrac{1}{\sqrt{2}}$
**10 a** $4x - y = 1$    **b** $y - x = 0$
**11 a** $2x + y = 7$    **b** $7x + y = 2$
**12 a** $x + 3y = 6$    **b** $2x - y = 3$
**13 a** $y = -x + 15$    **b** $y = 2x - 12$
**14 a** $y = -x$    **b** $y = -x + 3$
**15 a** $y = \dfrac{1}{3}x - \dfrac{1}{3}$    **b** $y = \dfrac{1}{16}x + \dfrac{81}{16}$
**16 a** $y = 2x - 3$    **b** $y = x + 1$
**17 a** $y = \dfrac{1}{4}x + 1$    **b** $y = -\dfrac{1}{16}x + \dfrac{3}{4}$
**18 a** $y = -\dfrac{1}{4}x + \dfrac{3}{4}$    **b** $y = -\dfrac{1}{32}x + \dfrac{7}{32}$
**19 a** $y = x - 1$    **b** $y = x - 1$
**20 a** $y \approx 2.77x - 1.55$    **b** $y \approx 1.10x + 1$
**21 a** $y = 14 - 4x$    **b** $y = 1 - x$
**22 a** $8x + 5y = 47$    **b** $4x + 11y = 126$
**23 a** $x + y \approx 2.10$    **b** $x + 2y \approx 13.4$
**24 a** $y = -x$    **b** $y = x - 2$
**25 a** $-0.984$    **b** $-0.894$
**26 a** 2    **b** 0.693
**27 a** $-3, -1$    **b** $0, -2$
**28 a** $4x^3 - 1$    **b** $-1$    **c** $y = x$
**29 a** $3x^2 - x^{-2}$    **b** 2    **c** $y = 2x$
**30** $y = \dfrac{11}{4}x - \dfrac{9}{4}$
**31** $y = 4x + \dfrac{17}{2}$
**32** $x = 0$
**33** $\left(-\dfrac{3}{2}, \dfrac{9}{4}\right)$
**34** $\left(-\dfrac{1}{8}, -8\right)$

**35** $y + 2x = -1$

**36** $y = \dfrac{1}{4}x + \dfrac{15}{4}$

**37** $y = \dfrac{4}{3} - \dfrac{x}{3}$ or $y = -\dfrac{4}{3} - \dfrac{x}{3}$

**38** $(2, 6), (-2, -6)$

**39** $(-3, -2), (1, 6)$

**40** $(2, 0.5)$

**41** $\dfrac{2}{3}$ or $2$

**42** $(1, 1), (3, 9)$

**44** $a = 20$

# Chapter 9 Mixed Practice

**1** a $8x - 1$      b $(2, 14)$

**2** a

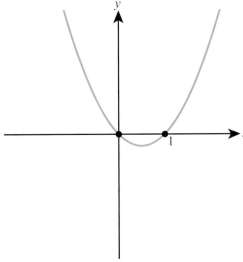

b $x > \dfrac{1}{2}$

c

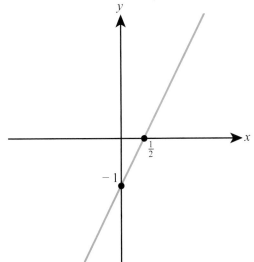

**3** a $f'(x) = 6x^2 + 10x + 4$

   b $f'(-1) = 0$      c $y = 2$

**4** $-0.5$

**5** $y = 27x - 58$

**6** a $12 + 10t$

   b $V(6) = 302$, $\dfrac{dV}{dt}(6) = 72$

    After 6 minutes, there is $302\,\text{m}^3$ of water in the tank, and the volume of water in the tank is increasing at a rate of $72\,\text{m}^3$ per minute.

   c $\dfrac{dV}{dt}(10) = 112$, so the volume is increasing faster after 10 minutes than after 6 minutes.

**7** a $2 - 2t$     b i $0.75$      ii $1$

   c $0 < t < 1$

**8** a $6$              b $x + 8x^{-2}$

   c $0$              d $y = 6$

   e, f

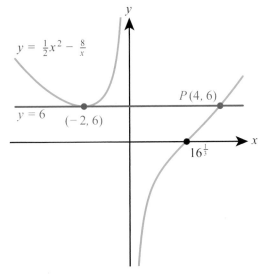

   g $(4, 6)$

**9** $(1, -3), (-1, 9)$

**10** a $(0, 0), (4, -32)$      b $y = -8x$

**11** a $y = 6x$, $y = -6x - 12$

   b $(-1, -6)$

**12** $p = -2$, $c = -3$

**13** a $30t - 3t^2$      b $72, -72$

   c The profit is increasing after 6 months, but decreasing after 12 months.

**14** a i $2$      ii $2x + 3$      b $-\dfrac{1}{2}$

c

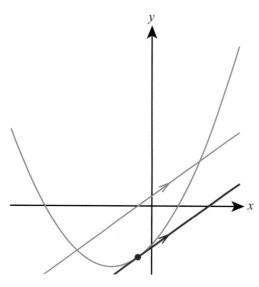

**15** **a** **i** $0 < x < \dfrac{4}{3}$

**ii**

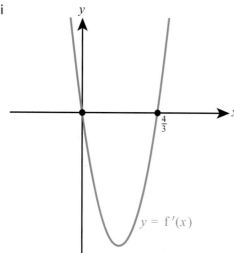

$y = f'(x)$

**b** **i** $x < 0$ and $0 < x < 2$

**ii**

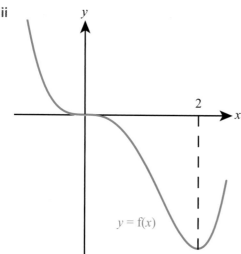

$y = f(x)$

**16** **a** 3        **b** $2x - 1$

   **c** $(2, 4)$        **d** $\left(\dfrac{1}{3}, \dfrac{16}{9}\right)$

   **e** $y = 5x - 7$        **f** $\left(\dfrac{1}{2}, \dfrac{7}{4}\right)$, gradient $= 0$

**17** **a** $2x + 1$        **b** 3 and $-2$

**18** $a = 2, b = -5$

**19** $a = -8, b = 13$

**20** $(1.5, 6)$

**21** $(0.701, 1.47)$

You need to make good use of technology in this question!

# Chapter 10
# Prior Knowledge

**1** $x^{-2} - 4x^{-3} + 4x^{-4}$        **2** $f'(x) = -12x^{-4} + 2$

# Exercise 10A

**1** **a** $f(x) = \dfrac{1}{4}x^4 + c$      **b** $f(x) = \dfrac{1}{6}x^6 + c$

**2** **a** $f(x) = -x^{-1} + c$

   **b** $f(x) = -\dfrac{1}{2}x^{-2} + c$

**3** **a** $f(x) = c$        **b** $f(x) = x + c$

**4** **a** $y = x^3 + c$        **b** $y = -x^5 + c$

**5** **a** $y = -\dfrac{7}{4}x^4 + c$      **b** $y = \dfrac{3}{8}x^8 + c$

**6** **a** $y = \dfrac{1}{4}x^6 + c$      **b** $y = -\dfrac{1}{6}x^{10} + c$

**7** **a** $y = x^{-3} + c$        **b** $y = -x^{-5} + c$

**8** **a** $y = -\dfrac{3}{2}x^{-4} + c$     **b** $y = 2x^{-2} + c$

**9** **a** $y = \dfrac{2}{5}x^{-1} + c$      **b** $y = -\dfrac{1}{4}x^{-7} + c$

**10** **a** $y = x^3 - 2x^2 + 5x + c$

    **b** $y = \dfrac{7}{5}x^5 + 2x^3 - 2x + c$

**11** **a** $y = \dfrac{1}{2}x^2 - \dfrac{2}{3}x^6 + c$

    **b** $y = \dfrac{3}{2}x^4 - \dfrac{5}{8}x^8 + c$

**12 a** $y = \frac{1}{4}x^3 + \frac{7}{6}x^2 + c$

**b** $y = \frac{4}{5}x - \frac{2}{15}x^5 + c$

**13 a** $y = -\frac{1}{2}x^{-4} + \frac{3}{2}x^2 + c$

**b** $y = 5x + 3x^{-3} + c$

**14 a** $y = -\frac{5}{2}x^{-1} + \frac{2}{7}x^{-5} + c$

**b** $y = -\frac{2}{3}x^{-2} + \frac{1}{15}x^{-6} + c$

**15 a** $\frac{1}{4}x^4 + \frac{5}{3}x^3 + c$   **b** $3x^2 - x^3 + c$

**16 a** $\frac{1}{3}x^3 - \frac{1}{2}x^2 - 6x + c$   **b** $\frac{5}{2}x^2 - \frac{1}{3}x^3 - 4x + c$

**17 a** $3x^3 - 6x - x^{-1} + c$

**b** $\frac{4}{3}x^3 - 4x^{-1} - \frac{1}{5}x^{-5} + c$

**18 a** $\frac{3}{2}x^2 - 2x + c$   **b** $\frac{5}{2}x^2 - 3x + c$

**19 a** $-x^{-2} + \frac{7}{6}x^{-3} + c$   **b** $-\frac{1}{2}x^{-1} + \frac{3}{8}x^{-2} + c$

**20 a** $y = x^3 + 6$   **b** $y = x^5 + 5$

**21 a** $y = \frac{1}{4}x^4 - 6x^2 + 10$   **b** $y = 3x - 2x^5 + 4$

**22 a** $y = -3x^{-2} + 2x^2 - \frac{5}{3}$

**b** $y = 3x^3 + 2x^{-1} - 4$

**23** $-\frac{4}{3t} + \frac{1}{2t^4} + c$   **24** $y = x^3 - 4x + 7$

**25** $y = -\frac{4}{x} - x^3 + 10$

**26** $\frac{3x^4}{4} - \frac{2x^3}{3} + \frac{3x^2}{2} - 2x + c$

**27** $\frac{z^4}{4} + \frac{z^2}{2} + c$   **28** $\frac{x^3}{9} + \frac{2}{3x} + c$

**29 a** $k = 0.2$   **b** 3 kg

**30 a** 40 litres   **b** 4   **c** no

# Exercise 10B

**1 a** 97.6   **b** 2.25

**2 a** −0.625   **b** −145.5

**3 a** 4   **b** 36

**4 a** 7.5   **b** 4.5

**5** 21.3

**6** 0.25

**7** 30

**8** 4.5

**9** $\frac{32}{3}$

**10** 18

**11 a** (2.5, 6.25)   **b** 18.2

**12** 12.7 litres

**13** 3600 g

**14 a** 58 g   **b** 60 g

**c** It suggests it takes forever for all the sand to fall through / sand is infinitely divisible.

**15** 18.75

**16** $12x^2 - \frac{8}{x^3}$

**17** $\frac{16}{3}$

**18** $af(a) - A$

# Chapter 10 Mixed Practice

**1** $\frac{x^4}{4} + \frac{3}{x} + c$

**2** $\frac{4}{3}x^3 - \frac{3}{2}x^2 + 5x + c$

**3** 0.661

**4** $y = x^3 - 4x^3 + 6$

**5** $a = 8$

**6** 5.7

**7 a** (2, 0), (5, 0)   **b** 6.75

**8 a** $y = 1.6x - 3.2$   **c** 1.07

**9 b** 1.5

**10** 1.5

**11** 1000 cm³

**12** 21.5 kg

**13** 47

**14 a** e.g. no energy lost to surroundings.

**b** 160 calories

**15** $y = \frac{5x^3}{3} + 3$

**16** 0

**17** $6x + \frac{2}{x^2}$

**18 a** (3, 0)   **b** 18

# Core SL content: Review Exercise

**1** a  4      b  83      c  900

**2** a  $\dfrac{2}{3}$      b  2      c  $3x + 2y = 31$

**3** a

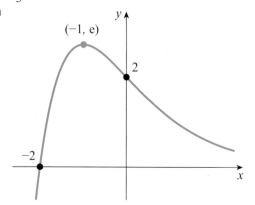

     b  $y = 0$      c  $y \leqslant 2.72$

**4** a  $a = 2.5; x \geqslant 2.5$      b  $f(x) \geqslant 0$

     c  18.5

**5** a  38.9°      b  95.5 cm²

**6** a  4.5      b  0.305

     c  28.2

**7** a  $(-2, 0), (2, 0)$      b  15.6

**8** a  4.5      b  2

     c  $\dfrac{7}{10}$

**9** a  9.56 cm³      b  88.9 g

     d  52.8°      e  30.9 cm²

**10** a and c

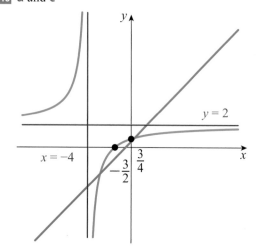

**b**  $x = -4$

**d**  $(-2.85078, -2.35078)$
     OR $(0.35078, 0.85078)$

**e**  1

**f**  $y = -x - 5$

**11**  $x - 11y + 112 = 0$

**12** a  1000      b  $-2$

**13** a  $8x^7 - 24x^{-2}$      b  $56x^6 + 48x^{-3}$

**14** a  $18r^3$      b  $\dfrac{18(1 - r^{15})}{1 - r}$

     c  0.315 or $-1.05$

**15**  3.6%

**16**  $\dfrac{60}{41}$

**17** a  12      b  0.0733

     c  0.764

**18** a  $-2$

     c  $2x + y = -3$      d  $(-3, 3)$

     e  $\sqrt{20}$      f  10

**19** a  i  1294.14 cm³      ii  6

     iii  I  431 cm³

          II  $4.31 \times 10^{-4}$ m³

     b  i  I  73.5°

        II  55.8 m

       ii  55.0 m      iii  217 m

**20** a  i  0.985

       ii  strong positive

     b  $y = 260x + 699$

     c  4077 USD

     d  e.g. $3952 < 4077$

     e  i  $304x$

       ii  $304x - (260x + 699)$

        iii  16

**21** a  0.159

     b  i  0.119

       ii  0.395

**22** a  3

     b  $\dfrac{7}{26}$

     c  $\dfrac{8}{15}$

     d  $\dfrac{172}{325}$

# Chapter 11 Prior Knowledge

1   $2x^2 + 7x - 4$

2   $\dfrac{23}{20}$

# Exercise 11A

**1**  $a = 4, b = 7$

**2**  $a = 6, b = 37$

**3**  $p = 4, q = \pm 5$

**8**  a and c

**9**  $a = 1, \ b = 3$

**10**  $p = 5, \ q = 2$

**13**  a  e.g. $x = 2$, $y = 1$      b  1

# Chapter 11 Mixed Practice

**1**  b and c

**2**  b  $k = 6$

**3**  b  $p = 2, \ q = 5$

**4**  $a = 4, b = 16$

**6**  $A = 1, \ B = 2$

**12**  a  $a = 2, b = 4$

# Chapter 12 Prior Knowledge

1  a  $15x^9 y^5$      b  $\dfrac{4d^2}{c}$      c  $9a^8 b^{-4}$

2  $y = 91$

3  $x = \dfrac{5}{3}$

4  $x = \dfrac{\ln 11 + 1}{2} \approx 1.70$

# Exercise 12A

1  a  7         b  5

2  a  2         b  3

3  a  4         b  5

4  a  4         b  8

5  a  125      b  25

6  a  $\dfrac{1}{10}$      b  $\dfrac{1}{10}$

7  a  $\dfrac{1}{4}$      b  $\dfrac{1}{9}$

8  a  $\dfrac{1}{4}$      b  $\dfrac{1}{1000}$

9  a  $x^{\frac{5}{2}}$                        b  $x^{\frac{4}{3}}$

10 a  $x^{\frac{5}{3}}$                       b  $x^{\frac{3}{2}}$

11 a  $4x^{-\frac{3}{2}}$                   b  $5x^{-\frac{5}{2}}$

12 a  $\dfrac{1}{5}x^{\frac{3}{2}}$             b  $\dfrac{1}{3}x^{\frac{2}{3}}$

13 a  $x^{\frac{5}{6}}$                       b  $x^{\frac{5}{3}}$

**14**  $\dfrac{1}{16}$

**15**  2

**16**  $\dfrac{8}{27}$

**17**  $x^{\frac{11}{12}}$

**18**  $3x^{-\frac{1}{3}} + 2x^{\frac{1}{2}}$

**19**  $\dfrac{1}{3}x^{-\frac{3}{2}}$

**20**  $\dfrac{1}{4}$

**21**  $\dfrac{25}{4}$

**22**  $x^{\frac{7}{6}}$

**23**  $x^{-\frac{3}{2}}$

**24**  $x^{\frac{8}{3}}$

**25**  $\dfrac{1}{8}x^{-\frac{3}{2}}$

**26**  $x^{\frac{1}{2}} + x^{-1}$

**27**  $x^{\frac{3}{2}} - x^{\frac{1}{2}}$

**28**  $\dfrac{1}{3}x^{-\frac{3}{2}}$

**29**  $\dfrac{1}{2}x^1 + \dfrac{3}{2}x^{-\frac{1}{2}}$

**30**  $y^4 = 16x^{\frac{8}{3}}$

**31**  $\sqrt[3]{y} = 3x^{\frac{1}{6}}$

**32**  $y = x^{\frac{3}{2}}$

**33**  $y = \dfrac{8}{27}x^{-\frac{3}{2}}$

**34**  $x = y^{\frac{2}{3}}$

**35**  $\pm 27$

**36**  64

# Exercise 12B

1  a  4                          b  2
2  a  −3                         b  −2
3  a  $\dfrac{1}{2}$             b  $\dfrac{1}{4}$
4  a  −1                         b  −1
5  a  0                          b  0
6  a  $\dfrac{1}{9}$             b  $\dfrac{1}{16}$
7  a  $\dfrac{9}{8}$             b  $\dfrac{19}{9}$
8  a  23                         b  61
9  a  3                          b  2
10 a  5                          b  −2
11 a  $2p - q$                   b  $3q - p$
12 a  $2p - 3q$                  b  $4q - 2p$
13 a  $\dfrac{3}{2}p + \dfrac{1}{2}q$   b  $2p + \dfrac{3}{2}q$
14 a  $2 + p - 2q$               b  $1 + 2p - 5q$
15 a  500                        b  2
16 a  9.5                        b  4
17 a  $\dfrac{17}{7}$            b  $\dfrac{5}{4}$
18 a  $\dfrac{3 + 5e^4}{1 - e^4}$   b  $\dfrac{e^3 + 2}{e^3 - 1}$
19 a  $\dfrac{\ln 7}{\ln 2}$     b  $\dfrac{\ln 8}{\ln 5}$
20 a  $\dfrac{\log_2 3}{\log_2 5}$   b  $\dfrac{\log_3 5}{\log_3 7}$
21 a  $\dfrac{\log_5 4}{\log_5 e}$   b  $\dfrac{\log_4 7}{\log_4 e}$
22 a  $\dfrac{\log_2 5}{4}$      b  $\dfrac{\log_3 2}{3}$
23 a  $\dfrac{2}{\log_5 2}$      b  $\dfrac{3}{\log_2 3}$
24 a  4 or $\dfrac{1}{4}$        b  25 or $\dfrac{1}{25}$
25 a  5 or $\dfrac{1}{5}$        b  2 or $\dfrac{1}{2}$
26 a  8 or $\dfrac{1}{8}$        b  256 or $\dfrac{1}{256}$
27 a  $\dfrac{4}{3}$             b  $-\dfrac{8}{3}$
28 a  $-\dfrac{9}{7}$            b  9
29 a  2.58                       b  0.528

30 a  0.490                      b  2.28
31 a  1.49                       b  −0.277
32 a  $\dfrac{\ln 2 + 2\ln 3}{\ln 3 - \ln 2}$   b  $\dfrac{2\ln 2 + \ln 3}{\ln 3 - \ln 2}$
33 a  $\dfrac{3\ln 2 + 5\ln 7}{2\ln 7 - \ln 2}$   b  $\dfrac{8\ln 2 - \ln 7}{3\ln 7 - \ln 2}$
34 a  $x + 4y$                   b  $2x + y - 5z$
   c  $3 + 2x + 3y$
35 a  $2 + \dfrac{1}{2}x$        b  $y - 1 - 5z$
36  $\ln(a^2 b^6)$
37  $\ln\left(\dfrac{\sqrt[3]{x}}{\sqrt{y}}\right)$
38 a  $-\dfrac{1}{2}$            b  $\dfrac{1}{3}$
39  5
40  42
41 a  1.43                       b  1.77
42  19.9
43  2
44  4
45 a  2.4                        b  $\log_2 2.4$
46 a  $\dfrac{1}{\ln x}$         b  $\sqrt[4]{e}$ or $\dfrac{1}{\sqrt[4]{e}}$
47  $-\dfrac{9}{7}$
48  $\dfrac{3\ln 5 + 5\ln 9}{\ln 9 - 2\ln 5}$
49 a  10                         b  6.58 days
50 a  i  1000                    ii  1210
   b  $\log_{1.1} 2 \approx 7.27$ hours
51  7.97 years
52  64 or $\dfrac{1}{64}$
53  $\dfrac{\ln 96}{\ln 72}$
54  $2\log_2 10$ years = 6.64 years (3 sf)
55 a  $-a^b$                     b  1

# Chapter 12 Mixed Practice

1  a  $\dfrac{27}{8}$            b  −3
2  $\ln 36$
3  5
4  $\log_{1.05} 2 \approx 14.2$

**5** 2

**6** $\dfrac{8}{5}$

**7** $e^{-4}$

**8** a $2a+b$ b $a-b-3c$ c $\dfrac{1}{2}c+\dfrac{3}{2}a$

**9** a 2 b 2 c −1

**10** a $x-\dfrac{1}{2}y$ b $2x-3y-3$

**11** a $x+y$ b $x+2y$ c $x-2y$

**12** $\log 6250$

**13** $\dfrac{1}{2}e^{\frac{5}{3}}$

**14** $\dfrac{1}{9}$ or 9

**15** $-\dfrac{13}{3}$

**16** $x=8$, $y=9$

**17** $\sqrt{2}$

**18** $\dfrac{\ln 2}{2\ln 3-1}$

**19** $\dfrac{\ln 5+3\ln 7}{\ln 7-2\ln 5}$

**20** $\dfrac{\log 12}{\log 48}$

**21** a 150 b 3397
  c 1.82 hours

**22** 16

**23** a 3 b 125

**24** a 5 b $p=5$, $q=-3$

**25** a 12 b −1 c 8

**26** $-\dfrac{\ln 2}{\ln 3}$

**27** 2

**28** $\dfrac{x}{2y}$

**29** a $2x-3$ b $\dfrac{2}{x}$

**30** $210\ln x$

**31** −4

# Chapter 13 Prior Knowledge

**1** $\dfrac{320}{3}\approx 107$

**2** 0

**3** $-1<x<1$

**4** $9-24x+16x^2$

# Exercise 13A

**1** a $\dfrac{1}{3}$ b 1

**2** a 1 b $\dfrac{5}{9}$

**3** a $\dfrac{16}{3}$ b 9

**4** a $\dfrac{1}{3}$ b $\dfrac{1}{4}$

**5** a $\dfrac{75}{8}$ b $\dfrac{64}{7}$

**6** a $1<x<3$ b $-4<x<-2$

**7** a $|x|<\dfrac{1}{2}$ b $|x|<\dfrac{1}{3}$

**8** a $|x|<2$ b $|x|<5$

**9** a $-5<x<-3$ b $0<x<2$

**10** a $|x|<\dfrac{2}{3}$ b $|x|<\dfrac{3}{4}$

**11** 4

**12** 4

**13** $\dfrac{3}{2}$

**14** $-\dfrac{1}{3}$

**15** $\dfrac{1}{4}$

**16** $2,\dfrac{2}{3},\dfrac{2}{9}$

**17** $\dfrac{1}{3},\dfrac{2}{3}$

**18** a $1,\dfrac{2}{5},\dfrac{4}{25}$ b $\dfrac{5}{3}$

**19** 3

**20** a $|x|<9$ b $\dfrac{27}{7}$

**21** a $2<x<4$ b $\dfrac{5}{4-x}$

**22** a $|x|<\dfrac{1}{2}$ b $\dfrac{2}{1-2x}$

**23** 6

**24** a $0<x<\dfrac{1}{2}$ b $\dfrac{x}{1-4x^2}$

**25** a $\dfrac{6}{5}$ b $|x|<2$

**26** 9

# Exercise 13B

1  a  $64 + 192x + 240x^2 + 160x^3$
   b  $128 + 448x + 672x^2 + 560x^3$
2  a  $243 - 405x + 270x^2 - 90x^3$
   b  $81 - 108x + 54x^2 - 12x^3$
3  a  $1 + 12x + 60x^2 + 160x^3$
   b  $1 + 14x + 84x^2 + 280x^3$
4  a  $1 - 20x + 150x^2 - 500x^3$
   b  $1 - 25x + 250x^2 - 1250x^3$
5  a  $1024 + 15\,360x + 103\,680x^2 + 414\,720x^3$
   b  $512 + 6912x + 41\,472x^2 + 145\,152x^3$
6  a  $32 - 240x + 720x^2 - 1080x^3$
   b  $64 - 576x + 2160x^2 - 4320x^3$
7  a  $a^7 + 14a^6b + 84a^5b^2 + 280a^4b^3$
   b  $a^8 + 16a^7b + 112a^6b^2 + 448a^5b^3$
8  a  $243a^5 - 810a^4b + 1080a^3b^2 - 720a^2b^3$
   b  $729a^6 - 2916a^5b + 4860a^4b^2 - 4320a^3b^3$
9  a  1120　　　　　　b  4032
10 a  $-10\,206$　　　　b  $-61\,236$
11 a  13 440　　　　　b  5376
12 a  24 634 368　　　b  7 185 024
13 a  760　　　　　　b  $-684$
14 a  $-24\,634\,368$　　b  3 421 440
15 a  $1 + 8x + 24x^2 + 32x^3 + 16x^4$
   b  $16 + 32x + 24x^2 + 8x^3 + x^4$
16 a  $27 + 27x + 9x^2 + x^3$
   b  $1 + 9x + 27x^2 + 27x^3$
17 a  $x^5 + 10x^4 + 40x^3 + 80x^2 + 80x + 32$
   b  $x^5 + 5x^4 + 10x^3 + 10x^2 + 5x + 1$
18 a  6　　　　　　　b  8
19 a  21　　　　　　b  15
20 a  84　　　　　　b  56
21 a  10　　　　　　b  13
22 a  105　　　　　　b  55
23 a  5, 9　　　　　b  7, 8
24 a  3, 17　　　　　b  5, 12
25 a  12　　　　　　b  18
26  $10\,000 - 12\,000x + 5400x^2 - 1080x^3 + 81x^4$
27  $32x^5 - 80x^4 + 80x^3 - 40x^2 + 10x - 1$
28  127 575
29  51 963 120

30  12
31  a  $1024 + 15\,360x + 103\,680x^2$
    b  1039
32  a  $19\,683 - 118\,098x + 314\,928x^2 - 489\,888x^3$
    b  18 533.0
33  a  $128 + 448x + 672x^2$　　b  $384 + 1216x + 1568x^2$
34  a  $32 - 240x + 720x^2$　　b  240
35  1040
36  a  $16x^3 + 64x$　　　　　b  0.640 016
37  12
38  10
39  $x^4 + 8x^2 + 24 + \dfrac{32}{x^2} + \dfrac{16}{x^4}$
40  $x^{10} + 15x^9 + 90x^8 + 270x^7 + 405x^6 + 243x^5$
41  12 285
42  $-1\,959\,552$
43  0

# Chapter 13 Mixed Practice

1  $16 + 32x + 24x^2 + 8x^3 + x^4$
2  $\dfrac{1}{2}$
3  $\dfrac{9}{2}$
4  3
5  14
6  a  $p = 5$, $q = 7$, $r = 5$ (or $r = 7$)
   b  $-24\,634\,368$
7  a  11　　　　　　b  $262\,440x^3$
8  a  $1 + 20x + 180x^2$　　b  1.02018
9  a  $\dfrac{1}{3} < x < 1$　　　b  $\dfrac{5}{9}$
10  $\dfrac{2}{3}$
11  13 440
12  a  4　　　　　　b  108
13  a  $n$
    b i  9　　　　　ii  144
14  $\pm\dfrac{1}{2}$
15  $\dfrac{19}{20}$
16  $\dfrac{15}{2}$

**17** b $\dfrac{e^{-x}}{1-e^{-x}}$  c $\ln\left(\dfrac{3}{2}\right)$

**18** $\dfrac{1}{3}$

**19** 7

**20** −96

**21** a ii $2^a$  iii $2^a(2^d)^{n-1}$

  b i $\dfrac{2^a(2^{nd}-1)}{2^d-1}$  ii $d<0$

  iii $\dfrac{2^a}{1-2^d}$  iv −1

  c $\ln\left[p^n q^{\frac{n(n-1)}{2}}\right]$

# Chapter 14 Prior Knowledge

1  11

2  $x \neq -3$

3  a  $f(x) \geqslant -3$

  b  Many-to-one

4

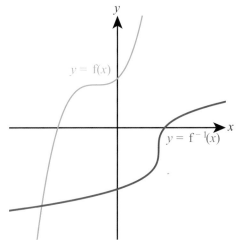

5  a  $x = \dfrac{y+1}{y-3}$  b  $x = \ln(y+2)$

# Exercise 14A

1  a  $3x^2 - 1$  b  $4x^2 + 2$
2  a  $(2x+1)^2$  b  $(3x-2)^2$
3  a  $12x^2 + 4x$  b  $20x^2 - 6x$
4  a  $3e^{2x+5}$  b  $4e^{3x+1}$
5  a  $12e^x + 1$  b  $6e^x + 5$
6  a  $e^{3x} - 2e^x$  b  $e^{3x} + 4e^x$

**7** a  $\dfrac{1}{3x+3}$  b  $\dfrac{1}{2x+3}$

**8** a  $\dfrac{3}{x} - \dfrac{2}{x^2}$  b  $\dfrac{4}{x} + \dfrac{3}{x^2}$

**9** a  $x > -8$  b  $x > -9$

**10** a  $x < 1$  b  $x < 0$

**11** a  $x \geqslant \dfrac{1}{2}$  b  $x \geqslant \dfrac{5}{3}$

**12** a  $x \leqslant \dfrac{7}{2}$  b  $x \leqslant \dfrac{5}{3}$

**13** a  $x \neq \dfrac{1}{4}$  b  $x \neq 1$

**14** a  $x \neq \ln 3$  b  $x \neq \ln 7$

**15** a  $11 - 9x$  b  $\dfrac{1}{3}$

**16** a  $x^4 + 2x^2 + 2$  b  $x = 1$

**17** $16x^9$

**18** a  $x \geqslant -2$  b  $\dfrac{19}{3}$

**19** a  $fg(8) = \ln 3,\ gf(8) = \ln 8 - 5$

  b  $e^8 + 5$

**20** a  i  2  ii  1

  b  4

**21** a  i  1  ii  1

  b  1

**22** a  $x > 3$  b  $e + 3$

  c  $e^4$  d  $\dfrac{3e^3}{e^3 - 1}$

**23** a  $\dfrac{x-3}{2x-5}$  b  $x \neq 2.5,\ 3$

  c  $\dfrac{7}{3}$

**24** $k = 16$

**25** a  $x \neq -\dfrac{2}{3}, -\dfrac{7}{6}$  b  $ff(x) \neq 0, \dfrac{1}{2}$

  c  $-\dfrac{5}{3}$

# Exercise 14B

1  a  $f^{-1}(x) = \dfrac{2x-1}{4}$  b  $f^{-1}(x) = \dfrac{5x-4}{3}$

2  a  $f^{-1}(x) = \dfrac{x+3}{4}$  b  $f^{-1}(x) = \dfrac{x-1}{5}$

3  a  $f^{-1}(x) = \dfrac{1}{4}\ln x$  b  $f^{-1}(x) = \dfrac{1}{3}\ln x$

4  a  $f^{-1}(x) = \ln\left(\dfrac{x}{3}\right) + 2$  b  $f^{-1}(x) = \ln\left(\dfrac{x}{2}\right) - 3$

**5 a** $f^{-1}(x) = \frac{1}{3}(2^x - 1)$  **b** $f^{-1}(x) = \frac{1}{4}(3^x + 1)$

**6 a** $f^{-1}(x) = x^2 + 2$  **b** $f^{-1}(x) = x^2 - 3$

**7 a** $f^{-1}(x) = \sqrt[3]{x} - 2$  **b** $f^{-1}(x) = \sqrt[3]{x} + 3$

**8 a** $f^{-1}(x) = \sqrt[3]{x+2}$  **b** $f^{-1}(x) = \sqrt[3]{x-5}$

**9 a** $f^{-1}(x) = \dfrac{2x+3}{x-1}$  **b** $f^{-1}(x) = \dfrac{3x+1}{x-1}$

**10 a** $f^{-1}(x) = \dfrac{2x+1}{3x-2}$  **b** $f^{-1}(x) = \dfrac{3x-1}{2x-3}$

**11 a** $x \geqslant 2$  **b** $x \geqslant -5$

**12 a** $x \leqslant -1$  **b** $x \leqslant 3$

**13 a** $x \geqslant 2$  **b** $x \geqslant 0$

**14 a** $-1 \leqslant x \leqslant 1$  **b** $-2 \leqslant x < 2$

**15 a** $x \leqslant -1$  **b** $x \leqslant 1$

**16 a** $x \geqslant -2$  **b** $x \geqslant 3$

**17 a** $x < -2$  **b** $x < -1$

**18 a** $\dfrac{5}{3}$  **b** $\dfrac{3x-4}{x}$ domain $x \neq 0$

**19** $\dfrac{1}{5}\ln\left(\dfrac{x}{3}\right)$ domain $x > 0$

**20 a**

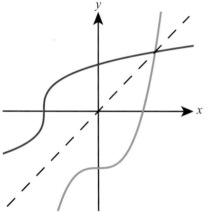

**b** $f^{-1}(x) = \sqrt[3]{5x+15}$

**21 a**

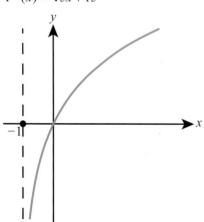

**b** $f^{-1}(x) = 2\ln(x+1),\ x > -1$

**22 a** $\sqrt{\dfrac{4x+1}{x-1}}$

**b** $f^{-1}(x) > 2$

**23** $\dfrac{1}{7}$

**24 a**

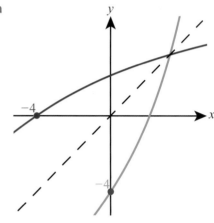

**b** $2\ln 5$

**25 a** $0$  **b** $-\sqrt{x-3}$

**26 a** $x \leqslant 5$

**b** $5 - \dfrac{1}{3}\sqrt{x}$

**27** $4$

# Chapter 14 Mixed Practice

**1 a** $f^{-1}(x) = \dfrac{x+1}{3}$

**2 a** $5$  **b** $-4$

**3** $0.462$

**4 a** $3$  **b** $1$

**5** $x = -3$

**6 a** $x > 2$

**b** $2 + e^{\frac{x}{3}},\ g^{-1}(x) > 2$

**7** $\sqrt[3]{\dfrac{x-1}{3}}$

**8 a** $2x^3 + 3$

**b** $-1.14$

**9 a** **i** $-1$

**ii** $0$

**b** $-3 \leqslant x \leqslant 3$

c

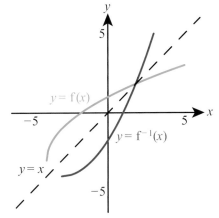

**10 a** $\dfrac{1}{x-1}-2, x\neq 1$

**11** $f^{-1}(x)=\dfrac{4x+1}{3-x}$ domain $x\neq 3$

**12 b** 9.39

**13 a** 3  **b** $-\dfrac{1}{4}$

**14 a** $e^x+2, h^{-1}(x)>2$  **b** $x-2$

**15 a** $a=-\dfrac{1}{2}$  **b** $\dfrac{-1-\sqrt{x}}{2}$

**16 a** 28  **b** $9-x^2$

  **c** **i** Reflection in the line $y=x$
    **ii** $\sqrt{x-3}$  **iii** $f^{-1}(x)>2\geqslant 1$

**17 a** **i** 22  **ii** $\dfrac{4x+11}{x-1}$
    **iii** $9x+4$
  **b** $f(x)$ can be 1, which is not in the domain of g
  **c** **ii** $f^{-1}(x)\neq 1$

**18 a** 9  **b** 5

**19 a** $\dfrac{x+2}{3}$
  **c** **i** 2.5
    **ii**

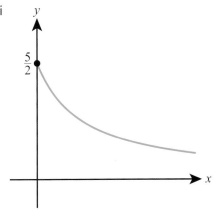

**d** **i** 2.5  **ii** $x=0$
  **e** 1

**21** $\dfrac{29}{4}$

**22 a** 0.577
  **b** Reflection in the line $y=x$
  **c** $\sqrt[3]{2}$

**23 a** **i** $\dfrac{1}{2x+3}, x\neq -\dfrac{3}{2}$
    **ii** $\dfrac{2}{x}+3, x\neq 0$
  **b** $(-1, 1)$

# Chapter 15 Prior Knowledge

**1** $3x^2+10x-8$
**2** $(x-10)(x+1)$
**3** $2\sqrt{3}$
**4** $x>\dfrac{7}{3}$

# Exercise 15A

**1 a i** 2  **ii** 1  **iii** 3
  **b i** 3  **ii** 2  **iii** 1
**2 a i** 1  **ii** 3  **iii** 2
  **b i** 1  **ii** 2  **iii** 3
**3 a i** 2  **ii** 3  **iii** 1
  **b i** 2  **ii** 1  **iii** 3
**4 a**

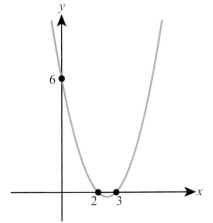

b

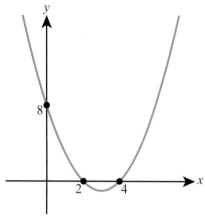

b

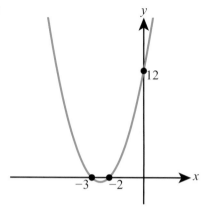

5 a

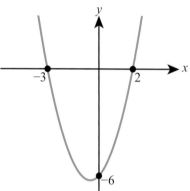

7 a

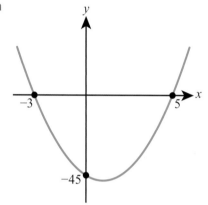

b

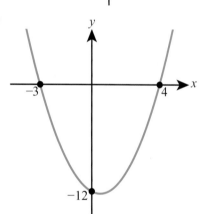

b

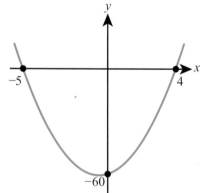

6 a

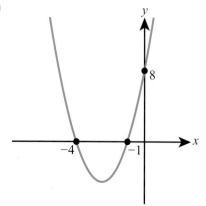

8 a

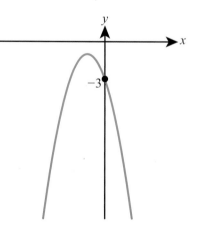

b

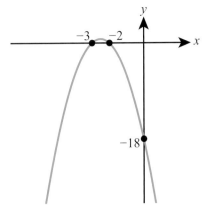

b

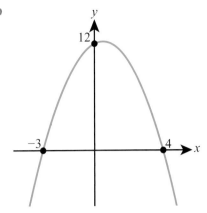

9 a

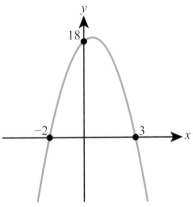

b

10 a

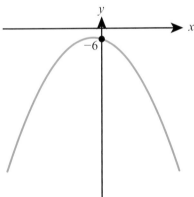

b

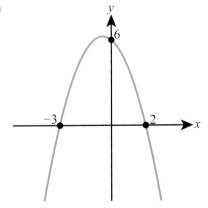

11 a $y = (x-3)(x+1)$    b $y = (x-4)(x+2)$

12 a $y = 2(x-1)(x-3)$    b $y = 3(x-1)(x-4)$

13 a $y = 3(x-2)(x+3)$    b $y = 2(x-1)(x+4)$

14 a $y = -3(x-1)(x-2)$

   b $y = -2(x-1)(x-3)$

15 a $y = -4(x-1)(x+3)$

   b $y = -3(x-3)(x+1)$

16 a $y = -(x-3)(x+2)$

   b $y = -(x-2)(x+3)$

17 a $(x-3)^2 + 6; (3, 6)$

   b $(x-4)^2 + 2; (4, 2)$

18 a $(x+2)^2 + 3; (-2, 3)$

   b $(x+3)^2 + 6; (-3, 6)$

19 a $(x+2)^2 - 3; (-2, -3)$

   b $(x+5)^2 - 20; (-5, -20)$

20 a $(x-2)^2 - 3; (2, -3)$

   b $(x-4)^2 - 13; (4, -13)$

21 a $(x+2)^2 - 9; (-2, -9)$

   b $(x+1)^2 - 9; (-1, -9)$

22 a $\left(x+\dfrac{3}{2}\right)^2 + \dfrac{11}{4}; \left(-\dfrac{3}{2}, \dfrac{11}{4}\right)$

   b $\left(x+\dfrac{5}{2}\right)^2 + \dfrac{7}{4}; \left(-\dfrac{5}{2}, \dfrac{7}{4}\right)$

23 a $\left(x-\dfrac{3}{2}\right)^2 - \dfrac{5}{4}; \left(\dfrac{3}{2}, -\dfrac{5}{4}\right)$

   b $\left(x-\dfrac{7}{2}\right)^2 - \dfrac{37}{4}; \left(\dfrac{7}{2}, -\dfrac{37}{4}\right)$

24 a $\left(x-\dfrac{1}{4}\right)^2 - \dfrac{25}{16}; \left(\dfrac{1}{4}, -\dfrac{25}{16}\right)$

   b $\left(x-\dfrac{1}{6}\right)^2 - \dfrac{13}{36}; \left(\dfrac{1}{6}, -\dfrac{13}{36}\right)$

**25 a** $2(x+2)^2 + 7; x = -2$

 **b** $2(x+3)^2 + 5; x = -3$

**26 a** $3(x-1)^2 + 7; x = 1$

 **b** $3(x-2)^2 - 7; x = 2$

**27 a** $2(x+1)^2 - 3; x = -1$

 **b** $2(x+2)^2 - 11; x = -2$

**28 a** $2\left(x-\frac{1}{2}\right)^2 + \frac{1}{2}; x = \frac{1}{2}$

 **b** $2\left(x-\frac{3}{2}\right)^2 + \frac{1}{2}; x = \frac{3}{2}$

**29 a** $3\left(x+\frac{1}{6}\right)^2 + \frac{23}{12}; x = -\frac{1}{6}$

 **b** $3\left(x-\frac{1}{6}\right)^2 + \frac{11}{12}; x = \frac{1}{6}$

**30 a** $-2(x-1)^2 + 3; x = 1$

 **b** $-2(x-2)^2 + 7; x = 2$

**31 a** $-(x+2)^2 + 2; x = -2$

 **b** $-(x+3)^2 + 12; x = -3$

**32 a** $y = (x-3)^2 + 2$  **b** $y = (x-2)^2 + 3$

**33 a** $y = (x+1)^2 + 2$  **b** $y = (x+2)^2 + 3$

**34 a** $y = 2(x-2)^2 - 3$  **b** $y = 3(x-3)^2 - 1$

**35 a** $y = -(x+2)^2 - 5$  **b** $y = -(x+1)^2 - 2$

**36 a** $y = -2(x-1)^2 - 5$  **b** $y = -3(x-1)^2 + 2$

**37 a** $y = -(x+3)^2 - 4$  **b** $y = -2(x+2)^2 - 5$

**38 a** $-24$  **b** $-3, 8$

**39 a** $2, 3$  **b** $-3$

 **c** $y = -3x^2 + 3x + 18$

**40 a** $\left(x-\frac{5}{2}\right)^2 - \frac{21}{4}$  **b** $x = \frac{5}{2}$

**41 a** $(0, 23)$  **b** $2(x+3)^2 + 5$

 **c** $(-3, 5)$

**42 a** $h = 2, k = 7$  **b** $-5$

 **c** $y = -5x^2 + 20x - 13$

**43 a** $3(x+1)^2 - 5$  **b** $-5$

**44 a** $5(x-1)^2 - 2$  **b** $f(x) \geqslant -2$

# Exercise 15B

**1 a** $2, 3$  **b** $3, 5$

**2 a** $-3, 2$  **b** $-3, 4$

**3 a** $3$  **b** $-5$

**4 a** $-1, -3$  **b** $-2, -1$

**5 a** $-2, 3$  **b** $-5, 4$

**6 a** $-3$  **b** $2$

**7 a** $0, 5$  **b** $0, -3$

**8 a** $0, -4$  **b** $0, 3$

**9 a** $4, -4$  **b** $7, -7$

**10 a** $2, -2$  **b** $4, -4$

**11 a** $3 \pm \sqrt{6}$  **b** $2 \pm \sqrt{6}$

**12 a** $-5 \pm 2\sqrt{7}$  **b** $-4 \pm 2\sqrt{3}$

**13 a** $\dfrac{3 \pm \sqrt{5}}{2}$  **b** $\dfrac{5 \pm \sqrt{33}}{2}$

**14 a** $\dfrac{-1 \pm \sqrt{21}}{2}$  **b** $\dfrac{-3 \pm \sqrt{13}}{2}$

**15 a** $\dfrac{6 \pm \sqrt{33}}{2}$  **b** $\dfrac{2 \pm \sqrt{5}}{2}$

**16 a** $\dfrac{-1 \pm \sqrt{13}}{4}$  **b** $\dfrac{-3 \pm \sqrt{5}}{4}$

**17 a** $\dfrac{-7 \pm \sqrt{37}}{2}$  **b** $\dfrac{-5 \pm \sqrt{13}}{2}$

**18 a** $\dfrac{-5 \pm \sqrt{33}}{2}$  **b** $\dfrac{-3 \pm \sqrt{29}}{2}$

**19 a** $\dfrac{5 \pm \sqrt{17}}{2}$  **b** $\dfrac{9 \pm \sqrt{65}}{2}$

**20 a** $-4 \pm \sqrt{19}$  **b** $-3 \pm \sqrt{13}$

**21 a** $\dfrac{-5 \pm \sqrt{17}}{4}$  **b** $\dfrac{-7 \pm \sqrt{33}}{4}$

**22 a** $\dfrac{7 \pm \sqrt{89}}{10}$  **b** $\dfrac{3 \pm \sqrt{89}}{10}$

**23 a** $\dfrac{5 \pm \sqrt{22}}{3}$  **b** $\dfrac{4 \pm \sqrt{10}}{3}$

**24 a** $\dfrac{-2 \pm \sqrt{7}}{2}$  **b** $\dfrac{-1 \pm \sqrt{2}}{2}$

**25 a** $3 < x < 7$  **b** $2 < x < 6$

**26 a** $-3 \leqslant x \leqslant 5$  **b** $-2 \leqslant x \leqslant 5$

**27 a** $x < -3 \text{ or } x > -2$  **b** $x < -4 \text{ or } x > -3$

**28 a** $x \leqslant 0 \text{ or } x \geqslant 6$  **b** $x \leqslant 0 \text{ or } x \geqslant 7$

**29 a** $(x+3)(x-4)$  **b** $-3, 4$

**30** a $(x-3)^2 - 11$  b $3 \pm \sqrt{11}$

**31** a $(x+5)^2 - 25$  b $-5 \pm 4\sqrt{2}$

**32** $1 < x < 5$

**33** $x \leq -b$ or $x \geq b$

**34** $-3, 7$

**35** $x > 4$

**36** $\dfrac{2 \pm \sqrt{7}}{3}$

**37** $x > q + p$ or $x < p - q$

**38** b $5, 8$ or $-8, -5$

**39** b $22\,\text{cm}$

**40** $(-2, 0)$ and $(3, 5)$

**41** $x = 3$

**42** a $2\,\text{s}$  b $5\,\text{m}$  c $\dfrac{4\sqrt{5}}{5}\,\text{s}$

**43** $\left(\dfrac{3 - \sqrt{23}}{2}, \dfrac{-3 - \sqrt{23}}{2}\right)$ and $\left(\dfrac{3 + \sqrt{23}}{2}, \dfrac{-3 + \sqrt{23}}{2}\right)$

**44** $x = \dfrac{5}{3}$ or $-2$

**45** $-3 < x < -2$ or $2 < x < 3$

**46** $1 < x \leq 3$

**47** $y = x$ or $y = 2x$

**48** $x = k$ or $x = k + 1$

**49** $x = 0,\ y = 1$

## Exercise 15C

**1** a $-24$, zero  b $17$, two

**2** a $0$, one  b $0$, one

**3** a $9$, two  b $49$, two

**4** a $56$, two  b $-60$, zero

**5** a $k < 16$  b $k < 9$

**6** a $k > \dfrac{25}{12}$  b $k > \dfrac{1}{4}$

**7** a $k = 12$  b $k = 20$

**8** a $k = \pm 12$
   b $k = \pm 20$

**9** a $-8 < k < 8$
   b $-6 < k < 6$

**10** a $\dfrac{16}{3}$  b $-\dfrac{4}{3}$

**11** $\pm 20$

**12** $k < \dfrac{9}{8}$

**13** $k > \dfrac{25}{24}$

**14** b $4$

**15** b $-1, -9$

**16** $k < -4$ or $k > 4$

**17** $a \leq 0$ or $a \geq 12$

**18** $\pm 6$

**19** $c > \dfrac{9}{8}$

**20** $0 < b < 8$

**21** $a < -1 - 4\sqrt{3}$ or $a > -1 + 4\sqrt{3}$

**22** $k < -3$ or $k > 1$

**23** $a \leq \dfrac{9}{4}$

**24** $-4 < a < 4$

**25** $a < -1$

**26** $k = \dfrac{27}{4}$

**27** $-5 < k < 5$

# Chapter 15 Mixed Practice

**1** a $(x+9)(x-2)$
   b $(-9, 0), (2, 0)$

**2** $\left(-\dfrac{3}{2}, 0\right), \left(\dfrac{3}{2}, 0\right)$

**3** a $-2(x-2)^2 + 5$
   b $(2, 5)$

**4** a $3$  b $1$  c $2$

**5** a $a = 3, b = 2$
   b $(3 - \sqrt{2}, 0), (3 + \sqrt{2}, 0)$

**6** a $(-3, 0), (5, 0)$
   b $x = 1$
   c $-16$

**7** a $41$  b two

**8** $k > \dfrac{4}{5}$

**9** **a** $x = 2$

**b**

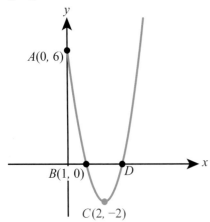

A(0, 6)

B(1, 0)   D

C(2, −2)

**c** 3

**10** **a** $3(x-1)^2 + 7$ **b** $f(x) \geqslant 7$

**11** **a** $-(x+2)^2 + 7$ **b** $f(x) \leqslant 7$

**12** $-2 \pm 2\sqrt{6}$

**13** **b** $9\,\mathrm{cm}^2$

**14** $k \leqslant -6\sqrt{2}$ or $k \geqslant 6\sqrt{2}$

**15** **a** **i** −5 **ii** $\dfrac{7}{8}$

**b**

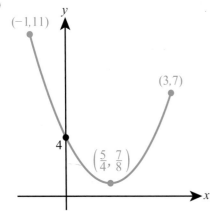

(−1, 11)

(3, 7)

4

$\left(\dfrac{5}{4}, \dfrac{7}{8}\right)$

**16** **a** $-10(x+4)(x-6)$ **b** $-10(x-1)^2 + 250$

**17** **b** $\pm 5$

**18** **a** $(x-3)^2 + 6$ **b** $\dfrac{1}{2}$

**19** $\pm\sqrt{15}$

**20** $a > 1$

# Chapter 16 Prior Knowledge

**1** **a** $x \geqslant -1$ **b** $f(x) \geqslant 0$

**2** $x = 1, y = 1$

**3** **a** $x^2$ **b** $4x$

**4**

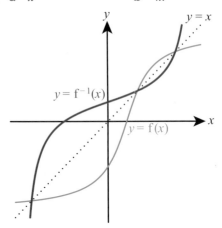

$y = x$

$y = f^{-1}(x)$

$y = f(x)$

# Exercise 16A

**1** **a**

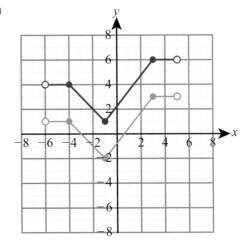

**b**

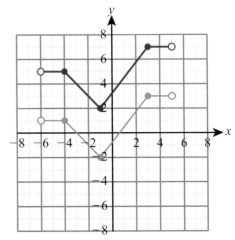

2  a

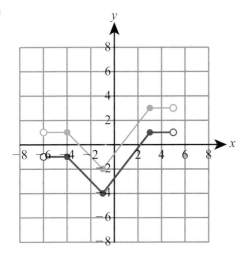

b

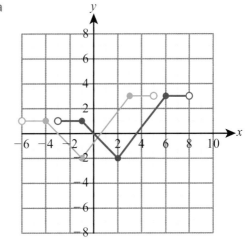

b

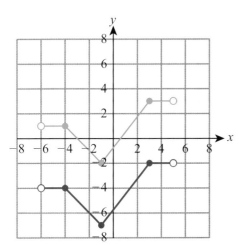

4  a

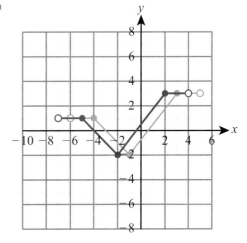

3  a

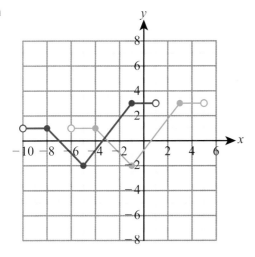

b

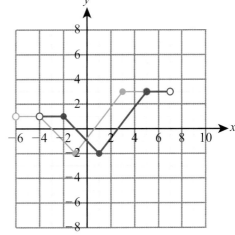

5  a  $y = 3x^2 + 3$          b  $y = 2x^3 + 5$

6  a  $y = 8x^2 - 7x - 4$     b  $y = 8x^2 - 7x - 1$

7  a  $y = 4(x - 3)^2$        b  $y = 3(x - 6)^3$

8  a  $y = (x+3)^2 + 6(x+3) + 2$

   b  $y = (x+2)^2 + 5(x+2) + 4$

9  a

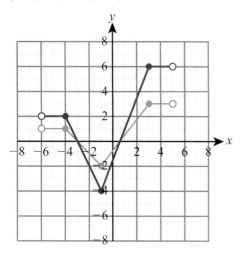

   b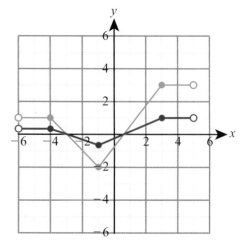

   b

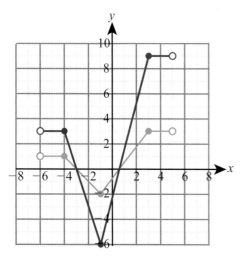

11 a

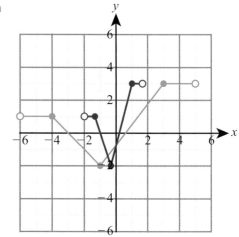

10 a

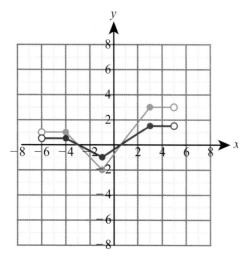

   b

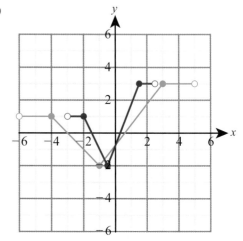

**12 a**

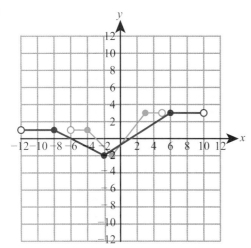

**b**

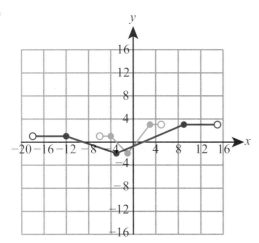

**13 a** vertical stretch, sf 3

　**b** vertical stretch, sf 2

**14 a** vertical stretch, sf $\frac{1}{2}$

　**b** vertical stretch, sf $\frac{1}{3}$

**15 a** horizontal stretch, sf $\frac{1}{2}$

　**b** horizontal stretch, sf $\frac{1}{3}$

**16 a** horizontal stretch, sf 2

　**b** horizontal stretch, sf 3

**17 a** $(2, -3)$ 　　　　　**b** $(5, -1)$

**18 a** $(-2, 4)$ 　　　　　**b** $(-3, 1)$

**19 a** $(-2, 3)$ 　　　　　**b** $(-1, 5)$

**20 a** $(2, -3)$ 　　　　　**b** $(5, -1)$

**21 a**

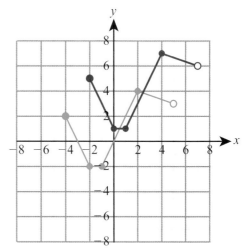

**b**

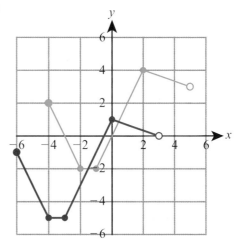

**22 a**

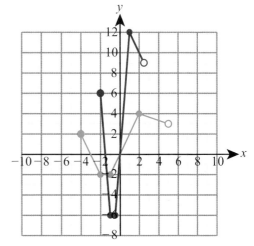

b

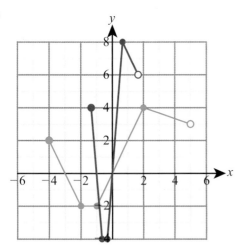

24 a

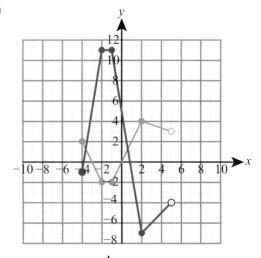

23 a

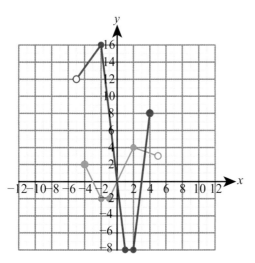

b

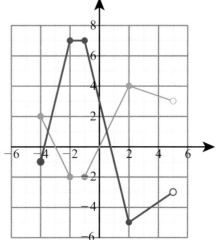

b

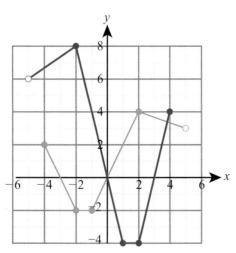

25 a  vertical stretch with scale factor 3, vertical translation −2 units

   b  vertical stretch scale factor 5, vertical translation 1 unit

26 a  horizontal stretch scale factor 2, vertical translation −1 unit

   b  horizontal stretch scale factor 2, vertical translation 3 units

27 a  horizontal translation 2 units, vertical translation 5 units

   b  horizontal translation −3 units, vertical translation −4 units

28 a  horizontal stretch scale factor $\frac{1}{3}$, reflection in $y$-axis

   b  horizontal stretch scale factor $\frac{1}{2}$, reflection in $y$-axis

**29 a**

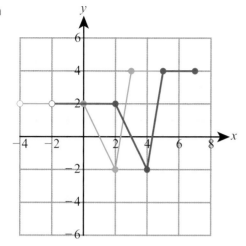

**b**

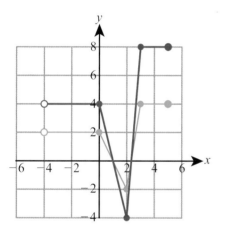

**c**

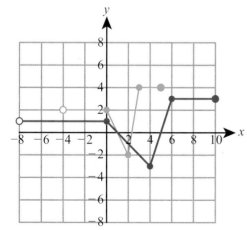

**30** $y = 12x^2 + 56x + 60$

**31** $y = 3e^{x-2}$

**32 a** $(x-5)^2 - 14$

  **b** horizontal translation 5 units, vertical translation −14 units

**33 a** $5(x+3)^2$

  **b** horizontal translation −3 units, vertical stretch scale factor 5

**34 a** $y = 9x^2$      **b** 3

**35 a** $y = 16x^3$      **b** $\dfrac{1}{2}$

**36** $y = 2x^3 + 5x^2$

**37 a** $y = 5f(x) + 3$

  **b** translation $\dfrac{3}{5}$ units, stretch sf 5

**38** $y = 2f(x) - 1$

# Exercise 16B

**1 a**

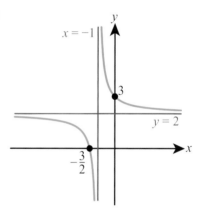

**b**

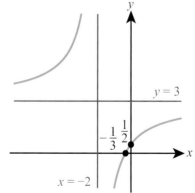

2 a

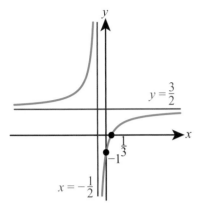

b

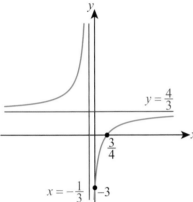

3 a

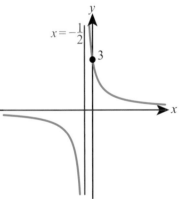

b

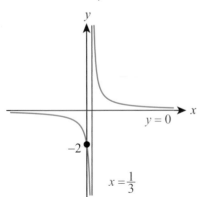

4 a

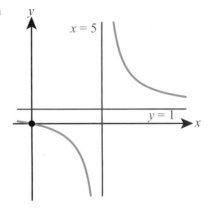

b

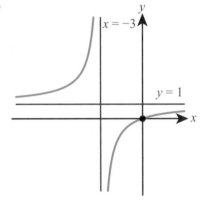

5 a $x = 0$, $y = 0$

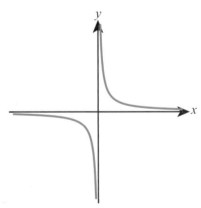

b vertical translation 2 units up

c $x = 0$, $y = 2$

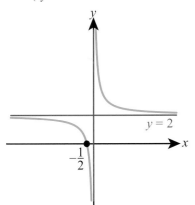

**6 a** $x = 0$, $y = 0$

**b** horizontal translation 3 units to the right

**c** $x = 3$, $y = 0$

**7 a** horizontal translation by $-2$ units

**b**

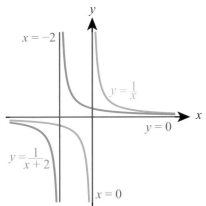

**8 a** $(-3, 0)$, $\left(0, -\dfrac{3}{2}\right)$

**b** $x = 2$, $y = 1$

**c**

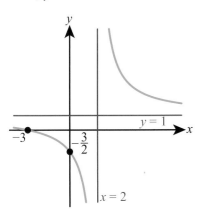

**9 a** $x = -1$, $y = 2$

**b** $-\dfrac{3}{2}$

**10**

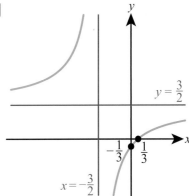

**11**

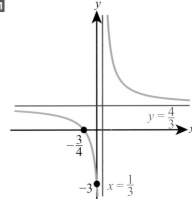

**12 a** $2 + \dfrac{5}{x}$

**b** vertical stretch with scale factor 5, vertical translation by 2 units

**c** $x = 0$, $y = 2$

**13 b** horizontal translation by 3 units, vertical translation by 2 units

**c**

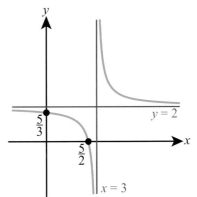

**14 a** $5 - \dfrac{1}{x}$

**b** reflection in the $x$-axis, vertical translation by 5 units

**c**

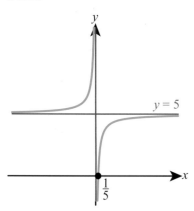

**15 a** $\dfrac{x}{x-2}$

**b**

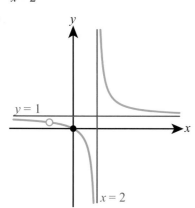

**16 a**

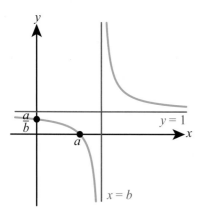

**b**

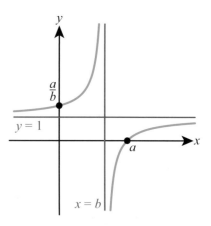

# Exercise 16C

**1 a**

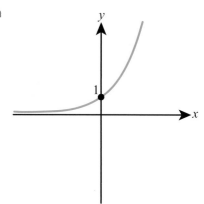

**b**

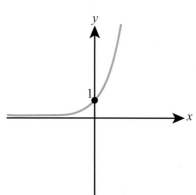

**2 a**

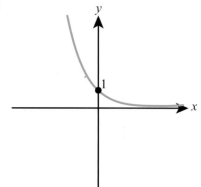

**b**

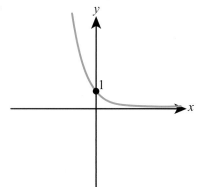

**b**

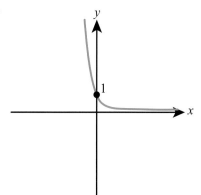

**3 a**

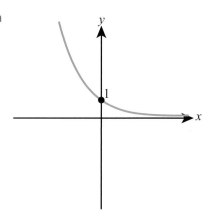

**5 a**

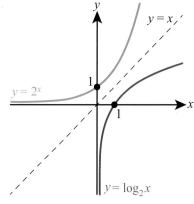

**b**

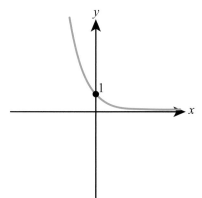

**b**

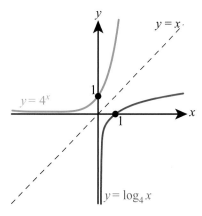

**4 a**

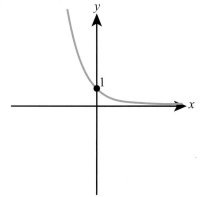

**6 a**

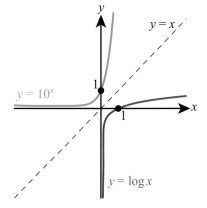

**b**

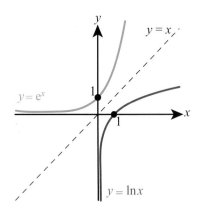

**7 a**

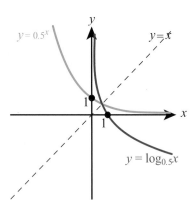

**b**

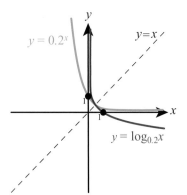

**8 a**

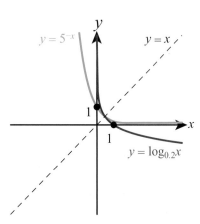

**b**

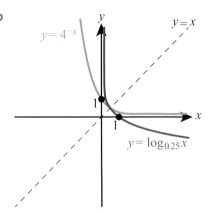

**9 a** $k = 0.742$      **b** $k = 0.531$

**10 a** $k = 1.44$      **b** $k = 1.63$

**11 a** $k = -0.511$      **b** $k = -0.693$

**12 a** $k = -1.10$      **b** $k = -1.39$

**13** A = ii, B = i, C = iii

**14**

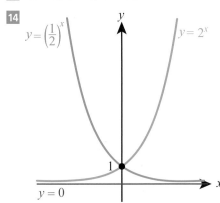

**15**

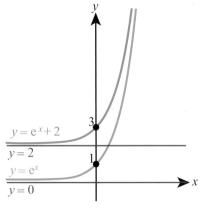

16

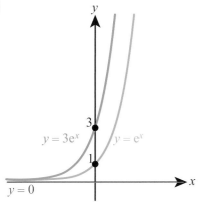

17

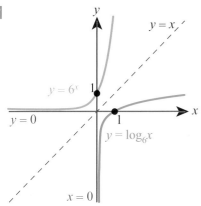

18  A = iii,  B = i,  C = ii
19  A = ii,  B = i,  C = iii
20

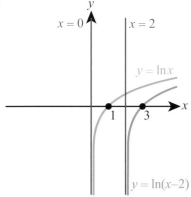

21  a   2                b   3
22  a   8                b   $y = 6$
23  a   $x = -5$

b

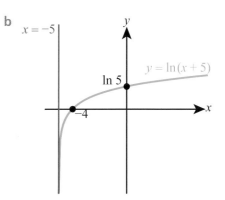

24  $C = 3$, $a = 4$
25  a   2                b   0.85
26  a   1.65
    b   horizontal stretch with scale factor 0.607
27  a   $f(x) > -2$

    b   $f^{-1}(x) = \ln\left(\dfrac{x+2}{3}\right)$, $x > -2$

    c

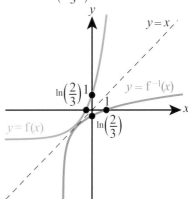

28  a   $y = \ln 3x$
    b   vertical translation by $\ln 3$ units
29  a   vertical stretch scale factor 5
    b   horizontal translation $-\ln 5$ units

# Chapter 16 Mixed Practice

1  a

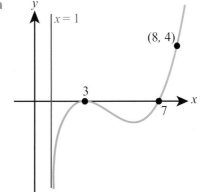

**b**

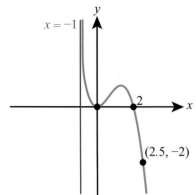

$x = -1$

$2$

$(2.5, -2)$

**c**

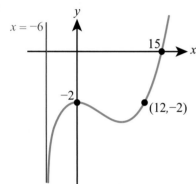

$x = -6$

$15$

$-2$

$(12, -2)$

**2** $y = 2x^3 - 18x^2 + 50x - 42$

**3 a** $(x + 2)^2 + 5$

 **b** horizontal translation by $-2$ units, vertical translation by $5$ units

**4 a** $y = \dfrac{3}{x + 2}$

 **b**

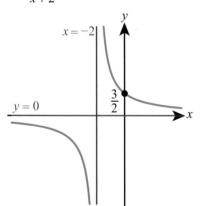

$x = -2$

$\dfrac{3}{2}$

$y = 0$

**5 b** translation 5 units to the right and 2 units up.

 **c** $x = 5$, $y = 2$

**6 a**

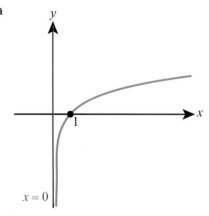

$1$

$x = 0$

**b**

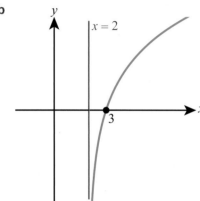

$x = 2$

$3$

**c**

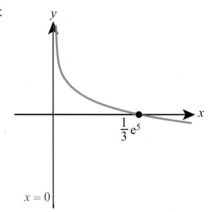

$\dfrac{1}{3}\mathrm{e}^5$

$x = 0$

**7** $a = 3$, $b = 4$

**8** vertical stretch scale factor $\dfrac{1}{3}$, translation 5 units to the left

**9** $y = \ln(\mathrm{e}^6(x - 2)^2)$

**10 a** $x = -\dfrac{7}{2}$, $y = 2$  **b** $\left(\dfrac{3}{4}, 0\right), \left(0, -\dfrac{3}{7}\right)$

c

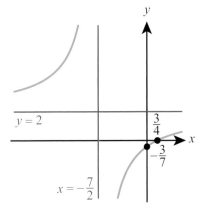

**11 a** $x = -5$, $y = 3$     **b** $x \neq -5$, $\mathrm{f}(x) \neq 3$

**12 a** translation 3 units to the left

   **b**

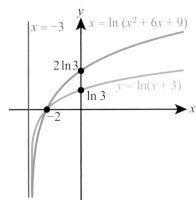

**13 a** $p = -4$, $q = 6$     **b** $y = -x^2 + 12x - 31$

**14 a** 3     **b** 7     **c** $y = 7$

**15 a** 800     **b** 2     **c** 40

**16** $\ln\left(\dfrac{e^2}{x-3}\right)$

**17 a** $x = 5$, $y = 2$     **b** $\alpha = 2$, $\beta = 1$

   **c** horizontal translation 5 to the right, vertical translation 2 up

   **d** $\dfrac{5x-9}{x-2}$     **e** reflection in $y = x$

**18** horizontal stretch factor 2

**19** vertical stretch factor $\dfrac{1}{\ln 10}$

**20**

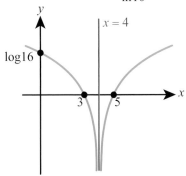

**21**

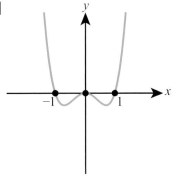

**22** $y = 2 - \mathrm{f}(x)$

# Chapter 17 Prior Knowledge

**1** $x = 0, \dfrac{2}{3}$

**2** $x = -4, 7$

**3** $x = \ln 7 - 2 \approx -0.0541$

**4** $x = \dfrac{1}{2}$

**5** £113.14

# Exercise 17A

**1 a** $0, \ln 4$     **b** $0, \ln 5$

**2 a** $0, e^2$     **b** $0, e^3$

**3 a** $1, \ln 2$     **b** $1, \ln 3$

**4 a** $0, \pm 1$     **b** $0, \pm 3$

**5 a** $0, 8$     **b** $0, 23$

**6 a** $0, 1, \ln 5$     **b** $0, -2, \ln\left(\dfrac{20}{3}\right)$

**7 a** $\pm 1, \pm 2$     **b** $\pm 2, \pm 3$

**8 a** $\sqrt[3]{2}, \sqrt[3]{3}$     **b** $1, \sqrt[3]{7}$

**9 a** $\pm 2$     **b** $\pm 3$

**10 a** $4, 25$     **b** $9, 16$

**11 a** $16$     **b** $25$

**12 a** $0, \ln 6$     **b** $0, \ln 3$

**13 a** $\ln 3, \ln 5$     **b** $\ln 3, \ln 4$

**14 a** $\ln 2$     **b** $\ln 2$

**15 a** $\ln 3$     **b** $\ln 4$

**16 a** no solutions     **b** no solutions

**17 a** 2     **b** 1

**18 a** 3     **b** 5

**19** $0, \pm\sqrt{5}$

20  $1, \dfrac{8}{5}$

21  $\pm 2, \pm\sqrt{6}$

22  $e^4, e^{-2}$

23  $\ln 3, \ln 7$

24  $\ln 3$

25  $\ln 2$

26  2

27  $-1, 27$

28  3

29  10

30  2, 3

31  $8, \dfrac{1}{2}$

32  $1 - \log_3 2$

33  $\pm 4, 8, 6$

# Exercise 17B

1  a  $-1.52$      b  1.24

2  a  $-1, 2.26$      b  $-3.04, 0.650$

3  a  $-0.836, 0.910, 1.84$

    b  $-1.71, -0.731, 0.689$

4  a  $0.144, 3.26$      b  $0.300, 2.36$

5  a  1.17      b  2.42

6  a  $-3.02, -0.175$      b  $0.191, 4.12$

7  a  $-1.75$      b  1.15

8  a  $1.59, 4.41$      b  $4.81, 10.2$

9  $-2.02, 0.379, 1.12$

10  $4.42, 11.0$

11  a

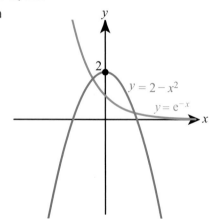

    b  two

12  $-1.84, 1.15$

13  a

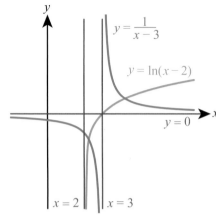

    b  two

14  $0.0922, 0.643, 13.8$

15  $-3.08 < k < 3.08$

16  0.667

17  $k > 2.15$

18  a  three      b  zero

# Exercise 17C

1  1.54 seconds

2  2.6 months

3  2.20 seconds

4  17.2 hours

5  $72\,cm^2$

6  12

7  $-0.257, 0.361, 0.896$

8  a  $600 \times 1.03^n$

9  58 years

10  a  12 m      b  60 km per hour

    c  50 km per hour

11  a  $\ln\left(2 \pm \sqrt{3}\right)$

12  b  29.4

13  b  $\ln 2$

# Chapter 17 Mixed Practice

1  $0, e^5 - 1$

2  $-1.81, 2.12$

3  a  42 units

    b  2.47 and 6.93 hours

    c  30 hours

**4** ±3

**5 b** 0.17 s

**6 a**

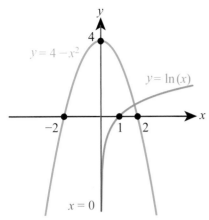

**b** one

**7 a** 10 500 **b** 13 400 **c** 33

**8** 1.51, 2.10

**9** 1, 25

**10** $\ln 2, \ln 5$

**11** 3

**12** 1

**13** $0 < k < 10.1$

**14 a** two **b** two **c** zero

**15** $x = -1.89$

**16** $x = \ln 3$

**17 a** $\dfrac{1}{3}$ **b** 78 732 **c** 10

**18** 3

**19** 2 or 3

**20** $\ln\left(k \pm \sqrt{k^2 - 1}\right)$

**21** $\dfrac{2}{3}$

**22 a** −0.796 **b** −0.898

**23 a** none **b** one

**25** $\sqrt{2} - 1$

# Chapter 18 Prior Knowledge

**1** $\theta = 21°$

**2** $x = -2, -3$

**3** $y = (x-3)^2 + 2(x-3) = x^2 - 4x + 3$

**4** $\dfrac{2x}{x^2 - 1}$

# Exercise 18A

**1 a** $\dfrac{\pi}{3}$ **b** $\dfrac{\pi}{4}$

**2 a** $\dfrac{5\pi}{6}$ **b** $\dfrac{2\pi}{3}$

**3 a** $\dfrac{\pi}{2}$ **b** $\dfrac{3\pi}{2}$

**4 a** 0.489 **b** 0.628

**5 a** 1.17 **b** 1.36

**6 a** 3.42 **b** 4.12

**7 a** 35.5° **b** 47.6°

**8 a** 72.2° **b** 77.3°

**9 a** 264.1° **b** 300.2°

**10 a** 36° **b** 22.5°

**11 a** 105° **b** 48°

**12 a** 420° **b** 330°

**13**

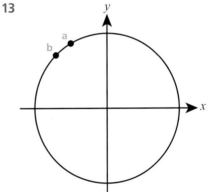

**14**

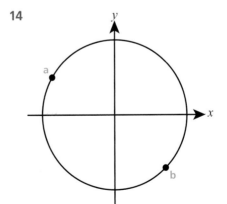

**15**

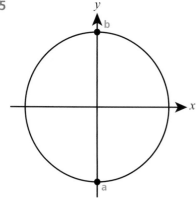

**16**

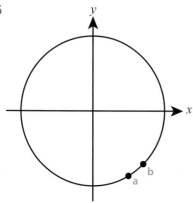

**17**

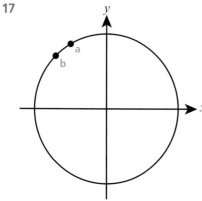

**18**

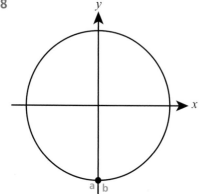

**19 a** 3.5 cm, 8.75 cm²     **b** 8.8 cm, 35.2 cm²

**20 a** 7.2 cm, 14.4 cm²     **b** 14.7 cm, 51.5 cm²

**21 a** 33.6 cm, 134 cm²     **b** 25.5 cm, 63.8 cm²

**22 a** 19.3 cm², 28.0 cm     **b** 4.73 cm², 15.9 cm

**23 a** 41.3 cm², 33.7 cm     **b** 15.5 cm², 19.2 cm

**24 a** 177 cm², 57.3 cm     **b** 8.56 cm², 13.2 cm

**25** per = 20.8 cm, area = 19.2 cm²

**26** per = 27.9 cm, area = 48.1 cm²

**27** 10.3 cm

**28 a** $\theta = 1.4$     **b** 17.5 cm²

**29** 0.699

**30** 15.5 cm

**31 a** 11.6 cm     **b** 38.2 cm

**32** area = 57.1 cm², per = 30.3 cm

**33** 6.84 cm

**34** 13.7 cm

**35** $12\pi$ cm

**36** 0.241 cm

**37** per = 14.3 cm, area = 2.55 cm²

**38 b** 1.74     **c** 13.1 cm

**39** 3

**40** $r = 23.4$ cm, $\theta = 2.15$

**41** per = 33.5 cm, area = 78.6 cm²

# Exercise 18B

**1 a** −0.6     **b** −0.6

**2 a** 0.6     **b** −0.6

**3 a** −0.6     **b** 0.6

**4 a** 0.6     **b** −0.6

**5 a** −0.940     **b** −0.940

**6 a** −0.940     **b** −0.940

**7 a** 0.940     **b** 0.940

**8 a** −0.940     **b** 0.940

**12 a** −0.73     **b** −0.73

**13 a** 0.73     **b** 0.73

**14 a** −0.73     **b** −0.73

**15 a** $\dfrac{1}{2}$     **b** $-\dfrac{1}{2}$

**16 a** $\dfrac{\sqrt{2}}{2}$     **b** $-\dfrac{\sqrt{2}}{2}$

**17 a** $-\sqrt{3}$     **b** $-\sqrt{3}$

18 a $\dfrac{\sqrt{3}}{2}$     b $-\dfrac{\sqrt{3}}{2}$

19 a $-\dfrac{\sqrt{2}}{2}$     b $-\dfrac{\sqrt{2}}{2}$

20 a $-\dfrac{1}{\sqrt{3}}$     b $\dfrac{1}{\sqrt{3}}$

21 a $64°, 116°$     b $49°, 131°$

22 a $55°$     b $57°$

23 a $59°, 121°$     b $66°, 114°$

24 a $57°$     b $50°$

25 $0$

27 $4 - 2\sqrt{3}$

28 $\dfrac{5}{6}\sqrt{3}$

29 $\cos x$

30 $\sqrt{61}$ cm

31 $52.7, 127$

32 $4.42\,\text{cm}, 6.37\,\text{cm}$

33 $18.2$ cm

34 $\sqrt{39}$

35 $y = \dfrac{x}{\sqrt{3}}$

36 c $-1$

# Exercise 18C

1 a $\pm\dfrac{\sqrt{15}}{4}$     b $\pm\dfrac{\sqrt{5}}{3}$

2 a $\pm\dfrac{2\sqrt{2}}{3}$     b $\pm\dfrac{\sqrt{7}}{4}$

3 a $-\dfrac{1}{2}$     b $-\dfrac{7}{9}$

4 a $\dfrac{1}{9}$     b $\dfrac{7}{8}$

5 a $\cos x = \pm\dfrac{\sqrt{21}}{5}, \tan x = \pm\dfrac{2}{\sqrt{21}}$

   b $\cos x = \pm\dfrac{\sqrt{7}}{4}, \tan x = \pm\dfrac{3}{\sqrt{7}}$

6 a $\sin x = \pm\dfrac{\sqrt{15}}{4}, \sin 2x = \mp\dfrac{\sqrt{15}}{8}$

   b $\sin x = \pm\dfrac{2\sqrt{6}}{5}, \sin 2x = \mp\dfrac{4\sqrt{6}}{25}$

7 a $\dfrac{4\sqrt{5}}{9}$     b $\dfrac{3\sqrt{7}}{8}$

8 a $-\dfrac{2}{\sqrt{21}}$     b $-\dfrac{1}{2\sqrt{2}}$

9 a $\dfrac{\sqrt{15}}{8}$     b $\dfrac{4\sqrt{2}}{9}$

10 a $-\dfrac{\sqrt{5}}{2}$     b $-\sqrt{3}$

11 a $\dfrac{\sqrt{2 - \sqrt{2}}}{2}$     b $\dfrac{\sqrt{2 + \sqrt{2}}}{2}$

12 a $\dfrac{\sqrt{2 - \sqrt{3}}}{2}$     b $\dfrac{\sqrt{2 + \sqrt{3}}}{2}$

13 a $\dfrac{\sqrt{2 - \sqrt{3}}}{2}$     b $\dfrac{\sqrt{2 + \sqrt{3}}}{2}$

14 a $\dfrac{\sqrt{2 - \sqrt{2}}}{2}$     b $\dfrac{\sqrt{2 + \sqrt{2}}}{2}$

15 a $-\dfrac{\sqrt{21}}{5}$     b $-\dfrac{4\sqrt{21}}{25}$

16 a $-\dfrac{2\sqrt{10}}{7}$     b $-\dfrac{3\sqrt{10}}{20}$

17 a $\dfrac{49}{81}$     b $\pm\dfrac{\sqrt{65}}{9}$

18 $7 - 4\sin^2 x$

19 $9\cos^2 x - 5$

23 a $\dfrac{\sqrt{2}}{2}$

24 $\dfrac{\sqrt{33}}{6}$

25 $\dfrac{\sqrt{3}}{3}$

26 $-\dfrac{79}{81}$

27 a $1 - 2\sin^2 x$

32 $\dfrac{\sqrt{14}}{4}$

33 a $\sqrt{\dfrac{2 - \sqrt{2}}{4}}$

# Exercise 18D

1 a

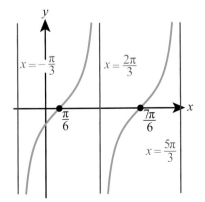

**b**

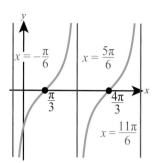

**2 a**

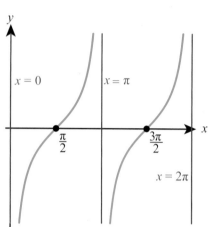

**b**

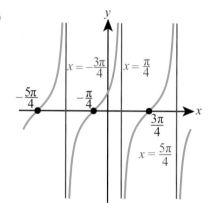

**3 a**

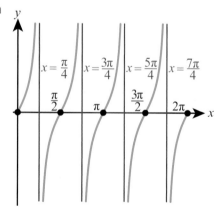

**b**

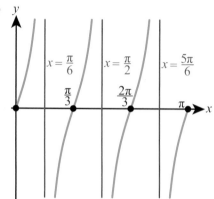

**4 a**

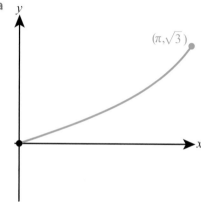

**b**

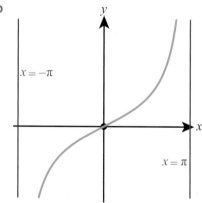

**5 a**

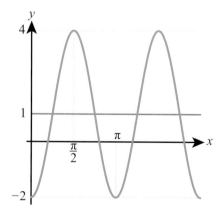

b

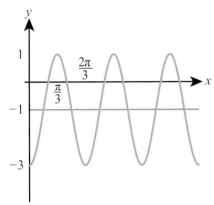

b

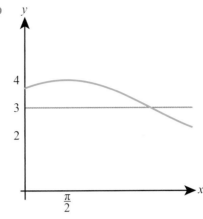

6 a

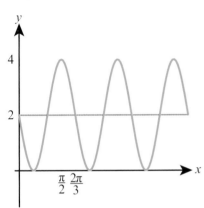

8 a

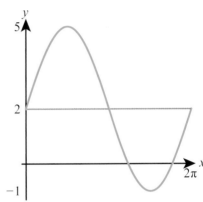

b

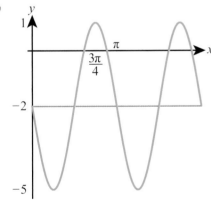

b

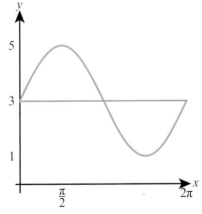

7 a

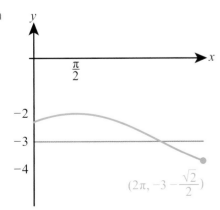

9 a $\quad a = 3,\ b = 2,\ d = 1$
  b $\quad a = 2,\ b = 3,\ d = 1$

10 a $\quad a = 2,\ c = \dfrac{\pi}{2},\ d = 2$

  b $\quad a = 3,\ c = \dfrac{\pi}{3},\ d = 1$

11 a $\quad a = 4,\ b = 2,\ d = -3$
  b $\quad a = 2,\ b = 3,\ d = 2$

12 a $\quad a = -2,\ c = \dfrac{\pi}{2},\ d = 2$

  b $\quad a = -1,\ c = \dfrac{\pi}{4},\ d = 3$

**13**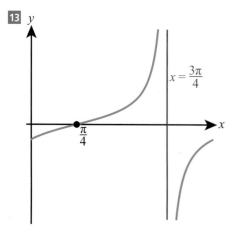

**14** $\dfrac{\pi}{2}$

**15** −17, 3

**16** 1.4 m

**17** a  6.6 m          b  6 a.m.

**18** a  8 minutes     b  11 m          c  8.6 m

**19** $a = 5,\ b = 2$

**20** 3

**21** a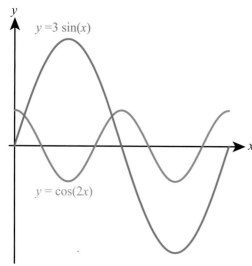

  b  two          c  six

**22** a  1.6 m         b  7
  c  0.279 seconds

**23** $\pi$

**24** a  no
  b  $\dfrac{3}{4}$, for $x = \dfrac{3\pi}{4}$

# Exercise 18E

**1**  a  0.386, 1.71, 2.48          b  0.912, 2.23

**2**  a  no solutions              b  no solutions

**3**  a  −0.795, 0.205            b  5.20, 11.5

**4**  a  $\dfrac{\pi}{6}, \dfrac{5\pi}{6}$          b  $\dfrac{\pi}{3}, \dfrac{2\pi}{3}$

**5**  a  $-\dfrac{7\pi}{4}, -\dfrac{5\pi}{4}, \dfrac{\pi}{4}, \dfrac{3\pi}{4}, \dfrac{9\pi}{4}, \dfrac{11\pi}{4}$

   b  $-\dfrac{3\pi}{2}, \dfrac{\pi}{2}, \dfrac{5\pi}{2}$

**6**  a  $\dfrac{4\pi}{3}, \dfrac{5\pi}{3}$          b  $\dfrac{5\pi}{4}, \dfrac{7\pi}{4}$

**7**  a  $-\dfrac{5\pi}{2}, -\dfrac{\pi}{2}$          b  $\dfrac{7\pi}{6}, \dfrac{11\pi}{6}$

**8**  a  60°, 300°                b  45°, 315°

**9**  a  −330°, −30°, 30°        b  −300°, −60°, 60°

**10** a  −135°, 135°             b  120°, −120°

**11** a  −180°, 180°             b  −360°, 0°, 360°

**12** a  $\dfrac{\pi}{3}, \dfrac{4\pi}{3}$          b  $\dfrac{\pi}{6}, \dfrac{7\pi}{6}$

**13** a  $-\dfrac{7\pi}{4}, -\dfrac{3\pi}{4}, \dfrac{\pi}{4}, \dfrac{5\pi}{4}$

   b  $-\dfrac{5\pi}{3}, -\dfrac{2\pi}{3}, \dfrac{\pi}{3}, \dfrac{4\pi}{3}$

**14** a  $\dfrac{5\pi}{6}, \dfrac{11\pi}{6}, \dfrac{17\pi}{6}$          b  $\dfrac{3\pi}{4}, \dfrac{7\pi}{4}, \dfrac{11\pi}{4}$

**15** a  $-\dfrac{7\pi}{3}, -\dfrac{4\pi}{3}, -\dfrac{\pi}{3}, \dfrac{2\pi}{3}$          b  −2π, −π, 0

**16** a  230°, 350°              b  265°, 355°

**17** a  65°, 155°               b  90°, 150°

**18** a  −π, 0, π                b  $-\dfrac{5\pi}{6}, \dfrac{\pi}{6}$

**19** a  $-\dfrac{13\pi}{18}, -\dfrac{11\pi}{18}, -\dfrac{\pi}{18}, \dfrac{\pi}{18}, \dfrac{11\pi}{18}, \dfrac{13\pi}{18}$

   b  $-\dfrac{5\pi}{6}, -\dfrac{\pi}{6}, \dfrac{\pi}{6}, \dfrac{5\pi}{3}$

**20** a  $\dfrac{5\pi}{12}, \dfrac{7\pi}{12}$          b  $\dfrac{7\pi}{24}, \dfrac{11\pi}{24}, \dfrac{19\pi}{24}, \dfrac{23\pi}{24}$

**21** a  22.5°, 112.5°           b  20°, 80°, 140°

**22** a  $\dfrac{\pi}{6}, \dfrac{\pi}{3}, \dfrac{7\pi}{6}, \dfrac{4\pi}{3}$          b  $\dfrac{\pi}{8}, \dfrac{3\pi}{8}, \dfrac{9\pi}{8}, \dfrac{11\pi}{8}$

**23 a** $105°, 165°, 285°, 345°$
  **b** $120°, 150°, 300°, 330°$

**24 a** $\dfrac{\pi}{3}, \dfrac{4\pi}{3}$    **b** $\dfrac{\pi}{6}, \dfrac{7\pi}{6}$

**25 a** $-135°, 45°$    **b** $-45°, 135°$

**26 a** $\dfrac{\pi}{6}, \dfrac{\pi}{2}, \dfrac{5\pi}{6}$    **b** $\dfrac{\pi}{6}, \dfrac{3\pi}{2}, \dfrac{5\pi}{6}$

**27 a** $45°, 135°$    **b** $0°, 135°, 180°$

**28 a** $\dfrac{2\pi}{3}, \dfrac{4\pi}{3}$    **b** $\dfrac{\pi}{3}, \dfrac{5\pi}{3}$

**29 a** $\dfrac{\pi}{6}, \dfrac{\pi}{2}, \dfrac{5\pi}{6}$    **b** $\dfrac{\pi}{6}, \dfrac{3\pi}{2}, \dfrac{5\pi}{6}$

**30 a** $0, \dfrac{2\pi}{3}, \dfrac{4\pi}{3}, 2\pi$    **b** $\dfrac{\pi}{3}, \pi, \dfrac{5\pi}{3}$

**31 a** $\dfrac{\pi}{2}$    **b** $\pi$

**32 a** $0, \pi, 2\pi$    **b** $\dfrac{\pi}{2}, \dfrac{3\pi}{2}$

**33** $0.702, 4.41$

**34** $30°, 150°$

**35** $\pm 45°, \pm 135°$

**36** $130°, 250°$

**37 b** $60°, 90°, 270°, 300°$

**38** $45°, 225°$

**39** $0, \dfrac{\pi}{3}, \pi$

**40** $30°, 150°, 270°$

**41** $\dfrac{-3 + \sqrt{57}}{8}$

**42** $-\dfrac{2\pi}{3}, 0, \dfrac{2\pi}{3}$

**43** $k \leqslant 0$

# Chapter 18 Mixed Practice

**1 a** $1.3\,\mathrm{m}$    **b** $2.51\,\mathrm{m}$

**3** $60°, 120°$

**4** $55.2°, 125°$

**5** $0.695, 3.54, 2.45, 5.88$

**6** $0, \dfrac{2\pi}{3}, \pi, \dfrac{5\pi}{3}, 2\pi$

**7 a** $-\dfrac{12}{13}$    **b** $-\dfrac{5}{12}$    **c** $-\dfrac{120}{169}$

**8 a i** $14\,\mathrm{m}$    **ii** $26\,\mathrm{m}$
  **b** $10$ seconds, $30$ seconds

**c**
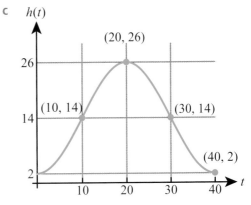

**d** $a = -12, b = \dfrac{\pi}{20}, c = 14$

**e i**
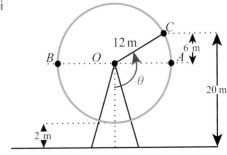

  **ii** $120°$    **iii** $13.3$ seconds

**9 a** $-\dfrac{\sqrt{7}}{4}$    **b** $-\dfrac{1}{8}$

**10** $\pm\dfrac{5\pi}{6}, \pm\dfrac{\pi}{6}$

**12** $5\left(\sqrt{6} - 1\right)$

**13** $17.8\ \mathrm{cm}^2$

**14 b** $0, \dfrac{\pi}{3}, \dfrac{2\pi}{3}, \pi$

**16** $a = 3, c = \dfrac{\pi}{4}$

**17 a** $5$    **b** $4$
  **c** $p = -\dfrac{\pi}{4}, q = 5$

**18** $\dfrac{\pi}{3}, \dfrac{5\pi}{3}$

**19** $3.7\,\mathrm{cm}$ and $3.6\,\mathrm{cm}$

**20 a** $44.0°$ ($0.769$ rad)    **b** $1.18\,\mathrm{cm}^2$

**23** $20\pi$

**24** $0 \leqslant \mathrm{f}(x) \leqslant 25$

**25** $0.207$ amps

**26** $6\pi$

**27** a

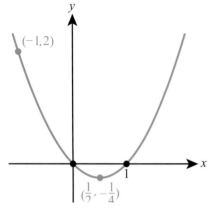

b  $-0.25 \leqslant k \leqslant 2$

**28** a  $82.3°$            b  $111\,cm$

**29** $\dfrac{31}{81}$

# Chapter 19 Prior Knowledge

**1** a  $y = -0.393x + 11.1$     b  $5.99$

**2** $\dfrac{3}{7}$

**3** $0.125$

**4** a  $0.106$               b  $x = 8.32$

# Exercise 19A

**1** a  $x = 1.66y - 0.0984$     b  $x = 0.952 + 4.02$

**2** a  $x = -1.31 + 9.81$      b  $x = -1.22 + 30.7$

**3** a  i  $x = 21.9$
    ii  reliable
  b  i  $x = -4.26$
    ii  not reliable (extrapolation)

**4** a  i  $x = 73.5$
    ii  not reliable (no correlation)
  b  i  $x = 87.9$
    ii  not reliable (no correlation)

**5** a  $x = -0.0620y + 15.5$
  b  $2.8\,km$
  c  Reliable; correlation is high and the values are within the range of the data

**6** $363$

**7** a  $65.6\,cm$            b  $149\,cm$

**8** $76$

**9** a  It predicts chemistry mark for a given mathematics mark
   b  $72$                    c  No, extrapolation

**10** a  $\$3845$             b  $\$4150, \$280$

**11** a  $-1$                 b  $x = -2y + 8$

**12** a  $6\,cm/minute$. The fact that the regression line has a non-zero estimate for $t = 0$ suggests that the model is not perfect, either due to measurement error or because the snail is not moving at a constant speed.
   b   $Y = 3.6T + 0.05$
   c   $0.6$

# Exercise 19B

**1** a  $0.5$                 b  $0.25$

**2** a  $0.8$                 b  $0.25$

**3** a  $0.4$                 b  $0.63$

**4** a  $0.21$                b  $0.06$

**5** a  $0.25$                b  $0.5$

**6** a  $0.625$               b  $0.5$

**7** a  no                    b  yes

**8** a  yes                   b  no

**9** a  yes                   b  yes

**10** a  no                   b  no

**11** a  no                   b  yes

**12** a  no                   b  no

**13** a  no                   b  yes

**14** a  yes                  b  no

**15** a  yes                  b  no

**16** a  $\dfrac{1}{3}$        b  $\dfrac{1}{4}$

**17** a  $0.15$               b  $0.3$

**18** a  $0.1$                c  $0.25$

**19** a  $P(A|B) = P(A)$       b  $0.15$
   c  $0.65$

**20** yes

**21** $x = 0.14, y = 0.39$  or  $x = 0.39, y = 0.14$

**22** $\dfrac{8}{45}$

**23** $\dfrac{8}{71}$

**24** $0.75$

**25** $\dfrac{5}{9}$

# Exercise 19C

1 a 1.33      b 0.25
2 a −1.5      b −2.38
3 a −1      b −3.25
4 a 0.833      b −0.476
5 a 71      b 143
6 a 3.59      b 0.11
7 a 5      b 9
8 a 133      b 8.75
9 a $P(Z < 1.5)$      b $P(Z < 1.4)$
10 a $P(-1.53 < Z < 0.2)$
   b $P(-1.67 < Z < -1.13)$
11 a $P(Z > 0.625)$      b $P(Z > 1.42)$
12 a $P(-0.286 < Z < 0.357)$
   b $P(1 < Z < 1.28)$
13 a 12      b 22
14 a 16      b 8.6
15 a 68      b 18
16 a 6.2      b 7.5
17 a $\mu = 21, \sigma = 2.5$      b $\mu = 13, \sigma = 1.6$
18 a $\mu = 5.5, \sigma = 0.82$      b $\mu = 6.2, \sigma = 0.95$
19 a $\mu = 35, \sigma = 12$      b $\mu = 40, \sigma = 15$
20 a $\mu = 28, \sigma = 8.5$      b $\mu = 24, \sigma = 7.3$
21 0.0918
22 2.42 cm
23 0.925
24 5.79
25 7.30
26 7.76
27 $\mu = 51.6, \sigma^2 = 126$
28 4.62%
29 99.7%
30 90.0

# Chapter 19 Mixed Practice

1 a 0.949      b 55
   c reliable; correlation is high and the value is within the range of the data
2 a low correlation
   b the line is $y$ on $x$
   c extrapolation

3 a $\dfrac{1}{10}$      b $\dfrac{1}{2}$
4 a 0.2      b no
5 a $\mu -1.645 \; \sigma = 22.5$      b $\mu = 24.7, \sigma = 1.31$
6 a $\dfrac{1}{10}$      b $\dfrac{17}{20}$      c no
7 a $y = 10.7x + 121$
   b i additional cost per box
      ii fixed costs
   c $760      d 13
   e i extrapolation
      ii the line is $y$ on $x$
9 6.14 cm
10 $\mu = 19.2, \sigma = 5.61$
11 20.5%
12 159
13 a 0.309      b 0.217
14 a 2.28%      b 10.1%
15 2630
16 $\mu = 766, \sigma = 10$
17 a 0.785      b 0.761
18 $\dfrac{1}{2}$
19 a $\dfrac{1}{2}$      b no
20 0.284
21 a −0.0209      b 0.983
   c $y = 2.07x^2 - 1.52$
22 a 0.819
   c 48.7
   d i 0.360      ii 0.924

# Chapter 20 Prior Knowledge

1 $f'(x) = 2x$
2 $y - 1 = 3(x - 1)$
3 $x = \ln 5 \approx 1.61$

# Exercise 20A

**1 a** $f'(x) = \frac{2}{3}x^{-\frac{1}{3}}$      **b** $f'(x) = \frac{3}{4}x^{-\frac{1}{4}}$

**2 a** $f'(x) = -\frac{1}{2}x^{-\frac{3}{2}}$      **b** $f'(x) = -\frac{4}{3}x^{-\frac{7}{3}}$

**3 a** $f'(x) = 9x^{\frac{1}{2}}$      **b** $f'(x) = 2x^{-\frac{3}{4}}$

**4 a** $f'(x) = -6x^{-\frac{5}{3}}$      **b** $f'(x) = -4x^{-\frac{7}{5}}$

**5 a** $f'(x) = x^{-\frac{1}{2}}$      **b** $f'(x) = \frac{3}{2}x^{-\frac{3}{4}}$

**6 a** $f'(x) = 4x^{-\frac{4}{3}}$      **b** $f'(x) = x^{-\frac{6}{5}}$

**7 a** $-3x^{-\frac{4}{3}} - \frac{4}{3}x^{-\frac{7}{3}}$      **b** $-x^{-\frac{3}{2}} + \frac{3}{2}x^{-\frac{5}{2}}$

**8 a** $3x^{\frac{1}{4}} - \frac{1}{20}x^{-\frac{3}{4}}$      **b** $2x^{\frac{4}{3}} + \frac{1}{21}x^{-\frac{2}{3}}$

**9 a** $-\frac{2}{3}x^{-\frac{4}{3}} - 2x^{-\frac{1}{3}}$      **b** $-\frac{5}{4}x^{-\frac{5}{4}} + 6x^{-\frac{1}{4}}$

**10 a** $f'(x) = 3\cos x$      **b** $f'(x) = -4\sin x$

**11 a** $f'(x) = -\frac{1}{2}\sin x - 5\cos x$

     **b** $f'(x) = \frac{3}{4}\cos x + 2\sin x$

**12 a** $\frac{dy}{dx} = \frac{3}{x}$      **b** $\frac{dy}{dx} = -\frac{4}{x}$

**13 a** $\frac{dy}{dx} = \frac{2}{x}$      **b** $\frac{dy}{dx} = -\frac{1}{x}$

**14 a** $y' = 5e^x$      **b** $y' = -6e^x$

**15 a** $y' = -\frac{e^x}{2}$      **b** $y' = \frac{3e^x}{4}$

**16** 0.5

**17** $-5\,\mathrm{m\,s^{-1}}$. Falcon is descending.

**18** (0,1)

**19** 5

**20** $y = -\frac{3x}{2} + \frac{\pi}{4} + \frac{3\sqrt{3}}{2}$

**21** $y = x + 3 - 3\ln 3$

**22** (9,3)

**23 a** $y = 2x - 5/2$

**24 a** 1 million      **b** 4 hours

     **c** 1.5 million/hour

     **d** no upper limit to the number of bacteria

**25** $x > 0.5$

**26** 0.5

**27 a** $\frac{\pi^2}{4}$      **b** $\frac{\pi^2}{4}$

**28** (4, 2)

**29** $\left(\frac{13\pi}{6}, 0\right)$

**30** 0.5

# Exercise 20B

**1 a** $\frac{dy}{dx} = 12(3x + 2)^3$

     **b** $\frac{dy}{dx} = 10(2x - 7)^4$

**2 a** $\frac{dy}{dx} = \frac{3}{2}x(x^2 + 3)^{-\frac{1}{4}}$

     **b** $\frac{dy}{dx} = -\frac{4}{3}x(4 - x^2)^{-\frac{1}{3}}$

**3 a** $\frac{dy}{dx} = \frac{1}{2}(2x^3 + x)^{-\frac{1}{2}}(6x^2 + 1)$

     **b** $\frac{dy}{dx} = \frac{1}{3}(5x - x^3)^{-\frac{2}{3}}(5 - 3x^2)$

**4 a** $f'(x) = 2\cos 2x$

     **b** $f'(x) = \frac{1}{3}\cos\frac{1}{3}x$

**5 a** $f'(x) = -\pi\sin \pi x$

     **b** $f'(x) = -5\sin 5x$

**6 a** $f'(x) = 3\cos(3x + 1)$

     **b** $f'(x) = -4\cos(1 - 4x)$

**7 a** $f'(x) = 3\sin(2 - 3x)$

     **b** $f'(x) = -\frac{1}{2}\sin\left(\frac{1}{2}x + 4\right)$

**8 a** $3e^{3x}$      **b** $\frac{1}{2}e^{\frac{x}{2}}$

**9 a** $-3x^2 e^{-x^3}$      **b** $8xe^{4x^2}$

**10 a** $\frac{1}{x}$      **b** $\frac{1}{x}$

**11 a** $\frac{2x}{x^2 + 1}$      **b** $-\frac{8x}{3 - 4x^2}$

**12 a** $y' = 6\sin 3x \cos 3x$

**b** $y' = -8\cos 4x \sin 4x$

**13 a** $y' = -2\sin 2x e^{\cos 2x}$ **b** $y' = 5\cos 5x e^{\sin 5x}$

**14 a** $y' = \dfrac{1}{2x}(\ln 3x)^{-\frac{1}{2}}$ **b** $y' = \dfrac{1}{3x}(\ln 2x)^{-\frac{2}{3}}$

**15** $-e^{-x}$

**16** 23

**17** $y = \dfrac{4x}{5} + \dfrac{9}{5}$

**18** $(3, 0)$

**19** $y = \dfrac{\pi^2 x}{4} - \dfrac{\pi}{2}$

**20** $y = \dfrac{x}{3} - \dfrac{5}{3} + \ln 3$

**21 a** $2\sin x \cos x$

**b** $\left(\dfrac{\pi}{2}, 1\right), (\pi, 0), \left(\dfrac{3\pi}{2}, 1\right)$

**25** $x = 3$

**26 b** no, e.g. $y = 3e^{2x}$

# Exercise 20C

**1 a** $\dfrac{dy}{dx} = x\cos x + \sin x$

**b** $\dfrac{dy}{dx} = x^{\frac{1}{2}}\cos x + \dfrac{1}{2}x^{-\frac{1}{2}}\sin x$

**2 a** $\dfrac{dy}{dx} = -x^2\sin x + 2x\cos x$

**b** $\dfrac{dy}{dx} = -x^{-1}\sin x - x^{-2}\cos x$

**3 a** $\dfrac{dy}{dx} = x^{-\frac{1}{2}}e^x - \dfrac{1}{2}x^{-\frac{3}{2}}e^x$

**b** $\dfrac{dy}{dx} = x^3 e^x + 3x^2 e^x$

**4 a** $\dfrac{dy}{dx} = x^{-\frac{1}{3}} + \dfrac{2}{3}x^{-\frac{1}{3}}\ln x$

**b** $\dfrac{dy}{dx} = x^3 + 4x^3 \ln x$

**5 a** $f'(x) = x(2x+1)^{-\frac{1}{2}} + (2x+1)^{\frac{1}{2}}$

**b** $f'(x) = \dfrac{9}{2}x^2(3x-4)^{\frac{1}{2}} + 2x(3x-4)^{\frac{3}{2}}$

**6 a** $f'(x) = 3x^2\cos 3x + 2x\sin 3x$

**b** $f'(x) = 2x^{\frac{3}{4}}\cos 2x + \dfrac{3}{4}x^{-\frac{1}{4}}\sin 2x$

**7 a** $f'(x) = -4x\sin(4x-1) + \cos(4x-1)$

**b** $f'(x) = -5x^{-1}\sin 5x - x^{-2}\cos 5x$

**8 a** $f'(x) = 4x^{\frac{1}{2}}e^{4x+5} + \dfrac{1}{2}x^{-\frac{1}{2}}e^{4x+5}$

**b** $f'(x) = -x^{-2}e^{1-x} - 2x^{-3}e^{1-x}$

**9 a** $f'(x) = \dfrac{2x^{-1}}{2x-3} - x^{-2}\ln(2x-3)$

**b** $f'(x) = -\dfrac{x^3}{5-x} + 3x^2\ln(5-x)$

**10 a** $y' = \dfrac{x^2 + 4x}{(x+2)^2}$ **b** $y' = \dfrac{-12}{(x-4)^2}$

**11 a** $y' = \dfrac{\dfrac{1}{2}x + 2}{(x+1)^{\frac{3}{2}}}$ **b** $y' = \dfrac{x-5}{(x-3)^3}$

**12 a** $y' = \dfrac{e^x(3x-5)}{(3x+1)^3}$ **b** $y' = \dfrac{e^x(2x-7)}{(2x-1)^4}$

**13** $y = \dfrac{4}{\pi} - \dfrac{4x}{\pi^2}$

**14** $y = \dfrac{2}{\pi}x - \dfrac{1}{2}$

**15** $7e^6$

**16** $e^x((x+1)\ln x + 1)$

**17** $e^{x\sin x}(\sin x + x\cos x)$

**18** $(\ln 2, 8 - 11\ln 2)$

**19** $\left(1, \dfrac{1}{e}\right)$

**20** $(e, e)$

**21** $\dfrac{k}{(k+x^2)^{1.5}}$

**22** $a = 3, b = 4$

**23** $a = 0, b = -2$

**24 a** $\dfrac{12e^{-n}}{(1-3e^{-n})^2}$

**b i** 1000 rabbits

   **ii** 750 rabbits per year

**c** 4

**d**

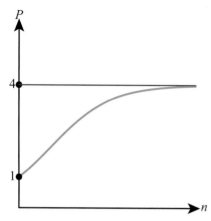

25 **b** $y = x$

# Exercise 20D

**1 a** 10        **b** 12

**2 a** 62        **b** 8

**3 a** $4e^{-4}$      **b** $18e^6$

**4 a** $-\dfrac{1}{4}$      **b** 1

**5 a**

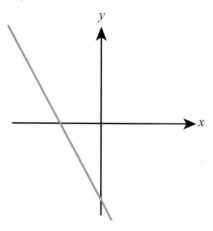

**b**

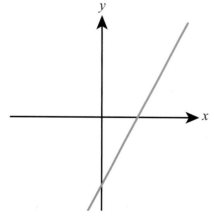

**6 a**

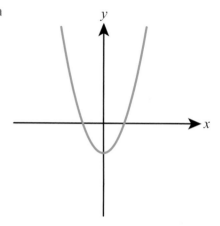

**b**

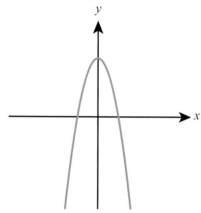

**7 a i** increasing, concave-down

     **ii** concave-up

     **iii** concave-down

   **b i** decreasing, concave-up

     **ii** increasing

     **iii** decreasing, concave-down

**8 a** $x < \dfrac{4}{3}$       **b** $x > -3$

**9 a** $x < -1$ or $x > 3$      **b** $0 < x < 3$

10 2

11 $a = 1$, $b = -1$

12 $(x + 2)e^x$

**13**

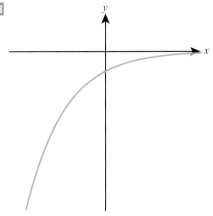

**14**

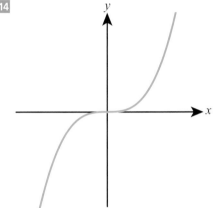

**15** $k = 3$

**16** $4, -3$

**20** $\left(\dfrac{1}{e}, -\dfrac{1}{e^2}\right)$

**21** $x < \dfrac{5}{3}$

**22 a** $-6$

 **b** $b > 12$

**23 a** $6$

**24** $5x^2 + 10x + 2$

# Exercise 20E

**1 a** $x = -1$ local max

 $x = 3$ local min

 **b** $x = 1$ local max

 $x = 2$ local min

**2 a** $x = \dfrac{1}{9}$ local max

 **b** $x = \dfrac{1}{4}$ local min

**3 a** $x = \dfrac{\pi}{3}$ local max

 $x = \dfrac{5\pi}{3}$ local min

 **b** $x = \dfrac{\pi}{6}$ local max

 $x = \dfrac{5\pi}{6}$ local min

**4 a** $x = \ln 5$ local min

 **b** $x = \ln\dfrac{1}{2}$ local max

**5 a** $x = \dfrac{3}{2}$ local max

 **b** $x = 4$ local min

**6** $b = -2, c = 3$

**7** $b = -4, c = 1$

**8** \$2.5 million

**9 a** $40\,\text{cm}$ **b** $100\,\text{cm}^2$

**10** $(0, 0)$ and $(-2, -4)$

**11** $(1, -1)$, local min

**12** $\left(\dfrac{\pi}{3}, \dfrac{\sqrt{3}}{2} - \dfrac{\pi}{6}\right)$

**13** $108\,\text{cm}^3$

**14 a** $\dfrac{1 - \ln x}{x^2}$ **b** $\left(e, \dfrac{1}{e}\right)$

 **d** Local max

**15** $\left(\dfrac{\pi}{2}, e\right)$, local max

 $\left(\dfrac{3\pi}{2}, \dfrac{1}{e}\right)$, local min

**16 a** $6\,\text{million}$ **b** $5.61\,\text{million}$

**17 a** \$7

 **b** It does not predict negative sales if $x > 10$.

 **c** \$9

**18 a** $\dfrac{800}{m} - \dfrac{200}{m^2}$ **b** $0.5\,\text{kg}$

**19** $f(x) \geqslant 10 - 3\ln 3$

**20** $f(x) \geqslant -30$

**21** (0, 3) local max

(ln 2, ln 16) local min

**22** $1.5\sqrt[3]{2}$ m

# Exercise 20F

**1 a** $x = 2$, non-stationary

**b** $x = 4$, non-stationary

**2 a** $x = 0$, stationary

$x = 3$, non-stationary

**b** $x = 0$, non-stationary

$x = 2$, stationary

**3 a** $x = 0$, stationary

$x = \pi$, non-stationary

$x = 2\pi$, stationary

**b** $x = \dfrac{\pi}{2}$, non-stationary

$x = \dfrac{3\pi}{2}$, stationary

**4 a** $x = \ln\dfrac{1}{2}$, non-stationary

**b** $x = 0$, non-stationary

**5** (−3, 50)

**6** (2, −288)

**7** $b = -3$, $c = 5$

**8** $a = 6$, $b = -3$

**9** $a = -3$, $b = 3$, $c = 2$

**10 a** (2, 4 + 8ln2)　　**b** non-stationary

**11 b** (1, 1)

**12 a, b, c**

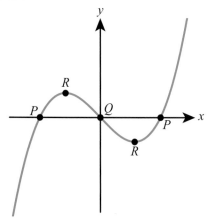

**d** No. $f'(x) \neq 0$ at these points.

**13 a** $x = -2, 1$

**b** $x = -2$ local minimum

$x = 1$ point of inflection

**14 a** $x = 0, 4$

**b** local minimum at $x = 0$

local maximum at $x = 4$

**15** $\left(\dfrac{\sqrt{2}}{2}, e^{-\frac{1}{2}}\right), \left(-\dfrac{\sqrt{2}}{2}, e^{-\frac{1}{2}}\right)$

**16** $\left(e^{-\frac{3}{2}}, -\dfrac{3}{2}e^{-3}\right)$

**17** (0, 2)

# Chapter 20 Mixed Practice

**1 a** $3ax^2 - 3$　　**b** −3　　**c** $\dfrac{1}{4}$

**2** (−2, 25)

**3** $b = -4$, $c = 7$

**4** $b = -6$, $c = 4$

**5** 0.25

**6** $y = x - 1$

**7** $y = 1 - \dfrac{x}{2}$

**8 a** $A(e^{-2}, 4e^{-2}), B(1, 0)$

**b** $(e^{-1}, e^{-1})$

**9** $\dfrac{\sqrt{3}}{3}$ mg l$^{-1}$

**10** $\dfrac{ds}{dt} = 25$

**11 a** 0

**b** $x > 0$

**c i** 2

**d** f(x)

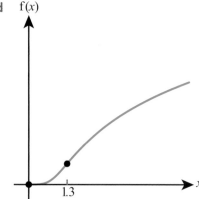

**12** b $\left(-1, \dfrac{1}{9}\right)$ c $\dfrac{1}{9} < k < 1$

**13** $y = x$

**14** $2e^{2x}(2x^2 + 4x + 1)$

**15** a $4.5\,\text{cm}^2$ b $6 + 3\sqrt{2}\,\text{cm}$

**16** $6e^5$

**17** $k = 12$

**18** $e^x((x+1)\sin x + x\cos x)$

**19** $a = 1,\ b = 8,\ c = 1.5$

**20** a $x < 2$ b $(2, 2e^{-2})$

**21** $\left(\dfrac{\pi}{4}, \dfrac{e^{-\frac{\pi}{4}}}{\sqrt{2}}\right)$

**22** a $5$ million $\text{m}^3$

  b $5 + \dfrac{2}{e}$ million $\text{m}^3$

  c 2 hours after the storm

**24** a $\dfrac{100}{51}$ b $34.3$

  c $0 < \text{f}(x) < 100$ e $5$

**25** b Not if $q < b < p$

**26** a $-2 \leqslant x \leqslant 2$

**27** a $0.5$ b $\pm\sqrt{2}$

# Chapter 21 Prior Knowledge

**1** $y = \dfrac{1}{3}x^3 + 2x + 4$

**2** $48.4$

**3** $\dfrac{dy}{dx} = 2e^{2x} + \cos x$

**4** $\text{f}'(x) = x(x^2 + 3)^{-\frac{1}{2}}$

# Exercise 21A

**1** a $\dfrac{3}{5}x^{\frac{5}{3}} + c$ b $\dfrac{4}{7}x^{\frac{7}{4}} + c$

**2** a $2x^{\frac{1}{2}} + c$ b $-3x^{-\frac{1}{3}} + c$

**3** a $4x^{\frac{5}{2}} + c$ b $4x^{\frac{5}{4}} + c$

**4** a $12x^{\frac{1}{3}} + c$ b $5x^{\frac{3}{5}} + c$

**5** a $\dfrac{3}{8}x^{\frac{4}{3}} + c$ b $\dfrac{5}{18}x^{\frac{6}{5}} + c$

**6** a $8x^{\frac{3}{4}} + c$ b $14x^{\frac{1}{2}} + c$

**7** a $y = 3\sin x + c$ b $y = -\sin x + c$

**8** a $y = 2\cos x + c$ b $y = -\dfrac{1}{2}\cos x + c$

**9** a $y = 5e^x + c$ b $y = -\dfrac{4}{3}e^x + c$

**10** a $y = 2\ln|x| + c$ b $y = \dfrac{1}{2}\ln|x| + c$

**11** a $\dfrac{1}{3}(2x+1)^{\frac{3}{2}} + c$ b $-\dfrac{3}{8}(1-2x)^{\frac{4}{3}} + c$

**12** a $-\dfrac{2}{5}(3-5x)^{\frac{1}{2}} + c$ b $\dfrac{2}{3}(2x-7)^{\frac{3}{4}} + c$

**13** a $-\dfrac{1}{3}\sin(2-3x) + c$ b $\dfrac{1}{4}\sin(4x+3) + c$

**14** a $-2\cos\left(\dfrac{1}{2}x - 5\right) + c$ b $\dfrac{1}{2}\cos(5-2x) + c$

**15** a $\dfrac{1}{5}e^{5x+2} + c$ b $-\dfrac{1}{3}e^{1-3x} + c$

**16** a $\dfrac{1}{4}\ln|4x-5| + c$ b $-\dfrac{1}{2}\ln|3-2x| + c$

**17** a $\dfrac{1}{4}(x^2+4)^4 + c$ b $\dfrac{2}{5}(x^4-2)^{\frac{5}{2}} + c$

**18** a $\dfrac{1}{4}\sin^4 x + c$ b $\dfrac{1}{3}\cos^3 x + c$

**19** a $\ln|x^3 + 2x| + c$ b $\ln|x^4 - 5x| + c$

**20** a $\dfrac{2}{9}(x^3+4)^{\frac{3}{2}} + c$ b $-\dfrac{3}{8}(1-x^2)^{\frac{4}{3}} + c$

**21** a $-2e^{-x^2} + c$ b $3e^{x^3} + c$

**22** a $\dfrac{1}{3}\ln|x^3 + 5| + c$ b $\dfrac{1}{4}\ln|x^4 - 3| + c$

**23** $y = 2x^{\frac{3}{2}} - 42$

**24** $2\sin x + 3\cos x + 2$

**25** $y = 2e^x - 5\ln|x| - 2e$

**26** $x^2 + \dfrac{3}{2}\ln|x| + c$

**27** $V = \dfrac{1}{2}t^2 + \dfrac{1}{2}\cos t + \dfrac{3}{2}$

**28**  $x = 5t - 2e^t + 7$

**29**  $-\dfrac{1}{14}(5 - 2x)^7 + c$

**30**  $-\dfrac{3}{2}\cos(2x) - \dfrac{2}{3}\sin(3x) + c$

**31**  $\dfrac{1}{6}(x^2 + 1)^6 + c$

**32**  $-\dfrac{1}{6}\cos(3x^2) + c$

**33**  $3\sqrt{x^2 + 2} + c$

**34**  $-\dfrac{1}{2(2x + 3)} + \dfrac{3}{2}$

**35**  $\dfrac{1}{3}(\ln x)^2 + 1$

**36**  $2\ln 3 + 5$

**37**  $-\cos^4 x + \dfrac{kx^3}{3} + x + c$

**38**  $e^x + \dfrac{1}{3}e^{-3x} + c$

**39**  $\dfrac{1}{3}(x^2 + 1)^{\frac{3}{2}} + c$

**40**  b   $\dfrac{x}{2} - \dfrac{1}{4}\sin 2x + c$

**41**  a   $1 - 2\sin^2 x$      b   $\dfrac{1}{2}x - \dfrac{1}{4}\sin 2x + c$

**42**  $-\ln|\cos x| + c$

**43**  $-\cos x + \dfrac{1}{3}\cos^3 x + c$

**44**  $\ln|\ln x| + c$

# Exercise 21B

**1**  a   $\dfrac{14}{3}$      b   12

**2**  a   $\dfrac{33}{5}$      b   $-\dfrac{1}{6}$

**3**  a   $\dfrac{26}{3}$      b   $-60$

**4**  a   $\dfrac{1}{2}$      b   1

**5**  a   2      b   5

**6**  a   $\ln 5$      b   $\ln\dfrac{3}{2}$

**7**  a   0.774      b   $-0.152$

**8**  a   1.12      b   3.02

**9**  a   $\dfrac{4}{3}$      b   $\dfrac{4}{3}$

**10**  a   $\dfrac{27}{2}$      b   $\dfrac{5}{2}$

**11**  a   1.63      b   9.77

**12**  a   2.83      b   1.83

**13**  a   36      b   $\dfrac{9}{2}$

**14**  a   $\dfrac{1}{3}$      b   9

**15**  a   $\dfrac{8}{3}$      b   $\dfrac{1}{3}$

**16**  $\dfrac{112}{9}$

**17**  3

**18**  3.29

**19**  25

**20**  a   $(-3, 0), (3, 0)$      b   36

**21**  a   $\dfrac{8}{3}, -\dfrac{5}{12}$      b   $\dfrac{37}{12}$

**22**  11.0

**23**  $\dfrac{52}{81}$

**24**  $-5\ln 3$

**26**  a   $\left(\dfrac{\pi}{4}, \dfrac{\sqrt{2}}{2}\right)$      b   $2 - \sqrt{2}$

**27**  a   $A(0, 1), B\left(1, \dfrac{1}{e}\right)$      b   0.115

**28**  a

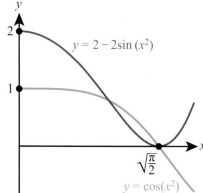

$y = 2 - 2\sin(x^2)$

$\sqrt{\dfrac{\pi}{2}}$

$y = \cos(x^2)$

b   0.0701

**29**  5.33

**30**  2

**31**  $\dfrac{7}{3}$

**32**  $\dfrac{3}{2}(1 - e^{-2})$

**33** $\frac{1}{2}\ln\left(\frac{11}{5}\right)$

**34** $\ln\left(\frac{2}{3}\right)$

**35** 27

**36** 17.5

**37 a** $\ln x + 1$      **b** 1

**38** −2

# Exercise 21C

**1 a** $v = 3t^2 + 6t$      **b** $v = 4t^3 - 5$
$a = 6t + 6$           $a = 12t^2$

**2 a** $v = -2\sin 2t$      **b** $v = \frac{1}{2}\cos\frac{t}{2}$

$a = -4\cos 2t$           $a = -\frac{1}{4}\sin\frac{t}{2}$

**3 a** $v = 3e^{3t}$      **b** $v = -4e^{-4t}$
$a = 9e^{3t}$         $a = 16e^{-4t}$

**4 a** $v = \frac{1}{t}$      **b** $v = \frac{1}{t}$

$a = -\frac{1}{t^2}$        $a = -\frac{1}{t^2}$

**5 a** 45      **b** 124

**6 a** 4      **b** $\frac{4}{5}$

**7 a** $3\sqrt{3}$      **b** 3

**8 a** $\ln\frac{5}{3}$      **b** $\ln\frac{16}{3}$

**9 a** 1.21      **b** 0.935

**10 a** 1.27      **b** 0.984

**11 a** 0.672      **b** 2.53

**12 a** 1.54      **b** 0.795

**13 a** 0.657      **b** 1.435

**14 a** 16 m      **b** $4\,\text{m s}^{-1}$
    **c** $-2\,\text{m s}^{-2}$

**15** $2 - 2e^{-1.5} \approx 15.5\,\text{m}$

**16 a** −48 m      **b** $\frac{176}{3} \approx 58.7\,\text{m}$

**17** $23\,\text{m s}^{-1}$

**18 a** $256\,\text{m s}^{-1}$      **b** $-32\,\text{m s}^{-2}$
    **c** 4 s         **d** 819.2 m

**19** 3.16 m

**20 a** 0 s, 6 s      **b** 0 s, 4 s
    **c** 4 s

**21 a** $\sin t + t\cos t$      **b** $0\,\text{m s}^{-1}$

    **c** $\frac{4 + \pi}{4\sqrt{2}}\,\text{m s}^{-1}$      **d** $2\,\text{m s}^{-2}$

    **e** 0.556 s, 1.57 s, 5.10 s
    **f** 13.3 m

**24 a** $-10\,\text{m s}^{-2}$      **b** 0.5 s, 61.25 m
    **c** 4 s          **d** 62.5 m
    **e** e.g. no obstructions to path of ball

**25 a** $18\,\text{m s}^{-1}$      **b** 3 s
    **c** $-8\,\text{m s}^{-2}$      **d** 108 m
    **e** −36 m        **f** 36 m

**26** $10\,\text{m s}^{-1}$

**27** $8\,\text{m s}^{-1}$

**28** 0.5 m

**29 a** 0 m      **b** 50 m

**30 a** $6.06\,\text{m s}^{-1}$      **b** 0

    **c** $\frac{71}{24} \approx 2.96\,\text{m s}^{-1}$

**31 a i** $v = \frac{At^2}{2} + u$      **ii** $s = \frac{At^3}{6} + ut$

**32** 6 seconds

**33** 0.9375 m

# Chapter 21 Mixed Practice

**1** 51

**2** $y = \frac{1}{2}\sqrt{x} + 2$

**3** $\frac{\sqrt{3} - 1}{2}$

**4** $\frac{3}{4}\ln|x| + \frac{1}{2x} + c$

**5** $v = -10.3\,\text{m s}^{-1}, \ a = -3.10\,\text{m s}^{-2}$

**6** $6\ln\left(\frac{2x - 5}{3}\right)$

**7 a** $-31.7\,\text{cm s}^{-1}$
    **b** 7.41 cm

**8** $2x^{\frac{3}{2}} + 4x^{-\frac{1}{2}} + c$

**9** $y = \sin^3 x + 2$

**10 a**

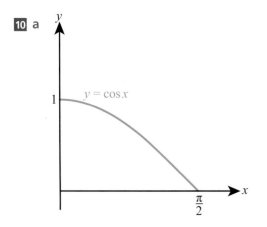

$y = \cos x$

**b** $A(0,1)$, $B\left(\dfrac{\pi}{2},0\right)$    **c** $y = -\dfrac{2}{\pi}x + 1$

**d** $1 - \dfrac{\pi}{4}$

**11** $\dfrac{4}{3}$

**12 a** $A(0,1)$, $B(2, 5)$, $C(7, 0)$

   **b** 17.2

**13 a** $-9\,\text{m s}^{-2}$

   **b** 18.5 m

**14 a**

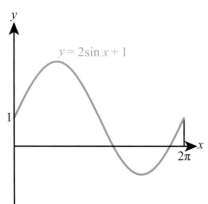

$y = 2\sin x + 1$

   **b** $2.20\,\text{m s}^{-1}$    **c** $6.28\,\text{m}$     **d** $9.02\,\text{m}$

**15** $\dfrac{1}{2}\ln 17$

**16** $\dfrac{5\pi}{3}$

**17 a** displacement = A, acceleration = B

   **b** $t = 3$

**18** 6

**19 a** $0 < x < d$

   **b** $a < x < b$ and $x > c$

   **c** 15

**21 a** $\dfrac{x}{2} - \dfrac{1}{12}\sin 6x + c$     **b** $\dfrac{x}{2} + \dfrac{1}{12}\sin 6x + c$

# Analysis and approaches SL: Practice Paper 1

**1 a** 13 cm     **b** 38     **c** $\dfrac{5}{8}$   [6]

**2 a** 15     **b** $(5, 0)$

   **c** $(2.5, 7.5)$                            [5]

**3 a** 80     **b** $\dfrac{1}{8}$     **c** $\dfrac{3}{5}$   [5]

**4 a** $-2$     **b** $8, \dfrac{1}{4}$           [5]

**5 a** 1     **b** decreasing

   **c** maximum $(1, 4)$, minimum $(4, -3)$

   **d** $(2, 1)$                            [6]

**6 a** $\dfrac{1}{2}$                                   [6]

**7** 4                                     [7]

**8 a** $(3, 0)$                               [1]

   **b i** $(5, 4)$

     **ii** $x^2 - 10x + 29$             [3]

   **c** $(5, 4)$                          [3]

   **d ii** $\dfrac{13}{4}$                     [6]

**9 a** $\dfrac{1}{2}(1 - p)$              [2]

   **b** $a = 2$, $b = 2$         [2]

   **c i** $(1 - p)^3 \left(\dfrac{1}{2}\right)^3$

     **ii** $(1 - p)^k \left(\dfrac{1}{2}\right)^k$      [3]

   **d** $\dfrac{2}{3}$                         [6]

**10 a** $f'(x) \geqslant 0$ for all $x > -1$   [1]

   **c** $f''(0) = 0$ and $f'(x) > 0$ on either side   [2]

   **d** $(\sqrt[3]{2}, \ln 3)$              [3]

   **e** $f^{-1}(x) = \sqrt[3]{e^x - 1}$, $x \in \mathbb{R}$, $f^{-1}(x) > -1$                         [4]

# Analysis and Approaches SL: Practice Paper 2

**1 a** 9.75 cm **b** 16.4 cm² [5]

**2 a** $f'(x) = \dfrac{2}{x}\ln x$

**c** $x = 0.340, 1.86$ [6]

**3** 44.8° or 135° [5]

**4 a**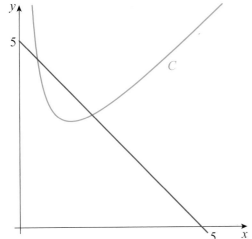

**b** 0.977 [6]

**5 a** $x \neq -1$ **b** $f(x) \neq 2$

**c** $f^{-1}(x) = \dfrac{3-x}{x-2}$ [6]

**6 a** $a = e^t \cos 2t - 2e^t \sin 2t$

**b** 4.76 m [6]

**7 a** 51.8 cm **b** 26.8 cm

**c** 5980 cm³ [7]

**8 a i** $-\dfrac{3}{2}$ **ii** $(0, 6)$ [3]

**b** $2x - 3y - 21 = 0$ [3]

**c** $(6, -3)$

**d i** $\sqrt{117} \approx 10.8$ **ii** $\sqrt{13} \approx 3.61$

**e** 19.5 [5]

**9 a ii** $\mu - 1.6449\sigma = 3$

**iii** $\mu = 12.0, \sigma = 5.47$ [7]

**b** 7.39 minutes [2]

**c i** 0.291 **ii** 0.135 [5]

**10 a i** $-\dfrac{1}{p^2}$ [5]

**b i** $Q(2p, 0), R\left(0, \dfrac{2}{p}\right)$

**ii** 2 [4]

**d** $2\sqrt{2} \approx 2.83$ [2]

# Glossary

**Amplitude** Half the distance between the maximum and minimum values of a periodic function

**Arithmetic sequence** A sequence with a common difference between each term

**Arithmetic series** The sum of the terms of an arithmetic sequence

**Asymptotes** Lines to which a graph tends but that it never reaches

**Axis intercepts** The point(s) where a graph crosses the axes

**Base** The number $b$ in the expression $b^x$

**Biased** A description of a sample that is not a good representation of a population

**Binomial coefficients** The constants of each term in the expansion of $(a + b)^n$ given by $^nC_r$

**Chain rule** A rule for differentiating composite functions

**Completing the square** The process of writing a quadratic in the form $a(x - h)^2 + k$

**Compound interest** The amount added to an investment or loan, calculated in each period as a percentage of the total value at the end of the previous period

**Concave-down** The part(s) of a curve where the second derivative is negative

**Concave-up** The part(s) of a curve where the second derivative is positive

**Continuous** Data that can take any value in a given range

**Definite integration** Integration with limits – this results in a numerical answer (or an answer dependent on the given limits) and no constant of integration

**Depreciate** A decrease in value of an asset

**Derivative** A function that gives the gradient at any point of the original function (also called the slope function or gradient function)

**Differentiation** The process of finding the derivative of a function

**Discrete** Data that can only take distinct values

**Discrete random variable** A variable with discrete output that depends on chance

**Discriminant** The expression $b^2 - 4ac$ for a quadratic

**Equation** A statement that two expressions are equal for certain values of the variable(s)

**Event** A combination of outcomes

**Exponent** The number $x$ in the expression $b^x$

**Exponential equation** An equation with the variable in the power (or exponent)

**Factorial** The product of all integers from 1 to $n$, denoted by $n!$

**Geometric sequence** A sequence with a common ratio between each term

**Geometric series** The sum of the terms of a geometric sequence

**Gradient function** See **derivative**

**Gradient of a curve** The gradient of the tangent to a curve at the given value

**Hyperbola** The shape of the graph of $y = \dfrac{1}{x}$

**Identity** A statement that two expressions are equal for all values of the variable(s)

**Identity function** A function that has no effect on any value in its domain

**Indefinite integration** Integration without limits – this results in an expression in the variable of integration (often $x$) and a constant of integration

**Inflation rate** The rate at which prices increase over time

**Integration** The process of reversing differentiation

**Intercept** A point at which a curve crosses one of the coordinate axes

**Interquartile range** The difference between the upper and lower quartiles

**Limit of a function** The value that $f(x)$ approaches as $x$ tends to the given value

**Limits of integration** The lower and upper values used for a definite integral

**Normal to a curve** A straight line perpendicular to the tangent at the point of contact with the curve

**Outcomes** The possible results of a trial

**Outlier** An extreme value compared to the rest of the data set: one that is more than 1.5 standard deviations above the upper quartile or below the lower quartile

**Parabola** The shape of the graph of a quadratic function

**Period** The smallest value of $x$ after which a function repeats

**Point of inflection** A point on a curve where the concavity changes

**Population** The complete set of individuals or items of interest in a particular investigation

**Product rule** A rule for differentiating a product of two functions

**Quartiles** The points one quarter and three quarters of the way through an ordered data set

**Quotient rule** A rule for differentiating a quotient of two functions

**Radians** An alternative measure of angle to degrees: $2\pi$ radians $= 360°$

**Range** The difference between the largest and smallest value in a data set

**Reflection** The mirror image of a curve in a given line

**Relative frequency** The ratio of the frequency of a particular outcome to the total frequency of all outcomes

**Roots of an equation** The solutions of an equation

**Sample** A subset of a population

**Sample space** The set of all possible outcomes

**Self-inverse function** A function that has an inverse the same as itself

**Simple interest** The amount added to an investment or loan, calculated in each period as a percentage of the initial sum

**Slope function** See **derivative**

**Standard deviation** A measure of dispersion, which can be thought of as the mean distance of each point from the mean

**Standard index form** A number in the form $a \times 10^k$ where $1 \leqslant a < 10$ and $k \in \mathbb{Z}$

**Standard normal distribution** The normal distribution with mean 0 and variance 1

**Stationary point** A point on a curve where the gradient is zero

**Stretch** Multiplication of the $x$ (or $y$) values of all points on a curve by a given scale factor

**Subtended** The angle at the centre of a circle subtended by an arc is the angle between the two radii extending from each end of the arc to the centre

**Sum to infinity** The sum of infinitely many terms of a geometric sequence

**Tangent to a curve** A straight line that touches the curve at the given point but does not intersect the curve again (near that point)

**Translation** The addition of a constant to the $x$ (or $y$) values of all points on a curve

**Trial** A repeatable process that produces results

**Value in real terms** The value of an asset taking into account the impact of inflation

**Variance** The square of the standard deviation

**Vertices of a graph** Points where the graph reaches a maximum or minimum point and changes direction

**Zeros of a function** Values of $x$ for which $f(x) = 0$

# Acknowledgements

The Publishers wish to thank Pedro Monsalve Correa for his valuable comments and positive review of the manuscript.

The Publishers would like to thank the following for permission to reproduce copyright material.

**Photo credits**

**p.xiii** © https://en.wikipedia.org/wiki/Principia_Mathematica#/media/File:Principia_Mathematica_54-43.png/ https://creativecommons.org/licenses/by-sa/3.0/; **p.xxiv** © R MACKAY/stock.adobe.com; **pp.2-3** *left to right* © 3dmentat - Fotolia; © GOLFCPHOTO/stock.adobe.com; © Janis Smits/stock.adobe.com; **pp.22-23** © Reineg/ stock.adobe.com; **p.26** © The Granger Collection/Alamy Stock Photo; **p.30** *left to right* © glifeisgood - stock. adobe.com; The Metropolitan Museum of Art, New York/H. O. Havemeyer Collection, Bequest of Mrs. H. O. Havemeyer, 1929/https://creativecommons.org/publicdomain/zero/1.0/; © Iarygin Andrii - stock.adobe.com; © Nino Marcutti / Alamy Stock Photo; **pp.48-9** *left to right* © Julija Sapic - Fotolia; © Getty Images/iStockphoto/ Thinkstock; © Marilyn barbone/stock.adobe.com; © SolisImages/stock.adobe.com; **pp.74-5 *left to right*** © Evenfh/stock.adobe.com; © Andrey Bandurenko/stock.adobe.com; © Arochau/stock.adobe.com; © Tony Baggett/stock.adobe.com; © Peter Hermes Furian/stock.adobe.com; **p.93** *left to right* © Shawn Hempel/stock. adobe.com; © Innovated Captures/stock.adobe.com; **p.133** © Bettmann/Getty Images; **pp.178-9** *left to right* © Lrafael/stock.adobe.com; © Comugnero Silvana/stock.adobe.com; © Talaj/stock.adobe.com © tawhy/123RF ; **pp.200-201** *left to right* © Daniel Deme/WENN.com/Alamy Stock Photo; © Maxisport/stock.adobe.com; © The Asahi Shimbun/Getty Images; **pp.222-3** *left to right* © Shutterstock/Pete Niesen; © Luckybusiness - stock. adobe.com; © Shutterstock/Muellek Josef; **p.225** © Caifas/stock.adobe.com; **p.230** © Dmitry Nikolaev/stock. adobe.com; **pp.256-7** *left to right* © Alena Ozerova/stock.adobe.com © Alena Ozerova/123RF; © Vesnafoto/ stock.adobe.com; © Drobot Dean/stock.adobe.com; **pp.278-9** *left to right* © Africa Studio/stock.adobe.com; © Norman/stock.adobe.com; © Hoda Bogdan/stock.adobe.com; © Artem Merzlenko/stock.adobe.com; **pp.286-7** *left to right* © Ollirg - Fotolia; © Cristian/stock.adobe.com; © Christian/stock.adobe.com; © Analysis121980/ stock.adobe.com; **pp.314-15** *left to right* © AntonioDiaz/stock.adobe.com; © AntonioDiaz/stock.adobe.com; © AntonioDiaz/stock.adobe.com; **pp.326-7** *left to right* © Pink Badger/stock.adobe.com; © Vchalup/stock.adobe. com; © Louizaphoto/stock.adobe.com; © Preserved Light Photography/Image Source/Getty Images; © Wajan/ stock.adobe.com; **pp.354-5** *left to right* © https://en.wikipedia.org/wiki/File:Wallpaper_group-p2-3.jpg/https:// creativecommons.org/publicdomain/mark/1.0/deed.en; https://en.wikipedia.org/wiki/File:Wallpaper_group-pmm-4.jpg/https://creativecommons.org/publicdomain/mark/1.0/deed.en; https://en.wikipedia.org/wiki/ File:Wallpaper_group-p4m-1.jpg/https://creativecommons.org/licenses/by-sa/3.0/; https://en.wikipedia.org/wiki/ File:Alhambra-p3-closeup.jpg/https://creativecommons.org/licenses/by-sa/3.0/deed.en; **pp.382-3** *left to right* © Granger Historical Picture Archive/Alamy Stock Photo; © Nahhan/stock.adobe.com; © Peter Hermes Furian/ stock.adobe.com; **pp.394-5** *left to right* © Vitaliy Stepanenko - Thinkstock.com; © L.R. Greening/Shutterstock. com; © Ingram Publishing Limited / Ultimate Lifestyle 06; © Howard Taylor/Alamy Stock Photo; **pp.440-1** *left to right* © Eyetronic/stock.adobe.com; © Seafarer81/stock.adobe.com; © Gorodenkoff/stock.adobe.com; **pp.458-9** *left to right* © Ipopba/stock.adobe.com; © Alekss/stock.adobe.com; © Inti St. Clair/stock.adobe.com; © Vegefox. com/stock.adobe.com; **pp.490-1** *left to right* © Andrey ZH/stock.adobe.com; © Maximusdn/stock.adobe.com; © Phonlamaiphoto/stock.adobe.com; © Alexey Kuznetsov/stock.adobe.com

**Text Credits**

All IB Past Paper questions are used with permission of the IB © International Baccalaureate Organization

**p.441** Data from https://commons.wikimedia.org/wiki/File:Population_curve.svg [accessed 16.7.2019]; **p.442** Data from GISTEMP Team, 2019: GISS Surface Temperature Analysis (GISTEMP), version 4. NASA Goddard Institute for Space Studies. Dataset accessed 2019-07-16 at https://data.giss.nasa.gov/gistemp/; **p.442** Data from: https://commons.wikimedia.org/wiki/File:FTSE_100_Index.png [accessed 16.7.2019]

Every effort has been made to trace all copyright holders, but if any have been inadvertently overlooked, the Publishers will be pleased to make the necessary arrangements at the first opportunity.

# Index